D1483588

QUANTUM MECHANICS

QUANTUM MECHANICS

by

JOHN L. POWELL

Department of Physics
University of Oregon

and

BERND CRASEMANN

Department of Physics
University of Oregon

ADDISON-WESLEY PUBLISHING COMPANY, INC.

READING, MASSACHUSETTS, U.S.A.

LONDON, ENGLAND

PREFACE

This book is based on courses in quantum mechanics which the authors have taught for several years. An attempt has been made to emphasize the physical basis of the subject, but without undue neglect of its mathematical aspects.

The early chapters of the book provide an introduction to the subject along essentially historical lines. Physical problems that led to the development of quantum mechanics are described in Chapter 1, and the "old quantum theory" is outlined. In Chapter 2, the foundations of wave mechanics are laid in terms of the wave-particle duality; the Schrödinger equation for a free particle is presented, and the Born interpretation of the wave function is discussed. After a discussion of wave packets and the uncertainty principle in Chapter 3, forces are introduced into the wave equation, and the analogy between wave mechanics and optics is traced. The theory of the Schrödinger equation is discussed in Chapter 5, and selected one-dimensional problems are worked out.

The formal structure of the subject is introduced in Chapter 6 through a discussion of linear operators, eigenfunctions, and commutation relations. Spherically symmetric one-particle systems are treated, and a chapter is devoted to the elementary theory of elastic scattering. The algebra of linear vector spaces is developed in Chapter 9, in a form suitable for application to quantum mechanics. The methods of matrix mechanics are then used in the development of the theory of angular momentum. Perturbation theory and the theory of radiative transitions are outlined in Chapter 11, and the book closes with a brief treatment of identical particles.

Throughout the book, the role of symmetry operations and the essentially algebraic structure of quantum-mechanical theory are emphasized. Thorough treatments of selected physical systems are given in the text. A large number of problems has been provided, calling for additional applications of the theory and supplementing the textual material.

The authors are much indebted to several of their colleagues for helpful suggestions, and to their students, on whom the material in the book has been tried out. Mr. Herschel Neumann, in particular, has carefully read one draft of the manuscript and suggested several corrections and clarifications. Sincere thanks are due to Mrs. June Powell and Mrs. Grace Lehrbach, for patiently and accurately typing several drafts of the manuscript.

J.L.P.

B.C.

CONTENTS

CHAPTER 1

HISTORICAL ORIGINS OF THE QUANTUM THEORY

1–1 Difficulties with classical models. Classical physics deals primarily with macroscopic phenomena. Most of the effects with which classical theory is concerned are either directly observable or can be made observable with relatively simple instruments. There is a close link between the world of classical physics and the world of sensory perception.

During the first decades of the present century, physicists turned their attention to the study of atomic systems, which are inherently inaccessible to direct observation. It soon became clear that the concepts and methods of classical macroscopic physics could not be applied directly to atomic phenomena. If classical laws of physics are to be applied at all to systems of atomic size, they can be considered only in connection with *models* of such systems. Models are usually conceived as systems of particles which interact with one another and with electromagnetic radiation according to assumed simple laws. One attempts to construct a microscopic model which will reproduce, as nearly as possible, observed macroscopic effects.

Many of the observed characteristics of atomic systems, however, are such that they cannot be reproduced by any model which behaves according to classical laws. The early development of atomic theory consisted of efforts to overcome these difficulties by modifying the laws of classical physics and the properties of the models to which they were applied. These efforts reached their successful conclusion in the period from 1925 to 1930, when an entirely new theoretical discipline, quantum mechanics, was developed by Schrödinger, Heisenberg, Dirac, and others. The present book is an introduction to this theory. The first chapter is a review of the history of the subject and provides a background for later work. Some of the problems which must be faced in an attempt to understand atomic phenomena will be pointed out.

The earliest evidence of the need for revision of classical concepts came from the field of chemistry. It had long been realized that the molecules of which a pure substance is composed are chemically identical, and that they retain their identity over long periods of time. A molecule of nitrogen, for example, consists of a number of positively and negatively charged particles which are held together by electrostatic forces. However, it is stated by Earnshaw's theorem that a system of charged particles cannot remain at rest in stable equilibrium under the influence of purely electrostatic forces. If a classical picture is adopted, these particles must, there-

1

fore, be in relative motion. Yet, this motion must be such as not to destroy the identity of the molecules, and must persist indefinitely. If the particles are confined to a restricted region of space, they must frequently or continuously change their direction of motion, i.e., be accelerated. It is, however, a well-known fact that accelerated charged particles radiate energy in the form of electromagnetic waves. Hence, the particles in a molecule should progressively change their state of motion in accordance with this loss of energy—a conclusion which does not agree with observation. This fact alone shows that the stability of molecules cannot be understood on the basis of a classical model.

The enormous range of electrical conductivities of solid materials is an example of a property of matter which cannot be reasonably explained in terms of classical ideas. For instance, the conductivity of silver is more than 10^{24} times larger than that of fused quartz. Electrical conduction presumably consists of a relative motion of the negatively and positively charged components of the material under the influence of an applied electric field. It is not possible to comprehend, in terms of a classical picture, how such motion occurs readily in silver, but not at all in quartz. It seems to be necessary to recognize the existence of a *principle of selection* which prohibits the motion in quartz, but not in silver. A similar situation is found in ferromagnetism: the magnetic susceptibility of iron is observed to be of the order of 10^9 times larger than that of other metals. Quite generally, any attempts to understand the chemical behavior of matter, as summarized in the periodic table of the elements, must take into account the fact that not all the states of motion permitted by a classical model are accessible to the systems of particles which comprise the molecules of a chemical substance.

1–2 Optical spectra. Light is emitted by substances which are raised to a high temperature or subjected to an electrical discharge. This light can be separated into its spectral components by means of a diffraction grating, and the wavelengths of the various components can be measured with great precision. Characteristic emission spectra, consisting of discrete frequencies or lines, are obtained in this way for each element.

The emission of this electromagnetic radiation must be associated with the accelerated motion of the charged particles in the atoms of the emitting substance. An attempt to construct a classical model reproducing the observed frequencies naturally leads to the conclusion that these must be the same as the frequencies of the periodic motions of the particles. The spectral lines should therefore fall into groups within which the frequencies ν are related by the harmonic law

$$\nu = n_1\nu_1 + n_2\nu_2 + n_3\nu_3 + \cdots,$$

we obtain for the average oscillator energy

$$\frac{\epsilon x/(1-x)^2}{1/(1-x)} = \frac{\epsilon x}{1-x} = \frac{\epsilon}{e^{\epsilon/kT}-1}. \tag{1-8}$$

In the limit $\epsilon \to 0$, or at very high temperature, this result reduces to kT, in agreement with the value obtained from the equipartition theorem.

At the time of Planck's work, Wien had already shown on thermodynamic grounds that the energy density u_λ of wavelength λ can depend upon the temperature only through a function of the product λT. This requirement is satisfied if one sets

$$\epsilon = h\nu, \tag{1-9}$$

where h is a universal constant. If the expression $8\pi\nu^2 \, d\nu/c^3$ is introduced for the number of modes in the interval $(\nu, \nu + d\nu)$, *Planck's law* results:

$$u_\nu = \frac{8\pi h\nu^3}{c^3} \frac{1}{e^{h\nu/kT}-1}. \tag{1-10}$$

This law is in agreement with experiment. The avoidance of the ultraviolet catastrophe on the basis of the quantum partition law is easily understood, since the available states are now widely separated in energy when ν is large, and can be reached only by the absorption of very high-energy quanta, a relatively rare occurrence.

The constant h (*Planck's constant*), as well as the Boltzmann constant k, can be evaluated by comparison with experiment. The total energy density, obtained by integrating Planck's distribution function over the whole spectrum, must depend on the temperature in accordance with *Stefan's law:*

$$u = \int_0^\infty u_\nu \, d\nu = \frac{4\sigma}{c} \, T^4. \tag{1-11}$$

If the integral is evaluated, using Eq. (1–10) for the energy density, one obtains for the value of Stefan's constant

$$\sigma = \frac{2\pi^5 k^4}{15h^3 c^2}. \tag{1-12}$$

Furthermore, Planck's law must yield a maximum energy density at the wavelength predicted by *Wien's law:*

$$\lambda_{\max} T = b. \tag{1-13}$$

The experimental values for the *Stefan-Boltzmann constant* σ and for *Wien's constant b* then determine the value of k and of Planck's constant,[1]

$$h = 6.625 \times 10^{-27} \text{ erg·sec.}$$

Additional insight into the significance of the quantum hypothesis is afforded by an alternative derivation of the radiation law, due to Einstein.[2]   Detailed reference to the electromagnetic radiation theory is avoided in this treatment, and the role played by statistical considerations is emphasized.   Suppose that two states of energy E_1 and E_2 ($E_2 > E_1$) are available to an atom bathed in radiation of density u_ν.   It is assumed that quantum jumps between these states can take place in three different ways: (1) *spontaneous* transitions, with the probability A_{12} that the atom jumps from state 2 to state 1, emitting a quantum, (2) *absorption*, with the probability $B_{21}u_\nu$ that the atomic state changes from 1 to 2 with absorption of a quantum, and (3) *induced emission*, with probability $B_{12}u_\nu$, in which the state changes from 2 to 1 under the influence of the incident radiation.   In thermal equilibrium, the probabilities that states 1 and 2 are occupied are proportional to $e^{-E_1/kT}$ and $e^{-E_2/kT}$, respectively. Therefore the equation

$$(A_{12} + B_{12}u_\nu)e^{-E_2/kT} = B_{21}u_\nu e^{-E_1/kT} \qquad (1\text{-}14)$$

expresses the condition of equilibrium, namely,

(number of transitions from 2 to 1) = (number of transitions from 1 to 2).

Now it is assumed that the *Einstein coefficients* A_{12}, B_{12}, B_{21} are properties of the atomic states 1 and 2, and therefore independent of u_ν. Consequently, if the temperature is sufficiently high and u_ν therefore very large, it may be assumed that

$$B_{12}u_\nu \gg A_{12}; \qquad (1\text{-}15)$$

hence $B_{21} = B_{12}$. This is the *principle of detailed balance* for induced transitions, that is, the probability that a transition is induced by the action of the electromagnetic field on the charged particles in the atom is the same for the transition $1 \to 2$ as for $2 \to 1$. With this substitution in Eq. (1-14) one obtains, on rearranging,

$$u_\nu = \frac{A_{12}/B_{12}}{e^{h\nu/kT} - 1}, \qquad (1\text{-}16)$$

---

[1] J. W. M. DuMond and E. R. Cohen, *Revs. Modern Phys.* **25**, 691 (1953).
[2] A. Einstein, *Verhandl. deut. physik. Ges.* **18**, 318 (1916) and *Physik. Z.* **18**, 121 (1917).

where $E_2 - E_1 = h\nu$. By comparison with Eq. (1–10), it is seen also that

$$\frac{A_{12}}{B_{12}} \propto \nu^3. \qquad (1\text{–}17)$$

The above argument, which is very much in the spirit of modern theory, is based upon the idea that quantum jumps take place in a random way, and that only *probabilities* for their occurrence can be defined. This application of the concept of probability is to be distinguished from its application in classical statistical mechanics. It is asserted that there is a certain probability per unit time that a random event may occur. This concept will receive further elaboration in later chapters. The Einstein coefficients depend upon the detailed structure of the atom and can be computed from quantum principles (cf. Chapter 11).

Einstein's derivation of Eq. (1–16) is essentially nonclassical. It contains as a basic element the assumption that the atom and radiation can coexist in a stable state, and that transitions from one state to another can be predicted only statistically.

1–4 The photoelectric effect. There is no satisfactory classical explanation for the detailed characteristics of the photoelectric effect. Apparatus for the study of this effect may consist of a vacuum chamber containing a metal plate opposite a collector (Fig. 1–4a). A retarding potential is maintained between the two electrodes. When a beam of ultraviolet light strikes the plate, electrons are emitted from the metal. Those electrons whose initial kinetic energy equals or exceeds the work that is required to overcome the retarding potential can reach the collector and contribute to the photoelectric current between the electrodes.

The lowest retarding potential for which the photoelectric current is zero is called the *stopping potential* (Fig. 1–4b). It is proportional to the maximum energy with which photoelectrons are emitted from the plate. According to classical considerations, this maximum energy should increase when the intensity of the incident light is increased; the force which the light exerts on electrons in the metal surface should be proportional to the magnitude of the electric vector ε of the incident light wave, and the magnitude of this vector increases when the light is made more intense. Contrary to this expectation, experiment shows that the maximum energy of the photoelectrons is independent of the intensity of the incident light. However, the energy of the photoelectrons is found to increase with the *frequency* of the incident light.

These properties of the photoelectric effect were first explained by Einstein.[1] Planck's earlier work on blackbody radiation, described in

---

[1] A. Einstein, *Ann. Physik* **17**, 132 (1905).

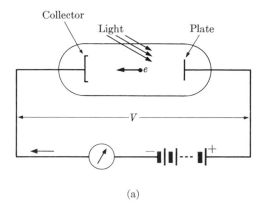

(a)

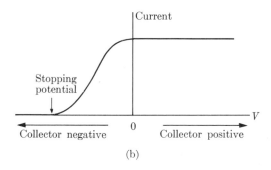

(b)

FIG. 1–4. Photoelectric effect. (a) Apparatus for the study of the photoelectric effect. (b) Typical variation of photoelectric current with potential between collector and plate.

the preceding section, had indicated that the exchange of energy between radiation and the walls of a cavity occurs in quanta of magnitude $h\nu$. Einstein extended this idea and suggested that, in inducing the emission of photoelectrons, light does not act like a wave, but like a stream of discrete quanta or photons. The energy $h\nu$ of one photon can be imparted to only one electron. Conservation of energy then leads to *Einstein's photoelectric law:*

$$E = h\nu - W. \tag{1–18}$$

Here, E is the kinetic energy with which the photoelectron leaves the metal surface, and W is the *work function*, which depends on the nature of the metal. Of the energy $h\nu$ of the incident quantum, a portion W is expended in freeing the electron from the surface, and the remainder, E, is imparted to the photoelectron as kinetic energy. Einstein's photo-electric law agrees closely with experiment and leads to the same value for h that is obtained from Planck's law.

1–5 The Franck-Hertz experiment. Planck's work on blackbody radiation and Einstein's study of the photoelectric effect showed that electromagnetic radiation when interacting with matter acts like an assemblage of discrete quanta of energy rather than like a continuous wave. The idea of naturally discontinuous processes of emission and absorption was new in physics, and it is understandable that further experimental confirmation was diligently sought. In 1914, Franck and Hertz[1] reported an unusually elegant experiment that proved that mechanical energy, like electromagnetic energy, is absorbed by atoms in discrete quanta. Because of its classic simplicity and its convincing nature, this experiment deserves somewhat detailed consideration.

The apparatus used by Franck and Hertz (Fig. 1–5) consisted of an electrically heated wire, located along the axis of a cylindrical grid, which was surrounded by a collector. The device was housed in an enclosure filled with mercury vapor. An accelerating voltage was applied between heater and grid, and a retarding voltage was maintained between grid and collector.

From the variation of collector current with retarding voltage, the energy of electrons that had passed through the mercury vapor could be determined. It was found that electrons whose total energy was less than

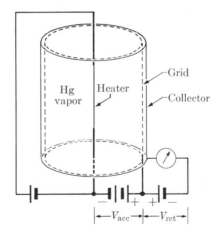

FIG. 1–5.   Apparatus for the experiment of Franck and Hertz.

[1] J. Franck and G. Hertz, *Verhandl. deut. physik. Ges.* **16**, 457 and 512 (1914).

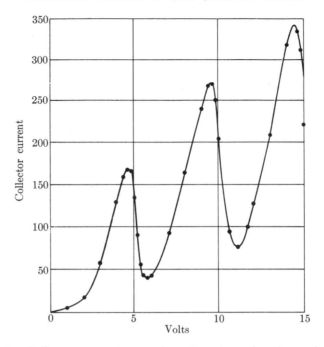

FIG. 1–6.   Collector current as a function of accelerating voltage in the Franck-Hertz experiment. [After J. Franck and G. Hertz, *Verhandl. deut. physik. Ges.* **16,** 457 (1914).]

4.9 ev[1] lost no detectable energy at all in collisions with mercury atoms. Any collisions between these low-energy electrons and atoms of the vapor must therefore have been perfectly elastic, imparting no energy to the atoms except for negligible recoil.   However, when the accelerating potential was increased above 4.9 volts, inelastic collisions occurred near the grid, in which electrons gave up their entire kinetic energy to mercury atoms.   After losing their energy in an inelastic collision, the electrons were no longer able to traverse the retarding field, and the collector current fell to a minimum.

A further increase in the accelerating voltage moved the region where electrons reached the critical energy of 4.9 ev closer to the central wire. After attaining the critical energy and losing it through collisions, electrons could now pick up new energy on their way to the grid, so that the collector current rose again.   A second current minimum was obtained with an accelerating potential of approximately 10 volts, due to a second region of

---

[1] One electron volt (ev) $= 1.6 \times 10^{-12}$ erg is the energy which an electron gains in falling through an accelerating potential difference of one volt.   Cf. also the list of constants and conversion factors, Appendix A–8.

inelastic collisions in the vicinity of the grid. The graph of collector current as a function of accelerating voltage is reproduced in Fig. 1–6. The experimental results show clearly that mercury atoms absorb mechanical energy in quanta of 4.9 ev.

The question immediately arose whether the atoms would reradiate quanta of the same energy. The frequency of the emitted radiation would then be $1.18 \times 10^{15} \text{ sec}^{-1}$, which corresponds to ultraviolet light of wavelength 2530 A. Franck and Hertz repeated their experiment in a quartz container, transparent to ultraviolet radiation, and photographed the emission spectrum from the mercury vapor. A strong line appeared at 2536 A, agreeing, within the limits of error, with the calculated value.

1–6 The Rutherford atom. Once the quantum nature of energy exchange was recognized, theoretical physicists were confronted with the task of suggesting a mechanism for the absorption and emission of quanta, in accordance with spectroscopic evidence. The mechanism would, of course, depend closely upon a satisfactory model of the atom.

The most important information about the structure of atoms was discovered by Rutherford in 1911. In a classical paper,[1] Rutherford analyzed experimental results of Geiger and Marsden on the scattering of alpha particles from thin metallic foils. The experiments had shown that a few incident alpha particles (about one in 20,000) were deflected through an average angle of 90° in passing through a thin (4×10^{-5} cm) gold foil. Rutherford assumed that such large deflections were produced in a single encounter of an alpha particle with an atom, since previous calculations based on multiple scattering did not give a satisfactory result. He showed that the experimental results could be explained if the atom was assumed to consist of a strong positive or negative central charge, concentrated within a distance of less than 3×10^{-12} cm and surrounded by a "sphere of electrification" of the opposite charge that extended throughout the remainder of the atom, i.e., to a distance of approximately 10^{-8} cm. The scattering could then be assumed to be due mainly to the central charge or *nucleus*, which would cause the alpha particle to describe a hyperbolic path with the center of the atom as one focus.

A calculation based on classical mechanics and the Coulomb force between alpha particle and atomic nucleus led Rutherford to his well-known formula for the number $n(\theta)$ of particles which are deflected into unit solid angle in the direction θ:

$$n(\theta) = n_0 N \left(\frac{ZZ'e^2}{2mv^2} \right)^2 \frac{1}{\sin^4 (\theta/2)}. \qquad (1\text{--}19)$$

---

[1] E. Rutherford, *Phil. Mag.* **21**, 669 (1911).

In this expression, m and v are the mass and velocity, respectively, of the incident alpha particle, N is the number of scattering centers per unit area, n_0 is the number of incident alpha particles per unit area, and Ze and $Z'e$ are, respectively, the charge of the nucleus and of the alpha particle.[1] The correctness of Rutherford's scattering formula, and therefore of his assumption of a nuclear atom, was borne out by further experiments,[2] during which more than 100,000 scintillations produced by alpha particles impinging upon a zinc sulfide screen were counted. In this work, Geiger and Marsden confirmed the $1/\sin^4(\theta/2)$-dependence of the number of scattered alphas on the scattering angle, as well as the direct proportionality of the thickness of the scattering foil to the number of alphas scattered in any given direction. They found also that the scattering per atom of foils of various materials varies approximately with the square of the atomic weight, and that the scattering by a given foil is inversely proportional to the fourth power of the velocity of the incident alpha particles. Finally, Geiger and Marsden were able to calculate from their data that the charge of the atomic nucleus is approximately equal to one-half the atomic weight.

It is interesting to note that a nuclear atom had already been considered mathematically by Nagaoka[3] in 1904, but it was Rutherford's analysis that established this concept as an experimental fact. The problem which immediately arose, however, concerned the stability of such a system. For reasons pointed out in Section 1–1, a classical model of a nuclear atom is unstable.

1–7 Stationary states of atoms. In 1913, Niels Bohr was able to resolve the question of the stability of the Rutherford atom and to take account of the absorption and emission of quanta. Bohr formulated a completely new theory of atomic structure, based on postulates that deviated fundamentally from the pattern of classical physics.[4] Bohr's work constitutes one of the most brilliant advances in modern physics and was basic in the development of the quantum theory.

In his first postulate, Bohr assumed the existence of discrete stationary states of the atom, with electrons moving about a positive nucleus in orbits which can be computed from classical theory. In the simplest case, that of the hydrogen atom, a single electron is assumed to describe a circle, or an ellipse with the nucleus at one focus. The total energy of the atom

---

[1] For a thorough discussion of this formula, see F. Rasetti, *Elements of Nuclear Physics*. New York: Prentice Hall, Inc., 1936, p. 53 ff.

[2] H. Geiger and E. Marsden, *Phil. Mag.* **25,** 604 (1913).

[3] H. Nagaoka, *Phil. Mag.* **7,** 445 (1904).

[4] N. Bohr, *Phil. Mag.* **26,** 1 (1913).

in such a stationary state remains constant; contrary to classical electro-dynamics, none is radiated.

As a second postulate, Bohr suggested that an atomic electron can make a transition from one stable orbit to another in a way which cannot be treated classically. If the energy E_m of the atom in the final state is lower than its energy E_n in the initial state, then the energy difference is radiated as a single photon. Since the energy of a photon of frequency ν is $h\nu$, conservation of energy requires that

$$E_n - E_m = h\nu. \tag{1-20}$$

The *Bohr frequency rule* follows:

$$\nu = \frac{E_n - E_m}{h}. \tag{1-21}$$

In the inverse process, a photon of the appropriate energy $h\nu$ is absorbed by the atom, raising an electron from an orbit of lower energy to one of higher energy.

Bohr showed also that the ionization potential of hydrogen and the frequencies of the lines of the *Balmer series* could be obtained if one assumed that the stationary states of the hydrogen atom were characterized by certain definite values of the angular momentum of the electron about the nucleus. According to this theory, the angular momentum is an integral multiple of Planck's constant, divided by 2π:

$$p_\phi = \frac{nh}{2\pi}, \qquad (n = 1, 2, 3, \ldots). \tag{1-22}$$

Bohr introduced his postulates of nonradiating, discrete stationary states and of transitions between states without justification other than that they explained a number of experimentally known facts. A deeper under-standing of Bohr's assumptions became possible only at a later time, with the advent of wave mechanics.

It will be recalled that the empirical Ritz combination principle states that the frequencies of the spectral lines of an atom can be obtained as differences between pairs of term values. In the light of Bohr's postulates, the set of characteristic term values of an atom is to be interpreted as the set of "allowed" energy values for the atomic system, divided by the velocity of light times Planck's constant. With this interpretation, the Ritz combination principle and the Bohr frequency rule become identical.

1-8 The correspondence principle. Bohr's postulates imply that atomic systems are not entirely governed by the laws of classical physics. How-ever, a basic condition can be stated, which can serve as a guide in the development of a more adequate theory. Whatever form it may take, the new theory must agree with classical physics in any of the very large number of situations in which classical physics has been found to provide the correct answer.

Newtonian mechanics and classical electrodynamics are based on a wealth of thoroughly established experimental evidence. Therefore, it must be demanded that the quantum theory yield, in every instance, results that become identical with those of classical physics if the masses and dimensions of the system under consideration are made to approach the masses and dimensions of classical systems. This fundamental idea is already apparent in Bohr's early work and was explicitly stated by him in 1923. It is known as the *correspondence principle* and is essential in the formulation of quantum mechanics.

1–9 The Bohr atom. The greatest triumph of the old quantum theory was its successful interpretation of the spectrum of hydrogen. The later wave-mechanical treatment of atoms differs in many respects from the early theory. Yet, Bohr's theory of the hydrogen atom retains sufficient interest and historical importance to justify its somewhat detailed presentation at this point.

In accordance with Kepler's first law, the electron orbit is, in general, an ellipse. However, the special case of a circular orbit is subject to particularly straightforward treatment and will be considered first. For further simplification, the nucleus will be assumed to be at rest. This is equivalent to assigning it infinite mass. A correction for the finite mass of the nucleus will be introduced later.

Consider an atom in which a single electron of mass m moves with velocity v in a circular orbit of radius a, centered at the nucleus. The charge of the electron is $-e$ and that of the nucleus, $+Ze$. The centripetal force is equal to the Coulomb force between electron and nucleus:

$$\frac{mv^2}{a} = \frac{Ze^2}{a^2}.$$ (1–23)

The kinetic energy of the electron, therefore, is

$$T = \tfrac{1}{2}mv^2 = \frac{Ze^2}{2a}.$$ (1–24)

Now according to Bohr, the stationary states of the system are characterized by definite values of the angular momentum, such that

$$p_\phi = mva = n\hbar \qquad (n = 1, 2, 3, \ldots),$$ (1–25)

where the symbol $\hbar$ ("h-bar" or "Dirac's h") is written for the quantity $h/2\pi$. Hence, the velocity of the electron in its nth orbit is

$$v = \frac{n\hbar}{ma}. \qquad (1\text{–}26)$$

When this value for the velocity is substituted in Eq. (1–24), the radius of the orbit is obtained:

$$a_n = \frac{n^2\hbar^2}{mZe^2}. \qquad (1\text{–}27)$$

Of particular interest is the radius of the orbit in the ground state of the hydrogen atom ($n = 1, Z = 1$):

$$a_0 = \frac{\hbar^2}{me^2} = 0.528 \times 10^{-8} \text{ cm} = 0.528 \text{ A}. \qquad (1\text{–}28)$$

This quantity is known as the *Bohr radius*.

The total energy of the atom is the sum of kinetic and potential energies:

$$E = T + V = \frac{Ze^2}{2a} - \frac{Ze^2}{a} = -\frac{Ze^2}{2a} = -T. \qquad (1\text{–}29)$$

The energy of the system in its nth state is therefore

$$E_n = -\frac{Ze^2}{2a_n} = -\frac{mZ^2e^4}{2n^2\hbar^2}. \qquad (1\text{–}30)$$

1–10 Spectroscopic series. The Balmer series of lines in the hydrogen spectrum is shown in Fig. 1–1. The series is named for J. J. Balmer, who in 1885 discovered the following empirical relation describing the wave numbers of the lines in this group:

$$\bar{\nu} = R_{\mathrm{H}}\left(\frac{1}{2^2} - \frac{1}{n^2}\right). \qquad (1\text{–}31)$$

Here, R_{H} is the *Rydberg constant* for hydrogen, with the numerical value[1] 109,677.58 cm^{-1}. Substitution of the integers 3, 4, 5, . . . for n in the Balmer formula gives the wave numbers of the lines in the Balmer series. These wave numbers have been verified experimentally to the very great accuracy attainable in spectroscopic measurements.

The Balmer formula follows from the Bohr frequency rule which, in terms of wave numbers, reads

$$\bar{\nu} = \frac{E_n - E_m}{ch}. \qquad (1\text{–}32)$$

---

[1] J. W. M. DuMond and E. R. Cohen, *Revs. Modern Phys.* **25**, 691 (1953).

Substituting the energy values from Eq. (1–30), we obtain

$$\bar{\nu} = \frac{2\pi^2 me^4}{ch^3}\left(\frac{1}{m^2} - \frac{1}{n^2}\right). \tag{1–33}$$

It is apparent that the Balmer series is produced by a group of transitions in which the electron falls from an outer orbit into the second innermost ($n = 2$) orbit. This is indicated in Fig. 1–7. Note that the indices m and n of the frequency rule appear in reversed order in the Balmer formula, because E_m and E_n are negative quantities. The Rydberg constant (for infinite nuclear mass) appears as a combination of universal constants,

$$R_\infty = \frac{2\pi^2 me^4}{ch^3} = 109{,}737.31 \text{ cm}^{-1}. \tag{1–34}$$

The frequency rule suggests the existence of other series in the spectrum of hydrogen. In fact, other series arise when the index m in Eq. (1–33) has values different from 2. In accordance with this prediction, four other series have been observed, none of which, however, lie in the visible range.

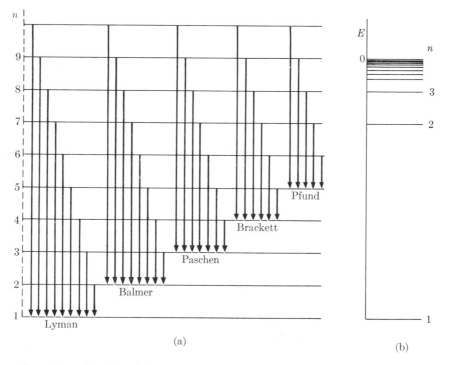

(a)

(b)

FIG. 1–7.   (a) Transitions producing the spectral series of hydrogen (energy not to scale). (b) Scale drawing of the energy levels of the hydrogen atom.

The five hydrogen series, named for their discoverers, are

$m = 1$ Lyman series   (ultraviolet)

$m = 2$ Balmer series   (visible)

$m = 3$ Paschen series   (infrared)

$m = 4$ Brackett series   (infrared)

$m = 5$ Pfund series   (infrared)

Representative transitions that produce lines of these series are also indicated in Fig. 1–7. The lines in each series fall closer together as the running index n is raised. In the limit of large n, the lines in each series converge toward the *series limit*, of wave number R/m^2.

1–11 Correction for finite mass of the nucleus. The theoretical value (1–34) for the Rydberg constant was obtained under the assumption that the nucleus has infinite mass and, therefore, remains at rest. Actually, both the nucleus and the electron move about their common center of mass. When the finite mass of the nucleus is taken into account, a slightly different value for the Rydberg constant is obtained, denoted by R_{H}. The correct energy levels do not differ greatly from those given by Eq. (1–30), since the ratio of nuclear mass to electron mass has the large value[1] 1836.13. One can account for the motion of the nucleus in the following manner: Newton's second law, as it applies to the electron, of mass m_1, moving with velocity $\mathbf{v}_1$ under the action of the Coulomb force $\mathbf{F}$, is

$$\mathbf{F} = m_1 \frac{d\mathbf{v}_1}{dt}. \qquad (1\text{--}35)$$

By Newton's third law, an equal and opposite force acts on the nucleus, of mass m_2, which has velocity $\mathbf{v}_2$:

$$-\mathbf{F} = m_2 \frac{d\mathbf{v}_2}{dt}. \qquad (1\text{--}36)$$

The relative acceleration of the electron with respect to the nucleus is

$$\frac{d}{dt}\mathbf{v} = \frac{d}{dt}(\mathbf{v}_1 - \mathbf{v}_2) = \frac{1}{m_1}\mathbf{F} + \frac{1}{m_2}\mathbf{F} = \frac{m_1 + m_2}{m_1 m_2}\mathbf{F}, \qquad (1\text{--}37)$$

so that the equation of motion for the electron with respect to the nucleus is

$$\mathbf{F} = \frac{m_1 m_2}{m_1 + m_2}\frac{d\mathbf{v}}{dt}. \qquad (1\text{--}38)$$

---

[1] J. W. M. DuMond and E. R. Cohen, *Revs. Modern Phys.* **25**, 691 (1953).

This result, familiar from mechanics, indicates that allowance for the finite mass of the nucleus can be made by substituting the *reduced mass* $m_1 m_2/(m_1 + m_2)$ in place of the electronic mass m_1 in the equation of motion. This substitution should be made in Eq. (1–24). Then, instead of m, the reduced mass will appear in Eq. (1–27) for the radius of the orbit and in Eq. (1–30) for the energy levels of the hydrogen atom; thus Eq. (1–30) becomes

$$E_n = - \frac{m_1 m_2 Z^2 e^4}{2(m_1 + m_2)\hbar^2 n^2}.$$ (1–39)

The Rydberg constant for finite nuclear mass is

$$R_H = \frac{2\pi^2 m_1 m_2 e^4}{(m_1 + m_2)ch^3} = R_\infty \frac{m_2}{m_1 + m_2},$$ (1–40)

and the wave numbers of the lines in the hydrogen spectrum are given by

$$\bar{\nu} = R_H \left(\frac{1}{m^2} - \frac{1}{n^2} \right) \qquad (n = m+1, m+2, \ldots).$$ (1–41)

Although the ratio of R_H to R_∞ is very nearly unity (approximately 1836/1837), the consequent small shift in wave numbers is within reach of spectroscopic observation. Of particular interest is the fact that Eq. (1–33), which contains R_∞ and hence does not depend upon the nuclear mass, predicts that certain spectral lines from singly ionized helium should coincide exactly with lines from hydrogen.[1] Such coincidences are not observed. The discrepancy caused Bohr in 1914 to take account of the nuclear mass, as has been done above.[2]

Another interesting consequence of the mass dependence of the Rydberg constant R_H is that the spectral lines of deuterium (H^2) are slightly shifted with respect to the corresponding lines of H^1. Through this shift the existence of deuterium was first proved by Urey and co-workers,[3] after it had been predicted by Birge and Menzel.[4]

---

[1] The factor $Z^2 = 4$ must be included in the expression for the energy levels of helium. For example, only alternate members of the Pickering series ($m = 4$) of He^+ are expected to coincide with lines of the Balmer series of H.

[2] N. Bohr, *Phil. Mag.* **27,** 506 (1914).

[3] H. C. Urey, F. G. Brickwedde, and G. M. Murphy, *Phys. Rev.* **40,** 1 (1932).

[4] R. T. Birge and D. H. Menzel, *Phys. Rev.* **37,** 1669 (1931).

1–12 Quantization of the phase integral. Bohr had shown that the stationary states of the hydrogen atom can be computed if it is assumed that the angular momentum of the system can have only discrete values, equal to multiples of a universal constant. A more general postulate, which includes Bohr's assumption and leads to the successful prediction of the allowed energy values for certain other systems as well, was discovered by W. Wilson[1] in 1915 and, independently, by A. Sommerfeld[2] the following year.

Let the motion of a periodic system of N degrees of freedom be described by the N coordinates q_i and the N canonically conjugate momenta p_i $(i = 1, 2, 3, \ldots, N)$. For example, the position of a mass point in three-dimensional space can be specified by the three cartesian coordinates $q_1 = x$, $q_2 = y$, and $q_3 = z$, and its velocity by the corresponding three momenta, $p_1 = p_x$, $p_2 = p_y$, and $p_3 = p_z$. Only systems whose motion is periodic will be considered. Then N *phase integrals* can be defined by

$$J_i = \oint p_i \, dq_i, \tag{1–42}$$

where the integration extends over one period of the variable q_i.

Wilson showed the following: If one postulates that the value of each phase integral must be equal to an integral multiple of Planck's constant h, then Planck's radiation law and the values for the energy levels of the hydrogen atom can be derived. It appeared, then, that periodic systems in nature obey the following rule:

$$\oint p_i \, dq_i = n_i h \qquad (n_i = 0, 1, 2, \ldots). \tag{1–43}$$

This relation, known as the *Wilson-Sommerfeld quantization rule*, was postulated because it led to results in agreement with experiment. In the following section, the rule will be applied to the computation of elliptic orbits in the hydrogen atom, and later, examples of its application to certain other simple systems will be studied.

In mechanics, it is often useful to think of the state of a system as represented by a point in *phase space*, i.e., in a space whose $2N$ coordinates are the coordinates q_i and the momenta p_i. A phase integral then represents an area in phase space. For example, consider a particle performing simple harmonic motion in one dimension. The motion is represented by a closed curve in a phase space that has the two coordinates x and p_x. The curve is an ellipse. The various stationary states of the oscillator, to be calculated in Section 1–16, can be represented by a set of ellipses, and the

¹ W. Wilson, *Phil. Mag.* **29**, 795 (1915).
² A. Sommerfeld, *Ann. Physik* **51**, 1 (1916).

quantization rule (1–43) demands that the area included between adjacent ellipses be equal to Planck's constant h. The dimensions of an area in phase space, as well as of the constant h, are those of *action* (energy $\times$ time or momentum $\times$ distance).

1–13 Elliptic electron orbits. Only circular electron orbits have been considered so far, and the theory of the hydrogen spectrum has been based entirely upon the result of quantization of the electron energy in such orbits. Circular orbits are, however, only special cases among those allowed by classical dynamics. Therefore, it is remarkable that the preceding calculations should have given all of the energy levels. In a sense, this is accidental, a consequence of the special form of the Coulomb law of force between electron and nucleus. The more general case of elliptic orbits will now be considered.[1]

From the discussion of Section 1–9 it is apparent that the radius a of a circular electron orbit in hydrogen is

$$a = \frac{p_\phi^2}{me^2}.\tag{1-44}$$

It is shown in mechanics that the elliptic orbits in the Coulomb field are described in general by

$$r = \frac{p_\phi^2/me^2}{1 - \epsilon \cos \phi},\tag{1-45}$$

where (r, ϕ) are the polar coordinates of the electron relative to the nucleus, measured in the plane of the orbit (Fig. 1–8), and

$$p_\phi = mr^2\dot{\phi}\tag{1-46}$$

is the constant value of the angular momentum. The number ϵ is the

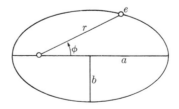

Fig. 1–8. Elliptic electron orbit.

---

[1] This problem was first solved by Sommerfeld in the same paper in which he introduced the quantization of the phase integral. Cf. A. Sommerfeld, *Ann. Physik* **51,** 1 (1916).

eccentricity of the ellipse which, in terms of p_ϕ and the total energy, is given by

$$\epsilon^2 = 1 + \frac{2p_\phi^2}{me^4}\, E. \tag{1–47}$$

For a circular orbit, the eccentricity is zero, so that Eq. (1–45) reduces to Eq. (1–44), and the energy becomes $E = -me^4/2p_\phi^2$.

The phase integrals for the system are

$$J_\phi = \oint p_\phi\, d\phi, \qquad J_r = \oint p_r\, dr, \tag{1–48}$$

where $p_r = m\dot{r}$ is the radial momentum of the electron. Since the angular momentum p_ϕ is constant, the first of the integrals (1–48) is

$$J_\phi = 2\pi p_\phi, \tag{1–49}$$

and the second can be evaluated most easily by observing that

$$p_r\, dr = m\frac{dr}{d\phi}\,\dot{\phi}\,\frac{dr}{d\phi}\, d\phi = p_\phi \left(\frac{1}{r}\frac{dr}{d\phi}\right)^2 d\phi; \tag{1–50}$$

whence, by Eq. (1–45),

$$J_r = p_\phi \int_0^{2\pi} \frac{\epsilon^2 \sin^2 \phi\, d\phi}{(1 - \epsilon \cos \phi)^2}. \tag{1–51}$$

Upon carrying out the integration, one obtains

$$J_r = J_\phi[(1 - \epsilon^2)^{-1/2} - 1]. \tag{1–52}$$

When ϵ^2 is eliminated between Eqs. (1–52) and (1–47), and the result solved for E, we find

$$E = -\frac{2\pi^2 me^4}{(J_r + J_\phi)^2}. \tag{1–53}$$

Now the Wilson-Sommerfeld quantization rule requires that

$$J_\phi = kh, \qquad J_r = n_r h \qquad (k,\, n_r \text{ integers}), \tag{1–54}$$

so that the quantized energy values are

$$E_n = -\frac{me^4}{2n^2\hbar^2}, \tag{1–55}$$

where $n = n_r + k$ is the *principal quantum number*.

When expressed in terms of the quantum numbers k and n, Eq. (1–47) for the eccentricity is

$$\epsilon^2 = 1 - \frac{k^2}{n^2},$$ (1–56)

which shows that the ratio of minor semiaxis to major semiaxis of the corresponding classical orbit is

$$\frac{b}{a} = \frac{k}{n}.$$ (1–57)

This provides a picture of the shape of the orbit belonging to given values of the quantum numbers.

It is now clear that several different orbits which are allowed by the quantization rules correspond to the same energy, namely, all orbits for which the sum $n_r + k$ has the same value n. This is a consequence of the fact that the energy of the system depends upon the phase integrals only through their sum, $J_r + J_\phi$, which is in turn an expression of the circumstance that the frequencies associated with the periodic motions of r and ϕ are the same, so that the classical orbit is a closed curve. Among all the orbits having a given n, however, there is always one which is circular, namely that for which $n_r = 0$, $k = n$. This explains why all the energy levels were found by consideration of the circular case alone.

If the quantum number k were zero, the electron would move along a straight line through the nucleus. For this reason, the value $k = 0$ was ruled out by Sommerfeld. The three electron orbits with $n = 3$ are shown in Fig. 1–9.

Whenever, in a dynamical system, there are several different states of motion which correspond to the same total energy, these states are said to be *degenerate* with respect to the energy. In the preceding example, the degeneracy arises because of the special nature of the Coulomb force, and is called *accidental*. It can be shown that a perturbation of the force, such as might be produced by the presence of other electrons, has the effect of removing the degeneracy, so that states with the same value of n but different values of k have different energies. Many of the properties of complex atoms can be understood in these terms and will be the subject of later discussion.

A more fundamental reason for the occurrence of degeneracy in the states of any isolated physical system is the fact that the energy does not depend upon the spatial orientation of the system. For example, the orientation of the plane of the electronic orbit is not important for the determination of the energy of the hydrogen states. A *fundamental* degeneracy of this kind is associated with the direction of the angular momentum of the system which, as will appear later, is also subject to quantum rules allowing

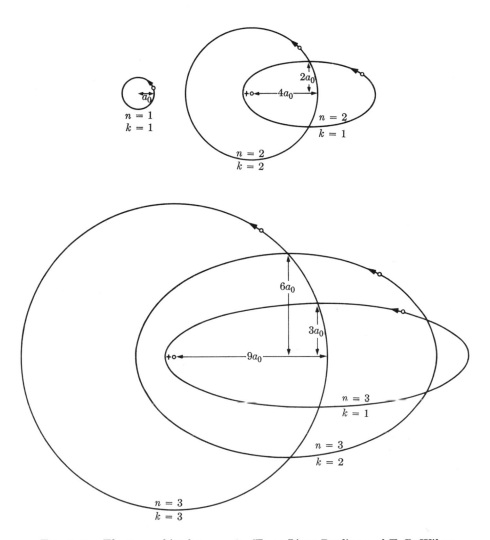

Fig. 1–9. Electron orbits for $n = 3$. (From Linus Pauling and E. B. Wilson, *Introduction to Quantum Mechanics*, McGraw-Hill, New York, 1935, by permission.)

only a discrete set of states instead of the continuum of classical physics. This point will be discussed in detail in connection with the quantum properties of angular momentum. It may be remarked here, however, that spatial degeneracy is removed only if the system is subjected to forces (e.g., a magnetic or electric field) which determine a direction in space in such a way that the energy is dependent upon the orientation of the system.

1–14 The particle in a box. It has been indicated in Section 1–12 that the Wilson-Sommerfeld quantization rule can be applied successfully to a number of periodic systems other than the hydrogen atom. A few such applications will be treated in the remaining sections of this chapter.

One of the simplest examples illustrating the quantization of the phase integral is that of a particle, of mass m, constrained to move in one dimension, x. At $x = 0$ and $x = a$, the particle undergoes perfectly elastic collisions with rigid "walls" [Fig. 1–10(a)]. No other forces act on the parti-

(a)

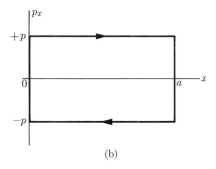

(b)

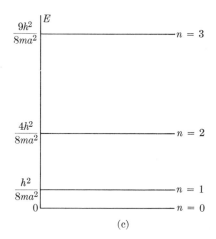

(c)

Fig. 1–10.   Particle in a box: (a) one-dimensional box; (b) phase diagram; (c) energy levels.

cle. The phase diagram is shown in Fig. 1–10(b). The phase integral for the periodic motion is

$$J = \oint p_x \, dx = p \oint dx = 2ap, \tag{1-58}$$

where p is the constant magnitude of the momentum. The quantization rule demands that

$$J = 2ap = nh. \tag{1-59}$$

The allowed values for the momentum follow from Eq. (1–59). The total energy of the system is equal to the kinetic energy:

$$E_n = \frac{p^2}{2m} = \frac{n^2 h^2}{8ma^2} \qquad (n = 0, 1, 2, \ldots). \tag{1-60}$$

According to classical mechanics, the particle can move with any speed, and consequently, its energy can have any positive value. For the continuous energy spectrum of classical physics, quantum mechanics substitutes a spectrum of discrete energy levels with values E_n given by Eq. (1–60). The system can exist only in one of the energy states E_n, which are plotted in Fig. 1–10(c).

The mass of the particle and the square of a appear in the denominator of the expression for the energy levels. Furthermore, the Planck constant h is a very small quantity. Consequently, the spacing of the energy levels E_n is very small for a system of macroscopic mass and dimensions. This can be seen by substituting one gram for m and one centimeter for a, for example. The energy difference between the first two levels of such a system is only 1.6×10^{-53} erg. This difference is so small that it would escape detection in any conceivable experiment. For a system of macroscopic size, the discrete quantum-mechanical spectrum becomes so finely structured as to be indistinguishable from the continuous spectrum of classical mechanics, in accordance with the demands of the correspondence principle.

However, if an electron is constrained to move in a "box" of atomic dimensions, the spacing of the energy levels is appreciable. For example, if $a = 10^{-8}$ cm and $m = 9.1 \times 10^{-28}$ gm, Eq. (1–60) shows that $E_1 = 0.6 \times 10^{-10}$ erg and $E_2 = 2.4 \times 10^{-10}$ erg. The difference between these two levels is 1.8×10^{-10} erg or 113 ev, which is large on the scale of atomic energies.

1–15 The rigid rotator. A rigid body constrained to rotate about a fixed axis is a system for which the phase integral can be expressed in terms of the angle of rotation ϕ and the angular momentum $p_\phi = I\dot{\phi}$. Here, I

is the moment of inertia of the system. There is no potential energy, so
that the total energy is given by

$$E = \tfrac{1}{2}I\dot{\phi}^2. \tag{1–61}$$

The angular momentum is a constant of the motion. Classically, the frequency of rotation is given by

$$\nu_c = \frac{\dot{\phi}}{2\pi} = \frac{1}{2\pi}\left(\frac{2E}{I}\right)^{1/2}. \tag{1–62}$$

The quantum-mechanical energy levels are found by quantizing the
phase integral:

$$J = \oint p_\phi \, d\phi = 2\pi p_\phi = nh; \qquad p_\phi = n\hbar. \tag{1–63}$$

Hence, the quantized energy levels of the rigid rotator are

$$E_n = \frac{p_\phi^2}{2I} = \frac{n^2\hbar^2}{2I}. \tag{1–64}$$

If the rotator carries a charge, classical electrodynamics predicts the
continuous emission of radiation. The frequency of this radiation is equal
to the frequency ν_c of the rotator, given by Eq. (1–62). On the other hand,
Bohr's first postulate of nonradiating, stationary states, which applies to
all periodic quantum-mechanical systems, indicates that the rotator does
not radiate so long as it remains in one of the energy states E_n. Radiation
from the quantum-mechanical charged rotator occurs only when the
system jumps from one energy level E_n to a lower level E_m. According to
the Bohr frequency rule, the radiation emitted in such a transition has
the frequency

$$\nu = \frac{E_n - E_m}{h} = (n^2 - m^2)\frac{h}{8\pi^2 I}. \tag{1–65}$$

Absorption of a photon of the same frequency by a rotator in state m will
raise the system to the state n.

In principle, the quantum-mechanical rotator might undergo a transition
from an initial level E_n to any other energy level E_m. Many other levels
may be available as a final state for the transition. Hence, a large number
of frequencies would be emitted or absorbed by the system unless the number of possible transitions were reduced by some subsidiary condition.
Such a condition, called a *selection rule*, follows from the correspondence
principle.

According to the correspondence principle, the quantum-mechanical frequency given by Eq. (1–65) must coincide, in the limit of large quantum numbers, with the classical frequency of Eq. (1–62). For large n and m, we have

$$(n^2 - m^2) = (n - m)(n + m) \approx \Delta n \cdot 2n \qquad (\Delta n \ll 2n),$$

so that Eq. (1–65) for the quantum-mechanical frequency can be re-written as

$$\nu = \Delta n \frac{nh}{4\pi^2 I} = \Delta n \frac{1}{2\pi} \left(\frac{n^2 h^2}{8\pi^2 I}\right)^{1/2} \left(\frac{2}{I}\right)^{1/2} \qquad (n \text{ large}). \qquad (1\text{--}66)$$

This expression contains the square root of the quantum-mechanical energy E_n as given by Eq. (1–64) and is equivalent to

$$\nu = \Delta n \frac{1}{2\pi} \left(\frac{2E}{I}\right)^{1/2} \qquad (n \text{ large}). \qquad (1\text{--}67)$$

The quantum-mechanical frequency becomes equal to the classical frequency of Eq. (1–62) if it is postulated that, among all possible transitions, only those between adjacent energy levels are *allowed:*

$$\Delta n = \pm 1. \qquad (1\text{--}68)$$

This restriction on the change of quantum number in a transition is the selection rule for the rigid rotator. The positive sign corresponds to transitions in which energy is absorbed by the system, while $\Delta n = -1$ characterizes transitions in which energy is emitted. The rule, which has been derived for the limiting case of large n, is found experimentally to hold throughout the entire range of quantum numbers. Transitions which would violate the selection rule do not occur and are said to be *forbidden.*

1–16 The harmonic oscillator. The one-dimensional harmonic oscillator consists of a particle of mass m bound to the origin by a restoring force $F = -kx$. The force constant k and angular frequency $\omega = 2\pi\nu$ are connected by the relation $k = m\omega^2$. The particle performs simple harmonic motion with amplitude a; its displacement from the origin at time t is $x = a \sin \omega t$.

The representative point in phase space, with coordinates x and p_x, describes an ellipse, as has been pointed out in Section 1–12. The energy of the oscillator is

$$E = \frac{p_x^2}{2m} + \frac{1}{2} kx^2. \qquad (1\text{--}69)$$

The semiaxis a of the ellipse is found by setting $p_x = 0$ in Eq. (1–69), and the semiaxis b is found by setting $x = 0$, so that

$$a = \left(\frac{2E}{k}\right)^{1/2}; \quad b = (2mE)^{1/2}. \tag{1-70}$$

The phase integral J is the area πab of the ellipse, or

$$J = \pi \left(\frac{2E}{k}\right)^{1/2} (2mE)^{1/2} = \frac{2\pi E}{\omega}. \tag{1-71}$$

Quantization according to Eq. (1–43) therefore gives the energy values

$$E_n = n\hbar\omega = nh\nu \quad (n = 0, 1, 2, \ldots). \tag{1-72}$$

Thus, the quantum-mechanical energy levels of the harmonic oscillator are equally spaced, with an interval

$$\Delta E = h\nu. \tag{1-73}$$

Classically, a charge that performs simple harmonic motion with frequency ν emits radiation of this frequency only, with no overtones. Many quantum-mechanical frequencies appear possible, and in order to satisfy the correspondence principle, a selection rule must be introduced. In the limit of large quantum numbers, the selection rule must allow only the quantum-mechanical frequency which corresponds to the classical frequency of the oscillator. This purpose is accomplished by the rule

$$\Delta n = \pm 1. \tag{1-74}$$

The correspondence principle demands agreement with classical theory only in the limit of highly excited states. However, it is an interesting, experimentally confirmed property of the harmonic oscillator that its quantum-mechanical frequency, $(E_n - E_{n-1})/h$, is equal to the classical frequency for all quantum numbers. The selection rule (1–74) holds for dipole radiation. Higher multipole orders are less likely, but not strictly forbidden (cf. Chapter 11).

The calculation of the oscillator energy levels is easily extended to more than one dimension. A particle oscillating in a plane and bound to the origin by a force $-k_x x$ acting in the x-direction, and a force $-k_y y$ in the y-direction, has a total energy $E = n_x h\nu_x + n_y h\nu_y$, where n_x and n_y are integers. If the oscillator is isotropic, the force constants k_x and k_y are equal, and $E_n = (n_x + n_y)h\nu = nh\nu$. There are $(n + 1)$ combinations of quantum numbers n_x and n_y that yield a given value of n and hence result in the same energy E_n. That is, for the two-dimensional oscillator, the energy levels for $(n + 1)$ different quantum states of the system coincide, and the energy level E_n is said to be $(n + 1)$-fold degenerate. Similarly, the nth energy level of the three-dimensional isotropic oscillator is $E_n = (n_x + n_y + n_z)h\nu$ and is $[(n + 1)(n + 2)/2]$-fold degenerate.

The harmonic oscillator can be used successfully as a model for many systems found in nature. For example, an atom in a molecule is bound by a potential which has a minimum for a certain interatomic distance. In the vicinity of the minimum, the potential-energy function can be approximated by the parabolic potential of the harmonic oscillator. This is equivalent to the assumption that, for small displacements from its equilibrium position, the atom experiences a restoring force that is proportional to the magnitude of the displacement. The vibrational energy levels of an oscillating molecule are therefore given approximately by an expression of the form

$$E_n = nh\nu. \tag{1–75}$$

In addition, a molecule is, in general, capable of rotation. The rigid rotator of Section 1–15 is a suitable model from which the rotational energy states of a diatomic molecule can be derived. If k is the rotational quantum number, the energy levels of a vibrating and rotating diatomic molecule are given by

$$E_{n,k} = nh\nu + \frac{k^2\hbar^2}{2I}, \tag{1–76}$$

which agrees approximately with spectroscopic evidence.

1–17 Shortcomings of the old quantum theory. The success of the old quantum theory in obtaining a solution of the problem of the hydrogen atom caused high hopes among the physicists of the day, who felt that the road to the solution of other problems in atomic physics might now be open. However, these hopes soon proved unjustified. Attempts to use Bohr's postulates and the Wilson-Sommerfeld quantization rule to calculate the energy levels of more complicated systems failed even when directed at the normal helium atom and the H_2 molecule. Even for the hydrogen atom, it was not possible to understand the relative intensities of the spectral lines.

Since the phase integral is defined only for periodic systems, the quantization of this integral could be carried out only for a limited number of cyclic processes. Moreover, new phenomena without classical analogues, connected with the spin of the electron and the Pauli exclusion principle, were yet to be discovered.

On a fundamental level, the basic inadequacy of the old quantum theory is that Bohr's postulate of discrete, nonradiating energy states was an empirical assumption, successful in its implications but without underlying physical foundation. Only the subsequent discovery of the wave nature of matter yielded a basis from which not only Bohr's postulates but also most of modern wave mechanics could be derived in a satisfactory manner.

REFERENCES

BORN, MAX, *Atomic Physics*. London: Blackie and Son Ltd., 1951. Very clear discussions of many of the experimental foundations of quantum mechanics.

HERZBERG, GERHARD, *Atomic Spectra and Atomic Structure*. New York: Dover Publications, 1944. Concise presentation of the basic principles of atomic spectroscopy.

PAULING, LINUS and WILSON, E. B., *Introduction to Quantum Mechanics*. New York: McGraw-Hill Book Company, Inc., 1935. Clear discussion of the old quantum theory including many applications. The hydrogen atom is covered in great detail.

RICHTMYER, F. K., and KENNARD, E. H., *Introduction to Modern Physics*. 5th ed. New York: McGraw-Hill Book Company, Inc., 1955. Extensive text which contains a thorough discussion of blackbody radiation. Valuable references to the original literature are included.

SEMAT, HENRY, *Introduction to Atomic Physics*. New York: Rinehart and Company, Inc., 1946. The book includes an elementary discussion of the hydrogen atom.

WHITE, H. E., *Introduction to Atomic Spectra*. New York: McGraw-Hill Book Company, Inc., 1934. A more complete and extensive, but less critical, text than Herzberg's book.

PROBLEMS

1-1. Consider a classical model of the hydrogen atom, with the electron moving in a circular orbit about the nucleus. It is shown in classical electrodynamics that a charge e, subjected to an acceleration $\ddot{r}$, radiates energy at the rate

$$\frac{dE}{dt} = -\frac{2e^2}{3c^3}\,(\ddot{r})^2.$$

Find the time during which the radius of the classical model of a hydrogen atom would shrink by a factor two.

1-2. Show that the Planck radiation law and the Rayleigh-Jeans law become identical if the size of the quantum ϵ is allowed to vanish, or if the temperature becomes very high.

1-3. Prove that Wien's displacement law requires that $\epsilon = h\nu$ in the Planck radiation law.

1-4. Calculate $\int_0^\infty u_\nu\,d\nu$ and derive the formula for σ given in Eq. (1-12).

1-5. Prove that the constant b in Wien's law is

$$b = \frac{hc}{4.965k},$$

where $x = 4.965$ is the positive root of $[1 - (x/5)]e^x = 1$.

1-6. Look up a recent compilation of the fundamental constants and examine the accuracy with which the relations of Problems 1–4 and 1–5 are verified by experiment. Describe one or two experiments which are basic for the determination of h.

1-7. The work function of zinc is 3.6 ev. What is the maximum energy of the photoelectrons ejected by ultraviolet light of wavelength 3000 A?

1-8. In an experiment on the photoelectric effect, a cesium plate is illuminated with ultraviolet light of wavelength 2000 A. The stopping potential is found to be 4.21 volts. What is the work function of the cesium surface?

1-9. In addition to the line $\lambda = 2536$ A mentioned in Section 1–5, the mercury spectrum has a second strong line at $\lambda = 1849$ A. In a Franck-Hertz experiment, at what voltage would one expect a current drop associated with the second line?

1-10. Look up Lord Rutherford's paper on alpha-particle scattering in *Phil. Mag.* **21**, 669 (1911), as well as Geiger and Marsden's report on the experimental verification of Rutherford's scattering formula, *Phil. Mag.* **25**, 604 (1913).

1-11. Verify Eq. (1-52) for J_r. [*Note:* The calculus of residues (Appendix A–1) can be applied to advantage in calculating this integral.]

1-12. Consider a neutral particle with the mass of an electron, which is bound by gravitational attraction to another neutral particle with the mass of a proton. Apply the Wilson-Sommerfeld quantization rule and calculate the energy levels and the elliptic orbits. Compute the dimensions of the system in its ground state.

1-13. Calculate the wavelength of the first three lines of the Balmer series of hydrogen and of deuterium and find the shift between corresponding lines.

1-14. Compute the wave numbers of the first four members of the Pickering series ($m = 4$) of singly ionized helium and calculate the shift of the even-numbered lines with respect to the corresponding lines of the Balmer series of hydrogen.

1-15. Make a scale drawing of the elliptic orbits of the hydrogen electron, corresponding to the principal quantum number $n = 4$.

1-16. Compute the first three energy levels (in electron volts) for a proton in a one-dimensional "box" of length 10^{-12} cm.

1-17. An electron is constrained to move in a box with sides 10^{-8}, 2×10^{-8}, and 3×10^{-8} cm. Find the two lowest energy levels for which all three quantum numbers are greater than zero.

1-18. Find the energy levels of the H^1Cl^{35} molecule, considered as a rigid rotator, and plot them (interatomic distance 1.27×10^{-8} cm). What is the frequency of the radiation that would be emitted in the transition

$$(n = 2) \rightarrow (n = 1)?$$

CHAPTER 2

FOUNDATIONS OF WAVE MECHANICS

2–1 Photons as particles: the Compton effect. The photoelectric effect and the Franck-Hertz experiment show that the energy of an electromagnetic wave is absorbed and emitted in discrete quanta. The Compton effect gives evidence of the corpuscular nature of radiation. The effect was first described by A. H. Compton in 1923 in two papers on the theoretical and experimental aspects of the scattering of x-rays from light elements.[1]

A classical argument concerning the phenomenon with which Compton's work dealt will be presented first. Compton used x-rays from molybdenum, with an energy of approximately 20 kev. On entering a scattering material, such an x-ray interacts with the atomic electrons. These electrons can be considered essentially free since they are bound to the atoms of the scatterer with an energy of the order 10 ev, which is very much smaller than the energy of the x-ray. The electric field of the incident electromagnetic wave train exerts a force on the atomic electrons and causes them to oscillate with the frequency of the x-ray. An electron that oscillates in simple harmonic motion will radiate as an electric dipole. The frequency of the radiation is that of the oscillations and therefore is initially equal to the frequency of the incident x-ray.

The incident wave train carries momentum E/c, where E is the energy of the wave train, and c the velocity of light. Since the x-ray gives some of its energy to the electron, it loses momentum. This momentum is imparted to the electron, which recoils in the direction of propagation of the incident wave train. As the electron recedes from the source of the x-ray, it no longer sees the original frequency, but a lower frequency, due to the Doppler effect. The electron then re-radiates this lower frequency.

This classical argument leads to the conclusion that the frequencies of scattered x-rays should have a continuous range of values. When the target electron is still at rest, the frequency of the scattered radiation, observed in the laboratory, should be that of the incident x-ray. The frequency should then decrease continuously to a final value, attained when the electron has gained its final velocity, after scattering the entire wave train.

This conclusion was not borne out by the results of Compton's experiments. Rather, the x-rays scattered by electrons at a particular angle are found to have just one sharply defined frequency, lower than that of the incident rays. This fact shows that the scattering is not a gradual process

---

[1] A. H. Compton, *Phys. Rev.* **21,** 483 (1923) and **22,** 409 (1923).

35

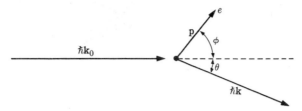

FIG. 2–1.   The Compton effect: momentum vectors.

during which the electron picks up momentum at a continuous rate, but that the interaction of x-ray and electron is instantaneous.

The fact that the x-ray transfers its energy and momentum instantaneously, in one packet, suggests that the scattering process can be treated as a collision between two particles: a photon and an electron.   Such instantaneous transfer does, indeed, occur in a two-particle collision. Figure 2–1 shows a diagram of the event. The incident "particle" is a photon of frequency ν and momentum $\hbar \mathbf{k}_0$, where $\mathbf{k}_0$ is the propagation vector of magnitude $1/\lambda_0 = 2\pi\nu_0/c$. The scattered photon leaves at the angle θ with momentum $\hbar \mathbf{k}$, and the electron recoils at the angle ϕ and carries off momentum $\mathbf{p}$.

The square of the total relativistic energy of the electron is

$$E^2 = m^2c^4 + p^2c^2. \tag{2-1}$$

Conservation of energy is expressed by

$$\hbar c k_0 + mc^2 = \hbar c k + E,$$

or

$$\hbar c(k_0 - k) = E - mc^2. \tag{2-2}$$

Squaring each side of this equation yields

$$\hbar^2 c^2 (k_0^2 + k^2 - 2kk_0) = E^2 + m^2c^4 - 2Emc^2. \tag{2-3}$$

Conservation of momentum requires

$$\hbar \mathbf{k}_0 - \hbar \mathbf{k} = \mathbf{p}. \tag{2-4}$$

The square of this relation is

$$\hbar^2 c^2 (k_0^2 + k^2 - 2\mathbf{k}_0 \cdot \mathbf{k}) = E^2 - m^2c^4, \tag{2-5}$$

in which p^2 has been eliminated by means of Eq. (2–1).   One-half the difference between Eq. (2–5) and Eq. (2–3) is

$$k_0 k - \mathbf{k}_0 \cdot \mathbf{k} = \frac{m}{\hbar^2} (E - mc^2),$$

which can be transformed, by use of Eq. (2–2), into

$$k_0 k - \mathbf{k}_0 \cdot \mathbf{k} = \frac{mc}{\hbar} (k_0 - k). \qquad (2\text{–}6)$$

Dividing by $k_0 k$, it is found that

$$1 - \cos \theta = \frac{mc}{\hbar} \left(\frac{1}{k} - \frac{1}{k_0} \right),$$

or

$$\lambda - \lambda_0 = \frac{\hbar}{mc} (1 - \cos \theta). \qquad (2\text{–}7)$$

The frequency ν of the scattered photon is therefore

$$\nu = \frac{\nu_0}{1 + (h\nu_0/mc^2)(1 - \cos \theta)}. \qquad (2\text{–}8)$$

Experiment shows that this equation is correct. The important theoretical fact is that it was obtained by considering the light quantum or photon as a particle.

If the Compton effect has the true features of a two-particle collision, then it must be possible to observe the scattered photon and the recoiling electron simultaneously. This was first tested in a classic experiment by Bothe and Geiger[1] in 1925. A beam of x-rays was scattered in a small volume of hydrogen at atmospheric pressure. The scattered photons and recoil electrons were detected by Geiger tubes, arranged opposite each other and at right angles to the x-ray beam. Simultaneity of the counts was determined by a photographic method. The center wires of the Geiger tubes were connected to two electrometers, and images of the electrometer leaves were projected side by side onto a film. By running the film at a controlled speed of 12.5 meters/sec, the resolving time of the device could be made to reach 10^{-4} second, and coincidences between photons and electrons were proved within this relatively long resolving time. More recent experiments have shown the existence of coincidences within the shortest resolving times that can be attained by modern electronic circuits.[2]

In summary, the details of the Compton effect show clearly that photons behave like particles when they collide with electrons. It is also true that light exhibits the well-known interference phenomena of physical optics, which can only be explained by the wave nature of electromagnetic radiation. In such phenomena, light consists of waves. This dual wave-particle nature of electromagnetic radiation is an experimental fact.

---

[1] W. Bothe and H. Geiger, *Z. Physik* **32**, 639 (1925).
[2] A. Bernstein and A. K. Mann, *Am. J. Physics* **24**, 445 (1956).

2–2 **Particle diffraction.** While light exhibits particle characteristics in its interaction with matter, as in the Compton effect, there also is evidence to indicate that particles exhibit wave-like behavior.

The first proof of the existence of "matter waves" was obtained in 1927 by Davisson and Germer[1] in an investigation which started from a laboratory accident. During an experiment on the angular distribution of electrons scattered from a nickel target, an explosion broke the vacuum and the target oxidized. The oxide layer was removed by lengthy heat treatment. In the process, the target happened to recrystallize into several large nickel crystals. When the experiment was finally continued, the angular distribution of electrons had changed markedly. Strong beams of electrons were observed coming from the target in certain directions, which were dependent upon the speed of the incident electrons. Davisson and Germer noticed that the strongest of the scattered electron beams corresponded accurately to diffraction maxima that would be expected in the diffraction of x-rays by the same crystal.[2] The angular distribution of the scattered electrons was analogous to optical diffraction patterns from a plane diffraction grating whose lines consisted of the rows of nickel atoms in the surface of the target crystal. Further experiments proved that the angular distribution of electrons scattered from a single nickel crystal actually represents an interference phenomenon—the diffraction of electron "matter waves" from the crystal lattice. The wavelength associated with the diffraction pattern is precisely that predicted somewhat earlier by L. de Broglie on the basis of theoretical considerations that will be discussed in Section 2–7. This wavelength which, according to de Broglie, is associated with a particle of momentum p, is

$$\lambda = h/p, \tag{2-9}$$

where h is Planck's constant.

The diffraction of electron waves by thin metal foils was first demonstrated by G. P. Thomson.[3] Properly treated metal foils contain a large number of crystals oriented at random and yield (similarly to crystalline powders) x-ray diffraction patterns that consist of a system of concentric rings. Thomson obtained ring-shaped diffraction patterns with 20- to 60-kev electrons that had passed through films of gold, platinum, alumi-

---

[1] C. Davisson and L. H. Germer, *Nature* **119**, 558 (1927) and *Phys. Rev.* **30**, 705 (1927).

[2] The possibility of electron diffraction had been foreseen by W. Elsasser [*Naturwiss.* **13**, 711 (1925)].

[3] G. P. Thomson, *Nature* **120**, 802 (1927) and *Proc. Roy. Soc., London*, **A117**, 600 (1928). Cf. also G. P. Thomson, *The Wave Mechanics of Free Electrons.* New York: McGraw-Hill Book Company, Inc., 1930.

FIG. 2-2. Diffraction pattern produced by electrons passing through a thin gold foil. (After G. P. Thomson, *The Wave Mechanics of Free Electrons*, Cornell University Press, Ithaca, N. Y., 1930, by permission.)

num, and other materials. One of the striking patterns that he obtained is reproduced in Fig. 2-2.

The size of a small-angle diffraction pattern is proportional to the wavelength. Hence, Thomson was able to verify de Broglie's wavelength equation (2-9) by showing that the diameters of the innermost diffraction rings are inversely proportional to the corresponding electron momenta. The results for gold are reproduced in Table 2-1. The last column contains the product of ring diameter D_1, proportional to wavelength, and of the quantity $V^{1/2}[1 + eV/(1200\ mc^2)]$, proportional to electron momentum (including a relativistic correction). The constancy of this product, within the experimental error, confirms de Broglie's hypothesis.

Since the early work, a large number of experiments on the wave

TABLE 2-1

DATA ON ELECTRON DIFFRACTION BY GOLD FOIL

(from G. P. Thomson's experiments)

Accelerating voltage V (volts)	Diameter of first diffraction ring D_1 (cm)	$D_1 V^{1/2}[1 + eV/(1200mc^2)]$
24,600	2.50	398
31,800	2.15	390
39,400	2.00	404
45,600	1.86	405
54,300	1.63	388
61,200	1.61	410

nature of particles has been conducted.   Diffraction effects have been
shown to occur on reflection and transmission not only of electrons, but
also of heavier particles.   In recent years, *neutron optics* has become an
important branch of nuclear physics.[1]   The de Broglie wavelength in
angstroms of a neutron of energy E electron volts is

$$\lambda = 0.287E^{-1/2}. \tag{2-10}$$

It follows that a 0.22-ev neutron has approximately the same wavelength (0.62 A) as a 20-kev x-ray.   A number of "optical" properties of
pile neutrons have been studied, including diffraction, refraction, reflection,
and polarization.

Neutron diffraction has become a useful technique in the study of
crystal structure.   In one of several methods, a beam of neutrons is sent
through a single crystal.   If d is the spacing of the lattice planes in the
crystal and θ the angle which the incident neutron beam makes with these
planes, then only neutrons will be reflected whose wavelengths satisfy
the Bragg condition

$$n\lambda = 2d \sin \theta. \tag{2-11}$$

A pattern of *Laue spots* is produced by the diffracted neutrons, and conclusions about the structure of the crystal can be drawn.   A Laue pattern
obtained from a sodium chloride crystal is reproduced in Fig. 2–3.

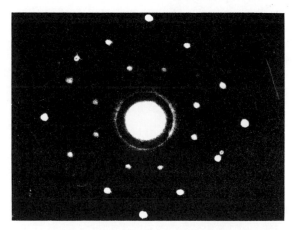

FIG. 2–3.   A Laue photograph of NaCl obtained by C. G. Shull and E. O.
Wollan at the Oak Ridge National Laboratory.   (From D. J. Hughes, *Pile
Neutron Research*, Addison-Wesley Pub. Co., Reading, Mass., 1953.)

---

[1] See, for example: D. J. Hughes, *Pile Neutron Research*. Reading, Mass.:
Addison-Wesley Publishing Co., Inc., 1953, Chapters 10 and 11.

The wave properties of neutrons make possible the construction of crystal monochromators which permit the selection of a very nearly monoenergetic beam from the neutron flux that is drawn from a pile.[1] A collimated beam of neutrons is allowed to fall on a large single crystal (e.g. lithium fluoride) at a small angle. Only those neutrons whose wavelength obeys the Bragg formula (2–11) for the given angle of incidence are reflected. This method is very useful in the investigation of neutron resonances and similar phenomena.

2–3 Elements of Fourier analysis. The experiments of Davisson and Germer and of Thomson clearly show that certain aspects of the motion of a very small material particle are wavelike. Associated with the motion of a particle is a wavelength, inversely proportional to the momentum of the particle, on the basis of which one can predict the observed diffraction effects. The recognition of this wave-particle duality has been essential in the development of modern physical theory.

The remainder of this book will be devoted to a description of quantum mechanics, and it will be shown how this discipline leads to a consistent picture of the phenomena. It will be helpful to have a mathematical structure adapted to the description of waves. Hence, it becomes advisable to digress temporarily to present a description of the essentials of Fourier analysis.

The principal theorems of Fourier analysis which are needed here can be derived in a heuristic way by considering the Fourier transform to be a suitable limit of a trigonometric series. Let a function $\psi(x)$ be given, which is defined in the interval $(-L/2, L/2)$, and let $\phi(x)$ be a periodic function, of period L, which coincides with ψ throughout the interval within which the latter is defined (Fig. 2–4). [The auxiliary function $\phi(x)$ is introduced temporarily.]

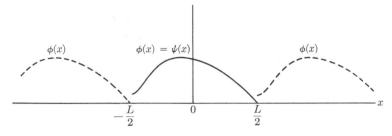

Fig. 2–4. The function $\psi(x)$, defined in $(-L/2, L/2)$, and the periodic function $\phi(x)$.

---

[1] D. J. Hughes, *op. cit.*, p. 159.

In the theory of Fourier series it is shown[1] that, provided $\phi(x)$ satisfies suitable (e.g., Dirichlet's) conditions, it can be expanded in harmonic series of the form

$$\phi(x) = \frac{a_0}{2} + \sum_{n=1}^{\infty} \left[a_n \cos \frac{2n\pi x}{L} + b_n \sin \frac{2n\pi x}{L} \right], \qquad (2\text{--}12)$$

and that at points of discontinuity the series on the right is equal to the mean value of ϕ. The coefficients a_n and b_n can be evaluated by means of the *orthogonality relations*

$$\frac{2}{L} \int_{-L/2}^{L/2} \cos \frac{2m\pi x}{L} \sin \frac{2n\pi x}{L} \, dx = 0, \qquad (2\text{--}13)$$

and

$$\frac{2}{L} \int_{-L/2}^{L/2} \cos \frac{2m\pi x}{L} \cos \frac{2n\pi x}{L} \, dx = \frac{2}{L} \int_{-L/2}^{L/2} \sin \frac{2m\pi x}{L} \sin \frac{2n\pi x}{L} \, dx = \delta_{mn}, \qquad (2\text{--}14)$$

where δ_{mn} is the Kronecker symbol, which is unity if $m = n$ and zero otherwise.

For example, if the expression for ϕ is multiplied by $(2/L) \cos (2m\pi x/L)$ and integrated over the interval $(-L/2, L/2)$, the result is

$$\frac{2}{L} \int_{-L/2}^{L/2} \phi(x) \cos \frac{2m\pi x}{L} \, dx = \frac{a_0}{2} \frac{2}{L} \int_{-L/2}^{L/2} \cos \frac{2m\pi x}{L} \, dx$$

$$+ \sum_{n=1}^{\infty} a_n \frac{2}{L} \int_{-L/2}^{L/2} \cos \frac{2m\pi x}{L} \cos \frac{2n\pi x}{L} \, dx, \quad (2\text{--}15)$$

whence, for $m = 0$,

$$a_0 = \frac{2}{L} \int_{-L/2}^{L/2} \phi(x) \, dx, \qquad (2\text{--}16)$$

and for $m \neq 0$,

$$a_m = \frac{2}{L} \int_{-L/2}^{L/2} \phi(x) \cos \frac{2m\pi x}{L} \, dx. \qquad (2\text{--}17)$$

By similar calculation with the multiplier $2/L$ [sin $(2m\pi x/L)$], the co-

---

1 W. Kaplan, *Advanced Calculus*. Reading, Mass.: Addison-Wesley Publishing Company, Inc., 1952, Chapter 7.

efficients b_n are obtained, and the results are summarized in the equations

$$a_n = \frac{2}{L} \int_{-L/2}^{L/2} \phi(x) \cos \frac{2n\pi x}{L}\, dx, \tag{2–18}$$

$$b_n = \frac{2}{L} \int_{-L/2}^{L/2} \phi(x) \sin \frac{2n\pi x}{L}\, dx. \tag{2–19}$$

[The series can be shown to converge uniformly in the interval $(-L/2,\, L/2)$, so that the term-by-term integration involved in the derivation of these equations is justified.]

The series (2–12) can be written in a more compact form by substituting

$$e^{i(2n\pi x/L)} = \cos \frac{2n\pi x}{L} + i \sin \frac{2n\pi x}{L}$$

in each of the trigonometric terms; thus

$$\phi(x) = \tfrac{1}{2}\left\{ a_0 + \sum_{n=1}^{\infty} [(a_n - ib_n)e^{i(2n\pi x/L)} + (a_n + ib_n)e^{-i(2n\pi x/L)}] \right\}. \tag{2–20}$$

Introducing the coefficients c_n:

$$c_n = \begin{cases} \tfrac{1}{2}(a_n - ib_n), & n > 0, \\ \tfrac{1}{2}a_0, & n = 0, \\ \tfrac{1}{2}(a_{-n} + ib_{-n}), & n < 0, \end{cases} \tag{2–21}$$

the Fourier series becomes

$$\phi(x) = \sum_{n=-\infty}^{\infty} c_n e^{i(2n\pi x/L)}, \tag{2–22}$$

where the coefficients are given by

$$c_n = \frac{1}{L} \int_{-L/2}^{L/2} \phi(x) e^{-i(2n\pi x/L)}\, dx. \tag{2–23}$$

Henceforth, the series will be used in this form.

2–4 Parseval's formula and the Fourier integral theorem. An essential concept associated with Fourier's series or, indeed, with any expansion in series of orthogonal functions. is that of *approximation in the mean*. To make this concept clear, suppose that a *trigonometric polynomial* has the form

$$S_N = \sum_{n=-N}^{N} d_n e^{i(2n\pi x/L)}, \qquad (2\text{–}24)$$

in which the d_n are given complex numbers, and that one wishes to examine the degree to which S_N affords an approximation to $\phi(x)$ in the interval $(-L/2, L/2)$. The quantity

$$\epsilon_N = \int_{-L/2}^{L/2} |\phi(x) - S_N|^2 \, dx \qquad (2\text{–}25)$$

is a positive number which measures the departure of S_N from $\phi(x)$ in the interval. That choice of the numbers d_n which corresponds to the smallest possible value of ϵ_N will give the best approximation to $\phi(x)$ "in the mean" or "in the sense of least squares." It is easy to see that the appropriate choice is $d_n = c_n$, that is, the Fourier coefficients; for multiplying out, integrating, and using Eq. (2–23), we obtain

$$\epsilon_N = \int_{-L/2}^{L/2} |\phi(x) - \sum_{n=-N}^{N} d_n e^{i(2n\pi x/L)}|^2 \, dx$$

$$= \int_{-L/2}^{L/2} |\phi(x)|^2 \, dx - \sum_{n=-N}^{N} \{ L c_n^* d_n + L d_n^* c_n - L d_n^* d_n \}$$

$$= \int_{-L/2}^{L/2} |\phi(x)|^2 \, dx - L \sum_{n=-N}^{N} |c_n|^2 + L \sum_{n=-N}^{N} |d_n - c_n|^2, \quad (2\text{–}26)$$

which clearly has its smallest value for $d_n = c_n$. In summary, the approximation S_N to $\phi(x)$ which is obtained by deleting from the Fourier series for $\phi(x)$ those terms for which $|n| > N$ is the best attainable in the "mean" sense.

Note also that ϵ_N is a positive number, whence

$$\sum_{n=-N}^{N} |c_n|^2 \leq \frac{1}{L} \int_{-L/2}^{L/2} |\phi(x)|^2 \, dx. \qquad (2\text{–}27)$$

The sum on the left, therefore, is a bounded sum of positive terms and hence convergent as $N \to \infty$. From this follows *Bessel's inequality*:

$$\sum_{n=-\infty}^{\infty} |c_n|^2 \leq \frac{1}{L} \int_{-L/2}^{L/2} |\phi(x)|^2 \, dx; \qquad (2\text{–}28)$$

and if the approximation in the mean is convergent, i.e., if

$$\lim_{N \to \infty} \epsilon_N = 0,$$

the equality sign holds, so that

$$\frac{1}{L} \int_{-L/2}^{L/2} |\phi(x)|^2 \, dx = \sum_{n=-\infty}^{\infty} |c_n|^2; \qquad (2\text{--}29)$$

this is called *Parseval's formula.*

Now suppose that a class of functions is given, such that Parseval's formula is true for each member of the class. Then the functions $e^{i(2n\pi x/L)}$ are a *complete set* of functions relative to the given class. In particular, they are complete with respect to the class of functions of period L which satisfy Dirichlet's conditions. It follows that there can be no function of this class which is not zero and is, at the same time, orthogonal to $e^{i(2n\pi x/L)}$ for every n, for such a function would have $c_n = 0$ for every n, but $\int_{-L/2}^{L/2} |\phi(x)|^2 \, dx \neq 0$, in contradiction to Parseval's formula.

Equation (2–29) can be regarded as the statement that the quantity on the left is equal to the area under a broken graph in which each rectangle has unit width, and height $|c_n|^2$ (Fig. 2–5). The Fourier expansion itself can be given a similar interpretation. If the variable $k = 2n\pi/L$ is introduced, so that an increase of n by $\Delta n = 1$ corresponds to $\Delta k = (2\pi/L) \, \Delta n$, then

$$\sum_{-\infty}^{\infty} |c_n|^2 \, \Delta n = \sum_{-\infty}^{\infty} |c_n|^2 \frac{L}{2\pi} \, \Delta k, \qquad (2\text{--}30)$$

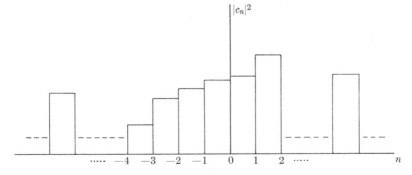

FIG. 2–5.   Graphical interpretation of Eq. (2–29).

or

$$\int_{-L/2}^{L/2} |\phi(x)|^2 \, dx = \sum_{-\infty}^{\infty} \left| \frac{Lc_n}{\sqrt{2\pi}} \right|^2 \Delta k. \qquad (2\text{-}31)$$

If we write

$$a_L(k) = \frac{Lc_n}{\sqrt{2\pi}} = \frac{1}{\sqrt{2\pi}} \int_{-L/2}^{L/2} \phi(x) e^{-ikx} \, dx \qquad (2\text{-}32)$$

and pass to the limit $L \to \infty$, $\Delta k \to 0$, Parseval's formula[1] becomes

$$\int_{-\infty}^{\infty} |\phi(x)|^2 \, dx = \int_{-\infty}^{\infty} |a(k)|^2 \, dk, \qquad (2\text{-}33)$$

where

$$a(k) = \lim_{L \to \infty} a_L(k) = \frac{1}{\sqrt{2\pi}} \int_{-\infty}^{\infty} \phi(x) e^{-ikx} \, dx. \qquad (2\text{-}34)$$

A similar transformation of the sum representing ϕ leads to

$$\phi(x) = \sum_{-\infty}^{\infty} c_n e^{i(2\pi nx/L)} = \frac{\sqrt{2\pi}}{L} \sum_{-\infty}^{\infty} a_L(k) e^{ikx} \frac{L}{2\pi} \Delta k, \qquad (2\text{-}35)$$

which approaches the limit

$$\phi(x) = \frac{1}{\sqrt{2\pi}} \int_{-\infty}^{\infty} a(k) e^{ikx} \, dk. \qquad (2\text{-}36)$$

This transition from an infinite series to an integral is not easily justified mathematically, but serves as a heuristic device to explain the formal content of the Fourier integral. Inasmuch as the interval is now infinite, it is no longer necessary to distinguish the function $\psi(x)$ from the auxiliary, periodic function $\phi(x)$ introduced in Section 2–3. In summary, the *Fourier transforms* can be written as

$$\psi(x) = \frac{1}{\sqrt{2\pi}} \int_{-\infty}^{\infty} a(k) e^{ikx} \, dk, \qquad (2\text{-}37)$$

$$a(k) = \frac{1}{\sqrt{2\pi}} \int_{-\infty}^{\infty} \psi(x) e^{-ikx} \, dx, \qquad (2\text{-}38)$$

---

[1] A more general version of this relation, important for many applications, is derived in Appendix A–2.

and Parseval's formula takes the form

$$\int_{-\infty}^{\infty} |\psi(x)|^2 \, dx = \int_{-\infty}^{\infty} |a(k)|^2 \, dk. \tag{2-39}$$

The pair of Fourier transforms can be combined into the single formula

$$\psi(x) = \frac{1}{2\pi} \int_{-\infty}^{\infty} \int_{-\infty}^{\infty} \psi(x') e^{ik(x'-x)} \, dx' \, dk, \tag{2-40}$$

known as *Fourier's integral theorem*.

Sufficient conditions for the validity of Fourier's theorem are that $\psi(x)$ be of limited total fluctuation, equal to its mean value at every point, and such that $\int_{-\infty}^{\infty} |\psi(x)| \, dx$ exists.[1]

2–5 Fourier transforms; examples. A physical interpretation of the Fourier transform of a function $\psi(x)$ is obtained if we note that

$$e^{ikx} = \cos kx + i \sin kx$$

is a linear combination of harmonic waves of wavelength $\lambda = 2\pi/k$, so that, by means of Eq. (2–37), $\psi(x)$ is represented as a linear combination of such waves, the amplitude of the component of wave number k being proportional to $a(k)$.

An easy example, which will be of importance in later work, is presented by the amplitude function of Fig. 2–6(a):

$$a(k) = \begin{cases} 1/\sqrt{\epsilon}, & -\epsilon/2 \le k \le \epsilon/2 \\ 0, & |k| > \epsilon/2. \end{cases} \tag{2-41}$$

This function has the Fourier transform

$$\psi(x) = \frac{1}{\sqrt{2\pi}} \int_{-\epsilon/2}^{\epsilon/2} \frac{1}{\sqrt{\epsilon}} e^{ikx} \, dk = \sqrt{2/\pi\epsilon} \, \frac{\sin(\epsilon x/2)}{x}, \tag{2 42}$$

illustrated in Fig. 2–6(b).

In this example, Parseval's formula is

$$\int_{-\infty}^{\infty} |a(k)|^2 \, dk = \int_{-\epsilon/2}^{\epsilon/2} \frac{1}{\epsilon} \, dk = 1 = \int_{-\infty}^{\infty} |\psi(x)|^2 \, dx, \tag{2-43}$$

[1] R. Courant and D. Hilbert, *Methods of Mathematical Physics*. New York: Interscience Publishers, Inc., 1953, Vol. I, Chapter II, Section 6.

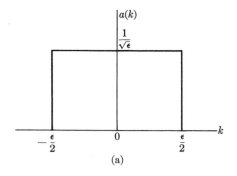

(a)

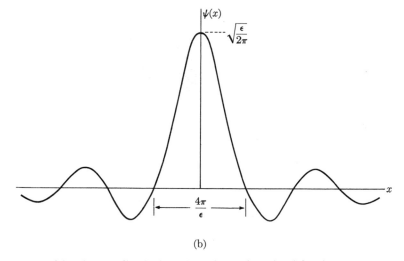

(b)

FIG. 2–6. (a) The amplitude function of Eq. (2–41). (b) The Fourier transform [Eq. (2–42)] of the amplitude function of Eq. (2–41).

which can be verified [1] as follows:

$$\int_{-\infty}^{\infty} |\psi(x)|^2 \, dx = \frac{2}{\pi\epsilon} \int_{-\infty}^{\infty} \frac{\sin^2(\epsilon x/2)}{x^2} \, dx = \frac{1}{\pi} \int_{-\infty}^{\infty} \frac{\sin^2 x}{x^2} \, dx$$

$$= \frac{1}{\pi} \int_{-\infty}^{\infty} \frac{\sin x}{x} \, dx = 1. \qquad (2\text{–}44)$$

---

[1] See "Mathematical Tables," *Handbook of Chemistry and Physics*, Cleveland: Chemical Rubber Publishing Co., 1947, 8th ed., p. 240, integral No. 338; or verify directly by means of the residue theorem.

The inverse transform is

$$a(k) = \frac{1}{\sqrt{2\pi}} \int_{-\infty}^{\infty} \sqrt{\frac{2}{\pi\epsilon}} \frac{\sin(\epsilon x/2)}{x} e^{-ikx} \, dx. \qquad (2\text{–}45)$$

This complex integral is most conveniently evaluated by means of the theory of residues. In Appendix A–1 it is shown that the integral is identical with the function $a(k)$, introduced above.

It is interesting to note that the graph of the function $a(k)$ has the width ϵ, while the graph of $\psi(x)$ has a width of order $1/\epsilon$, so that the product $\Delta k \, \Delta x$ is of order unity. This relation is significant in connection with the *uncertainty principle*, which will be referred to in Section 2–7 and discussed further in Chapter 3.

As a second example, important in later applications, consider the gaussian function

$$\psi(x) = \frac{1}{\sqrt{\sigma\sqrt{\pi}}} e^{-(x^2/2\sigma^2)}, \qquad (2\text{–}46)$$

for which the Fourier transform is

$$a(k) = \frac{1}{\sqrt{2\pi}} \frac{1}{\sqrt{\sigma\sqrt{\pi}}} \int_{-\infty}^{\infty} e^{-(x^2/2\sigma^2)} e^{-ikx} \, dx$$

$$= \frac{e^{-\sigma^2 k^2/2}}{\sqrt{2\pi\sigma\sqrt{\pi}}} \int_{-\infty}^{\infty} e^{-(1/2\sigma^2)(x+ik\sigma^2)^2} \, dx. \qquad (2\text{–}47)$$

The integral has the value $\sqrt{2\pi}\sigma$ (see Appendix A–1) and therefore

$$a(k) = \sqrt{\sigma/\sqrt{\pi}} \, e^{-\sigma^2 k^2/2}. \qquad (2\text{–}48)$$

The functions $\psi(x)$ and $a(k)$ are exhibited graphically in Fig. 2–7. The

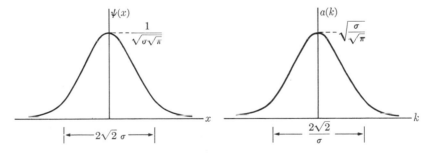

FIG. 2–7. The gaussian function $\psi(x)$ [Eq. (2–46)] and its Fourier transform $a(k)$ [Eq. (2–47)], which is also gaussian.

inverse can be calculated in a similar way, and the reader may verify that $\psi(x)$ is obtained in its original form. Note that the product $\Delta x \, \Delta k$ is again of order unity. In this example, Parseval's formula is

$$\frac{1}{\sqrt{\pi}\sigma} \int_{-\infty}^{\infty} e^{-x^2/\sigma^2} \, dx = \frac{\sigma}{\sqrt{\pi}} \int_{-\infty}^{\infty} e^{-\sigma^2 k^2} \, dk = 1. \qquad (2\text{-}49)$$

The Fourier transform is easily generalized, making it applicable to functions of more than one variable. The function $\psi(x, y, z)$, for example, can be considered as a function of the single variable x, with y and z being treated as parameters. Then,

$$\psi(x, y, z) = \frac{1}{\sqrt{2\pi}} \int_{-\infty}^{\infty} a_1(k_x, y, z) e^{ik_x x} \, dk_x, \qquad (2\text{-}50)$$

where

$$a_1(k_x, y, z) = \frac{1}{\sqrt{2\pi}} \int_{-\infty}^{\infty} \psi(x, y, z) e^{-ik_x x} \, dx \qquad (2\text{-}51)$$

is simply the Fourier theorem for the single variable x. The function $a_1(k_x, y, z)$ can now be analyzed in a similar way with respect to the variable y:

$$a_1(k_x, y, z) = \frac{1}{\sqrt{2\pi}} \int_{-\infty}^{\infty} a_2(k_x, k_y, z) e^{ik_y y} \, dk_y; \qquad (2\text{-}52)$$

$$a_2(k_x, k_y, z) = \frac{1}{\sqrt{2\pi}} \int_{-\infty}^{\infty} a_1(k_x, y, z) e^{-ik_y y} \, dy. \qquad (2\text{-}53)$$

The process can be repeated with respect to the variable z, and the final result is

$$\psi(x, y, z) = \frac{1}{(\sqrt{2\pi})^3} \int_{-\infty}^{\infty} \int_{-\infty}^{\infty} \int_{-\infty}^{\infty} a(k_x, k_y, k_z) e^{i(k_x x + k_y y + k_z z)} \, dk_x \, dk_y \, dk_z, \qquad (2\text{-}54)$$

$$a(k_x, k_y, k_z) = \frac{1}{(\sqrt{2\pi})^3} \int_{-\infty}^{\infty} \int_{-\infty}^{\infty} \int_{-\infty}^{\infty} \psi(x, y, z) e^{-i(k_x x + k_y y + k_z z)} \, dx \, dy \, dz. \qquad (2\text{-}55)$$

To simplify the writing of these formulae, it is customary to introduce the vectors

$$\mathbf{r} = (x, y, z) \qquad \text{and} \qquad \mathbf{k} = (k_x, k_y, k_z),$$

and to write the volume elements as

$$d\mathbf{r} = dx \, dy \, dz; \qquad d\mathbf{k} = dk_x \, dk_y \, dk_z.$$

In these terms,

$$\psi(\mathbf{r}) = \frac{1}{(\sqrt{2\pi})^3} \int a(\mathbf{k}) e^{i\mathbf{k}\cdot\mathbf{r}} \, d\mathbf{k}, \qquad (2\text{--}56)$$

$$a(\mathbf{k}) = \frac{1}{(\sqrt{2\pi})^3} \int \psi(\mathbf{r}) e^{-i\mathbf{k}\cdot\mathbf{r}} \, d\mathbf{r}, \qquad (2\text{--}57)$$

and it is understood that the integration is to be extended over the entire space in the indicated variables.

For example, the function

$$\psi = \frac{1}{(\sqrt{\sigma\sqrt{\pi}})^3} e^{-r^2/2\sigma^2} \qquad (r^2 = x^2 + y^2 + z^2) \qquad (2\text{--}58)$$

represents a three-dimensional wave packet which is concentrated mainly within a spherical volume of dimensions of order σ in the neighborhood of the origin. Its Fourier transform is

$$a(\mathbf{k}) = \left(\sqrt{\frac{\sigma}{\sqrt{\pi}}}\right)^3 e^{-k^2\sigma^2/2} \qquad (k^2 = k_x^2 + k_y^2 + k_z^2). \qquad (2\text{--}59)$$

It is readily verified that

$$\int |\psi(\mathbf{r})|^2 \, d\mathbf{r} = \int |a(\mathbf{k})|^2 \, d\mathbf{k} = 1. \qquad (2\text{--}60)$$

The first equality of Eq. (2–60) is, of course, of general validity since it expresses Parseval's formula for the 3-dimensional case.

2–6 Superposition of plane waves; time dependence.   The function $\exp(i\mathbf{k}\cdot\mathbf{r})$ represents a plane harmonic wave with propagation vector $\mathbf{k}$ and wavelength $\lambda = 2\pi/k$; for if $\mathbf{r} + \lambda(\mathbf{k}/k)$ is substituted in place of $\mathbf{r}$ in the expression for the phase, we obtain, by definition of the wavelength,

$$(\mathbf{k}\cdot\mathbf{r} + 2\pi) = \mathbf{k}\cdot\left(\mathbf{r} + \lambda\frac{\mathbf{k}}{k}\right),$$

or,

$$2\pi = \lambda k. \qquad (2\text{--}61)$$

Now consider the function $\exp[i(\mathbf{k}\cdot\mathbf{r} - \omega t)]$, where t is the time. It can be written

$$\exp\left[ik\left(\frac{\mathbf{k}}{k}\cdot\mathbf{r} - \frac{\omega}{k}t\right)\right]$$

and is recognized as a wave propagated in the direction of the vector **k** with *phase velocity* ω/k.

The most general superposition of such waves can be expressed, by Fourier's theorem, in the form

$$\psi(\mathbf{r}, t) = \frac{1}{(\sqrt{2\pi})^3} \int a(\mathbf{k}) \exp[i(\mathbf{k} \cdot \mathbf{r} - \omega t)] \, d\mathbf{k}. \qquad (2\text{-}62)$$

Note that $a(\mathbf{k})e^{-i\omega t}$ now takes the place of the $a(\mathbf{k})$ of the earlier discussion, and that $a(\mathbf{k})$, which is *independent of t*, is given by

$$a(\mathbf{k})e^{-i\omega t} = \frac{1}{(\sqrt{2\pi})^3} \int \psi(\mathbf{r}, t) \exp[-i\mathbf{k} \cdot \mathbf{r}] \, d\mathbf{r}, \qquad (2\text{-}63)$$

or

$$a(\mathbf{k}) = \frac{1}{(\sqrt{2\pi})^3} \int \psi(\mathbf{r}, t) \exp[-i(\mathbf{k} \cdot \mathbf{r} - \omega t)] \, d\mathbf{r}. \qquad (2\text{-}64)$$

It is an essential fact that, although t appears explicitly in the right-hand side of Eq. (2–64), $a(\mathbf{k})$ does not depend upon t, provided ψ is indeed a superposition of plane waves, as is assumed. Consequently,

$$\frac{\partial a(\mathbf{k})}{\partial t} = 0, \qquad (2\text{-}65)$$

indicating that

$$\int \left\{ \frac{\partial \psi}{\partial t} + i\omega\psi \right\} \exp[-i(\mathbf{k} \cdot \mathbf{r} - \omega t)] \, d\mathbf{r} = 0 \qquad (2\text{-}66)$$

is an equation which must be satisfied identically by every $\psi(\mathbf{r}, t)$ of the given form. If the law of dispersion is known, this condition can in many cases be reduced to a partial differential equation (e.g. Schrödinger's equation) for $\psi(\mathbf{r}, t)$.

2–7 Wave packets and the Einstein-de Broglie relations. With the mathematical preparation of the foregoing sections, we now take up the subject of matter waves and develop a theoretical description of the wave-particle duality. If a wave is to be associated with a particle in a consistent way, then it is reasonable to expect that the wavelike phenomena are localized in the neighborhood of the particle. It will therefore be necessary to deal with *wave packets*. The formulation of such packets will be considered first.

A simple harmonic wave of frequency $\nu = \omega/2\pi$ and wavelength $\lambda = 2\pi/k$ can be represented by the expression

$$e^{i(kx - \omega t)}, \qquad (2\text{-}67)$$

where only one coordinate (x) is considered for the moment. The speed of propagation of such a wave is the speed associated with the motion of a point at which the phase $(kx - \omega t)$ has a constant value. Such a point has the position

$$x = \text{constant} + \frac{\omega}{k}\, t,$$

and evidently moves with the velocity ω/k, which is the *phase velocity* of the wave:

$$v_{\text{ph}} = \omega/k. \tag{2–68}$$

A wave packet can be constructed by forming a group of plane waves, of the kind described in Section 2–6:

$$\psi(x, t) = \int_{-\infty}^{\infty} a(k) e^{i(kx - \omega t)}\, dk. \tag{2–69}$$

The behavior in time of such a wave group is determined by the way in which the angular frequency ω depends upon the wave number k, that is, by the *law of dispersion*. In the simplest case, ω and k are proportional, and the phase velocity is the same for each harmonic component of ψ. If we write[1] $\omega = c|k|$ and divide the interval of integration into two parts, the wave group is

$$\psi(x, t) = \int_{0}^{\infty} a(k) e^{ik(x - ct)}\, dk + \int_{0}^{\infty} a(-k) e^{-ik(x + ct)}\, dk. \tag{2–70}$$

The two integrals represent two parts of the wave group, traveling toward $+x$ and $-x$, respectively, with speed c and without change in shape.[2] *Undispersed* wave motion of this type is exemplified by the familiar vibrations of a tightly stretched string or by the propagation of light waves in empty space.

In the more general case of *dispersive propagation*, the connection between ω and k is not linear (for example, waves in water or light waves in an optical medium like glass). If the propagation is dispersive, the velocities of the individual components of ψ depend upon the wavelength; because of the change of relative phase of the components in time, the wave packet is no longer propagated without change in its shape. If,

---

[1] For mathematical convenience (cf. Section 2–4), the definition of k is extended to include negative real numbers. However, we adopt the convention that ω is positive.

[2] If the graph of the function $y = \psi(x)$ is given, the graph of the function $y = \psi(x - a)$ can be obtained by translating the former a distance a in the direction of the positive x-axis. The function $y = \psi(x - ct)$ represents, therefore, a graph of unchanging shape which is translated continuously with speed c.

however, the range of values of k for which the components have significant amplitude is limited, it is possible to assign an average velocity to the wave packet, as will be shown: On the assumption that $a(k)$ is negligible except when k lies in a small interval Δk, the wave packet can be written

$$\psi(x, t) = \int_{\Delta k} a(k) e^{i(kx - \omega t)} \, dk, \qquad (2\text{-}71)$$

where the integration extends only over the significant interval. If we assume that ω varies slowly with k, a good approximation is

$$\omega = \omega_0 + \left(\frac{d\omega}{dk}\right)_{k=k_0} (k - k_0) + O[(k - k_0)^2], \qquad (2\text{-}72)$$

in which $\omega_0 = \omega(k_0)$, and k_0 is some fixed value of k within Δk. To this approximation,[1]

$$\psi(x, t) \approx e^{i(k_0 x - \omega_0 t)} \int_{\Delta k} a(k) e^{i[x - (d\omega/dk) t](k - k_0)} \, dk \qquad (2\text{-}73)$$

represents a wave of wavelength $2\pi/k_0$ and frequency $\omega_0/2\pi$, which is modulated by the integral appearing as a factor. This factor depends upon x and t only in the combination $x - (d\omega/dk)t$ and thus represents a wave packet which moves with the *group velocity*

$$v_g = \frac{d\omega}{dk}. \qquad (2\text{-}74)$$

From another point of view, the expression for the group velocity can be obtained by noting that, outside the wave packet, the component wave trains are out of phase and hence interfere destructively, so that their contribution to ψ is small. In the neighborhood of a point within the wave packet, however, the phase

$$S = kx - \omega t \qquad (2\text{-}75)$$

is stationary with respect to k, that is,

$$\frac{\partial S}{\partial k} = x - \frac{d\omega}{dk} t = 0, \qquad (2\text{-}76)$$

---

[1] Note that ω enters ψ through the product ωt, and the terms of order $(k - k_0)^2$ are therefore multiplied by t. Hence, the approximation will fail after a time interval long enough to render the omitted terms no longer negligible. This is an expression of the fact that every superposition of harmonic waves will become dispersed in a sufficiently long time. It follows that the concept of group velocity is associated only with a wave packet localized both in space and time.

so that the various components stand in constant phase relation to one another. This point of constant relative phase is seen to move with the group velocity, Eq. (2–74).

The concept of group velocity relates to a wave packet that is localized in space. Furthermore, the existence of a group velocity depends upon the condition that the wave numbers of the component waves in the packet are comprised within a relatively small interval Δk. However, these two conditions, i.e., limited spatial extent Δx and limited wave number range Δk, cannot be fulfilled independently. Several examples in Section 2–5 have demonstrated that the intervals in coordinate and in wave number are connected by a relation of the form

$$\Delta x \, \Delta k \approx 1. \qquad (2\text{--}77)$$

Consequently, a very narrow packet, for which Δx is small, contains a large frequency interval in its Fourier decomposition. The concept of group velocity is correspondingly limited for this narrow packet, because of the rapid dispersal of components whose frequencies differ widely. On the other hand, a small range in k implies a large spatial extent of the packet, so that the group velocity is not clearly identified as the speed of propagation of a localized disturbance in space. This complementary relation between intervals in x and k is of great importance in wave mechanics.

So far, the present section has been devoted to a discussion of wave packets and some of their properties. The next step consists in linking wave packets with material particles in order to describe the wavelike properties of matter. The key to the relationship between particles and waves was provided by Prince Louis de Broglie in his doctoral thesis[1] and in subsequent work. It has already been indicated in Section 2–2 that de Broglie suggested the relation

$$\lambda = h/p \qquad (2\text{--}78)$$

between wavelength λ and momentum p of a particle, and that experiments on particle diffraction bear out this relationship. It follows that momentum and wave number are connected by

$$p = \hbar k. \qquad (2\text{--}79)$$

A relation between energy and frequency, on the other hand, was established by Planck's treatment of blackbody radiation and by Einstein's

¹ L. de Broglie, *Thèse de doctorat.* Paris: Masson, 1924. L. de Broglie, *J. phys. et radium* **7**, 1 and 321 (1926). L. de Broglie and L. Brillouin, *Selected Papers on Wave Mechanics.* London: Blackie and Son Ltd., 1928.

explanation of the photoelectric effect (Sections 1–3 and 1–4):

$$E = \hbar\omega. \tag{2–80}$$

De Broglie pointed out that, if ω and k are connected with the energy and momentum of a particle according to Eqs. (2–80) and (2–79), the group velocity of the associated wave packet is

$$v_g = \frac{d\omega}{dt} = \frac{dE}{dp}. \tag{2–81}$$

The quantity on the right, however, is just the classical Hamiltonian expression for the velocity of the particle. Thus, for a free nonrelativistic particle,

$$E = \frac{p^2}{2m}; \qquad \frac{dE}{dp} = \frac{p}{m} = v. \tag{2–82}$$

Similarly, in the relativistic case,

$$E^2 = p^2c^2 + m^2c^4; \qquad \frac{dE}{dp} = \frac{pc^2}{E}, \tag{2–83}$$

and the familiar relations among energy, momentum, and velocity, namely,

$$E = \frac{mc^2}{\sqrt{1 - v^2/c^2}} \qquad \text{and} \qquad p = \frac{mv}{\sqrt{1 - v^2/c^2}} \tag{2–84}$$

again lead to the relation

$$\frac{dE}{dp} = v. \tag{2–85}$$

Thus, the quantum hypothesis of Einstein and Planck and the wavelength-momentum relationship introduced by de Broglie lead to the conclusion that the velocity of a particle is to be associated with the group velocity of the corresponding wave packet. The consistency of this view is confirmed by the fact that the pairs of variables (x, t) and (k, ω) behave similarly when transformed from one Lorentz frame to another, so that the group velocity and particle velocity can be associated with one another in a relativistically covariant way. It was emphasized by de Broglie that the theory satisfies, at least to this degree, the requirements of special relativity.

2–8 Wave functions for a free particle; the Schrödinger equation. On the basis of the preceding discussion, a nonrelativistic free particle, of energy $E = \frac{1}{2}mv^2$ and momentum $\mathbf{p} = m\mathbf{v}$, is associated with a wave, of frequency $\nu = E/h$ and wavelength $\lambda = h/p$. Hence, the propagation vector is $\mathbf{k} = \mathbf{p}/\hbar$. Momentum and frequency are related through the law of dispersion, that is,

$$\omega = \frac{E}{\hbar} = \frac{p^2}{2m\hbar}. \tag{2-86}$$

A suitable plane wave is

$$\exp\left[\frac{i}{\hbar}\,(\mathbf{p}\cdot\mathbf{r} - Et)\right], \tag{2-87}$$

or, if we restrict the discussion temporarily to one dimension,

$$\exp\left[\frac{i}{\hbar}\left(px - \frac{p^2}{2m}\,t\right)\right]. \tag{2-88}$$

By superposition of a set of such waves, a wave packet can be constructed, as shown in Section 2–6:

$$\psi(x, t) = \frac{1}{\sqrt{2\pi\hbar}} \int_{-\infty}^{\infty} a(p) e^{(i/\hbar)(px - Et)}\, dp. \tag{2-89}$$

The amplitude of the component of momentum p is given by the Fourier transform

$$a(p) = \frac{1}{\sqrt{2\pi\hbar}} \int_{-\infty}^{\infty} \psi(x, t) e^{-(i/\hbar)(px - Et)}\, dx. \tag{2-90}$$

Since $a(p)$ is independent of the time (cf. Section 2–6), it is also given by the Fourier transform of $\psi(x, 0)$:

$$a(p) = \frac{1}{\sqrt{2\pi\hbar}} \int_{-\infty}^{\infty} \psi(x, 0) e^{-(i/\hbar)px}\, dx. \tag{2-91}$$

Also, by Parseval's formula,

$$\int_{-\infty}^{\infty} |\psi(x, t)|^2\, dx = \int_{-\infty}^{\infty} |a(p)|^2\, dp = \text{constant.} \tag{2-92}$$

The group velocity of the packet is given by Eq. (2–81):

$$v_g = \frac{dE}{dp} = \frac{p}{m} = v. \tag{2-93}$$

The phase velocity is, by Eq. (2–68),

$$v_{\text{ph}} = \frac{\omega}{k} = \frac{E}{p} = \frac{p}{2m} = v_g/2. \tag{2-94}$$

The phase velocity is without physical significance, apart from the fact that ω/k is the velocity of propagation of points of constant phase for a single harmonic component of ψ.

Generalization to three-dimensional wave packets yields

$$\psi(\mathbf{r}, t) = \frac{1}{(\sqrt{2\pi\hbar})^3} \int a(\mathbf{p}) \exp\left[\frac{i}{\hbar} (\mathbf{p} \cdot \mathbf{r} - Et)\right] d\mathbf{p}, \qquad (2\text{--}95)$$

representing a group of plane waves for which the wavelength and frequency are connected with energy and momentum according to the Einstein-de Broglie relations. The integration extends over the region within which ψ is defined. The group velocity is

$$\mathbf{v}_g = \left[\frac{\partial E}{\partial p_x}, \frac{\partial E}{\partial p_y}, \frac{\partial E}{\partial p_z}\right] = \operatorname{grad}_p E = \mathbf{v}, \qquad (2\text{--}96)$$

and, as before, the Parseval formula is

$$\int |\psi|^2 \, d\mathbf{r} = \int |a(\mathbf{p})|^2 \, d\mathbf{p} = \text{constant}. \qquad (2\text{--}97)$$

The function $\psi(\mathbf{r}, t)$, as defined in Eq. (2–95), satisfies a partial differential equation which is of central importance in wave mechanics. This equation, first constructed by E. Schrödinger[1] in 1926, is

$$-\frac{\hbar^2}{2m} \nabla^2 \psi + \frac{\hbar}{i} \frac{\partial \psi}{\partial t} = 0. \qquad (2\text{--}98)$$

The derivatives occurring in this equation are

$$\frac{\partial \psi}{\partial t} = \frac{1}{(\sqrt{2\pi\hbar})^3} \int \frac{-iE}{\hbar} a(\mathbf{p}) \exp\left[\frac{i}{\hbar} (\mathbf{p} \cdot \mathbf{r} - Et)\right] d\mathbf{p} \qquad (2\text{--}99)$$

and

$$\nabla^2 \psi = \frac{1}{(\sqrt{2\pi\hbar})^3} \int \frac{-p^2}{\hbar^2} a(\mathbf{p}) \exp\left[\frac{i}{\hbar} (\mathbf{p} \cdot \mathbf{r} - Et)\right] d\mathbf{p}, \qquad (2\text{--}100)$$

whence the expression on the left in Schrödinger's equation is

$$\frac{1}{(\sqrt{2\pi\hbar})^3} \int \left(\frac{p^2}{2m} - E\right) a(\mathbf{p}) \exp\left[\frac{i}{\hbar} (\mathbf{p} \cdot \mathbf{r} - Et)\right] d\mathbf{p}, \qquad (2\text{--}101)$$

which vanishes because of the relation $E = p^2/2m$.

---

[1] E. Schrödinger, *Ann. Physik* **79**, 361 and 489 (1926), **80**, 437 (1926), and **81**, 109 (1926).

The Schrödinger equation for the wave function of a free particle arises directly from the Einstein-de Broglie relations and is, in a sense, equivalent to them. However, until suitable boundary conditions and requirements concerning the continuity of solutions are imposed, the properties of the wave function are not completely described by the Schrödinger equation. Indeed, the application of Fourier's theorem that has been made above is only justified for functions for which the integrals involved are convergent, and this obviously implies that $\psi(\mathbf{r}, t)$ should become zero at a very large distance from the center of the wave packet. Moreover, it will usually be true that ψ is a continuous function of x, y, and z. In a very general sense, any continuous solution of the Schrödinger equation which vanishes at infinity in such a way that the integral $\int |\psi|^2 \, d\mathbf{r}$ exists, can be represented as a superposition of plane waves, and the two descriptions of the wave function are entirely equivalent. This is a matter of great practical importance, because many of the properties of ψ which are of physical interest are brought out more clearly by the Schrödinger equation than by the direct representation of ψ in terms of its harmonic components.

2-9 Physical interpretation of the Schrödinger wave function. In the foregoing discussion of the dual, wave-particle nature of physical systems, no direct connection has been established between the wave function ψ and the observable properties of the associated system. It is, however, an essential element of the quantum theory that the wave function associated with a physical system contains all relevant information about the behavior of the system and thus describes it completely. In other words, any meaningful question about the result of an experiment performed upon the system can be answered if the wave function is known. It is apparent, therefore, that postulates are required which will permit one to deduce the result of an experiment from knowledge of ψ.

The fundamental postulate relating to this question was formulated by Max Born[1] and states that the quantity $\psi^*\psi = |\psi|^2$ is to be interpreted as a *probability density* for a particle in the state ψ.[2] More precisely, if a particle is described by a wave function ψ, then, in a measurement of the position of the particle, the probability $P(\mathbf{r}) \, d\mathbf{r}$ of finding it in a volume element $d\mathbf{r} = dx \, dy \, dz$ at the point $\mathbf{r}$ is proportional to $|\psi(\mathbf{r})|^2 \, d\mathbf{r}$.

The probabilistic hypothesis of Born and the generalizations developed by him, Bohr, Dirac, Heisenberg, and others in the period from 1925 to

---

[1] M. Born, Z. *Physik* **37**, 863 (1926) and **38**, 803 (1926). Also M. Born, *Atomic Physics.* 5th ed. London: Blackie and Son Ltd., 1951, p. 93.

[2] The wave function gives a complete description of the state of the physical system; it is therefore natural to speak of "the state described by the function ψ," or briefly, of "the state ψ."

1930, are a precise formulation of a fundamentally new concept of natural processes which has emerged from the study of systems of atomic size. The point of view adopted by Einstein in his discussion of Planck's radiation law (cf. Section 1–3) is typical of this probabilistic interpretation of natural laws. A full appreciation of this hypothesis can only be attained through careful study of the operational meaning of physical measurements. Familiarity with the formal aspects of the subject is prerequisite to successful investigation in this field,[1] which lies beyond the scope of the present work. Only the consequences of Born's interpretation will be developed.

The probability density $P(\mathbf{r})$ is, at least in principle, an observable quantity. One may imagine, for example, a large number (i.e., an *ensemble*) of identical and noninteracting systems, each consisting of a free particle in three-dimensional configuration space, and each described by the same wave function ψ. In each of these systems, an experiment can be performed to determine whether the particle is within some given finite region of space, of volume V. The probability that one of these measurements will give a positive result is

$$\int_V P(\mathbf{r})\, d\mathbf{r}, \qquad (2\text{-}102)$$

where the integration extends over the volume in question. The quantity $P(\mathbf{r})$ can thus be determined with good accuracy by choosing a sufficiently small volume V.

Situations are frequently encountered where the particle is bound by forces to a limited region. Thus, it may be confined to a box with impenetrable walls or, as in the case of an atomic electron, it may be held close to the position of a heavy nucleus by the electrical force of attraction. In such cases, the particle is certainly to be found if the entire space is investigated, and therefore

$$\int P(\mathbf{r})\, d\mathbf{r} = 1, \qquad (2\text{-}103)$$

where the integration extends over all space. The proportionality between $P(\mathbf{r})$ and $\psi^*\psi$, postulated by Born, admits a constant which can be determined from Eq. (2–103). The relation (2–103) will clearly be fulfilled if one sets

$$P(\mathbf{r}) = \frac{\psi^*\psi}{\int \psi^*\psi\, d\mathbf{r}}. \qquad (2\text{-}104)$$

---

1 On this point, the reader may consult the works of Born, Bohr, and Dirac, listed at the end of this chapter.

This expression for $P(\mathbf{r})$ remains unchanged if ψ is multiplied by any constant factor; hence, two wave functions differing only by a constant factor describe identical physical systems. It is convenient to introduce a factor that makes the denominator on the right-hand side of Eq. (2–104) equal to unity. This is accomplished by forming

$$\psi_1 = \frac{1}{\sqrt{N}}\,\psi; \qquad N = \int \psi^* \psi \, d\mathbf{r}, \tag{2–105}$$

whence $P(\mathbf{r}) = \psi_1^* \psi_1$. The function ψ_1, defined in this way, satisfies the equation

$$\int \psi_1^* \psi_1 \, d\mathbf{r} = 1, \tag{2–106}$$

and is said to be *normalized*.

Not all wave functions, however, can be treated in this way. One frequently has to deal with functions, such as $\exp[i(\mathbf{k}\cdot\mathbf{r} - \omega t)]$, which cannot be normalized. In these cases, the quantity $\psi^*\psi$ must be interpreted as a *relative probability density*, in the sense that the ratio of the magnitudes of $\int \psi^*\psi \, d\mathbf{r}$ in two different regions of space determines the relative probability that these regions are occupied by the particle. In the example just cited, $\psi^*\psi = 1$, so that the particle is found with equal likelihood in every region of given volume, whatever the location of the region in space may be. Since only the relative magnitude of ψ at different points of space is of significance here, it is again true that two wave functions differing only by a multiplicative constant will describe identical physical situations. The choice of the constant multiplier, i.e., the "normalization" of the wave function, is then arbitrary and may be made in any way that is convenient for calculation. Furthermore, since the physical results are to be deduced from the quantity $|\psi|^2$, it is clear that an arbitrary multiplier $e^{i\gamma}$ of absolute magnitude unity is always allowed and has no effect upon the physical interpretation.

A less abstract idea of the interpretation of ψ can be obtained by considering a single region of space that contains a large number of *noninteracting* particles, each of which is described by the same wave function. No interaction between the particles means that the motion of one is unaffected by the presence of the others. This situation is equivalent to the statistical ensemble of isolated systems considered above. The common wave function of all the particles can be regarded as belonging to the assemblage, and, in this interpretation, the quantity $|\psi(\mathbf{r})|^2$ represents the average number of particles which will be found in unit volume at the point $\mathbf{r}$, when an experiment is performed to detect them. Hence, one may say that the absolute square of the wave function represents the *average particle density at* $\mathbf{r}$.

A practical situation of this kind arises, for example, if a beam of neutrons is produced by a chain-reacting pile or by a target bombarded with charged particles. The force between two neutrons is of very short range ($\sim 10^{-13}$ cm), and, in any realizeable beam, the distance between neutrons is larger by many orders of magnitude. Thus, collisions are extremely rare, and the ideal case of a system of noninteracting particles is very nearly attained. This is also true for beams of charged particles (because of their very low density) even though the Coulomb interaction has a much longer range. No detectable error is made when it is assumed that interactions are absent. For example, if all the particles in a beam have the same velocity $\mathbf{v}$, then the flux, or number of particles crossing unit area perpendicular to $\mathbf{v}$ per second, is accurately proportional to $\mathbf{v}\psi^*\psi$.

In conclusion, we shall prove an important theorem concerning the normalization of the wave function. This theorem asserts that the integral

$$N = \int_{\text{all space}} \psi^*\psi \, d\mathbf{r} \tag{2-107}$$

is *independent of the time* for every wave function ψ that is a solution of Schrödinger's equation [Eq. (2–98)]:

$$-\frac{\hbar^2}{2m} \nabla^2\psi + \frac{\hbar}{i} \frac{\partial\psi}{\partial t} = 0.$$

To prove the theorem, note that the complex conjugate of the Schrödinger equation is

$$-\frac{\hbar^2}{2m} \nabla^2\psi^* - \frac{\hbar}{i} \frac{\partial\psi^*}{\partial t} = 0. \tag{2-108}$$

By differentiation under the integral sign, the time derivative of N is obtained as

$$\frac{dN}{dt} = \int (\dot{\psi}^*\psi + \psi^*\dot{\psi}) \, d\mathbf{r} \tag{2-109}$$

which, by substitution from the Schrödinger equation and its complex conjugate, becomes

$$\frac{dN}{dt} = \int \frac{i\hbar}{2m} (\psi^*\nabla^2\psi - \psi\nabla^2\psi^*) \, d\mathbf{r}. \tag{2-110}$$

This integral can be transformed by Green's second identity[1] into

---

[1] P. M. Morse and H. Feshbach, *Methods of Theoretical Physics*. New York: McGraw-Hill Book Co., Inc., 1953, Vol. I, paragraph 7–2, p. 803.

$$\frac{dN}{dt} = \frac{i\hbar}{2m} \int_S \left(\psi^* \frac{d\psi}{dn} - \psi \frac{d\psi^*}{dn} \right) da \qquad (2\text{–}111)$$

in which the surface S encloses the volume of integration. In general, any continuous function ψ for which the integral N exists (which has been assumed) will vanish at large distances rapidly enough, so that the surface integral of Eq. (2–111) approaches zero as the surface of integration recedes to infinity. Hence,

$$\frac{dN}{dt} = 0,$$

and, consequently, the normalization of ψ will not change in time.[1]

2–10 Expectation of a dynamical quantity. The Born interpretation of ψ in terms of a probability density immediately permits the calculation of the "average" or "expected" result of a measurement, performed on a particle in the state ψ, of the position of the particle or, more generally, of any quantity which depends upon the coordinates of the particle. The *expectation* of a function $f(\mathbf{r}) = f(x, y, z)$ of the coordinates of the particle is, by definition,

$$\langle f \rangle = \int P(\mathbf{r}) f(\mathbf{r}) \, d\mathbf{r} = \frac{\int f|\psi|^2 \, d\mathbf{r}}{\int |\psi|^2 \, d\mathbf{r}}. \qquad (2\text{–}112)$$

The quantity $\langle f \rangle$ is thus a weighted average of the possible values of the function $f(\mathbf{r})$, the weight being the probability that the particle occupies that position in space at which the function takes on the value $f(\mathbf{r})$.

While the expectation of any quantity which depends only upon the position of the particle can be obtained from Eq. (2–112), nothing has been said as yet about other quantities of dynamical interest, such as momentum or energy. In the classical description of the motion of a free particle, position and momentum constitute a fundamental pair of dynamical quantities, in terms of which all properties of the motion can be expressed. It is therefore of interest to ask how momentum is to be defined in quantum theory. The correspondence principle (Section 1–8) is helpful in finding the answer to this question: The average motion of a well-defined wave packet [Eq. (2–89)] must coincide with the classical motion of the particle. On this basis, one may attempt to construct a

---

[1] The theorem has been proved only for a free particle, not interacting with any other particles. It will be seen later (Section 4–5) that only a trivial modification of the proof is required to establish the result in more general cases.

definition of momentum such that the classical rule

$$p = m \frac{dx}{dt},$$

connecting momentum and position, will have as its quantum counterpart

$$\langle p \rangle = m \frac{d}{dt} \langle x \rangle, \tag{2-113}$$

where $\langle x \rangle$ is given by Eq. (2-112) as

$$\langle x \rangle = \frac{\int x \psi^* \psi \, dx}{\int \psi^* \psi \, dx}. \tag{2-114}$$

The rule (2-113) leads directly to an expression for $\langle p \rangle$ in terms of ψ; substitution of $\langle x \rangle$ from Eq. (2-114) (we assume that ψ is normalized) gives

$$\langle p \rangle = m \frac{d}{dt} \int x \psi^* \psi \, dx = m \int x (\dot{\psi}^* \psi + \psi^* \dot{\psi}) \, dx = 2m \, \mathrm{Re} \int x \psi^* \frac{\partial \psi}{\partial t} \, dx, \tag{2-115}$$

where Re is an abbreviation for "the real part of."[1]   By means of the Schrödinger equation, the time derivative can be eliminated:

$$\langle p \rangle = \mathrm{Re} \, i\hbar \int_{-\infty}^{\infty} x \psi^* \frac{d^2 \psi}{dx^2} \, dx. \tag{2-116}$$

Integration by parts, with ψ assumed to be zero at $x = \pm\infty$, reduces Eq. (2-116) to

$$\langle p \rangle = \mathrm{Re} \, \frac{\hbar}{i} \int_{-\infty}^{\infty} \left(\psi^* + x \frac{d\psi^*}{dx} \right) \frac{d\psi}{dx} \, dx. \tag{2-117}$$

The quantity $(\hbar/i) \int_{-\infty}^{\infty} x |d\psi/dx|^2 \, dx$ is purely imaginary and can be dropped; whence

$$\langle p \rangle = \mathrm{Re} \int_{-\infty}^{\infty} \psi^* \frac{\hbar}{i} \frac{d\psi}{dx} \, dx. \tag{2-118}$$

---

[1] Note that t is contained only in ψ.

This integral is purely real, for its imaginary part is given by

$$2i \operatorname{Im} \int_{-\infty}^{\infty} \psi^* \frac{\hbar}{i} \frac{d\psi}{dx}\, dx = \int_{-\infty}^{\infty} \psi^* \frac{\hbar}{i} \frac{d\psi}{dx}\, dx + \int_{-\infty}^{\infty} \psi \frac{\hbar}{i} \frac{d\psi^*}{dx}\, dx$$

$$= \frac{\hbar}{i} \int_{-\infty}^{\infty} \frac{d}{dx}(\psi^*\psi)\, dx = \frac{\hbar}{i}(\psi^*\psi)\Big|_{-\infty}^{\infty} = 0. \quad (2\text{–}119)$$

Thus the symbol Re in Eq. (2–118) can be discarded and

$$\langle p \rangle = \int_{-\infty}^{\infty} \psi^* \frac{\hbar}{i} \frac{d\psi}{dx}\, dx. \quad (2\text{–}120)$$

If ψ were not normalized, we would obtain

$$\langle p \rangle = \frac{\displaystyle\int_{-\infty}^{\infty} \psi^*(\hbar/i)(d\psi/dx)\, dx}{\displaystyle\int_{-\infty}^{\infty} \psi^*\psi\, dx}. \quad (2\text{–}121)$$

Equations (2–120) and (2–121) can be accepted as the quantum-mechanical definition of $\langle p \rangle$. By means of this definition, one can deduce a probability distribution function $P(p)$ such that

$$\langle p \rangle = \int pP(p)\, dp. \quad (2\text{–}122)$$

Assuming again that ψ is a normalized wave packet of the form described by Eq. (2–89), we can transform the defining equation (2–122) for $\langle p \rangle$ as follows:

$$\langle p \rangle = \int_{-\infty}^{\infty} \psi^* \frac{\hbar}{i} \frac{d\psi}{dx}\, dx = \int_{-\infty}^{\infty} \psi^* \frac{\hbar}{i} \frac{1}{\sqrt{2\pi\hbar}} \int_{-\infty}^{\infty} \frac{i}{\hbar}\, p\, a(p) e^{(i/\hbar)(px-Et)}\, dp\, dx$$

$$= \frac{1}{\sqrt{2\pi\hbar}} \int_{-\infty}^{\infty} \int_{-\infty}^{\infty} \psi^* e^{(i/\hbar)(px-Et)}\, dx\, p\, a(p)\, dp. \quad (2\text{–}123)$$

The integral over x is now recognized as $a^*(p)$ [cf. Eq. (2–90)], whence

$$\langle p \rangle = \int_{-\infty}^{\infty} a^*(p)\, p\, a(p)\, dp = \int_{-\infty}^{\infty} p|a(p)|^2\, dp, \quad (2\text{–}124)$$

whence $P(p) = |a(p)|^2$. The Fourier transform plays the same formal role for p that $\psi(x)$ does for x.

Inasmuch as $\psi(x)$ and $a(p)$ are connected in a one-to-one way by the Fourier transform relations, either function gives a complete description of the quantum state of the system, and each formal relation involving one of the functions has its counterpart in terms of the other. It is usual to refer to $a(p)$ as the wave function in momentum space and to $\psi(x)$ as the wave function in coordinate space. Equations (2–120) and (2–124) are an example of the transformation from coordinate to momentum space. Similarly, one can transform the relation

$$\langle x \rangle = \int \psi^* x \psi \, dx \tag{2-125}$$

to

$$\langle x \rangle = \int a^*(p) i\hbar \frac{d}{dp} a(p) \, dp, \tag{2-126}$$

as the reader may verify. (Note the symmetry between the variables x and p.) The normalization of the momentum wave function $a(p)$ follows from that of ψ; this assures the consistency of the interpretation of $|a(p)|^2$ as a probability density in momentum space and is an expression of Parseval's formula [Eq. (2–39)]:

$$\int |\psi(x)|^2 \, dx = \int |a(p)|^2 \, dp. \tag{2-127}$$

This relation, incidentally, provides an immediate proof of the theorem that the integral on the left is independent of time.

REFERENCES

BOHR, NIELS, "Can Quantum-Mechanical Description of Physical Reality be Considered Complete?" in *Phys. Rev.* **48,** 696 (1935). Compare this with A. Einstein, B. Podolsky, and N. Rosen, "Can Quantum-Mechanical Description of Physical Reality be Considered Complete?" in *Phys. Rev.* **47,** 777 (1935).

BOHR, NIELS, *Atomic Theory and the Description of Nature.* Cambridge: Cambridge University Press, 1934. An account, by one of the originators of the quantum theory, of the experimental and theoretical basis of the subject.

BORN, MAX, *Atomic Physics.* London: Blackie and Son Ltd., 1951. This text contains an excellent discussion of wave-particle dualism in Chapter IV, Section 7.

DIRAC, P. A. M., *The Principles of Quantum Mechanics.* Oxford: The Clarendon Press, 1935, 2nd ed. (Because of a difficult notational system, the third edition is much harder to read than the second.) Every serious student should study this book.

KEMBLE, E. C., *The Fundamental Principles of Quantum Mechanics*. New York: McGraw-Hill Book Company, Inc., 1937. This early text contains discussions of the probabilistic interpretation of wave mechanics from the point of view adopted by von Neumann in his *Mathematical Foundations of Quantum Mechanics*, and includes a more detailed treatment of the time dependence of wave packets.

RICHTMYER, F. K., and KENNARD, E. H., *Introduction to Modern Physics*. 5th ed. New York: McGraw-Hill Book Company, Inc., 1955. A good reference on the Compton effect and on matter waves.

RUARK, A. E., and UREY, H. C., *Atoms, Molecules, and Quanta*. New York: McGraw-Hill Book Company, Inc., 1930. This book is out of date for the student of modern quantum theory. Nevertheless, it contains many valuable references to the experimental background of wave mechanics and its theoretical interpretation.

PROBLEMS

2–1. Show that, in the Compton effect, the relation between the electron recoil angle ϕ and the photon scattering angle θ is

$$\tan \phi = \frac{1}{1 + h\nu_0/mc^2} \operatorname{ctn}\left(\frac{\theta}{2}\right).$$

Derive an expression for the kinetic energy of the recoil electron.

2–2. Show that the electron energy in a Compton scattering event is a maximum in the forward direction, equal to

$$\hat{E}_e = \frac{h\nu_0}{mc^2/2h\nu_0 + 1}.$$

2–3. At what velocity is the de Broglie wavelength of an alpha-particle equal to that of a 1-kev x-ray?

2–4. Verify Eq. (2–10) for the de Broglie wavelength of a neutron.

2–5. Look up a description of a microwave spectrum analyzer (or better yet, experiment with the instrument itself) and study the way in which this device provides a spectral (i.e., Fourier) decomposition of a microwave pulse.

2–6. Calculate the inverse transform of (2–48) and show that it is equal to (2–46).

2–7. Derive Eq. (2–59) and verify Eq. (2–60).

2–8. The law of dispersion for gravitational waves in deep water is

$$v_{\text{ph}} = \left(\frac{\lambda g}{2\pi}\right)^{1/2}.$$

What is the group velocity?

2-9. A particle on a straight line is described by

$$\psi(x) = \frac{1 + ix}{1 + ix^2}.$$

(a) Normalize this wave function. (b) Make a graph of the probability distribution in x. (c) Where is the particle most likely to be found?

2-10. The normalized wave function of a particle on a straight line is given by

$$\psi(x) = \frac{1}{\sqrt{\sigma\sqrt{\pi}}} e^{-(x^2/2\sigma^2)} e^{i(P/\hbar)x}.$$

(a) Where is the particle most likely to be found? (b) What is the average momentum of the particle?

2-11. The state of an oscillator of angular frequency ω is represented by

$$\psi(x) = e^{-m\omega x^2/\hbar}.$$

What are the expectations of the momentum and the position?

2-12. In the preceding problem, find the probability that the magnitude of the momentum is larger than $(m\hbar\omega)^{1/2}$.

CHAPTER 3

WAVE PACKETS AND THE UNCERTAINTY PRINCIPLE

3–1 Uncertainty of position and momentum. The language of everyday life, which is the language in which we think, has been developed to describe the world of direct experience. Mainly because of sensory limitations, our experience is restricted to a rather small field of objects and events. Phenomena that lie outside the restricted region of direct perception can be studied through indirect reasoning, but we cannot, in general, expect that analogies with subjects of common experience will provide a correct picture of events in the atomic realm.

Quantum mechanics provides a mathematical formalism that describes atomic events, and it is difficult to resist the temptation of trying to visualize the results of wave-mechanical theory. These attempts are useful since they facilitate the intuitive grasp of the theory. But the limitations of these pictures must be borne in mind, and it is often necessary to accept seemingly contradictory features. One of the most novel, and not easily visualized, concepts in wave mechanics concerns the uncertainty which is naturally inherent in the simultaneous measurement of certain pairs of dynamical variables.

In quantum mechanics, a particle is described by a wave packet. The wave packet surrounds the position of the classical particle, and the "center of gravity" of the packet follows the classical particle trajectory. According to Born's postulate, the particle may be found anywhere within the region where the amplitude of the wave function is different from zero. This implies that the position of the particle is indeterminate within the limits of the wave packet. The question arises whether a packet can be constructed that is very much localized in space, so that the position of the particle is confined to a small volume. It can be expected that a narrow packet moves in such a way that the average position of the particle conforms to classical laws.

The construction of a localized wave packet is indeed possible, at least for a limited time. The fact that the packet will spread with time will be considered below. The Fourier integral expression for a wave packet has been given in Eq. (2–89). At time $t = 0$, a one-dimensional packet can be written

$$\psi(x, 0) = \frac{1}{\sqrt{2\pi\hbar}} \int a(p)e^{(i/\hbar)px} \, dp, \qquad (3\text{--}1)$$

where

$$a(p) = \frac{1}{\sqrt{2\pi\hbar}} \int \psi(x)e^{-(i/\hbar)px} \, dx. \qquad (3\text{--}2)$$

These equations can be used to construct a wave packet localized within a region Δx. For the sake of simplicity, one may choose the shape indicated in Fig. 3–1; that is, the wave function is zero everywhere except within a region of width Δx, where it has the value $1/\sqrt{\Delta x}$.[1] The Fourier transform can be constructed by analogy to the similar example in Section 2–5; in terms of $p = \hbar k$, it is

$$a(p) = \sqrt{\frac{\Delta x}{2\pi\hbar}} \ \frac{\sin{(p \, \Delta x/2\hbar)}}{p \, \Delta x/2\hbar}. \tag{3-3}$$

The physical significance of the momentum wave function $a(p)$ is that the absolute square of this quantity is proportional to the probability that the particle has the momentum p (cf. Section 2–10). For this reason, it is interesting to consider the behavior of $a(p)$ as a function of p. Figure 3–2 is a graph of Eq. (3–3). The maximum value of $a(p)$ occurs at $p = 0$. The function goes to zero on both sides of this maximum, then oscillates, decreasing rapidly. The main contribution to the momentum probability comes from the narrow momentum interval centered at the origin, which has a width of order $\hbar/\Delta x$. The relative widths of the coordinate and momentum functions are therefore related by

$$\Delta x \, \Delta p \approx \hbar. \tag{3-4}$$

A precise definition of Δx and Δp will be introduced later, and it will be shown that the "uncertainty product" $\Delta x \, \Delta p$ is always limited by a relation of this kind, irrespective of the shape of the packet. Equation (3–4) is an expression of *Heisenberg's uncertainty principle:* If the coordinate x of a particle represented by a wave packet is defined with an accuracy Δx, then the conjugate momentum p_x of the particle is defined only with an accuracy $\Delta p \approx \hbar/\Delta x$. It is not possible to build into the wave packet more exact information as to both position and momentum (or velocity) of the particle.

The relation $\Delta p \, \Delta x \approx \hbar$ follows from the mathematical properties of a wave packet. The question whether this uncertainty condition corresponds to physical reality remains open to discussion. Evidently, only two possibilities exist: (1) There are in nature particles of completely defined position and momentum; in that case, they cannot be represented by wave packets, and wave mechanics would fail. (2) All particles can be represented by wave packets; in that case, it must be physically im-

---

[1] The wave function of Fig. 3–1 is an idealization introduced because of its mathematical simplicity. Since its derivative is discontinuous, this function cannot represent an actual physical situation. For the same reason, certain averages in momentum space do not exist [e.g., Δp as given by Eq. (3–7)].

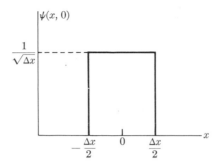

FIG. 3–1.   Rectangular wave packet.

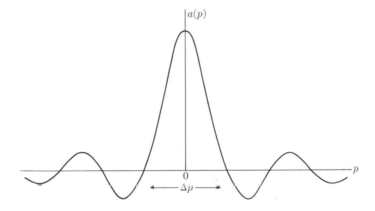

FIG. 3–2.   The Fourier transform $a(p)$ of the wave function illustrated in Fig. 3–1.

possible to measure simultaneously position and momentum of a particle with a higher degree of accuracy than the uncertainty relation allows. In view of the experimental evidence for the wave nature of matter, the second alternative must be true.   A precise expression of Heisenberg's relation for a one-dimensional wave packet of arbitrary form will now be derived.

3–2 Exact statement and proof of the uncertainty principle for wave packets.   A definition of uncertainty is the first step in a mathematical proof of the uncertainty principle.   By analogy with the standard deviation of statistics, it is customary to define the uncertainty in x by the relation

$$\Delta x = \langle (x - \langle x \rangle)^2 \rangle^{1/2}. \tag{3–5}$$

As stated before, $|\psi(x)|^2 \, dx$ is the probability that the position of a particle lies between x and $x + dx$. The average or expectation of x is defined by the relation

$$\langle x \rangle = \int \psi^*(x) x \psi(x) \, dx. \tag{3-6}$$

Similarly, the uncertainty in momentum is defined as

$$\Delta p = \langle (p - \langle p \rangle)^2 \rangle^{1/2}, \tag{3-7}$$

where the expectation of the momentum is given by

$$\langle p \rangle = \int \psi^*(x) \left(\frac{\hbar}{i} \frac{\partial}{\partial x} \right) \psi(x) \, dx. \tag{3-8}$$

The purpose of the present section is to prove that, for any wave function ψ,

$$\Delta x \, \Delta p \geq \tfrac{1}{2} \hbar. \tag{3-9}$$

We shall first present the proof for a wave function for which $\langle x \rangle = \langle p \rangle = 0$. It is then an easy step to generalize the statement for functions with finite $\langle x \rangle$ and $\langle p \rangle$. To obtain Heisenberg's result, we consider the integral

$$\int -\frac{\hbar}{i} \frac{d\psi^*}{dx} x\psi \, dx = \frac{\hbar}{i} \int \psi^* \psi \, dx + \int \frac{\hbar}{i} \frac{d\psi}{dx} x\psi^* \, dx, \tag{3-10}$$

in which the second form is obtained from the first through integration by parts. This equation can be rewritten in the form

$$2i \operatorname{Im} \int -\frac{\hbar}{i} \frac{d\psi^*}{dx} x\psi \, dx = \frac{\hbar}{i} \int |\psi|^2 \, dx. \tag{3-11}$$

Thus the square of the modulus of the integral on the right satisfies the inequality

$$\hbar^2 \left| \int |\psi|^2 \, dx \right|^2 = 4 \left| \operatorname{Im} \int -\frac{\hbar}{i} \frac{d\psi^*}{dx} x\psi \, dx \right|^2 \leq 4 \left| \int -\frac{\hbar}{i} \frac{d\psi^*}{dx} x\psi \, dx \right|^2; \tag{3-12}$$

this follows since the magnitude of the imaginary part of a complex number cannot exceed the modulus of the number itself. By the inequality of Schwarz (Appendix A–3), the right-hand member in the inequality (3–12) is in turn smaller than

$$4 \int -\frac{\hbar}{i} \frac{d\psi^*}{dx} \frac{\hbar}{i} \frac{d\psi}{dx} \, dx \cdot \int x\psi^* x\psi \, dx,$$

whence

$$\int x^2 |\psi|^2 \, dx \cdot \int \left| \frac{\hbar}{i} \frac{d\psi}{dx} \right|^2 dx \geq \frac{\hbar^2}{4} \left| \int |\psi|^2 \, dx \right|^2, \qquad (3\text{-}13)$$

or

$$\frac{\int x^2 |\psi|^2 \, dx}{\int |\psi|^2 \, dx} \cdot \frac{\int |(\hbar/i)(d\psi/dx)|^2 \, dx}{\int |\psi|^2 \, dx} \geq \frac{\hbar^2}{4}. \qquad (3\text{-}14)$$

The first factor on the left is recognized as the definition of $\langle x^2 \rangle$, and by an analysis similar to that leading to Eq. (2–121) for $\langle p \rangle$, it can be shown that (Problem 3–2)

$$\langle p^2 \rangle = \frac{\int p^2 |a(p)|^2 \, dp}{\int |a(p)|^2 \, dp} = \frac{\int |(\hbar/i)(d\psi/dx)|^2 \, dx}{\int |\psi|^2 \, dx}. \qquad (3\text{-}15)$$

Consequently, we have proved that

$$\langle x^2 \rangle \langle p^2 \rangle \geq \frac{\hbar^2}{4}, \qquad (3\text{-}16)$$

or, since we have assumed that $\langle x \rangle = \langle p \rangle = 0$,

$$\Delta x \, \Delta p \geq \frac{\hbar}{2}. \qquad (3\text{-}17)$$

The modification of the proof required to remove the restriction $\langle x \rangle = \langle p \rangle = 0$ is left for the reader (Problem 3–3).

3–3 An example of position-momentum uncertainty. The uncertainty principle sets a limit to the precision with which certain pairs of dynamical variables, such as position and momentum, can be defined simultaneously. There is no theoretical limit on the accuracy attainable in defining a single variable by a wave packet; but if one such variable is known accurately, then a measurement of the conjugate variable disturbs the state of the system sufficiently so that the uncertainty relation holds.

The physical implications of Heisenberg's principle are well illustrated by analysis of certain "thought experiments." Such illustrations show that the imperfect definition of a wave packet, caused by diffraction effects, can be understood to imply the relation $\Delta x \, \Delta p \approx \hbar$. An illuminating example, due to Heisenberg, is the "gamma-ray microscope:"[1]

A beam of monoenergetic electrons can be obtained experimentally. We assume that the velocity of the electrons is known exactly. Hence,

---

[1] W. Heisenberg, *The Physical Principles of the Quantum Theory*. New York: Dover Publications, Inc., 1930, p. 21.

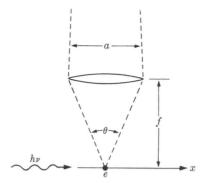

FIG. 3–3. Heisenberg's gamma-ray microscope.

according to the uncertainty relation, nothing can be known about their position. In principle, a microscope could be used to obtain information on the position of an electron in the beam. The resolving power of a microscope is inversely proportional to the wavelength of the light used. Thus, to obtain as accurate a position measurement as possible, light of short wavelength will be chosen, perhaps gamma rays. The accuracy of the measurement is given by physical optics as

$$\Delta x = \frac{\lambda}{\sin \theta} \approx \frac{\lambda f}{a},$$

(3–18)

where a is the aperture, λ the wavelength, and f the distance from the electron to the lens (Fig. 3–3).

For the electron to be observed, at least one photon must be scattered into the microscope, and this photon will interact with the electron, causing it to recoil. In the scattering process, momentum is transferred from the photon to the electron, according to the rules of the Compton effect. A quantum of wavelength λ has momentum h/λ; this is the order of magnitude of the recoil momentum imparted to the electron. The x-component of the recoil momentum, however, which is of interest, cannot be known exactly; we know only that the photon was scattered into the microscope, and hence the uncertainty of its direction is of the order of magnitude of the angle θ. Consequently, the uncertainty of the x-component of the electron's recoil momentum is of the order

$$\Delta p \approx \frac{h}{\lambda} \sin \theta,$$

(3–19)

so that

$$\Delta x \, \Delta p \approx h,$$

(3–20)

as predicted by Heisenberg's principle.

3–4 Energy-time uncertainty. In its mathematical aspect the uncertainty relation has been presented as a theorem in Fourier analysis. As such, it applies also, in terms of the function $e^{-(i/\hbar)Et}$, to the harmonic analysis of a wave packet which has a limited duration in time. A wave packet of duration Δt must be comprised of plane-wave components whose energies extend over a range ΔE, where

$$\Delta E \, \Delta t \approx \hbar. \tag{3–21}$$

This relation expresses the Heisenberg uncertainty principle as it applies to energy and time. A similar connection between frequency and time, $\Delta \nu \, \Delta t \approx 1$, is well known to electrical engineers, who are familiar with the fact that, e.g., a one-megacycle band width is required to reproduce a one-microsecond pulse.

Another version of the energy-time uncertainty relation emerges from the following consideration: The velocity of a wave packet can be determined by observing the position of the particle at two different times. If the packet has width Δx, then the momentum is uncertain within Δp, as given by Eq. (3–4), and the energy is indeterminate within

$$\Delta E = \frac{p}{m} \, \Delta p = v \, \Delta p. \tag{3–22}$$

Now the uncertainty in time is of order $\Delta x/v = \Delta t$, whence

$$\Delta E \, \Delta t \approx v \, \Delta p \, \Delta t \approx \Delta x \, \Delta p \approx \hbar. \tag{3–23}$$

It should be noted that, in this argument, Δt is the time required for a packet of width Δx to pass a given point in space, whereas in relation (3–21) Δt was the interval in time required to define the energy of a particle within ΔE.

The energy-time uncertainty relation will be encountered again in connection with the decay of quasi-stationary states. It will then receive still another interpretation.

3–5 Monochromatic waves. It was pointed out in the beginning of this chapter that a large number of component frequencies is required to construct a wave packet which is concentrated in space. The width of the band of frequencies or wave numbers is less for a widely spread packet, in conformity with the relation

$$\Delta x \, \Delta p \geq \frac{\hbar}{2}. \tag{3–24}$$

We wish to consider the two limiting cases which this relation admits.

One limit is $\Delta p \to 0$, $\Delta x \to \infty$, which corresponds to the familiar plane wave

$$\psi(x) = \frac{1}{\sqrt{2\pi\hbar}}\, e^{ikx}, \tag{3-25}$$

where $1/\sqrt{2\pi\hbar}$ is a normalization factor. The plane wave has a single wave number k and is spread out over all space. This corresponds to a state of sharply defined momentum $\hbar k$, but of completely undefined position, so that the uncertainty in x is infinite.

The other set of limiting values of Δp and Δx occurs for a wave packet associated with a particle whose position is exactly known, so that nothing can be said about its momentum. In this case, $\Delta k = \infty$, $\Delta x = 0$. These limiting cases are most efficiently dealt with by Fourier representations in terms of $\psi(x)$ and $a(p)$, in conformity with Eqs. (2–89) and (2–90). A mathematical difficulty is encountered, however, since the momentum function corresponding to (3–25), that is,

$$a(p) = \frac{1}{2\pi\hbar} \int e^{ikx} e^{-(i/\hbar)px}\, dx, \tag{3-26}$$

is presented as an integral in x which does not exist. The formal validity of Eq. (3–26) can, nevertheless, be restored by introducing the *Dirac delta function* (see Appendix A–4), in terms of which we write

$$a(p) = \delta(p - \hbar k). \tag{3-27}$$

In this way, a momentum representation for the single-frequency wave packet (3–25) is obtained. In an entirely symmetrical manner, the momentum function

$$a(p) = \frac{1}{\sqrt{2\pi\hbar}}\, e^{(i/\hbar)px_0} \tag{3-28}$$

corresponds, in the coordinate representation, to

$$\psi(x) = \frac{1}{2\pi\hbar} \int e^{(i/\hbar)(x-x_0)p}\, dp = \delta(x - x_0). \tag{3-29}$$

This expression represents the wave function for a particle located exactly at the point $x = x_0$. The wave functions (3–25) and (3–29) can be regarded as the limits of very narrow wave packets in the momentum and coordinate spaces, respectively.

3–6 The gaussian wave packet. According to the exact statement (3–9) of the uncertainty principle, the expression

$$\Delta x \, \Delta p \geq \tfrac{1}{2}\hbar$$

holds for any wave packet. A problem of special interest is that of finding the shape of the wave packet for which the uncertainty product attains its theoretical minimum value,[1] so that $\Delta x \, \Delta p = \tfrac{1}{2}\hbar$. This problem will now be solved. If we re-examine the proof of the uncertainty principle in Section 3–2, we see that the sign of equality can hold only if the functions entering the Schwarz inequality are proportional, i.e., if

$$\frac{\hbar}{i} \frac{d\psi}{dx} = Cx\psi, \tag{3–30}$$

where C is a suitable multiplier. Furthermore, the relation (3–12) will not admit the equality sign unless the integral

$$\int -\frac{\hbar}{i} \frac{d\psi^*}{dx} \, x\psi \, dx = C^* \int x\psi^* x\psi \, dx \tag{3–31}$$

is purely imaginary. Hence C must be a purely imaginary number. For convenience, we write

$$C = \frac{i\hbar}{C'}, \tag{3–32}$$

and thus

$$\frac{d\psi}{dx} = -\frac{x}{C'} \, \psi. \tag{3–33}$$

This differential equation is easily solved (Problem 3–11), yielding

$$\psi = Ne^{-x^2/2C'} ; \tag{3–34}$$

the necessity of $\int |\psi|^2 \, dx$ being convergent makes C' positive. Let us say that $C' = \sigma^2$, then

$$\psi = Ne^{-x^2/2\sigma^2}. \tag{3–35}$$

The form of the result shows that the minimum wave packet has the shape of a gaussian curve. It follows that a gaussian wave packet represents a particle whose position and momentum are simultaneously determined, as closely as the uncertainty principle permits.

---

[1] E. H. Kennard, Z. *Physik* **44,** 326 (1927).

A similar calculation, not involving the assumption $\langle x \rangle = \langle p \rangle = 0$, leads to the result (Problem 3–12)

$$\psi = N \exp\left[-\frac{(x - \langle x \rangle)^2}{2\sigma^2}\right] \exp\left(\frac{i}{\hbar} \langle p \rangle x\right). \qquad (3\text{-}36)$$

This wave packet moves with uniform average momentum $\langle p \rangle$. At time $t = 0$, for which the calculation has been carried out, the center of the packet is located at $x = \langle x \rangle$. Figure 3–4 is a graph of the function $|\psi(x)|$. The quantity σ measures the width of the packet. The constant N is determined by normalization, as follows:

$$\int_{-\infty}^{\infty} \psi^* \psi \, dx = |N|^2 \int_{-\infty}^{\infty} e^{-(x - \langle x \rangle)^2/\sigma^2} \, dx = \sqrt{\pi} \, \sigma |N|^2 = 1. \qquad (3\text{-}37)$$

Hence,

$$\psi(x, 0) = \frac{1}{\sqrt{\sigma \sqrt{\pi}}} \exp\left[-\frac{(x - \langle x \rangle)^2}{2\sigma^2}\right] \exp\left(i \frac{\langle p \rangle}{\hbar} x\right). \qquad (3\text{-}38)$$

Generalized to three dimensions, the gaussian packet is

$$\psi(\mathbf{r}, 0) = \frac{1}{(\sigma \sqrt{\pi})^{3/2}} \exp\left[-\frac{|\mathbf{r} - \langle \mathbf{r} \rangle|^2}{2\sigma^2}\right] \exp\left[i \frac{\langle \mathbf{p} \rangle \cdot \mathbf{r}}{\hbar}\right], \qquad (3\text{-}39)$$

where, as usual, the vectors $\mathbf{r}$ and $\mathbf{p}$ have the components (x, y, z) and (p_x, p_y, p_z), respectively.

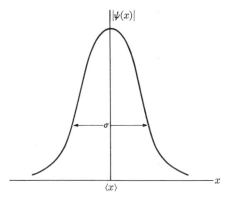

FIG. 3–4. The gaussian wave packet.

3–7 Spread of the gaussian packet with time. In the preceding section, it has been shown that a gaussian wave packet describes a particle for which the uncertainty relation has its minimum value, $\Delta x\,\Delta p = \hbar/2$, at a given time $t = 0$. The change of the gaussian packet with time will now be computed, and it will become apparent that the packet spreads out as time progresses. In general, the problem is that of finding $\psi(x, t)$ when $\psi(x, 0)$ is given. The calculation is carried through in two steps: First, we compute the momentum wave function which, according to Eq. (2–90), is given by

$$a(p) = \frac{1}{\sqrt{2\pi\hbar}} \int \psi(x, 0)e^{-ikx}\, dx; \qquad (3\text{–}40)$$

then we perform a second Fourier transformation, using Eq. (2–89), and obtain

$$\psi(x, t) = \frac{1}{\sqrt{2\pi\hbar}} \int a(p)e^{(i/\hbar)\,[px-(p^2/2m)\,t]}\, dp. \qquad (3\text{–}41)$$

The substitution $E = p^2/2m$ has been made. The width of the packet is conveniently described in terms of

$$(\Delta x)^2 = \langle x^2 \rangle - \langle x \rangle^2. \qquad (3\text{–}42)$$

If the origin is chosen at the center of the packet at $t = 0$, then $\langle x \rangle = 0$, whence[1]

$$(\Delta x)^2\big|_{t=0} = \langle x^2 \rangle = \frac{1}{\sigma\sqrt{\pi}} \int_{-\infty}^{\infty} x^2 e^{-x^2/\sigma^2}\, dx = \frac{\sigma^2}{2}, \qquad (3\text{–}43)$$

and

$$\Delta x\big|_{t=0} = \frac{\sigma}{\sqrt{2}}. \qquad (3\text{–}44)$$

Since $\Delta x\,\Delta p = \hbar/2$, the width in momentum space is

$$\Delta p = \frac{\hbar}{\sqrt{2}\,\sigma}. \qquad (3\text{–}45)$$

Inasmuch as $a(p)$ does not depend on t, the value of Δp remains constant as the wave packet moves.

---

[1] The general formula

$$\int_{-\infty}^{\infty} x^{2n} e^{-x^2}\, dx = \frac{1 \cdot 3 \cdot 5 \cdots (2n-1)\sqrt{\pi}}{2^n}$$

will find frequent application.

The Fourier transform of $\psi(x, 0)$ is

$$a(p) = \frac{1}{\sqrt{2\pi\sigma\hbar\sqrt{\pi}}} \int_{-\infty}^{\infty} e^{-x^2/2\sigma^2} e^{-(i/\hbar)(p-<p>)x}\, dx. \qquad (3\text{-}46)$$

By combining the exponents in the integrand and completing the square in x, we find, with the aid of Appendix A–1, that

$$a(p) = \sqrt{\frac{\sigma}{\hbar\sqrt{\pi}}}\, e^{-(\sigma^2/2\hbar^2)(p-<p>)^2}. \qquad (3\text{-}47)$$

The wave function in momentum space is therefore also a gaussian function, but it is time-independent. By substituting Eq. (3–47) into Eq. (3–41), we obtain the time-dependent wave function in coordinate space:

$$\psi(x, t) = \sqrt{\frac{\sigma}{2\pi\hbar^2\sqrt{\pi}}} \int e^{-(\sigma^2/2\hbar^2)(p-<p>)^2} e^{(i/\hbar)[px-(p^2/2m)t]}\, dp. \qquad (3\text{-}48)$$

The integral appearing here is of the same form as that encountered in Eq. (3–46). We arrive at

$$\psi(x, t) = \frac{1}{\sqrt{\sigma\sqrt{\pi}}} \frac{\sigma}{\alpha} \exp\left\{-\frac{[x-(\langle p\rangle/m)t]^2}{2\alpha^2}\right\} \exp\left[\frac{i}{\hbar}\langle p\rangle\left(x-\frac{\langle p\rangle}{2m}t\right)\right],$$
$$(3\text{-}49)$$

where $\alpha^2 = \sigma^2 + i\hbar t/m$.

It is evident that $\langle x\rangle = (\langle p\rangle/m)t$, whence the packet travels with the group velocity $v = \langle p\rangle/m$, which is the velocity of a classical particle of momentum $\langle p\rangle$. Furthermore, the second exponential indicates that the phase velocity is $\langle p\rangle/2m = v/2$, as shown in general in Section 2–8. The probability density is obtained as the absolute square of ψ:

$$|\psi(x, t)|^2 = \frac{\sigma}{|\alpha|^2\sqrt{\pi}} \exp\left\{-\frac{\sigma^2[x-(\langle p\rangle/m)t]^2}{|\alpha|^4}\right\}; \qquad (3\text{-}50)$$

and hence the width of the packet[1] at time t is

$$\Delta x = \frac{\sigma}{\sqrt{2}} \sqrt{1 + \frac{\hbar^2 t^2}{\sigma^4 m^2}}. \qquad (3\text{-}51)$$

The quantity $\Delta x(t)/\Delta x(0) = \sqrt{2}\,\Delta x/\sigma$ is shown as a function of time in Fig. 3–5. The effect of the dispersion is to spread out the packet; ultimately, the width increases in direct proportion to the time. Also, since Δp is constant, the uncertainty product $\Delta x\,\Delta p$ grows, i.e., ψ represents a mini-

---

[1] Note: When $|\psi|^2$ has the form $e^{-(x^2/\sigma^2)}$, then $\Delta x = \sigma/\sqrt{2}$.

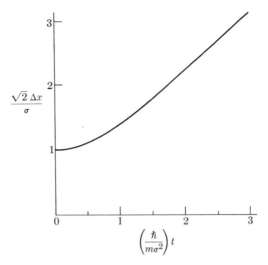

FIG. 3–5. Spread of the gaussian wave packet with time.

mum packet only at the initial instant. The time scale is given by the quantity $T = m\sigma^2/\hbar$, in terms of which

$$\Delta x = (\Delta x|_{t=0})\sqrt{1 + (t/T)^2}. \qquad (3\text{–}52)$$

The packet changes shape inappreciably as long as $t \ll T$, but at a time $t \gg T$ its original form is completely changed. To understand this effect, note that the width of the momentum packet is $\hbar/\sigma$, so that the spread Δv in the velocities of the component waves is approximately $\hbar/m\sigma$. The form of the wave packet is completely changed in a time during which the velocity difference Δv causes a change of order Δx in the relative position of two component waves. This time is

$$t \approx \frac{\Delta x}{\Delta v} \approx \frac{m\sigma^2}{\hbar} = T. \qquad (3\text{–}53)$$

For example, an electron wave packet initially confined within a region of size $\hbar/mc$ will approximately double in width in a time[1]

$$T = \frac{m}{\hbar}\left(\frac{\hbar}{mc}\right)^2 = \frac{\hbar}{mc^2} \approx 10^{-21} \text{ sec.}$$

If, however, the initial size of the packet is 1 cm, the characteristic time T is of order one second.

---

[1] The phase velocity in this example is of order c. Since the calculation is not relativistic, the result must be regarded as only an order-of-magnitude estimate.

3–8 General solution for time dependence of ψ; causality. The procedure outlined above for determining $\psi(x, t)$ from its value at an earlier time can be generalized. The calculation leads to a formulation of the wave-mechanical properties of a free particle. The relativistic generalization of this formulation is the starting point for much of the modern theoretical work on the physics of fundamental particles.[1] We shall work in three dimensions, following the steps of the preceding section without, however, specifying the form of ψ. We have seen that the expression for $\psi(\mathbf{r}, t)$ in terms of its Fourier components is

$$\psi(\mathbf{r}, t) = \frac{1}{(2\pi\hbar)^{3/2}} \int a(\mathbf{p}) \exp\left[\frac{i}{\hbar}\left(\mathbf{p} \cdot \mathbf{r} - \frac{p^2}{2m} t\right)\right] d\mathbf{p}, \qquad (3\text{-}54)$$

where the quantity $a(\mathbf{p})$ is independent of time, as emphasized previously. The momentum wave function can therefore be calculated by Fourier inversion of ψ at any other time t':

$$a(\mathbf{p}) = \frac{1}{(2\pi\hbar)^{3/2}} \int \psi(\mathbf{r}', t') \exp\left[-\frac{i}{\hbar}\left(\mathbf{p} \cdot \mathbf{r}' - \frac{p^2}{2m} t'\right)\right] d\mathbf{r}'. \qquad (3\text{-}55)$$

Substituting into Eq. (3–54), we obtain the result

$$\psi(\mathbf{r}, t) = \int K(\mathbf{r}, t; \mathbf{r}', t')\psi(\mathbf{r}', t')\, d\mathbf{r}', \qquad (3\text{-}56)$$

in which the function K is

$$K(\mathbf{r}, t; \mathbf{r}', t') = \frac{1}{(2\pi\hbar)^3} \int \exp\left\{\frac{i}{\hbar}\left[\mathbf{p} \cdot (\mathbf{r} - \mathbf{r}') - \frac{p^2}{2m} (t - t')\right]\right\} d\mathbf{p}. \qquad (3\text{-}57)$$

In Eq. (3–56), the wave function at the time t is expressed linearly in terms of its values at the earlier time t'. This is an explicit expression of the *principle of causality*, according to which knowledge of the behavior of a system at one instant is sufficient for the prediction of its future behavior. It is the quantum analogue of the classical principle that the future behavior of a particle can be predicted when its position and velocity are known at a given instant. Thus, in spite of the statistical nature of quantum theory, the causal relationship of events is not invalidated; rather, causality is now presented in a form which is consistent with the principle that knowledge of the behavior of any system is limited by the effects of measurements designed to determine its state at a given time. A complete description is embodied in the wave function: If the wave function is

---

[1] R. P. Feynman, *Phys. Rev.* **76**, 749 (1949), particularly footnote 4, p. 750.

known, the future behavior of the system it describes is predicted by the laws of quantum theory.

Mathematically, the principle of causality makes itself evident by the fact that the Schrödinger equation contains the time only in the first derivative. The Schrödinger equation, therefore, gives the rate of change of ψ in terms of its value at a given instant. The wave function at a later instant can therefore be calculated.

The integral of Eq. (3–57) can be evaluated and an explicit form for K derived. Since this computation is an instructive exercise in the treatment of integrals in three-dimensional momentum space, we shall indicate the steps. So far as the integration is concerned, the vector $\mathbf{r} - \mathbf{r}'$ is a constant; hence spherical polar coordinates in the $\mathbf{p}$-space can be introduced, with the polar axis in the direction of this vector. The integrand then does not depend upon the azimuthal angle, and the volume element is $d\mathbf{p} = 2\pi \sin \theta \, d\theta \, p^2 \, dp$, where $p = |\mathbf{p}|$, and θ is the polar angle measured from the direction of $\mathbf{r} - \mathbf{r}'$. The limits of integration are $(0, \infty)$ for p and $(0, \pi)$ for θ, or $(1, -1)$ for $\mu = \cos \theta$. With these transformations,

$$ K = \frac{1}{(2\pi\hbar)^3} \int_0^\infty \int_{-1}^1 \exp \frac{i}{\hbar} \left[p|\mathbf{r} - \mathbf{r}'|\mu - \frac{p^2}{2m}(t - t') \right] 2\pi \, d\mu p^2 \, dp. \tag{3–58} $$

The integration with respect to μ is immediate; that is,

$$ \int_{-1}^1 \exp \left[\frac{i}{\hbar} p|\mathbf{r} - \mathbf{r}'|\mu \right] d\mu = \frac{\hbar}{i} \frac{1}{p|\mathbf{r} - \mathbf{r}'|} \exp \left[\frac{i}{\hbar} p|\mathbf{r} - \mathbf{r}'|\mu \right] \Bigg|_{-1}^1 $$

$$ = \frac{2\hbar}{p|\mathbf{r} - \mathbf{r}'|} \sin \frac{p}{\hbar} |\mathbf{r} - \mathbf{r}'|, \tag{3–59} $$

and K becomes

$$ K = \frac{1}{(2\pi\hbar)^2} \int_{-\infty}^\infty \frac{\sin (p/\hbar)|\mathbf{r} - \mathbf{r}'|}{|\mathbf{r} - \mathbf{r}'|} \exp \left[-\frac{i}{\hbar} \frac{p^2}{2m}(t - t') \right] p \, dp, \tag{3–60} $$

in which a factor 2 has been absorbed by doubling the interval of integration. Further elementary transformations (Problem 3–5) yield finally[1]

$$ K(\mathbf{r}, t; \mathbf{r}', t') = \left[\frac{2\pi i\hbar}{m}(t - t') \right]^{-3/2} \exp \left[i \frac{m}{2\hbar} \frac{|\mathbf{r} - \mathbf{r}'|^2}{(t - t')} \right]. \tag{3–61} $$

[1] W. Pauli, *Die Allgemeinen Prinzipien der Wellenmechanik.* Ann Arbor: J. W. Edwards, 1947, p. 104.

REFERENCES

HEISENBERG, WERNER, *The Physical Principles of the Quantum Theory*. New York: Dover Publications, Inc., 1930. The discussion of the uncertainty principle, written by its originator, will repay careful study.

HEISENBERG, WERNER, "The Development of the Interpretation of the Quantum Theory," in *Niels Bohr and the Development of Physics*. New York: McGraw-Hill Book Co., Inc., 1955. A historical account containing a critique of modern attempts to elucidate the meaning of quantum concepts.

KEMBLE, E. C., *The Fundamental Principles of Quantum Mechanics*. New York: McGraw-Hill Book Co., Inc., 1937, Chapter II and Chapter VII, Section 33.

KRAMERS, H. A., *Die Grundlagen der Quantentheorie*. Leipzig: Akademische Verlagsgesellschaft m.b.H., 1938. English translation: *The Foundations of Quantum Theory*. Amsterdam: North-Holland Publishing Co., 1957, Chapter I. This clearly written text is one of the masterworks of mathematical physics.

PROBLEMS

3–1. Verify Eq. (3–3).

3–2. Prove the relation (3–15).

3–3. Modify the proof of (3–17) to remove the restriction $\langle x \rangle = \langle p \rangle = 0$. (Hint: The origin of coordinates in the configuration and momentum spaces can be transferred to the centers of the respective packets by means of a linear substitution.)

3–4. Derive the general formula

$$\int_{-\infty}^{\infty} x^{2n} e^{-x^2}\, dx = \frac{1 \cdot 3 \cdot 5 \cdots (2n - 1)\sqrt{\pi}}{2^n}.$$

3–5. Carry out in detail the steps leading from Eq. (3–57) to Eq. (3–61). (Hint:

$$\int_{-\infty}^{\infty} (\sin ap)e^{-ibp^2} p\, dp = -\frac{\partial I}{\partial a}, \qquad \text{where} \quad I = \int_{-\infty}^{\infty} (\cos ap)e^{-ibp^2}\, dp.$$

I can be evaluated by the methods of Appendix A–1 or directly, through use of a table of definite intergrals.

3–6. If $a(p)$ is real and the origin is chosen so that $\langle x \rangle$ is initially zero, show that the formula

$$(\Delta x)^2 = (\Delta x)^2 \big|_{t=0} + \frac{(\Delta p)^2 t^2}{m^2}$$

is true for a wave packet of arbitrary shape. (Hint: Judicious use of the representation in momentum space is helpful in solving this problem.)

3–7. Normalize the momentum function

$$a(\mathbf{p}) = N \exp\left(-\frac{\alpha}{\hbar}|\mathbf{p}|\right)$$

and show that the corresponding $\psi(\mathbf{r})$ is

$$\psi(\mathbf{r}) = \frac{1}{\pi}(2\alpha)^{3/2}\frac{\alpha}{(r^2 + \alpha^2)^2}.$$

Also, calculate Δx and Δp_x for this wave packet and evaluate the product $\Delta x \, \Delta p_x$.

3–8. Use the integral equation (3–56) and the kernel K to show that if

$$\psi(\mathbf{r}, 0) = \exp\left(\frac{i}{\hbar}\mathbf{p} \cdot \mathbf{r}\right),$$

then

$$\psi(\mathbf{r}, t) = \exp\left[\frac{i}{\hbar}\left(\mathbf{p} \cdot \mathbf{r} - \frac{p^2}{2m}t\right)\right].$$

3–9. Discuss the time behavior of the wave packet whose (unnormalized) form is, initially,

$$\psi(x, 0) = \frac{1}{x}\sin\frac{x}{a}\, e^{(i/\hbar)p_0 x}.$$

3–10. The 2s-state of hydrogen has the wave function

$$\psi(\mathbf{r}) = \frac{1}{\sqrt{32\pi}}(r - 2)e^{-r/2},$$

in which r is measured in units of $\hbar^2/me^2$. Find the momentum representation for this state.

3–11. Solve the differential equation (3–33).

3–12. Derive expression (3–36) for the gaussian packet with $\langle x \rangle \neq 0$, $\langle p \rangle \neq 0$.

CHAPTER 4

THE SCHRÖDINGER EQUATION

4–1 Interaction among particles. In the preceding chapters, the quantum-mechanical description of free particles has been considered. However, the theory would be of no value if it could not be extended to include the interactions among particles. The effects of such interactions do, in fact, play a much more fundamental role in quantum mechanics than in classical newtonian physics; it has already been pointed out how interactions, introduced by the process of measurement, have a finite, nonnegligible influence on the quantum state of an atomic system.

Four fundamental and closely related questions arise when the interactions among particles are considered: (1) How can the concept of *state* be extended to systems that contain more than one particle? Classically, the state of a system at a particular instant is known if the position and the velocity of each of its component particles are given at that instant. In quantum theory, an appropriate generalization of the one-particle wave function must be found, such that the results of observations can be calculated by rules analogous to those set forth in Section 2–10.

(2) What are the intrinsic properties of the particles found in nature, and what types of interactions exist between them? ("Intrinsic" properties are the mass, electric charge, spin, and other quantum-mechanical properties which characterize the wave function of a "free" particle of given type.) The answer to this question is obviously a matter of experiment. However, modern relativistic quantum theory provides a structure that permits us to predict the characteristics of possible types of particles and to classify them in a clear and elegant way.

It is remarkable that only two fundamental types of interactions were known prior to 1930, namely, the newtonian gravitational force and electrical forces, typified by the Coulomb force of electrostatics. The former does not enter into the discussion of atomic systems because it is about 10^{40} times smaller than the Coulomb force between two elementary charged particles. In atomic physics, therefore, we are concerned with systems of charged particles which interact according to the familiar law of electrostatics,

$$F = \frac{ZZ'e^2}{r^2}, \tag{4–1}$$

which is the force acting between two particles, of charges Ze and $Z'e$, respectively, along the line of length r joining their positions ($e = -4.8 \times$

10^{-10} esu is the fundamental electronic charge, and Z and Z' are the "atomic numbers").

Nuclear interactions, which come into play only at much smaller distances than those of interest for atomic systems ($<10^{-12}$ cm), are of a fundamentally different kind. It has not yet been possible to achieve a clear theoretical picture of these forces, comparable to the understanding of electrical forces in terms of the theory of Maxwell and Faraday. However, there is evidence to support the view that the problem of nuclear forces will be solved when a correct description of the interaction between nuclear particles and mesons has been developed.

In recent times, a large number of "fundamental" particles[1] have been discovered, and other interactions have become known, including those associated with the beta decay of nuclei and the interactions among the various mesons. In any experiment dealing with these effects, the energy of the particles is comparable to their rest energy, and a relativistic treatment is required. These phenomena will not be considered in the present book.

(3) How are the interactions between particles propagated? In electromagnetic theory, the mode of propagation is described classically by Maxwell's equations for the electromagnetic field which is propagated in empty space with the velocity of light. The speed of the electrons within atoms is usually small relative to the velocity of light, and effects ascribable to the finite time of propagation of the field are, in general, of secondary importance. The static law (4–1) can therefore be used as a good approximation in formulating a theory of atoms. Apart from this, however, it is believed that every correct physical theory must conform to the requirements of special relativity; the association of a ψ-field with the particles allows a relativistically correct, and hence more profoundly satisfying, picture of interacting fields than any model that could have been constructed on classical foundations. The field theory of quantum phenomena had its origins in theoretical studies of the interactions between atomic systems and light[2] and in the justification of the assumptions which Einstein used in his derivation of the radiation law (Section 1–3). Quantum field theory is the basic discipline in terms of which modern ideas in fundamental-particle physics are expressed.

(4) How do we include interactions in the wave-mechanical formalism? We must formulate equations which express the modifications of the wave function resulting from interactions among the particles described. Obviously, this problem is closely associated with the answer to our first question, concerning the extension of the concept of state, because at least

[1] A. M. Shapiro, *Revs. Modern Phys.* **28**, 164 (1956). J. D. Jackson, *The Physics of Elementary Particles*. Princeton: Princeton University Press, 1958.

[2] W. Heisenberg and W. Pauli, *Z. Physik* **9**, 338 (1931). See also E. Fermi, *Revs. Modern Phys.* **1**, 87 (1932).

two particles are involved in every interaction. However, in many situations, the influence of interaction on a single particle may be described without reference to the motions of others, provided the concept of potential energy is applicable. The particle is conceived to move in a fixed field of force that arises from its surroundings. At each point of space, the force on the particle is then given by

$$\mathbf{F} = -\nabla V, \tag{4-2}$$

where V is the potential energy, which depends only upon the position of the particle.[1] For a two-particle system, this description can be introduced rigorously by considering the relative motion of the two particles in the frame of reference in which the center of mass is at rest,[2] and we shall see in Chapter 7 that this classical device can be adapted without change to the quantum treatment. More complicated systems can also be treated in an approximate way whenever one of the particles can be considered to move in the average field produced by its interactions with the others. This single-particle model provides a valuable starting point for understanding the states of complex atomic and nuclear systems.

The description of the force on a particle by means of a potential function is based, of course, on the fundamental assumptions concerning the field of force which are requisite for the definition of V.[3] In the case of the Coulomb force (4–1), these requirements are satisfied, and one has

$$V = \frac{ZZ'e^2}{r}. \tag{4-3}$$

In other instances (e.g. nuclear forces), however, one can properly question the validity of Eq. (4–2) on the basis that the forces do not admit the definition of a potential energy. An important example arises when the force is dependent upon the velocity of the particle as well as upon its position. Much attention has been given to the question whether such forces may be of importance in nuclear interactions.[4] The Lorentz force

$$\mathbf{F} = \frac{e}{c} \mathbf{v} \times \mathfrak{B}, \tag{4-4}$$

---

[1] G. Joos, *Theoretical Physics*. New York: G. E. Stechert and Co., 1934, Chapter V, Section 4.

[2] Cf. Chapter 1, Section 1–11.

[3] G. Joos, *loc. cit.*

[4] J. A. Wheeler, *Phys. Rev.* **50**, 643 (1936). Cf. also the discussions in R. G. Sachs, *Nuclear Theory*. Reading, Mass.: Addison-Wesley Publishing Co., Inc., 1953, Section 4–2. Also, J. M. Blatt and V. F. Weisskopf, *Theoretical Nuclear Physics*. New York: John Wiley and Sons, 1952, Chapter III, Section 3.

which acts upon a particle of charge e, moving with velocity **v** in a magnetic field of intensity $\mathfrak{B}$, is explicitly velocity-dependent, and therefore not describable by Eq. (4–2). It may, however, be expressed in terms of the vector potential **A** for the magnetic field and thus brought into the single-particle theory. For atomic systems, the force (4–4) is smaller than the Coulomb force by a factor of order v^2/c^2, approximately, and hence makes a small, although important, contribution to the energies of atomic states.

4–2 Geometrical optics. The wave theory must be modified in order to include forces which can be expressed by means of a potential energy. This modification can be conjectured from the correspondence principle, by analogy with the well-known parallelism between geometrical optics and newtonian mechanics, developed by Hamilton. The laws of geometrical optics can be deduced from the laws of wave optics, i.e., from Maxwell's equations; they represent an approximation which holds provided that the wavelength of the light is very short in comparison to the size of the refracting or reflecting objects in the field. It will now be shown that, by analogy, newtonian mechanics can be considered to be a similar approximation, derived from wave mechanics and subject to the same restriction, that is, the de Broglie wavelength must be small compared to the dimensions of the physical objects involved. The diagram of Fig. 4–1 is helpful in understanding these similarities between optics and mechanics. It is instructive to consider in detail the connections indicated in the figure; we shall, therefore, digress briefly and describe the transition from wave optics to geometrical optics.

According to Huygens' principle, the course of a beam of light traversing an isotropic optical medium of varying index of refraction can be

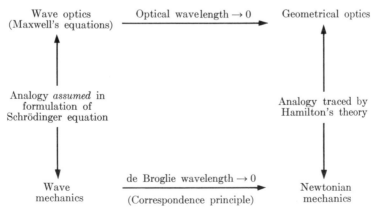

FIG. 4–1.   Analogy between optics and mechanics.

traced by means of the geometrical construction shown in Fig. 4–2. The wavefront Σ is the orthogonal trajectory of the optical rays in the medium; it is related to the neighboring wavefront Σ', occupied by the disturbance at a short time (Δt) later, in the following way: At each point P of Σ, one constructs a sphere of radius $v\,\Delta t$, where v is the velocity of light in the medium in the neighborhood of that point, i.e.,

$$v = \frac{c}{\mu}\,; \qquad \mu = \text{index of refraction at } P. \tag{4–5}$$

The wavefront Σ' is the envelope of the family of spheres constructed in this manner at each point of Σ. By repeating this process and passing to the limit $\Delta t \to 0$, one obtains a system of wave surfaces and their orthogonal trajectories, or rays[1] (Problem 4–1).

Now consider a system of wavefronts and rays generated by an initial surface Σ_0. The time of propagation of light along a ray P_0P from Σ_0 to Σ is, by construction,

$$t = \int_{P_0}^{P} \frac{ds}{v} = \frac{1}{c}\int_{P_0}^{P} \mu\,ds \tag{4–6}$$

(see Fig. 4–3). It is easy to show that this time is shorter than the time (computed by the same formula) for any other path joining these two

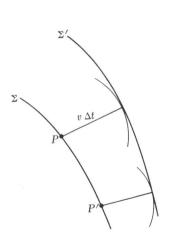

FIG. 4–2. Huygens' construction.

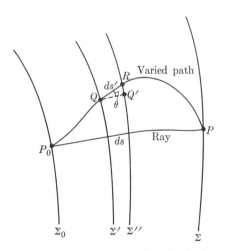

FIG. 4–3. Construction for the derivation of Fermat's principle of least time.

[1] J. L. Synge, *Geometrical Optics, an Introduction to Hamilton's Method.* Cambridge: Cambridge University Press, 1937.

points. To prove this, consider two nearby wave surfaces, Σ' and Σ'', between which the integral of Eq. (4-6) receives a contribution $(\mu/c)\,ds$. By construction, this quantity is the same for every normal line (e.g. QQ') joining these surfaces; but for a path QR, which is not normal, it is $\mu\,ds' = \mu\,ds/\cos\theta$, where θ is the angle between QR and the normal. Since $\cos\theta \leq 1$, it follows that

$$\frac{1}{c}\int \mu\,ds' \geq \frac{1}{c}\int \mu\,ds. \qquad (4\text{-}7)$$

This is Fermat's *principle of least time*, summarized in the statement that the time of propagation of light between two points in an optical medium is smaller than the time which would be computed for any other path in the medium joining the same two points. In the notation of the calculus of variations,[1] Fermat's principle can be written

$$\delta \frac{1}{c}\int \mu\,ds = 0, \qquad \text{(constant frequency)}. \qquad (4\text{-}8)$$

The index of refraction in a dispersive medium depends, of course, upon the frequency of the light, and the variation implied in Eq. (4-8) is to be made for light of a given constant frequency.

The wave surfaces can be defined by the equation

$$S = ct = \int_{\Sigma_0}^{\Sigma} \mu\,ds = \text{constant}, \qquad (4\text{-}9)$$

where S is the *optical path length* for the system of rays in question. It is easy to demonstrate, by the methods of the calculus of variations or by geometrical arguments similar to those just made, that the content of Eq. (4-8) is also expressed by the differential equation[2]

$$(\nabla S)^2 = \mu^2, \qquad (4\text{-}10)$$

which must be satisfied by every function S that describes a system of rays in a medium of index μ (Appendix A-5). It will now be shown how Eq. (4-10) can be regarded as an approximate expression of the behavior of light waves in the limit of *small* wavelength.

---

1 G. Joos, *Theoretical Physics*. New York: G. E. Stechert and Co., 1934, Chapter IV.

2 P. Frank and R. v. Mises, *Die Differential- und Integralgleichungen der Mechanik und Physik*. Braunschweig: Friedrich Vieweg und Sohn, 1935, Volume II, Chapter 1, Section 1.

It follows from Maxwell's equations[1] that the relation

$$\nabla^2\phi + k^2\phi = 0 \qquad (4\text{–}11)$$

is satisfied by each component of the electromagnetic field vectors in a light wave, provided that the local wavelength $\lambda = 2\pi/k$ is small in comparison with a distance in which the index of refraction undergoes an appreciable change, i.e., provided that $\lambda|\nabla\mu| \ll 1$. Under these circumstances, it is reasonable to expect that the light will be propagated locally as if the index of refraction were constant, and that the effect of the variation of μ will be felt slowly, that is, only after a distance of many wavelengths has been traversed. These considerations lead us to an approximate solution of Eq. (4–11) in the form

$$\phi = \phi_0 e^{ik_0 S}. \qquad (4\text{–}12)$$

The amplitude factor ϕ_0 is expected to be a slowly varying function of position, describing the effect on the intensity of the varying divergence of the rays from place to place in the medium. The quantity $k_0 = 2\pi/\lambda_0$ is the wave number for waves of the same frequency in free space, and the function S is the optical path length, expressing the effect of the refractive medium on the phase of the wave disturbance. It is readily shown that the quantity S satisfies the differential equation (4–10) and is, therefore, identical with the optical path length of the geometrical theory. Differentiation of Eq. (4–12) yields

$$\nabla^2\phi = [ik_0\nabla S \cdot (\nabla\phi_0 + ik_0\phi_0\nabla S) + \nabla^2\phi_0 + ik_0\nabla\phi_0 \cdot \nabla S$$
$$+ ik_0\phi_0\nabla^2 S]e^{ik_0 S}; \qquad (4\text{–}13)$$

and by substituting Eq. (4–13) into Eq. (4–11), cancelling the exponential factor, and equating real and imaginary parts,[2] we obtain

$$(\nabla S)^2 = \mu^2 + \frac{1}{k_0^2}\frac{\nabla^2\phi_0}{\phi_0} \qquad (4\text{–}14)$$

and

$$\nabla^2 S + \nabla S \cdot \nabla \ln \phi_0^2 = 0. \qquad (4\text{–}15)$$

In accordance with the approximations described, the second term in the right-hand member of Eq. (4–14) is vanishingly small in the limit $\lambda \to 0$,

---

[1] J. A. Stratton, *Electromagnetic Theory.* New York: McGraw-Hill Book Co., Inc., 1941, page 342, Examples 10 and 11. Cf. also Appendix A–6.

[2] It is a good exercise for the reader to justify the assumption that the functions ϕ_0 and S can be assumed to be *real*.

and S therefore satisfies Eq. (4–10). Equation (4–15) relates to the intensity of the light and can also be derived from the geometrical theory.[1] The approximation just described is a very useful one in the theory of waves. It will be encountered again in the discussion of the WKB method of solving the Schrödinger equation, which provides a link between wave mechanics and the older quantization rule of Bohr and Sommerfeld.

4–3 Analogy between optics and mechanics. With the preceding sketch of optical theory, we are prepared to describe Hamilton's analogy between geometrical optics and the classical mechanics of a particle moving in a field of force. It is shown in works on dynamics[2] that the orbit of a particle moving in accordance with Newton's laws can be deduced from the *principle of least action:*

$$\delta \int p(x, y, z, E) \, ds = 0, \qquad E \text{ constant.} \qquad (4\text{–}16)$$

According to this principle, the path followed by a particle in a field of force described by the potential energy function V is such that the value of the *action* $\int p \, ds$ is stationary when compared to other paths which have the same end points and are traversed with the same total energy. The momentum of a particle of mass m and total energy E is given by

$$p = \sqrt{2m(E - V)}. \qquad (4\text{–}17)$$

The formal similarity between Eq. (4–16) and Eq. (4–8) is at once apparent: The problems of geometrical optics and of particle dynamics become formally identical if the index of refraction is regarded as the optical analogue of the momentum of the particle. This is *Hamilton's analogy.*
Equation (4–8) can also be written in the form

$$\delta \int \frac{1}{\lambda(x, y, z, \nu)} \, ds = 0, \qquad \nu \text{ constant,} \qquad (4\text{–}18)$$

in which $\lambda = (1/\mu)(c/\nu)$ is the local wavelength. The quantum relations

$$p = \frac{h}{\lambda} \quad \text{and} \quad E = h\nu, \qquad (4\text{–}19)$$

connecting the wavelength and frequency of matter waves with the momentum and energy of the corresponding particle, impart a much deeper

---

[1] P. Frank and R. v. Mises, *loc. cit.* Cf. Problem 4–2.
[2] H. Goldstein, *Classical Mechanics.* Reading, Mass.: Addison-Wesley Publishing Co., Inc., 1950, Section 7–5. L. Page, *Introduction to Theoretical Physics.* 2nd ed. New York: D. Van Nostrand Co., Inc., 1935, Chapter IV, Section 55.

significance to the parallelism between Eqs. (4–16) and (4–18). It is natural to assume that an equation of the form (4–18) holds also for a quantum-mechanical system and provides an approximation to the wave theory if the wavelength is very short; i.e., precisely in the situation in which the newtonian laws must follow from the more complete theory.

The final step in this heuristic reasoning is made if we assume that the matter waves, expressed by the wave function ψ, are correctly described by an equation analogous to Eq. (4–11), in which the wave number is determined by Eqs. (4–19) and (4–17):

$$k^2 = \left(\frac{2\pi}{\lambda}\right)^2 = \frac{2m}{\hbar^2}(E - V). \tag{4–20}$$

The result is

$$\nabla^2\psi + \frac{2m}{\hbar^2}(E - V)\psi = 0, \tag{4–21}$$

which is Schrödinger's equation.

It will now be assumed that the Schrödinger equation is exactly satisfied[1] by the wave function for a particle in the field V, and its consequences will be developed on the basis of the Born interpretation.

First, it is noted that the Schrödinger equation (4–21) applies to a system that consists of a single particle of mass m in a field of potential energy V, the total energy being given and constant. The time dependence of the wave function is not specified. However, in the special case of a *free* particle ($V = 0$), one has

$$-\frac{\hbar^2}{2m}\nabla^2\psi = E\psi; \tag{4–22}$$

this expression is the same as the Schrödinger equation for a free particle obtained in Eq. (2–98), provided that

$$-\frac{\hbar}{i}\frac{\partial\psi}{\partial t} = E\psi. \tag{4–23}$$

This differential equation in t implies that the time behavior of ψ is described by

$$\psi(\mathbf{r}, t) = \psi(\mathbf{r}, 0)\exp\left[-\frac{i}{\hbar}Et\right]. \tag{4–24}$$

---

[1] Note that Eq. (4–11) is *not* exactly satisfied by the Maxwell field, so that the analogy to optical phenomena is not (even formally) complete. An analogy of this kind exists for any pair of wave fields. Cf. C. Eckart, "The Approximate Solution of One-Dimensional Wave Equations," *Revs. Modern Phys.* **20**, 399 (1948).

The assumption is made that this equation gives the time behavior of *every* wave function belonging to a system of fixed energy, so that Eq. (4–21) may equally well be written

$$-\frac{\hbar^2}{2m}\nabla^2\psi + V\psi = -\frac{\hbar}{i}\frac{\partial\psi}{\partial t}.$$ (4–25)

This is the *Schrödinger equation containing the time*. It is the generalization of Eq. (2–98). In this simple case, it provides an answer to the fourth question raised in Section 4–1, i.e. to the question of how interactions can be included in the wave-mechanical formalism.

Finally, to interpret ψ, we assume that the formulae for the expectations of physical quantities, which were derived in the free-particle case (Section 2–10), hold without change. Specifically,

$$\langle \mathbf{r}\rangle = \frac{\int \psi^* \mathbf{r}\psi\,d\mathbf{r}}{\int \psi^*\psi\,d\mathbf{r}}\,;\qquad \langle \mathbf{p}\rangle = \frac{\int \psi^*(\hbar/i)\nabla\psi\,d\mathbf{r}}{\int \psi^*\psi\,d\mathbf{r}}.$$ (4–26)

The time-dependent Schrödinger equation (4–25) is the central relation of nonrelativistic quantum mechanics and forms the basis for most of the work presented in the remainder of this book.

4–4 Principle of superposition of states.[1] The discussion leading to Eq. (4–25) was founded on the wave theory for a particle of fixed energy. However, since the Schrödinger equation is a linear equation, its solutions obey the *principle of superposition*. Hence, if two solutions ψ_1 and ψ_2 (belonging, perhaps, to different values of E) are known, other solutions can be constructed, of the form

$$\psi = a_1\psi_1 + a_2\psi_2,$$ (4–27)

with arbitrary choice of the constants a_1 and a_2. Furthermore, the function ψ will satisfy the same conditions of continuity and integrability that are satisfied by ψ_1 and ψ_2. In addition, we assume as a fundamental principle that *every* allowable solution of Eq. (4–25) corresponds to a realizable state of the system. Indeed, this principle of superposition played a basic role in the discussion of wave packets in Chapter 2: The representation of ψ in Eq. (2–89) is a linear combination of functions $\exp[(i/\hbar)(px - Et)]$, each of which satisfies Eq. (4–25) and is a state in its own right (cf. Section 3–5).

---

[1] P. A. M. Dirac, *The Principles of Quantum Mechanics*. 3rd ed. Oxford: The Clarendon Press, 1947, Chapter I.

Quantum mechanics is, therefore, a linear theory, and many of the classical methods which have been developed for the treatment of linear problems (such as the vibrating string, optical waves, etc.) are immediately adaptable. It must be borne in mind, however, that "the superposition which occurs in quantum mechanics is of an essentially different nature from any occurring in the classical theory,"[1] and that analogies to the interpretation of results in classical physics are very likely to be misleading. There is no analogue, in classical dynamics, to the principle of super-position of states.

4–5 Probability current. The remainder of this chapter will be devoted to a discussion of certain general properties of the solutions of the Schrö-dinger equation (4–25); these properties hold for all single-particle wave functions, independently of special assumptions as to the form of the potential energy. Since V is a real quantity, the complex conjugate of Eq. (4–25) is

$$-\frac{\hbar^2}{2m} \nabla^2 \psi^* + V\psi^* = \frac{\hbar}{i} \frac{\partial \psi^*}{\partial t} ; \qquad (4\text{–}28)$$

V can be eliminated between Eqs. (4–25) and (4–28), and we obtain

$$\frac{\hbar^2}{2m} (\psi^* \nabla^2 \psi - \psi \nabla^2 \psi^*) = \frac{\hbar}{i} \left(\psi^* \frac{\partial \psi}{\partial t} + \frac{\partial \psi^*}{\partial t} \psi \right), \qquad (4\text{–}29)$$

which is the same as

$$-\frac{\hbar}{2im} \nabla \cdot (\psi^* \nabla \psi - \psi \nabla \psi^*) = \frac{\partial}{\partial t} (\psi^* \psi). \qquad (4\text{–}30)$$

This relation has the form of the *equation of continuity*, i.e.,

$$\frac{\partial \rho}{\partial t} + \nabla \cdot \mathbf{S} = 0, \qquad (4\text{–}31)$$

in which $\rho = \psi^* \psi$ is the probability density, and $\mathbf{S}$ the *probability current:*

$$\mathbf{S} = \frac{\hbar}{2im} (\psi^* \nabla \psi - \nabla \psi^* \psi) = \frac{\hbar}{m} \text{Im} (\psi^* \nabla \psi). \qquad (4\text{–}32)$$

Equation (4–31) arises in any theory in which an extensive quantity (e.g., mass, charge, or heat energy) is known to satisfy a law of conserva-tion. If, for example, ρ represents the density of a compressible fluid, and $\mathbf{S} = \rho \mathbf{v}$ is the current of fluid crossing unit area normal to the direc-

---

[1] P. A. M. Dirac, *op. cit.*, page 14.

tion of the fluid velocity **v**, then the equation (4–31) expresses the conservation law; i.e., a change in the total amount of fluid contained within any small fixed volume element is accounted for by flow through the surface of the volume element. Using Gauss's divergence theorem, the equation of continuity can be transformed to the integral form

$$\frac{d}{dt} \int_V \rho \, d\mathbf{r} = \int_\Sigma \mathbf{S} \cdot \hat{\mathbf{n}} \, da, \tag{4–33}$$

in which $\hat{\mathbf{n}}$ is the unit outward normal to the surface Σ which encloses a region of volume V. The interpretation of Eq. (4–33) is similar to that of Eq. (4–31).

The interpretation of the quantity $\psi^*\psi$ as probability density leads, therefore, to the concept of the probability current [Eq. (4–32)]. The decrease of the probability of finding a particle within V can be described in terms of an outward flow of probability current through the surface of V. The decrease of probability arises, of course, because of the change of ψ with time.

If the wave function is regarded as belonging to a collection of noninteracting particles (Section 2–9), then the expectation of the vector **S** can be considered to be the average particle current. The wave function $\psi = \exp\,(i/\hbar)\mathbf{p} \cdot \mathbf{r}$ leads, for example, to

$$\mathbf{S} = \frac{\hbar}{m} \operatorname{Im}\left[\exp\left(-\frac{i}{\hbar}\,\mathbf{p} \cdot \mathbf{r}\right) \cdot \frac{i}{\hbar}\,\mathbf{p}\exp\left(\frac{i}{\hbar}\,\mathbf{p} \cdot \mathbf{r}\right)\right] = \frac{\mathbf{p}}{m} = \mathbf{v}, \tag{4–34}$$

which is just the current for a beam of particles of unit density ($\psi^*\psi = 1$) and velocity **v**.

It is assumed that the equation of continuity holds at every point of space, and that ψ and its derivative are continuous everywhere. Hence the possibility of sources for ρ is excluded, as must be the case if ψ is to be normalizable independently of the time.[1] It can be noted, incidentally, that Eq. (4–33), when applied to the entire volume of space, generalizes the normalization theorem of Section 2–9 to the case in which the particle is not free. The absence of sources for the ψ-field results from the homogeneous character of the Schrödinger equation.

In any state for which $\nabla \cdot \mathbf{S}$ is zero, the probability density is constant in time [Eq. (4–31)]. Such states are called *stationary*. The most important case arises for states of fixed energy, for which the time dependence of ψ is given by Eq. (4–24), and $\psi(\mathbf{r}, 0)$ satisfies therefore the time-independent Schrödinger equation (4–21). Except for the presence of ψ itself, however,

---

[1] The integral $\int|\psi|^2 \, d\mathbf{r}$ would be constant if there were a distribution of sources and sinks for ρ, of equal and opposite total strengths. However, no such distribution is permitted.

this equation is entirely real. Consequently, if $\psi(\mathbf{r}, 0)$ is a complex solution of Eq. (4–21), it can be separated into real and imaginary parts, each of which is itself a solution. It is possible that these two solutions are linearly independent of each other; if this is so, we have a case of degeneracy in which more than one physically distinguishable state belongs to the same value of the total energy. However, if the state in question is nondegenerate, then the real and imaginary parts of $\psi(\mathbf{r}, 0)$ must describe the same state, and therefore differ from $\psi(\mathbf{r}, 0)$ only by a multiplicative (complex) constant. The value of this constant is at our disposal; hence the function $\psi(\mathbf{r}, 0)$ in Eq. (4–24) can always be chosen to be real, provided the state in question is nondegenerate. Thus, according to Eq. (4–32), the current is zero for nondegenerate stationary states.

There are no nondegenerate stationary states for a free particle. Indeed, the state $\exp[(i/\hbar)(\mathbf{p} \cdot \mathbf{r})]$ is a free-particle state of energy E for every vector $\mathbf{p}$ of magnitude $\sqrt{2mE}$. It follows that the above argument concerning nondegenerate states applies only to bound states. The subject of degeneracy of quantum states will be discussed further and illustrated by examples later in the text.

4–6 Motion of wave packets. The principle of correspondence between the motion of a wave packet and the motion of a classical particle has been of fundamental significance in the development of quantum mechanics and is essential in the interpretation of the theory. A calculation due to Ehrenfest[1] will now be made, which demonstrates that the newtonian laws of motion, in the form

$$\text{(a)} \quad \frac{d\mathbf{r}}{dt} = \frac{\mathbf{p}}{m}, \qquad \text{(b)} \quad \frac{d\mathbf{p}}{dt} = -\nabla V, \qquad (4–35)$$

are satisfied exactly by the average motion of a wave packet described by a wave function ψ which is a solution of the Schrödinger equation. For simplicity, we shall deal with the components of $\mathbf{r}$ and $\mathbf{p}$ in the direction of the x-axis, and assume that ψ is a normalized wave function. The average or expectation of x for the packet is

$$\langle x \rangle = \int \psi^* x \psi \, d\mathbf{r}, \qquad (4–36)$$

and hence differentiation (cf. Section 2–9) yields

$$\frac{d\langle x \rangle}{dt} = \int \left[\frac{\partial \psi^*}{\partial t} x\psi + \psi^* x \frac{\partial \psi}{\partial t} \right] d\mathbf{r}. \qquad (4–37)$$

[1] P. Ehrenfest, Z. Physik **45**, 455 (1927).

This equation can be transformed, by means of Eq. (4–25) and its complex conjugate Eq. (4–28), into

$$\frac{d\langle x \rangle}{dt} = \frac{i}{\hbar} \int \left[-\frac{\hbar^2}{2m} (\nabla^2 \psi^* \psi - \psi^* \nabla^2 \psi)x \right] d\mathbf{r}$$

$$= -\frac{\hbar}{m} \operatorname{Im} \int x\psi^* \nabla^2 \psi \, d\mathbf{r}. \tag{4–38}$$

The vector identity

$$\nabla \cdot (x\psi^* \nabla \psi) = x\psi^* \nabla^2 \psi + x\nabla \psi^* \cdot \nabla \psi + \psi^* \frac{\partial \psi}{\partial x} \tag{4–39}$$

can now be used to obtain

$$\frac{d\langle x \rangle}{dt} = -\frac{\hbar}{m} \operatorname{Im} \int \left[\nabla \cdot (x\psi^* \nabla \psi) - \psi^* \frac{\partial \psi}{\partial x} \right] d\mathbf{r} \tag{4–40}$$

$$= \frac{1}{m} \int \psi^* \frac{\hbar}{i} \frac{\partial \psi}{\partial x} \, d\mathbf{r}. \tag{4–41}$$

In obtaining Eq. (4–40), we have omitted the second term in Eq. (4–39) since it is purely real. The integral of the divergence in Eq. (4–40) can be shown to vanish by application of Gauss's theorem and of the boundary conditions on ψ. It follows from Eq. (4–26) that Eq. (4–41) is just the quantum counterpart of the classical relation between velocity and position:

$$\frac{d\langle x \rangle}{dt} = \frac{\langle p_x \rangle}{m}. \tag{4–42}$$

In a similar way, we have

$$\langle p_x \rangle = \int \psi^* \frac{\hbar}{i} \frac{\partial \psi}{\partial x} \, d\mathbf{r};$$

$$\frac{d}{dt} \langle p_x \rangle = \frac{\hbar}{i} \int \left[\frac{\partial \psi^*}{\partial t} \nabla \psi + \psi^* \nabla \frac{\partial \psi}{\partial t} \right]_x d\mathbf{r}$$

$$= \int \left\{ -\frac{\hbar^2}{2m} [\nabla^2 \psi^* \nabla \psi - \psi^* \nabla(\nabla^2 \psi)] + V(\psi^* \nabla \psi) - \psi^* \nabla(V\psi) \right\}_x d\mathbf{r}$$

$$= -\frac{\hbar^2}{2m} \int \left[\nabla^2 \psi^* \frac{\partial \psi}{\partial x} - \psi^* \nabla^2 \frac{\partial \psi}{\partial x} \right] d\mathbf{r} - \int \psi^* \left(\frac{\partial V}{\partial x} \right) \psi \, d\mathbf{r}. \tag{4–43}$$

By Green's second identity the first integral is zero, and according to the general definition of expectation, the second integral is $-\langle \partial V/\partial x \rangle$, whence

$$\frac{d\langle p_x \rangle}{dt} = \left\langle -\frac{\partial V}{\partial x} \right\rangle ; \qquad (4\text{-}44)$$

this is the quantum equivalent of the x-component of Eq. [4–35(b)]. Thus the expectations of position, momentum, and force obey Newton's second law of motion exactly. This is an expression of the correspondence principle for wave packets which has now been founded on the Schrödinger equation.

REFERENCES

DE BROGLIE, L., and BRILLOUIN, L., *Selected Papers on Wave Mechanics.* London: Blackie and Son Limited, 1928. Especially: "On the Parallelism between the Dynamics of a Material Particle and Geometrical Optics."

DIRAC, P. A. M., *The Principle of Quantum Mechanics*, 3rd ed. Oxford: The Clarendon Press, 1947. Chapter I emphasizes the fundamental importance of the principle of superposition.

ECKART, C., "The Approximate Solution of One-Dimensional Wave Equations," *Revs. Modern Phys.* **20**, 399 (1948). A general review of approximate methods related to the classical approximation. Wave systems in other fields of physics are discussed.

JOOS, G., *Theoretical Physics.* New York: G. E. Stechert and Co., 1934. Chapter IV is a clear, concise summary of the calculus of variations.

PROBLEMS

4–1. Show that the rays and wave surfaces obtained by Huygens' construction are mutually orthogonal provided the medium is isotropic.

4–2. The intensity of the light in an optical medium can be described in terms of the vector $\mathbf{I} = I\hat{\mathbf{e}}$, where I is the energy which crosses unit area perpendicular to a ray in one second, and $\hat{\mathbf{e}}$ is a unit vector parallel to the ray. Show that $\nabla \cdot \mathbf{I} = 0$, and hence

$$\nabla^2 S + \nabla S \cdot \nabla \ln\left(\frac{I}{\mu}\right) = 0.$$

[Hint: In terms of S, $\hat{\mathbf{e}}$ is given by $\hat{\mathbf{e}} = (1/\mu)\nabla S$.] Compare this result with Eq. (4–15) and show that the two expressions are consistent in view of the approximations made in obtaining the latter.

4–3. Show that, for a normalized one-dimensional wave packet,

$$\int_{-\infty}^{\infty} S \, dx = \frac{\langle p \rangle}{m}.$$

Generalize to the three-dimensional case and comment on the physical meaning of this relation.

4–4. Calculate the probability current corresponding to the wave function

$$\psi = \frac{e^{ikr}}{r},$$

where $r^2 = x^2 + y^2 + z^2$. Examine $\mathbf{S}$ for large values of $\mathbf{r}$ and interpret the result.

101

CHAPTER 5

PROBLEMS IN ONE DIMENSION

5–1 Potential step. The Schrödinger equation for the wave function of a particle constrained to move on a straight line (the x-axis) is

$$-\frac{\hbar^2}{2m}\frac{\partial^2\psi}{\partial x^2} + V\psi = -\frac{\hbar}{i}\frac{\partial\psi}{\partial t}, \tag{5-1}$$

where $V(x)$ is the potential energy. If the total energy of the particle has the fixed value E, the time dependence of ψ is given by

$$\psi(x, t) = \psi(x, 0)e^{-(i/\hbar)Et}, \tag{5-2}$$

and therefore, ψ also satisfies the time-independent Schrödinger equation

$$\frac{d^2\psi}{dx^2} + \frac{2m}{\hbar^2}(E - V)\psi = 0. \tag{5-3}$$

In this chapter, we shall study the solutions of this equation for several forms of the function $V(x)$. The first example is the potential step represented by the function

$$V(x) = \begin{cases} 0 & (x < 0), \\ V_0 & (x > 0), \end{cases} \tag{5-4}$$

shown in Fig. 5–1. The time-independent Schrödinger equation is

$$\frac{d^2\psi}{dx^2} + k_0^2\psi = 0, \qquad k_0 = \frac{\sqrt{2mE}}{\hbar} \qquad (x < 0), \tag{5-5}$$

$$\frac{d^2\psi}{dx^2} + k^2\psi = 0, \qquad k = \frac{\sqrt{2m(E - V_0)}}{\hbar} \qquad (x > 0). \tag{5-6}$$

This equation can be solved immediately to yield

$$\psi = Ae^{ik_0x} + Be^{-ik_0x} \qquad (x < 0), \tag{5-7}$$

$$\psi = Ce^{ikx} + De^{-ikx} \qquad (x > 0), \tag{5-8}$$

where A, B, C, D are integration constants. The functions e^{ik_0x} and e^{-ik_0x}, when multiplied by the factor $\exp[(-i/\hbar)Et]$, represent waves moving toward the right and toward the left, respectively. Hence, ψ is a

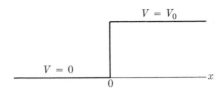

$$V = V_0$$

$$V = 0$$

$$x$$

$$0$$

FIG. 5–1.   Square potential step.

linear combination of plane waves in both intervals of the x-axis, the wave
number corresponding in each case to the kinetic energy, according to the
general formula

$$k = \frac{1}{\hbar} \sqrt{2m[E - V(x)]}. \tag{5–9}$$

It has been pointed out in Section 4–5 that the wave function and its
derivative must be continuous everywhere; otherwise, sources of the
probability density $\psi^*\psi$ could exist, and it would be impossible to normalize
ψ independently of the time.   Because of the discontinuity in V in the
present example, the form of the Schrödinger equation changes discon-
tinuously at $x = 0$.   The constants of integration must be chosen so that
ψ and its derivative are continuous at this point.   Continuity of ψ demands
that the wave functions (5–7) and (5–8) must be equal at $x = 0$, whence

$$A + B = C + D. \tag{5–10}$$

When the derivatives of (5–7) and (5–8) are set equal at $x = 0$, the
further condition

$$k_0(A - B) = k(C - D) \tag{5–11}$$

is obtained.   These two linear equations in the coefficients are easily
solved to give

$$C = \frac{2k_0}{k_0 + k} A - \frac{k_0}{k_0 + k} \frac{k}{} D, \qquad B = \frac{k_0 - k}{k_0 + k} A + \frac{2k}{k_0 + k} D, \tag{5–12}$$

and, by substitution in Eqs. (5–7) and (5–8), we have

$$\psi = A\psi_1 + D\psi_2, \tag{5–13}$$

where

$$\psi_1 = \begin{cases} e^{ik_0x} + \dfrac{k_0 - k}{k_0 + k} e^{-ik_0x} & (x < 0), \\[3mm] \dfrac{2k_0}{k_0 + k} e^{ikx} & (x > 0), \end{cases} \tag{5–14}$$

and

$$\psi_2 = \begin{cases} \dfrac{2k}{k_0 + k}\, e^{-ik_0 x} & (x < 0), \\[2ex] e^{-ikx} - \dfrac{k_0 - k}{k_0 + k}\, e^{ikx} & (x > 0). \end{cases} \tag{5-15}$$

The conditions of continuity at $x = 0$ are satisfied by both ψ_1 and ψ_2 and therefore by the linear combination (5–13). However, a further condition to be imposed upon the wave function is that ψ must remain finite as $|x|$ becomes infinite, and hence the remote parts of the x-axis must be examined. The behavior of the functions (5–14) and (5–15) for large values of $|x|$ is governed by the relative magnitudes of E and V_0. We distinguish three cases:

I. $E > V_0$. The numbers k_0 and k are real and may be assumed to be positive. Since the two functions ψ_1 and ψ_2 are finite everywhere, both are admissible solutions. In this case, the general solution (5–13) is a linear combination of the two *linearly independent*[1] solutions ψ_1 and ψ_2. Thus the solution ψ is doubly degenerate. The function ψ_1 represents a state in which a wave $\exp{(ik_0 x)}$ is incident from the direction of the negative x-axis. The second term in the first of Eqs. (5–14) represents a reflected wave traveling back from the step, while the part of the solution belonging to $x > 0$ represents a transmitted wave propagated beyond the step toward the right. A similar interpretation of ψ_2 can be made in which the term $\exp{(-ikx)}$ represents a wave incident upon the step from the right.

The special solution ψ_1 can be written

$$\psi_1 = \begin{cases} e^{ik_0 x} + Re^{-ik_0 x} & (x < 0), \\ Te^{ikx} & (x > 0), \end{cases} \tag{5-16}$$

in which $R = (k_0 - k)/(k_0 + k)$ and $T = 2k_0/(k_0 + k)$ may be called the *reflection* and *transmission coefficients*, respectively. R and T measure the (complex) amplitudes of the reflected and transmitted waves relative to the amplitude of the incident wave. The probability current [Eq. (4–32)] is

$$S = \frac{\hbar}{m}\, \mathrm{Im}\left(\psi_1^* \frac{d\psi_1}{dx}\right),$$

---

[1] Two functions ψ_1 and ψ_2 are linearly independent if the equation $A\psi_1 + D\psi_2 = 0$ is not satisfied identically in x for any choice of the complex numbers A and D except $A = D = 0$.

whence, for $x < 0$,

$$S_{x<0} = \frac{\hbar}{m} \text{Im} \left[(e^{-ik_0 x} + R^* e^{ik_0 x}) ik_0 (e^{ik_0 x} - R e^{-ik_0 x}) \right]$$

$$= \frac{\hbar k_0}{m} (1 - |R|^2), \quad (5\text{-}17)$$

and, for $x > 0$,

$$S_{x>0} = \frac{\hbar}{m} \text{Im} \left(T^* e^{-ikx} \cdot ik T e^{ikx} \right) = \frac{\hbar}{m} |T|^2 \text{Re } k. \quad (5\text{-}18)$$

In case I, k is real, and it is easily verified that $S_{x<0} = S_{x>0}$. The expressions (5-17) and (5-18) are forms of S which are immediately associated with the character of the wave functions from which they are derived. In terms of the index of refraction

$$\mu = \sqrt{1 - \frac{V_0}{E}}, \quad (5\text{-}19)$$

we have

$$|R|^2 = \left(\frac{1 - \mu}{1 + \mu} \right)^2 \quad \text{and} \quad |T|^2 = \left(\frac{2}{1 + \mu} \right)^2. \quad (5\text{-}20)$$

The ratio of the transmitted current to the incident current is

$$\frac{k}{k_0} |T|^2 = \mu \left(\frac{2}{1 + \mu} \right)^2. \quad (5\text{-}21)$$

These relationships are presented graphically in Fig. 5-2. A similar description of the state ψ_2 can easily be made (Problem 5-1). It may be remarked at this point that, because of the discontinuous behavior assumed

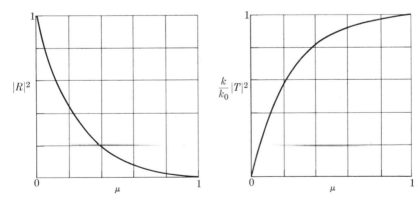

FIG. 5-2. Ratio of reflected and transmitted current to incident current for the square potential step, as a function of the index of refraction $\mu = \sqrt{1 - V_0/E}$.

for $V(x)$, there is no correspondence limit of the kind discussed in Section 4–2, i.e., there is *no* value of k_0 such that $\lambda|d\mu/dx| \ll 1$.

II. $0 < E < V_0$. The wave number k_0 is real, as before, but k is now purely imaginary. We write

$$k = \frac{i}{\hbar}\sqrt{2m(V_0 - E)} = i\kappa. \qquad (5\text{–}22)$$

The transmitted wave in the solution ψ_1 is now attenuated with the relaxation length $1/\kappa$. Also,

$$R = \frac{k_0 - i\kappa}{k_0 + i\kappa}, \qquad (5\text{–}23)$$

whence $|R|^2 = 1$. The incident and reflected waves carry equal currents toward and away from the step, and the total current is zero. Intuitively, this may be expected, since a classical particle would be "totally reflected" at the step. The wave function, however, *is not zero to the right of the step*, but has the value

$$\psi_1 = \frac{2k_0}{k_0 + i\kappa}e^{-\kappa x}. \qquad (5\text{–}24)$$

There is a finite probability that the particle is found in a region which is classically inaccessible. This nonclassical phenomenon of *barrier penetration* is fundamental to the understanding of certain phenomena in atomic and nuclear physics (Ramsauer effect, alpha decay). Its importance will be illustrated further in the following sections. The current on the right of the step is zero since Re $k = 0$ [Eq. (5–18)].

For $x > 0$, the function ψ_2 contains the term $\exp(-ikx) = \exp(\kappa x)$ which is not bounded for $x > 0$. Consequently, ψ_2 is not an admissible solution for $E < V_0$. In this case, therefore, the wave function is unique except for normalization, and the state is nondegenerate. It will be seen frequently in future examples that the condition of boundedness plays a decisive role in limiting the number of admissible solutions of the Schrödinger equation and hence the number of possible states.

III. $E < 0$. The wave numbers k_0 and k are both purely imaginary, and the argument of the preceding paragraph shows that there is no solution in this case. This is always true when the kinetic energy is negative everywhere. (Note that, in case II, the kinetic energy is negative in the region $x > 0$, but is positive to the left of the barrier.)

Finally, it is of interest to mention the limiting case $V_0 \to \infty$, in which the step is classically impenetrable. The solution for case II is the only one of interest; it reduces in the limit $\kappa \to \infty$ to

$$\psi_1 = \begin{cases} e^{ik_0x} - e^{-ik_0x} = 2i\sin k_0x & (x < 0), \\ 0 & (x > 0). \end{cases} \qquad (5\text{–}25)$$

The wave function is zero beyond the step, and although ψ_1 is continuous at $x = 0$, $d\psi_1/dx$ is not. The discontinuity of slope at $x = 0$ results from the unphysical assumption that the potential step is infinitely high. Whenever this idealization is employed, the slope of the wave function is discontinuous at the step.[1]

5–2 Potential barrier. The potential function (Fig. 5–3)

$$V(x) = \begin{cases} 0, & (x < 0), \\ V_0, & (0 < x < a), \\ 0, & (a < x), \end{cases} \tag{5–26}$$

admits a double degeneracy of the wave function for every positive value of E. From symmetry, it is clear that the wave can be incident on the barrier either from the right or from the left. In the latter case, the wave function has the form

$$\psi = \begin{cases} e^{ik_0x} + Re^{-ik_0x} & (x < 0), \\ Ae^{ikx} + Be^{-ikx} & (0 < x < a), \\ Te^{ik_0x} & (a < x), \end{cases} \tag{5–27}$$

in which $\hbar k_0 = \sqrt{2mE}$, and $\hbar k = \sqrt{2m(E - V_0)}$. The constants R, A, B, T are to be evaluated by requiring ψ and $d\psi/dx$ to be continuous at $x = 0$ and $x = a$. The conditions are:

$$1 + R = A + B,$$

$$ik_0(1 - R) = ik(A - B),$$

$$Te^{ik_0a} = Ae^{ika} + Be^{-ika}, \tag{5–28}$$

$$ik_0Te^{ik_0a} = ik(Ae^{ika} - Be^{-ika}).$$

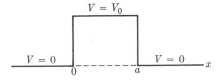

FIG. 5–3. Square potential barrier

[1] L. I. Schiff, *Quantum Mechanics*, 2nd ed. New York: McGraw-Hill Book Co., Inc., 1955, p. 29.

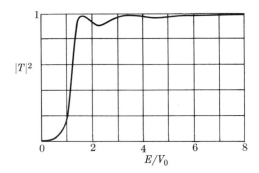

FIG. 5–4. Transmission of a square potential barrier as a function of energy.

Solving for the constants, we obtain

$$R = \frac{(1 - \mu^2)\sin ka}{(1 + \mu^2)\sin ka + 2i\mu\cos ka},$$

$$T = e^{-ik_0 a}\frac{2i\mu}{(1 + \mu^2)\sin ka + 2i\mu\cos ka},$$

$$A = e^{-ika}\frac{i(1 + \mu)}{(1 + \mu^2)\sin ka + 2i\mu\cos ka},$$ (5–29)

$$B = e^{ika}\frac{-i(1 - \mu)}{(1 + \mu^2)\sin ka + 2i\mu\cos ka},$$

where

$$\mu = \frac{k}{k_0} = \sqrt{1 - \frac{V_0}{E}}.$$ (5–30)

If $E \geq V_0$, k and μ are real, and the absolute square of the amplitude of the transmitted wave is given by

$$|T|^2 = \frac{(2\mu)^2}{(1 + \mu^2)^2\sin^2 ka + (2\mu)^2\cos^2 ka},$$ (5–31)

which has the value $[1 + (k_0 a/2)^2]^{-1}$ at $E = V_0$ and increases with E to the value unity at $E = V_0[1 + \pi^2/(2mV_0 a^2/\hbar^2)]$, after which it oscillates slightly, becoming asymptotically equal to unity for values of E that are large compared to V_0. In analogy to the well-known phenomenon of total transmission of light through a thin refracting layer, perfect transmission occurs for $ka = n\pi$, that is, whenever the barrier width a is an integral number of half wavelengths.

For $E < V_0$, the wave number k is purely imaginary, and we write

$$k = i\kappa = \frac{i}{\hbar}\sqrt{2m(V_0 - E)}\,, \qquad (5\text{–}32)$$

whence

$$|T|^2 = \frac{(2k_0/\kappa)^2}{(1 - k_0^2/\kappa^2)^2 \sinh^2\kappa a + [2(k_0/\kappa)]^2 \cosh^2\kappa a}\,. \qquad (5\text{–}33)$$

The transmission is therefore zero at $E = 0$ and increases steadily with E, joining smoothly to the value given by Eq. (5–31) for $E = V_0$ (Fig. 5–4).

5–3 Rectangular potential well. In the foregoing examples, the Schrödinger equation was found to have solutions for every positive value of the energy so that the states of the system form a continuum. This is true whenever the classical motion is not confined to a finite region of space. However, if the particle is bound, and hence the classical motion periodic, then stationary states of the quantum-mechanical system exist only for certain discrete values of the energy. In the early quantum theory, discussed in Chapter 1, these discrete energy levels were defined by the quantization rules of Bohr and Sommerfeld. We shall now see how the discreteness of bound states arises naturally from the rules governing the wave function. It will be found that the old quantization rules are approximations to the correct quantum-mechanical results.

A one-dimensional rectangular potential well is defined by

$$V(x) = \begin{cases} 0 & (x < 0), \\ -V_0 & (0 < x < a), \\ 0 & (a < x), \end{cases} \qquad (5\text{–}34)$$

in which V_0, the *depth of the well*, is a positive number (Fig. 5–5).

The wave function for a particle of positive energy can be obtained immediately by changing the sign of V_0 in the example of the preceding section, and again there is a continuum of doubly degenerate states. The details are left as an exercise (Problem 5–5).

If, however, E is negative, but larger than $-V_0$, the situation is quite different. The solutions of the Schrödinger equation outside the potential

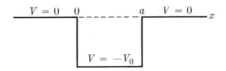

FIG. 5–5.   Rectangular potential well.

well are now $e^{\kappa x}$ and $e^{-\kappa x}$, where $\hbar\kappa = \sqrt{-2mE}$ is real. The function $e^{\kappa x}$, however, is not bounded as x becomes large and positive; hence, it must not appear in ψ for $x > a$. Similarly, ψ must not contain $e^{-\kappa x}$ for $x < 0$. The (unnormalized) wave function is, therefore, of the form

$$\psi = \begin{cases} e^{\kappa x} & (x < 0), \\ Ae^{ikx} + Be^{-ikx} & (0 < x < a), \\ Ce^{-\kappa x} & (a < x), \end{cases} \tag{5-35}$$

where $\hbar k = \sqrt{2m(E + V_0)}$. (For the sake of simplicity, we have chosen the arbitrary constant multiplier such that the coefficient of $e^{\kappa x}$ is unity.)

The continuity conditions at $x = 0$ and $x = a$ are found in the usual way, i.e.,

$$\begin{aligned} 1 &= A + B, \\ \kappa &= ik(A - B), \\ Ce^{-\kappa a} &= Ae^{ika} + Be^{-ika}, \\ -\kappa Ce^{-\kappa a} &= ik(Ae^{ika} - Be^{-ika}). \end{aligned} \tag{5-36}$$

The three constants A, B, C are therefore subject to *four* conditions, and the equations (5–36) cannot be satisfied for arbitrary values of E. By elimination of A, B, C from the equations (5–36), it is easy to show (Problem 5–7) that the condition of compatibility of these equations is

$$2 \cot ka = \frac{k}{\kappa} - \frac{\kappa}{k}, \tag{5-37}$$

and that, if this condition is satisfied, the solution is

$$A = B^* = \tfrac{1}{2}\left(1 - i\frac{\kappa}{k}\right), \qquad C = \tfrac{1}{2}\left(\frac{k}{\kappa} + \frac{\kappa}{k}\right)e^{\kappa a} \sin ka. \tag{5-38}$$

The solution of the transcendental Eq. (5–37), which can be done graphically, is facilitated by introducing the quantities

$$\gamma = \sqrt{\frac{2mV_0a^2}{\hbar^2}}, \qquad \alpha = \frac{ka}{\gamma} = \sqrt{1 + \frac{E}{V_0}}, \tag{5-39}$$

in terms of which Eq. (5–37) can be expressed (Problem 5–8) as

$$\gamma\alpha = (n - 1)\pi + 2\cos^{-1}\alpha, \qquad (n = 1, 2, \ldots). \tag{5-40}$$

Figure 5–6 illustrates this relation graphically for the three cases $\gamma = 1$, 4, 12. The abscissae of the points of intersection (denoted by dots) of the

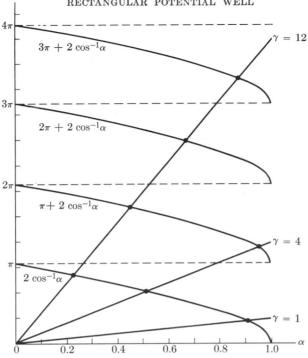

FIG. 5–6. Graphical solution of Eq. (5–40).

two graphs give the energies of the corresponding states according to the formula

$$E = -V_0(1 - \alpha^2). \tag{5–41}$$

The number of states for a given well depth is clearly the greatest integer contained in the quantity $(\gamma/\pi + 1)$. This quantity increases as V_0 is made larger. The values of E corresponding to the stationary states decrease as the well depth increases, and a new level appears at zero energy each time γ assumes the value $n\pi$. Energy-level diagrams for the three cases of Fig. 5–6 are drawn in Fig. 5–7. The broken lines connect the levels with the same value of n. The energy scale in this figure is distorted for the sake of convenient plotting.

The number n has an important meaning relative to the wave function. If the relations (5–38) are inserted into Eqs. (5–35), the wave function has the form

$$\psi = \begin{cases} e^{\kappa x} & (x < 0), \\ \dfrac{1}{\alpha}\sin\left[k\left(x - \dfrac{a}{2}\right) + \dfrac{n\pi}{2}\right] & (0 < x < a), \\ (-)^{n+1}e^{-\kappa(x-a)} & (a < x). \end{cases} \tag{5–42}$$

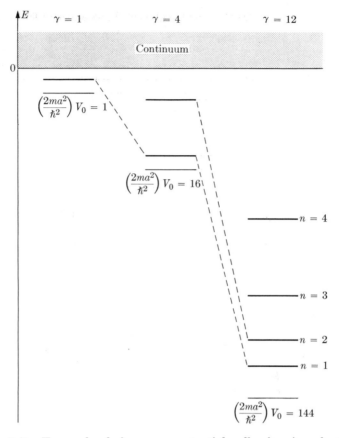

FIG. 5–7.   Energy levels for square potential wells of various depths.

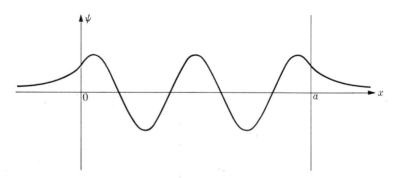

FIG. 5–8.   Wave function of a particle in a square potential well, for $n = 5$.

This wave function is shown on an arbitrary scale in Fig. 5–8. It is clear from the form of Eq. (5–42) that ψ is zero at $x = \pm\infty$ and at $n - 1$ points within the region $0 < x < a$, so that the total number of zeros of ψ is $n + 1$. It is a general property of bound-state solutions of the one-dimensional Schrödinger equation that the number of zeros of ψ is $n + 1$, where n is the *quantum number* ranking the energy levels in increasing order (see Section 5–7).

In the limit that the well becomes very deep, the straight line of Fig. 5–6 approaches the vertical axis, and the points of intersection approach the values

$$ka = n\pi \qquad (n = 1, 2, \ldots). \tag{5–43}$$

Hence, in this limit, the energy levels are

$$\frac{\hbar^2 k^2}{2m} = \frac{n^2 \pi^2 \hbar^2}{2ma^2}, \tag{5–44}$$

which agrees with Eq. (1–60), Chapter 1, obtained from the Wilson-Sommerfeld quantization rule, Eq. (1–43). Note, however, that the old quantum theory would have given the same result (5–44) *independently of the well depth* since the classical motion described in Section 1–14 is not affected by the value of V_0 as long as the particle is bound. Hence, in this example, rule (1–43) is correct only for large n, in which case the wavelength is very short compared to a, and the classical limit is approached.

In Section 5–13, it will be shown that this situation is general, in that a slightly modified form of Eq. (1–43) represents, for large n, an asymptotic approximation to the quantum-mechanical solution.

5–4 Degeneracy. Qualitative characteristics of the wave function. The character of the solutions of the Schrödinger equation (5–1), illustrated by the foregoing examples, depends upon the relative magnitudes of the total and potential energies. The energy spectrum can be qualitatively determined by inspection of the graph of $V(x)$. Thus, in Fig. 5–9(a), the spectrum belonging to a finite potential barrier [cf. Section 5–2] is continuous and doubly degenerate for every positive value of E. The double degeneracy corresponds to the fact that the wave can be incident on the barrier either from the right or from the left. The potential of Fig. 5–9(b) allows a continuum of states for every positive value of E; the states are doubly degenerate for $E > V_0$, but nondegenerate for $E < V_0$. In the latter case, the wave function is exponential in character to the right of the potential step and, as in the example of Section 5–1, uniquely determined by the condition that ψ be bounded for $x \to \infty$. In

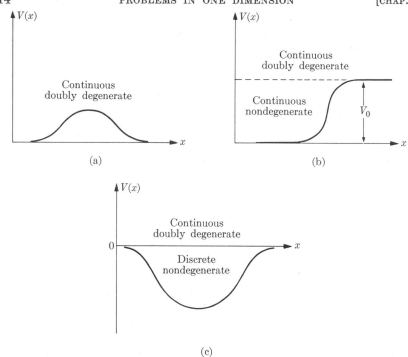

FIG. 5–9. Three potential functions and the types of energy spectra produced by these potentials.

Fig. 5–9(c), the states for $E > 0$ form a continuum of twofold degenerate states. However, if $E < 0$, the classical motion is periodic, and the continuity conditions for ψ can be satisfied only for particular values of E, and then in only one way, leading to a *discrete* spectrum of nondegenerate states. It is clear that every combination of continuous degenerate, continuous nondegenerate, and discrete spectra can arise for suitable forms of the potential-energy function (Problem 5–10).

It is interesting to consider the origin of discrete spectra in more detail. The points x_1, x_2 (Fig. 5–10), which correspond to the classical limits of the motion, are the points for which $E = V(x)$. They are called the *classical turning points* because the periodic classical motion would be confined to the interval (x_1, x_2). Within this interval, E is greater than $V(x)$, and the Schrödinger equation can be written

$$\frac{d^2\psi}{dx^2} = -k^2\psi, \tag{5–45}$$

where $k^2 = (2m/\hbar^2)(E - V)$ is a positive function of x. The second derivative $d^2\psi/dx^2$ measures the curvature of the graph of ψ against x.

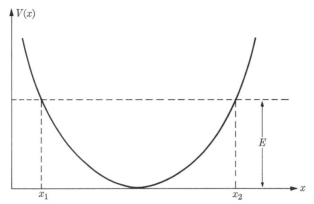

FIG. 5–10.   Potential function showing classical turning points x_1, x_2 for a particle of energy E.

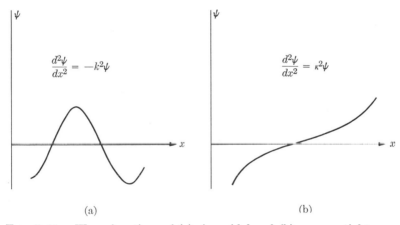

FIG. 5–11.   Wave functions of (a) sinusoidal and (b) exponential type.

Equation (5–45) shows, therefore, that the curvature is negative if ψ is positive, and positive if ψ is negative. Therefore, the graph turns toward the x-axis [Fig. 5–11(a)]. This behavior is characteristic of the sine curve, and solutions of Eq. (5–45) are therefore said to be of *sinusoidal type*. Outside the classical interval, E is smaller than $V(x)$, and the Schrödinger equation becomes

$$\frac{d^2\psi}{dx^2} = \kappa^2\psi, \qquad \kappa^2 = \frac{2m}{\hbar^2}(V - E); \qquad (5\text{–}46)$$

now the graph of ψ bends away from the x-axis, and ψ changes monotonically with x [Fig. 5–11(b)]. The exponential function is the prototype for this case, and a solution of Eq. (5–46) is of *exponential type*. It is

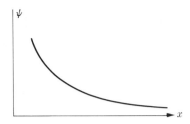

FIG. 5–12.   Bounded solution of exponential type.

intuitively clear that a function of exponential type can be zero at only one point. The graph of the solution must recede from the x-axis both to the left and to the right of a zero. Hence, if the exponential region extends to infinity and the solution is to be bounded, there must not be a zero at any finite x. A bounded solution must approach zero monotonically as $x \to \infty$ (Fig. 5–12).

Let us suppose that, for energy E, we have to the left in Fig. 5–10 a solution of exponential type which vanishes properly at $x = -\infty$. At the turning point x_1 it can be joined smoothly to a linear combination of two linearly independent solutions of sinusoidal type. This combination is uniquely determined by the requirement that ψ and $d\psi/dx$ be continuous. The interior solution must now be joined at $x = x_2$ to a solution of exponential type on the right. It is easily seen that it is not possible, for every value of E, to join the interior solution to a solution of decreasing exponential type at x_2, but that the increasing, nonintegrable solution will, in general, also be required if both ψ and $d\psi/dx$ are to be continuous. The wave function will be integrable only for special values of E. This is the origin of the quantization of levels for bound states.

5–5 Theory of the Schrödinger equation. Linear independence. The one-dimensional Schrödinger equation (5–3) can be written conveniently in the form

$$\frac{d^2\psi}{dx^2} + [\lambda - u(x)]\psi = 0, \qquad (5\text{–}47)$$

where $\lambda = 2mE/\hbar^2$ and $u(x) = 2mV(x)/\hbar^2$. For brevity, we shall also denote differentiation with respect to x by a prime:

$$\psi' = \frac{d\psi}{dx}, \qquad \psi'' = \frac{d^2\psi}{dx^2}, \qquad \text{etc.}$$

We first observe that ψ'' and ψ have the same sign if $\lambda < u(x)$, and opposite signs if $\lambda > u(x)$, corresponding to the exponential and sinusoidal types mentioned above. The points for which $\lambda = u(x)$ (turning points) or $\psi(x) = 0$ are points of inflection.[1]

The following theorem, the proof of which is found in mathematical texts,[2] is fundamental to the subject: *In any interval in x containing the point x_1, and within which u(x) is an analytic function of x, a unique analytic solution of Eq. (5–47) exists, satisfying the conditions*

$$\psi(x_1) = a, \qquad \psi'(x_1) = b,$$

where a and b are arbitrary (complex) numbers. In other words, the solution is determined uniquely when the ordinate and slope of the graph of $\psi(x)$ are given at any point. In particular, the only solution which satisfies

$$\psi(x_1) = 0, \qquad \psi'(x_1) = 0$$

is the trivial solution $\psi(x) = 0$. Consequently, the graph of any nontrivial ψ must cross the x-axis with a finite slope. Since ψ is analytic, its zeros are *isolated* and can be numbered in order from left to right.

Two solutions of Eq. (5–47) (or, more generally, any two functions of x) are *linearly independent* if the equation

$$C_1\psi_1 + C_2\psi_2 = 0 \qquad (5\text{–}48)$$

cannot be satisfied identically in x for any choice of the constants C_1 and C_2 except $C_1 = C_2 = 0$. If such nonzero constants exist, ψ_1 and ψ_2 are *linearly dependent.* In the latter case, differentiation of Eq. (5–48) yields also

$$C_1\psi_1' + C_2\psi_2' = 0. \qquad (5\text{–}49)$$

The two expressions (5–48) and (5–49) can be considered to be equations for C_1 and C_2; they are linear and homogeneous in these unknowns. Consequently, if a nontrivial solution exists, the determinant (called the *Wronskian* of ψ_1 and ψ_2)

$$W = \psi_1'\psi_2 - \psi_1\psi_2' \qquad (5\text{–}50)$$

must vanish. If, on the other hand $W \neq 0$, nontrivial solutions for C_1 and C_2 are impossible, and ψ_1 and ψ_2 are linearly independent. In sum-

---

[1] The quantities λ and $u(x)$ are both real; therefore, any complex solution of Eq. (5–47) is a linear combination of two real solutions, representing the real and imaginary parts. Hence, no generality is lost if the discussion is limited to real solutions of Eq. (5–47).

[2] Cf., for example, E. T. Whittaker and G. N. Watson, *A Course of Modern Analysis.* 4th ed., Cambridge: Cambridge University Press, 1927, Section 10–2.

mary, *the necessary and sufficient condition for ψ_1 and ψ_2 to be linearly independent is that their Wronskian does not vanish.*[1]

The Wronskian determinant formed from two solutions of Eq. (5–47) is of special significance because it is a *first integral* of this equation. This can be demonstrated by multiplying the equations

$$\psi_1'' + [\lambda - u(x)]\psi_1 = 0,$$
$$\psi_2'' + [\lambda - u(x)]\psi_2 = 0,$$

(5–51)

by ψ_2 and ψ_1, respectively, and subtracting, with the result

$$\psi_1''\psi_2 - \psi_1\psi_2'' = \frac{d}{dx}(\psi_1'\psi_2 - \psi_1\psi_2') = 0,$$

(5–52)

whence

$$W = \psi_1'\psi_2 - \psi_1\psi_2' = \text{a constant.}$$

(5–53)

The Wronskian is independent of x, and the linear independence of ψ_1 and ψ_2 can therefore be tested if the values of these functions and their derivatives are known at any point. Moreover, if one solution, say ψ_1, is known, then a second linearly independent solution can be found by solving the first-order differential equation (5–53). This is the origin of the term "first integral."

Two linearly independent solutions, ψ_1 and ψ_2, are a *complete set* in the sense that every solution of Eq. (5–47) can be expressed as a linear combination of ψ_1 and ψ_2:[2]

$$\psi = C_1\psi_1 + C_2\psi_2.$$

(5–54)

According to the fundamental theorem stated at the beginning of this section, ψ is uniquely determined by the quantities $\psi(x_1)$ and $\psi'(x_1)$, where x_1 is any particular value of x. The constants C_1 and C_2 are therefore

---

[1] A thorough understanding of the elementary theory of linear algebraic equations is essential to the study of quantum mechanics. The reader may consult any standard textbook in algebra. The excellent summary in R. Courant and D. Hilbert, *Methods of Mathematical Physics*, New York: Interscience Publishers, Inc., 1953, Vol. I, Chapter I, is immediately adaptable to applications in physics.

[2] The existence of a linearly independent set of two solutions can be inferred from the fundamental theorem by constructing ψ_1 and ψ_2 such that

$$\psi_1(0) = 0, \quad \psi_1'(0) = 1; \quad \psi_2(0) = 1, \quad \psi_2'(0) = 0,$$

so that $W = 1 \cdot 1 + 0 \cdot 0 = 1.$

determined by the equations

$$\begin{aligned}
\psi(x_1) &= C_1\psi_1(x_1) + C_2\psi_2(x_1), \\
\psi'(x_1) &= C_1\psi_1'(x_1) + C_2\psi_2'(x_1),
\end{aligned} \tag{5–55}$$

which are a linear inhomogeneous set of equations in C_1 and C_2. Their determinant, which is the Wronskian, does not vanish. Therefore, a unique solution satisfying Eq. (5–54) can always be found.

5–6 Properties of the zeros of ψ; Sturm's theorem. A general property of a linearly independent set of solutions of the Schrödinger equation is expressed by Sturm's theorem: *If $\psi_1(x)$ is a solution of Eq.* (5–47) *which has consecutive zeros at x_1 and x_2, then every linearly independent solution $\psi_2(x)$ vanishes once and only once in the interval* (x_1, x_2). The proof of Sturm's theorem follows from the properties of the Wronskian. If ψ_1 vanishes at $x = x_1$ and $x = x_2$ (see Fig. 5–13), then the Wronskian, which is independent of x, is

$$W = \psi_1'(x_1)\psi_2(x_1) = \psi_1'(x_2)\psi_2(x_2). \tag{5–56}$$

Clearly, $\psi_1'(x_1)$ and $\psi_1'(x_2)$ have opposite signs; hence by Eq. (5–56), the signs of $\psi_2(x_1)$ and $\psi_2(x_2)$ are also opposite. However, ψ_2 is a continuous function and must therefore vanish at least once in an interval within which it has both positive and negative values. It cannot vanish more than once, for if it did, it would have consecutive zeros in an interval within which the linearly independent function ψ_1 does not vanish. This relationship between linearly independent solutions is referred to as the *interlacing of the zeros:* The zeros of ψ_1 and ψ_2 are encountered alternately as x increases.

Other properties of the zeros which depend upon the relative magnitudes of λ and $u(x)$ are easily established. For example, we can prove that ψ can have at most one zero in any interval within which $\lambda < u(x)$. This property of the solutions in the nonclassical region has already been

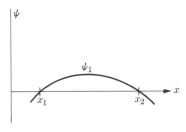

FIG. 5–13. Wave function used in the proof of Sturm's theorem.

mentioned in the qualitative discussion of Section 5–4. To prove this, we note that the Schrödinger equation in the form

$$\psi'' = (u - \lambda)\psi \qquad (5\text{--}57)$$

can be integrated at once to give, for any two points x_1, x_2,

$$\psi'(x_2) - \psi'(x_1) = \int_{x_1}^{x_2} (u - \lambda)\psi \, dx. \qquad (5\text{--}58)$$

If ψ is assumed to be zero at x_1 and x_2 and positive in the interval between these points, a contradiction follows: Figure 5–13 shows that $\psi'(x_1) > 0$ and $\psi'(x_2) < 0$. The left-hand member of Eq. (5–58) is therefore negative. However, since $\lambda < u$ by hypothesis, the right-hand member is positive. Because of this contradiction, the assumption that $\psi(x)$ has more than one zero cannot be upheld.

By means of Eq. (5–58) it is also easy to prove that if $\lambda < u(x)$, then ψ is monotonic in any interval to the right of a zero. For if $\psi(x_1) = 0$ and x is any point to the right of x_1, we have

$$\psi'(x) - \psi'(x_1) = \int_{x_1}^{x} (u - \lambda)\psi \, dx = p(x). \qquad (5\text{--}59)$$

If we now assume that $\psi'(x_1)$ is positive, then ψ is positive in the interval of integration, whence $p(x)$ is also positive. A second integration of Eq. (5–59) between x_2 and x_3 results in

$$\psi(x_3) = \psi(x_2) + \psi'(x_1)(x_3 - x_2) + \int_{x_2}^{x_3} p(x) \, dx. \qquad (5\text{--}60)$$

Thus, if x_1, x_2, x_3 are in increasing order, $\psi(x_3)$ is greater than $\psi(x_2)$, that is, ψ increases monotonically. The proof can be easily modified to show also that ψ decreases monotonically to the right of a zero at which $\psi'(x_1) < 0$. All of these properties of solutions in the nonclassical region are summarized in the statement that, in this region, ψ is of exponential type.

In any classically accessible interval $[\lambda > u(x)]$, the properties of ψ are quite different. If x_1 and x_2 are stationary points for the function ψ, that is, if $\psi'(x_1) = \psi'(x_2) = 0$, then Eq. (5–58) becomes

$$0 = \int_{x_1}^{x_2} (u - \lambda)\psi \, dx. \qquad (5\text{--}61)$$

The factor $(u - \lambda)$ has, by hypothesis, the same sign throughout the interval of integration, and it follows that ψ must have at least one zero within (x_1, x_2). The stationary points of ψ are therefore interlaced with its zeros, which is characteristic of solutions of sinusoidal type.

In order to avoid frequent repetition of the procedure employed in deriving Eq. (5–53), we shall now apply it to obtain a very general formula of great usefulness in the study of the Schrödinger equation. Let ψ_1 and ψ_2 be solutions of two different Schrödinger equations, belonging to different potential-energy functions $u_1(x)$ and $u_2(x)$ and to different energies λ_1 and λ_2:

$$\psi_1'' + (\lambda_1 - u_1)\psi_1 = 0,$$
$$\psi_2'' + (\lambda_2 - u_2)\psi_2 = 0. \tag{5-62}$$

Multiplying the first of these equations by ψ_2, the second by ψ_1, and subtracting, we obtain

$$\frac{d}{dx}(\psi_1'\psi_2 - \psi_1\psi_2') + (\lambda_1 - u_1)\psi_1\psi_2 - (\lambda_2 - u_2)\psi_1\psi_2 = 0. \tag{5-63}$$

Integration between the limits x_1, x_2 and rearrangement yield

$$(\psi_1'\psi_2 - \psi_1\psi_2')\Big|_{x_1}^{x_2} = (\lambda_2 - \lambda_1)\int_{x_1}^{x_2}\psi_1\psi_2\,dx - \int_{x_1}^{x_2}(u_2 - u_1)\psi_1\psi_2\,dx. \tag{5-64}$$

This formula is the source of many theorems concerning the properties of wave functions. For example, let us compare two functions belonging to the same energy but for different potentials, with $u_2 > u_1$ in the interval under consideration. Furthermore, suppose that ψ_2 has consecutive zeros at x_1 and x_2 and that it is positive between these points (Fig. 5–14). Then, since $\lambda_1 = \lambda_2$, Eq. (5–64) becomes

$$-\psi_1(x_2)\psi_2'(x_2) + \psi_1(x_1)\psi_2'(x_1) = -\int_{x_1}^{x_2}(u_2 - u_1)\psi_1\psi_2\,dx. \tag{5-65}$$

Hence ψ_1 must have a zero in (x_1, x_2); if it did not have a zero in this interval and were assumed to be positive, then the left-hand member of Eq. (5–65) would be positive and the right-hand member negative $(u_2 > u_1)$, so that a contradiction would be obtained.

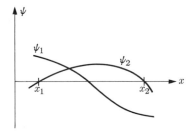

FIG. 5–14.   The wave functions ψ_1 and ψ_2 of Eq. (5–65).

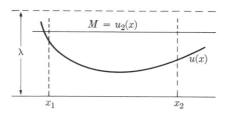

FIG. 5–15.   The potential functions $u(x)$ and $u_2(x)$.

We have thus proved the following theorem: *If ψ_1 and ψ_2 are solutions belonging to the same energy but different potential energy such that $u_2 > u_1$, then ψ_1 has at least one zero between two consecutive zeros of ψ_2.* Qualitatively, the kinetic energy is larger for ψ_1 than for ψ_2; consequently, ψ_1 oscillates more rapidly than ψ_2.

This theorem can be used to show that, as λ approaches infinity, the number of zeros in the classical region becomes infinite. In the above argument, let $u_1 = u$ be a given potential-energy function and consider an interval (x_1, x_2) within which $\lambda > u$. Let M be an upper bound of u in the interval and define $u_2 = M$ (Fig. 5–15). By the theorem just proved, there is at least one zero of $\psi = \psi_1$ between consecutive zeros of any solution of

$$\psi_2'' + (\lambda - M)\psi_2 = 0. \tag{5–66}$$

But such a solution is $\psi_2 = \sin[\sqrt{\lambda - M}\,(x - C)]$, where C is an arbitrary constant. This function has arbitrarily many zeros in (x_1, x_2) for sufficiently large λ, and hence the same is true of ψ.

Incidentally, it may be noted that, for a given λ, the greatest interval between zeros of ψ is not as large as $\pi/\sqrt{\lambda - M}$. This provides an estimate of the frequency of oscillation of ψ.

A simple variation of the preceding argument (Problem 5–15) can be made to show that the least interval between the zeros of ψ is not smaller than $\pi/\sqrt{\lambda - m}$, where m is a lower bound of u in (x_1, x_2). Thus, there is a finite number of zeros of ψ in any finite interval; i.e., as long as $u(x)$ is a bounded function, the frequency of oscillation of ψ is limited.

5–7 Bound states. When the classical motion is confined to a finite interval in x, that is, when there are classical turning points x_1 and x_2 such that

$$\lambda < u(x) \qquad \text{for} \qquad x < x_1 \quad \text{or} \quad x > x_2,$$

then, as we have seen, the Schrödinger equation admits integrable solutions only for discrete values of λ. The wave functions belonging to such discrete states can be studied in an elegant way by a method devised by Milne.[1]

We introduce the special solutions ψ_s and ψ_c defined by the conditions

$$\psi_s(0) = 0, \qquad \psi_c(0) = 1, \qquad \psi_s'(0) = 1, \qquad \psi_c'(0) = 0, \qquad (5\text{–}67)$$

and the functions $v(x)$ and $\phi(x)$ such that

$$\psi_s = v \sin \phi, \qquad \psi_c = v \cos \phi, \qquad (5\text{–}68)$$

that is,

$$v = \sqrt{\psi_s^2 + \psi_c^2}, \qquad \phi = \tan^{-1} \frac{\psi_s}{\psi_c}. \qquad (5\text{–}69)$$

The functions ψ_s and ψ_c are linearly independent, for by Eqs. (5–67), the Wronskian formed from them is

$$\psi_s'\psi_c - \psi_s\psi_c' = 1. \qquad (5\text{–}70)$$

Hence, any solution of the Schrödinger equation can be expressed as a linear combination of ψ_s and ψ_c [cf. Eq. (5–54)]:

$$\psi = C_1\psi_s + C_2\psi_c. \qquad (5\text{–}71)$$

If the constants C_1 and C_2 are written as

$$C_1 = C \cos \theta, \qquad C_2 = -C \sin \theta, \qquad (5\text{–}72)$$

and the relations (5–68) are substituted for ψ_s and ψ_c, Eq. (5–71) becomes

$$\psi = Cv \sin (\phi - \theta). \qquad (5\text{–}73)$$

In terms of v and ϕ, the Wronksian (5–70) is

$$W = 1 = v^2\phi', \qquad (5\text{–}74)$$

or,

$$\phi' = \frac{1}{v^2}. \qquad (5\text{–}75)$$

Now the functions ψ_s and ψ_c are linearly independent and cannot simultaneously vanish. Therefore, by Eqs. (5–69), v is not zero, and by

---

[1] W. E. Milne, *Trans. Am. Math. Soc.* **30**, 797 (1928) and *Phys. Rev.* **35**, 863 (1930). Cf. also V. Rojansky, *Introductory Quantum Mechanics*. New York: Prentice-Hall, Inc., 1946, Section 32.

Eq. (5–75), ϕ is a monotonically increasing function of x. Also, it follows from Eqs. (5–67) that

$$\phi(0) = 0, \qquad \phi'(0) = 1, \qquad v(0) = 1, \qquad v'(0) = 0. \qquad (5\text{–}76)$$

The function v satisfies the differential equation (Problem 5–16)

$$v'' + (\lambda - u)v = \frac{1}{v^3}. \qquad (5\text{–}77)$$

Now let ψ represent a bounded solution of the Schrödinger equation (5–47), and hence, an allowed energy state. ψ cannot have a zero outside the classical region, for it would then diverge. Every wave function which satisfies the boundary conditions at $x = \pm\infty$ is therefore nonzero in the nonclassical region, and

$$\psi \to 0 \qquad \text{as} \qquad |x| \to \infty. \qquad (5\text{–}78)$$

The function v is positive everywhere, and ϕ increases steadily with x. However, ϕ cannot be unbounded, since, according to Eq. (5–73), ψ would then have infinitely many zeros in the finite interval (x_1, x_2), which has been shown to be impossible. It can be concluded that there are finite numbers ϕ_1 and ϕ_2 such that

$$\phi \to \phi_1 \qquad \text{as} \qquad x \to -\infty; \qquad \phi \to \phi_2 \qquad \text{as} \qquad x \to \infty. \qquad (5\text{–}79)$$

Furthermore, it follows that $\phi' = 1/v^2 \to 0$ as $|x| \to \infty$, whence

$$v \to \infty \qquad \text{as} \qquad |x| \to \infty. \qquad (5\text{–}80)$$

Finally, the derivative of v also becomes infinite as $|x| \to \infty$. The general character of the functions v and ϕ is illustrated in Fig. 5–16.

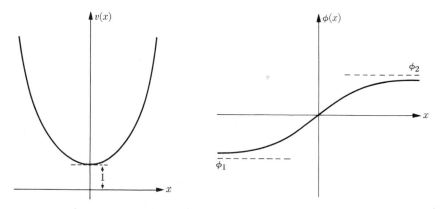

FIG. 5–16.   The functions $v(x)$ and $\phi(x)$ of Milne's method.

Now, by suitable choice of the constants C_1, C_2, C_3, C_4, any solution of Eq. (5–47) can be written in either of the forms

$$\psi = C_1 v \sin(\phi - \phi_1) + C_3 v \cos(\phi - \phi_1),$$
$$\psi = C_2 v \sin(\phi - \phi_2) + C_4 v \cos(\phi - \phi_2),$$

(5–81)

since each is a linear combination of the solutions ψ_s and ψ_c. If ψ is to be bounded, however, the cosine terms in Eqs. (5–81) cannot be present, because the first becomes infinite as $x \to -\infty$, and the second does so as $x \to +\infty$. The sine terms, however, approach zero as $|x| \to \infty$ (Problem 5–17) and thus satisfy the boundary condition (5–78). Consequently, the bounded solution is

$$\psi = C_1 v \sin(\phi - \phi_1) = C_2 v \sin(\phi - \phi_2),$$

(5–82)

from which it is readily shown that

$$\phi_2 - \phi_1 = n\pi,$$

(5–83)

where n is a positive integer. The constants C_1 and C_2 are related by $C_1 = (-)^n C_2$.

Equation (5–83) is the quantum condition; it implies that bounded solutions of Eq. (5–47) exist only for discrete values of λ. The quantum number n measures the number of zeros of the corresponding wave function. There are $n + 1$ zeros, of which two are at $x = \pm\infty$, and $n - 1$ are in the classical region. Corresponding to each integer $n = 1, 2, 3, \ldots$, there is a value λ_n of λ, that is, an energy level of energy E_n. Furthermore, $\lambda_1 < \lambda_2 < \lambda_3 < \cdots$, i.e., the quantum number n lists the energy levels in increasing order. To see this, we resort to Eq. (5–64) and compare two functions ψ_1 and ψ_2 for which $u_1 = u_2$ and $\lambda_2 > \lambda_1$:

$$\left. (\psi_1' \psi_2 - \psi_1 \psi_2') \right|_{x_1}^{x_2} = (\lambda_2 - \lambda_1) \int_{x_1}^{x_2} \psi_1 \psi_2 \, dx.$$

(5–84)

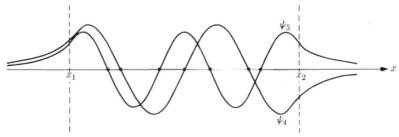

FIG. 5–17. The wave functions ψ_4 and ψ_5, belonging to the quantum numbers 4 and 5.

If x_1 and x_2 are consecutive zeros of ψ_1, then ψ_1 has a definite sign in (x_1, x_2), and by the usual *reductio ad absurdum*, ψ_2 must have at least one zero in (x_1, x_2). In a similar manner, it can be shown that the finite zeros of ψ_1 must all be included between the largest and the smallest finite zero of ψ_2. The character of the two solutions ψ_4 and ψ_5 belonging to the quantum numbers 4 and 5 is illustrated in Fig. 5–17.

We can now examine the behavior of wave functions for bound states as λ varies from $-\infty$ to $+\infty$. If λ is smaller than the least value of $u(x)$, then $\lambda < u(x)$ everywhere, and no solution has more than one zero. Hence, no energy level can exist, and $\phi_2 - \phi_1 < \pi$. As λ increases, the quantity

$$\phi_2 - \phi_1 = \int_{-\infty}^{\infty} \frac{dx}{v^2} \tag{5–85}$$

can be shown to increase steadily (Problem 5–21), passing through the values $\pi, 2\pi, 3\pi, \ldots$ in succession; there is a discrete, nondegenerate set of levels, each wave function having one more zero than the preceding one. If $u(x)$ increases to infinity as $|x| \to \infty$, we obtain an infinite number of these discrete levels. If $u(x)$ is bounded, however, the number of discrete levels is finite, and a continuum of states exists for values of λ such that $\lambda - u(x)$ is positive at $x = \infty$ or $x = -\infty$.

5–8 Orthogonality. The wave functions belonging to the *characteristic numbers* λ_n are called *characteristic solutions* of the Schrödinger equation, or more commonly, *eigenfunctions*. In this terminology, the wave function associated with λ_n is said to be "an eigenfunction belonging to the *eigenvalue* λ_n (or E_n) of the energy." A very important theorem will now be proved, concerning the integral

$$\int_{-\infty}^{\infty} \psi_m \psi_n \, dx, \tag{5–86}$$

in which ψ_m and ψ_n are eigenfunctions belonging to different eigenvalues λ_m and λ_n. Substituting $\psi_1 = \psi_m, \psi_2 = \psi_n, \lambda_1 = \lambda_m, \lambda_2 = \lambda_n, u_1 = u_2,$ $x_1 = -\infty, x_2 = \infty$ in Eq. (5–64), we have

$$(\lambda_n - \lambda_m) \int_{-\infty}^{\infty} \psi_m \psi_n \, dx = 0. \tag{5–87}$$

(Since ψ_m and ψ_n are eigenfunctions, they are zero at $\pm \infty$). It follows that the integral (5–86) must vanish for every pair of eigenfunctions belonging to different eigenvalues ($\lambda_m \neq \lambda_n$). If two functions ψ_m and ψ_n have this property, they are *orthogonal*. The corresponding integral for $n = m$ is a positive number and can be made equal to unity by normalization. Hence, if each ψ is normalized,

$$\int_{-\infty}^{\infty} \psi_m \psi_n \, dx = \delta_{mn}. \tag{5–88}$$

Functions which satisfy this equation are normalized and orthogonal, or more simply, *orthonormal*.

The eigenfunctions ψ_n are characteristic of the potential-energy function in the Schrödinger equation. It is of interest to study the dependence of the eigenvalues on $u(x)$. If ψ_1 and ψ_2 are eigenfunctions belonging to potential-energy functions u_1 and u_2, respectively, then for the infinite interval Eq. (5–64) is

$$(\lambda_2 - \lambda_1) \int_{-\infty}^{\infty} \psi_1 \psi_2 \, dx = \int_{-\infty}^{\infty} (u_2 - u_1) \psi_1 \psi_2 \, dx. \tag{5–89}$$

If ψ_1 and ψ_2 are the *lowest* states ($n = 1$) for u_1 and u_2, their only zeros are at $\pm\infty$, and both functions may be considered to be positive. Hence if $u_2 \geq u_1$, then $\lambda_2 \geq \lambda_1$; that is, if the potential energy is increased, the energy of the lowest state is raised. This result, which is of great importance in practical applications, can be generalized (Problem 5–23) as follows: *If $u_2 > u_1$ and ψ_1 and ψ_2 are eigenfunctions having the same number of zeros ($n_1 = n_2$), then $\lambda_2 > \lambda_1$.*

Equation (5–89) is of special significance when the difference $u_2 - u_1 = \delta u$ is *small*. In this case, ψ_1 and ψ_2 are expected to be nearly the same, and hence the difference in energy $\delta\lambda = \lambda_2 - \lambda_1$ is given, to the first order of small quantities, by

$$\delta\lambda = \frac{\int \delta u \psi^2 \, dx}{\int \psi^2 \, dx} = \langle \delta u \rangle: \tag{5–90}$$

The change in the energy produced by a small change (perturbation) of the potential-energy function is equal to the expectation of the change in potential energy, evaluated for the unperturbed state. This theorem, which has its counterpart[1] in classical mechanics, is central in perturbation theory (Chapter 11).

5–9 The linear harmonic oscillator. The Schrödinger equation for a harmonic oscillator, i.e., for a particle of mass m with the potential energy $V = (1/2)kx^2$, is

$$\frac{d^2\psi}{dx^2} + \frac{2m}{\hbar^2}\left(E - \tfrac{1}{2}kx^2\right)\psi = 0. \tag{5–91}$$

---

[1] J. H. Van Vleck, "Quantum Principles and Line Spectra," *Bull. Nat. Res. Council* **54** (1926), p. 205.

The solution of this important problem provides an example of the general theory developed in the preceding section.

The classical frequency of the oscillator is given by $\omega = \sqrt{k/m}$. For convenience, we measure the energy in units of $(1/2)\hbar\omega$:

$$E = \tfrac{1}{2}\lambda\hbar\omega. \tag{5-92}$$

In the classical motion, the energy is $E = (1/2)m\omega^2 a^2$, where a is the amplitude of oscillation. The amplitude of a classical oscillator of energy $(1/2)\hbar\omega$ is therefore

$$a = \sqrt{\frac{\hbar}{m\omega}}. \tag{5-93}$$

We shall choose a as the unit of distance for the problem and write

$$x = \sqrt{\frac{\hbar}{m\omega}}\, x'. \tag{5-94}$$

In terms of λ and x', the Schrödinger equation (5–91) becomes

$$\frac{m\omega}{\hbar}\left[\frac{d^2\psi}{dx'^2} + (\lambda - x'^2)\psi\right] = 0. \tag{5-95}$$

Only the variable x' will be used in the discussion of this equation, and the prime will be dropped for simplicity. When necessary, the appropriate distinction between x and x' is easily made. The resulting equation,

$$\psi'' + (\lambda - x^2)\psi = 0, \tag{5-96}$$

is of the form of Eq. (5–47), with $u(x) = x^2$ (Fig. 5–18).

Clearly, all eigenfunctions of the system belong to bound states of positive energy and must vanish for $|x| \to \infty$. To determine the approximate

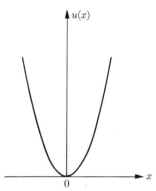

FIG. 5–18.   The potential function $u(x) = x^2$ of the linear harmonic oscillator.

behavior of $\psi(x)$ for large x, we note that if λ is negligible compared to x^2, Eq. (5–96) becomes

$$\psi'' \sim x^2\psi, \qquad |x| \gg \sqrt{\lambda}, \tag{5–97}$$

which has the approximate (asymptotic) solution (Problem 5–24)

$$\psi \sim x^n e^{-(1/2)x^2}, \tag{5–98}$$

where n is any constant. One is led, therefore, to expect that Eq. (5–96) has a solution of the form

$$\psi(x) = e^{-(1/2)x^2}\phi(x). \tag{5–99}$$

The differential equation for the function $\phi(x)$ is found by substituting Eq. (5–99) into Eq. (5–96):

$$\phi'' - 2x\phi' + (\lambda - 1)\phi = 0. \tag{5–100}$$

This differential equation can be solved by the method of series.[1] We wish to express the function $\phi(x)$ by a power series

$$\phi = \sum_{k=0}^{\infty} a_k x^{k+\alpha}, \tag{5–101}$$

in which the coefficients a_k are chosen so that Eq. (5–100) is formally satisfied. Substituting Eq. (5–101) into Eq. (5–100) and collecting terms, one finds

$$\sum_{k=0}^{\infty} \{(k+\alpha)(k+\alpha-1)a_k x^{k+\alpha-2} - [2(k+\alpha)-(\lambda-1)]a_k x^{k+\alpha}\} = 0. \tag{5–102}$$

For this equation to be satisfied identically in x, the coefficient of each power of x must vanish. The lowest power of x is $\alpha - 2$, and the corresponding coefficient is $\alpha(\alpha - 1)a_0$; hence we require

$$\alpha(\alpha - 1)a_0 = 0. \tag{5–103}$$

Therefore, the constant α must be 0 or 1. If $\alpha = 0$ is chosen, Eq. (5–102) becomes

$$\sum_{k=0}^{\infty} \{k(k-1)a_k x^{k-2} - [2k-(\lambda-1)]a_k x^k\} = 0. \tag{5–104}$$

---

[1] Cf., for example: E. D. Rainville, *Intermediate Course in Differential Equations.* New York: John Wiley and Sons, Inc., 1943, Chapter IV.

Setting the coefficients of the various powers of x equal to zero, we obtain

$$x^{-2}: \quad 0(-1)a_0 = 0, \tag{5-105}$$

$$x^{-1}: \quad 1(0)a_1 = 0, \tag{5-106}$$

$$\vdots$$

$$x^k: \quad (k+2)(k+1)a_{k+2} - [2k - (\lambda - 1)]a_k = 0. \tag{5-107}$$

Equation (5–105) is an identity resulting from $\alpha = 0$. Equation (5–106) is also an identity, irrespective of the value of a_1, so that a_1 is arbitrary. Equation (5–107), which must be satisfied for every k larger than 0, yields

$$a_{k+2} = \frac{2k - (\lambda - 1)}{(k+1)(k+2)}\, a_k. \tag{5-108}$$

This is a recurrence relation for the remaining coefficients, which can be calculated successively when the values of a_0 and a_1 are given.[1] It is clear that two solutions are obtained in this way:

one for $a_1 = 0$,

$$\phi_0 = a_0 + a_2 x^2 + a_4 x^4 + \cdots, \tag{5-109}$$

and one for $a_0 = 0$,

$$\phi_1 = a_1 x + a_3 x^3 + a_5 x^5 + \cdots, \tag{5-110}$$

where ϕ_0 is an even, and ϕ_1 an odd function of x; the two are therefore linearly independent (Problem 5–25). Consequently, (5–109) and (5–110) are a complete set of solutions of Eq. (5–100), in terms of which every other solution can be expressed. Hence, we need not be concerned further with the choice $\alpha = 1$, which also satisfies Eq. (5–103) (Problem 5–26).

The ratio of consecutive terms in either of these series is

$$\frac{a_{k+2} x^{k+2}}{a_k x^k} = \frac{2k - (\lambda - 1)}{(k+1)(k+2)}\, x^2, \tag{5-111}$$

and as $k \to \infty$, this ratio approaches zero. The series are therefore convergent for all values of x and define analytic functions which are easily shown to be solutions of Eq. (5–100). Moreover, when k is very large

---

[1] Note: A power-series solution for ψ, of the form of Eq. (5–101), might have been attempted directly by substitution into Eq. (5–96). This procedure, however, would have led to a recurrence relation among three of the a_k, and the simple solution (5–108) would not have been possible. This difficulty is avoided by the substitution (5–99), which is therefore necessary, independently of considerations as to the asymptotic behavior of ψ.

compared to λ, the ratio (5–111) is approximately $2x^2/k$. Now the ratio of consecutive terms in the series expansion of the function e^{x^2}, namely,

$$e^{x^2} = \sum_{\substack{0 \\ (k \text{ even})}}^{\infty} \frac{x^k}{(k/2)!}, \qquad (5\text{--}112)$$

is also $2x^2/k$, and it can be demonstrated that the functions ϕ_0 and ϕ_1 behave like e^{x^2} when x is large.[1] Consequently, the function ψ defined by Eq. (5–99) is, in general, not bounded for large x and is not admissible as a wave function. If, however, one of the coefficients a_k is zero, all the succeeding coefficients also vanish by Eq. (5–108), and the corresponding function ϕ is a polynomial. In this special case, the wave function is

$$\psi = (\text{polynomial}) \cdot e^{-x^2/2},$$

which is obviously bounded. Inspection of Eq. (5–108) shows that the series for ϕ is broken off and becomes a polynomial if and only if

$$\lambda = \lambda_n = 2n + 1, \qquad (5\text{--}113)$$

where n is a non-negative integer. If n is even, the solution ϕ_0 is a polynomial of degree n; if n is odd, ϕ_1 is a polynomial of degree n. The values of the energy corresponding to (5–113) are

$$E_n = \tfrac{1}{2}\lambda_n \hbar\omega = (n + \tfrac{1}{2})\hbar\omega. \qquad (5\text{--}114)$$

Thus, $E_{n+1} - E_n = \hbar\omega$, in agreement with the result of the old quantum theory (Section 1–16).

The polynomials obtained by evaluating the coefficients a_k are called *Hermite polynomials* and are designated by the symbol $H_n(x)$. If the constants a_0 and a_1 are chosen in such a way that the coefficient of the highest power of x in $H_n(x)$ is 2^n, one finds that

$$H_0(x) = 1, \qquad H_1(x) = 2x, \qquad H_2(x) = 4x^2 - 2,$$

$$H_3(x) = 8x^3 - 12x, \ldots \qquad (5\text{--}115)$$

---

[1] If x is large, the dominant terms in the series (5–101) are those for which k is large, i.e., they occur far out in the series. Heuristically, we conclude that the ratio a_{k+2}/a_k for these most important terms is approximately the same as the corresponding ratio for the series representation of e^{x^2}, and hence, that ϕ and e^{x^2} behave similarly for large x. Rigorous justification of this reasoning is difficult. However, the conclusion is confirmed by a closer study of the analytic properties of the solutions of Eq. (5–100). (Cf. P. M. Morse and H. Feshbach, *Methods of Theoretical Physics*. New York: McGraw-Hill Book Co., Inc., 1953, Part II, p. 1640.)

5–10 Hermite polynomials. The nth Hermite polynomial $H_n(x)$ is a solution of the differential equation

$$H_n'' - 2xH_n' + 2nH_n = 0; \qquad (5\text{–}116)$$

it is a polynomial of degree n in which the coefficient of the term in x^n is 2^n. By differentiation of Eq. (5–116), we obtain

$$(H_n')'' - 2x(H_n')' - 2H_n' + 2nH_n' = 0, \qquad (5\text{–}117)$$

whence the function $H_n'(x) = dH_n/dx$ satisfies the differential equation

$$\phi'' - 2x\phi' + 2(n - 1)\phi' = 0. \qquad (5\text{–}118)$$

This equation, however, is also satisfied by H_{n-1}, and since it has only one polynomial solution, we conclude that

$$H_n' = CH_{n-1}. \qquad (5\text{–}119)$$

By comparison of the terms in x^{n-1}, C can be evaluated:

$$n2^n x^{n-1} = C2^{n-1} x^{n-1}.$$

The constant C is therefore $2n$, and

$$H_n' = 2nH_{n-1}, \qquad (5\text{–}120)$$

which is a recurrence formula. By further differentiation and manipulation of Eqs. (5–120) and (5–116), the relations

$$H_{n+1} - 2xH_n + 2nH_{n-1} = 0, \qquad (5\text{–}121)$$

$$H_{n+1} = 2xH_n - H_n' \qquad (5\text{–}122)$$

can be derived (Problem 5–27). Equation (5–122) can be used to construct the polynomials in consecutive order, beginning with $H_0 = 1$.

The *generating function*, defined by

$$g(x, h) = \sum_{n=0}^{\infty} H_n(x)\, \frac{h^n}{n!}, \qquad (5\text{–}123)$$

is of great usefulness in calculations that involve the Hermite functions. By means of the recurrence formulae for H_n, a closed expression for $g(x, h)$ can be derived. Differentiating partially with respect to h, one obtains

$$\frac{\partial g}{\partial h} = \sum_{n=0}^{\infty} nH_n \frac{h^{n-1}}{n!} = \sum_{1}^{\infty} H_n \frac{h^{n-1}}{(n-1)!} = \sum_{0}^{\infty} H_{n+1} \frac{h^n}{n!}$$

$$= \sum_{n=0}^{\infty} (2xH_n - 2nH_{n-1}) \frac{h^n}{n!} = 2xg - 2 \sum_{1}^{\infty} H_{n-1} \frac{h^n}{(n-1)!}$$

$$= 2xg - 2h \sum_{0}^{\infty} H_n \frac{h^n}{n!} = (2x - 2h)g. \tag{5-124}$$

This partial differential equation in h can be solved, with the result

$$g = C(x)e^{2xh-h^2}, \tag{5-125}$$

where the constant of integration may depend upon x. However, the substitution $h = 0$ in Eqs. (5–123) and (5–125) yields

$$g(x, 0) = C(x) = H_0(x) = 1,$$

and, consequently,

$$g(x, h) = e^{2xh-h^2}. \tag{5-126}$$

Since Eq. (5–123) is a Taylor series in h, the formula

$$H_n(x) = \frac{\partial^n}{\partial h^n} \left(e^{2xh-h^2} \right) \Big|_{h=0} \tag{5-127}$$

is obviously true. It can be simplified to the form

$$H_n(x) = (-)^n e^{x^2} \frac{d^n}{dx^n} \left(e^{-x^2} \right) \tag{5-128}$$

(Problem 5–28). This is an explicit formula from which $H_n(x)$ can be found by direct, although laborious, calculation.

It is known from the general theory that the oscillator wave functions

$$\psi_n = A_n e^{-x^2/2} H_n(x) \tag{5-129}$$

are orthogonal, that is,

$$\int_{-\infty}^{\infty} \psi_n^* \psi_m \, dx - A_n A_m \int_{-\infty}^{\infty} e^{-x^2} H_n(x) H_m(x) \, dx = 0, \quad (m \neq n), \tag{5-130}$$

and they will be orthonormal, provided the constants A_n are chosen so that

$$|A_n|^2 \int_{-\infty}^{\infty} e^{-x^2} H_n^2(x) \, dx = 1. \tag{5-131}$$

The normalization integral (5–131) can be calculated by integrating the equation

$$e^{-x^2}g^2(x, h) = \sum_{m,n} e^{-x^2}H_m H_n \frac{h^{m+n}}{m!n!} \qquad (5\text{–}132)$$

between the limits $-\infty$ and $+\infty$. According to Eq. (5–130), the terms for which $m \neq n$ disappear, and we have

$$\int_{-\infty}^{\infty} e^{-x^2}g^2(x, h)\,dx = \int_{-\infty}^{\infty} e^{-(x^2-4xh+2h^2)}\,dx = \sqrt{\pi}\,e^{2h^2}$$

$$= \sum_{n=0}^{\infty} \int_{-\infty}^{\infty} e^{-x^2}H_n^2\,dx \cdot \frac{h^{2n}}{(n!)^2}. \qquad (5\text{–}133)$$

On writing

$$\sqrt{\pi}\,e^{2h^2} = \sqrt{\pi}\sum_{n=0}^{\infty}\frac{2^n h^{2n}}{n!} \qquad (5\text{–}134)$$

and comparing the terms in h^{2n} in the two Taylor's series, we obtain

$$\int_{-\infty}^{\infty} e^{-x^2}H_n^2\,dx = 2^n n!\sqrt{\pi}. \qquad (5\text{–}135)$$

The normalization constant is therefore[1]

$$A_n = \frac{1}{\sqrt{2^n n!\sqrt{\pi}}}, \qquad (5\text{–}136)$$

and the orthonormal eigenfunctions of the harmonic oscillator are

$$\psi_n = \frac{1}{\sqrt{2^n n!\sqrt{\pi}}}\,e^{-x^2/2}H_n(x). \qquad (5\text{–}137)$$

In terms of the functions ψ_n, the recurrence relations (5–120), (5–121), and (5–122) become

$$\sqrt{n+1}\,\psi_{n+1} - \sqrt{2}\,x\psi_n + \sqrt{n}\,\psi_{n-1} = 0, \qquad (5\text{–}138)$$

$$x\psi_n + \psi_n' = \sqrt{2n}\,\psi_{n-1}, \qquad (5\text{–}139)$$

$$x\psi_n - \psi_n' = \sqrt{2(n+1)}\,\psi_{n+1}. \qquad (5\text{–}140)$$

Equation (5–138) is of fundamental importance in determining the selection and intensity rules for radiation by a harmonic oscillator and will be referred to in this connection in Chapter 11.

---

[1] The arbitrary phase of A_n has been chosen to be zero (cf. Section 2–9).

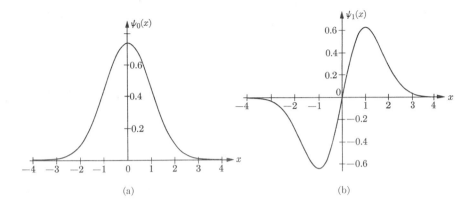

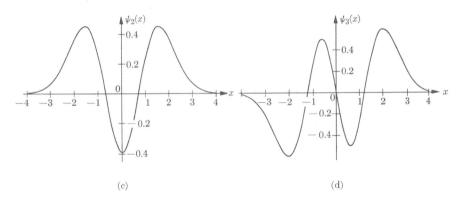

Fig. 5–19. Normalized harmonic-oscillator wave functions for the quantum numbers 0, 1, 2, and 3. [Cf. J. B. Russell, "A Table of Hermite Functions," *J. Math. Phys.* **12**, 291 (1933).]

5–11 Oscillator wave functions.

The eigenfunctions (5–137) for $n = 0$, 1, 2, 3 are shown in Fig. 5–19. The reader should study these graphs carefully in the light of the general theory of Sections 5–4 and 5–5.

The form of ψ_n for large n is of interest, because states of high quantum number are separated by an energy interval that is small compared to the total energy, and the classical motion is approached. In the classical limit, the local wavelength of the eigenfunctions can be expected to be very small, so that ψ_n oscillates very rapidly. Indeed, n zeros[1] of ψ_n are contained within the classical interval of length $2\sqrt{2E_n/k}$, and the average

---

[1] Note that the energy levels for the harmonic oscillator are conventionally labelled by $n = 0, 1, 2, \ldots$, whereas the enumeration $n = 1, 2, 3, \ldots$ was used in Section 5–7.

interval between zeros is therefore

$$2 \sqrt{\frac{2\dot{E}n}{k}} \Big| n \sim \frac{2}{\sqrt{n}} \sqrt{\frac{2\hbar\omega}{k}}, \qquad \text{approximately,} \qquad (5\text{-}141)$$

which approaches zero as $n \to \infty$. The probability density for $n = 11$ is shown in Fig. 5–20.

In any experiment measuring the probability distribution $P(x)$, a finite interval in x is involved whose size is determined by the refinement of the apparatus. For a macroscopic system, therefore, this interval is large compared to the quantum-mechanical wavelength. Consequently, the result of a measurement performed on a classical system will be the average value of $P(x)$, taken over a large number of periods of ψ.

The result to be expected in such a measurement is easily deduced from the classical motion: Since the limits of the motion are given by $x = \pm\sqrt{\lambda}$, the position of the particle is

$$x = \sqrt{\lambda} \sin \omega t,$$

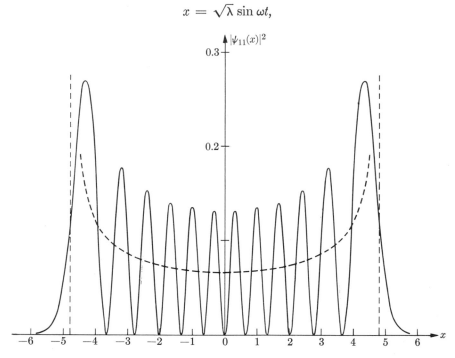

FIG. 5–20.   Probability density for the harmonic oscillator in the state ψ_{11}. The broken curve represents the classical probability distribution, as given by Eq. (5–143). [Cf. J. B. Russell, "A Table of Hermite Functions," *J. Math. Phys.* **12**, 291 (1933), and E. R. Smith, "Zeros of the Hermitian Polynomials," *Am. Math. Monthly* **43**, 354 (1936).]

and the speed is $\omega\sqrt{\lambda}\cos\omega t$. The fraction of the total time spent by the particle in the interval dx is therefore

$$\frac{dt}{T} = \frac{1}{T}\frac{2\,dx}{\omega\sqrt{\lambda}\cos\omega t} = \frac{1}{T}\frac{2\,dx}{\omega\sqrt{\lambda - x^2}}, \qquad (5\text{--}142)$$

in which $T = 2\pi/\omega$ is the period of oscillation. This quantity is the probability that, in a random observation, the particle will be found in the interval dx. Classically, therefore, we obtain

$$P(x)\,dx = \frac{dx}{\pi\sqrt{\lambda - x^2}}. \qquad (5\text{--}143)$$

Now it can be shown that if n is large, the Hermite polynomial is approximated by[1]

$$H_n(x) \sim \frac{2^{n+1}(n/2e)^{n/2}}{\sqrt{2\cos\alpha}}\,e^{n\alpha^2}\cos\left[(2n + \tfrac{1}{2})\alpha - \frac{n\pi}{2}\right], \qquad (5\text{--}144)$$

where α is the smallest positive angle whose sine is $x/\sqrt{2n}$. The probability function for the harmonic oscillator is therefore

$$|\psi_n|^2 = \frac{1}{2^n n!\sqrt{\pi}}\,e^{-x^2}H_n^2(x) \sim \frac{2}{\pi\sqrt{2n - x^2}}\cos^2\left[(2n + \tfrac{1}{2})\frac{x}{\sqrt{2n}} - \frac{n\pi}{2}\right], \qquad (5\text{--}145)$$

in which the approximation $n \gg 1$ has been used to simplify Eq. (5–144). The average value of the factor $\cos^2[\ldots]$ over many periods is $1/2$, and in this sense the mean value of $|\psi|^2$ (since $2n \sim \lambda$) is

$$P(x) = \frac{1}{\pi\sqrt{\lambda - x^2}},$$

in exact agreement with Eq. (5–143).

This example illustrates the asymptotic, and mathematically rather subtle, nature of the correspondence principle. The behavior described is typical of the classical limit and will be encountered in a more general form in the section on the WKB approximation (Section 5–13).

---

[1] The approximation implied by the symbol $\sim$ is *asymptotic* in the sense that, by choosing a sufficiently large n, the ratio of $H_n(x)$ to the expression (5–144) can be made arbitrarily close to unity. Cf. G. Szegö, *Orthogonal Polynomials*. New York: American Mathematical Society Colloquium Publications, 1939, Vol. XXIII. See also Problem 5–32.

The recurrence relation (5–138), in conjunction with the orthogonality relations

$$\int_{-\infty}^{\infty} \psi_m \psi_n \, dx = \delta_{mn},$$ (5–146)

can be employed in evaluating certain integrals. Thus, when Eq. (5–138) is multiplied by ψ_n and integrated, the result is

$$\langle x \rangle = \int_{-\infty}^{\infty} \psi_n x \psi_n \, dx = 0,$$ (5–147)

which is also obvious from the fact that ψ_n^2 is an even function of x. The same procedure with the multiplier ψ_{n+1} yields

$$\sqrt{n+1} - \sqrt{2} \int_{-\infty}^{\infty} \psi_{n+1} x \psi_n \, dx = 0,$$

or

$$\int_{-\infty}^{\infty} \psi_{n+1} x \psi_n \, dx = \sqrt{\frac{n+1}{2}}.$$ (5–148)

Replacing n by $n - 1$ in this expression, we also have

$$\int_{-\infty}^{\infty} \psi_{n-1} x \psi_n \, dx = \sqrt{\frac{n}{2}}.$$ (5–149)

Finally, if $m \neq n \pm 1$, a similar calculation results in

$$\int_{-\infty}^{\infty} \psi_m x \psi_n \, dx = 0 \qquad (m \neq n \pm 1).$$ (5–150)

In Chapter 11, the integral $\int \psi_m x \psi_n \, dx$ will be shown to determine the probability that a quantum of energy $\hbar\omega$ is emitted in a transition between the states m and n. Since the integral vanishes for $m \neq n \pm 1$, we have the selection rule

$$\Delta n = \pm 1,$$ (5–151)

which, in Section 1–16, was conjectured on the basis of the correspondence principle.

If we repeat the above procedure with the multiplier $x \psi_n$, and use the preceding results, we obtain

$$\langle x^2 \rangle = (\Delta x)^2 = n + \tfrac{1}{2}.$$ (5–152)

Also, from the differential equation (5–96),

$$\psi_n'' + (2n + 1 - x^2)\psi_n = 0,$$ (5–153)

one obtains, on multiplying by ψ_n and integrating,

$$-\int_{-\infty}^{\infty} \psi_n \psi_n'' \, dx = n + \tfrac{1}{2}. \tag{5-154}$$

If the original variables [Eqs. (5–92) and (5–94)] are restored,[1] these formulae become

$$(\Delta x)^2 = (n + \tfrac{1}{2})\frac{\hbar}{m\omega}, \tag{5-155}$$

and

$$\langle p^2 \rangle = (\Delta p)^2 = \int_{-\infty}^{\infty} \psi_n(-\hbar^2 \psi_n'') \, dx = (n + \tfrac{1}{2})\hbar m\omega, \tag{5-156}$$

whence

$$\Delta x \, \Delta p = (n + \tfrac{1}{2})\hbar. \tag{5-157}$$

For the lowest state $(n = 0)$, $\Delta x \, \Delta p = \hbar/2$; this is already known to be true since the wave function is the gaussian function $\exp(-\tfrac{1}{2}x^2)$ (cf. Section 3–6).

5–12 Parity. The potential-energy function $(1/2)kx^2$ for the harmonic oscillator is an even function of x. Consequently, the Schrödinger equation

$$\frac{d^2\psi}{dx^2} + [\lambda - u(x)]\psi = 0,$$

which involves x explicitly only in $u(x)$ and $\psi'' = d^2\psi/dx^2$, is not changed in form by the substitution

$$x' = -x, \tag{5-158}$$

that is, it becomes

$$\frac{d^2\psi}{dx'^2} + [\lambda - u(x')]\psi = 0.$$

The boundary conditions, $\psi \to 0$ as $|x| \to \infty$, also remain unchanged by this substitution. This fact, which results from the symmetry in x, is expressed by the statement that the problem is *invariant to the symmetry transformation* $x \to -x$.

---

[1] In returning to the original variables, it is noted that x is measured in units $\sqrt{\hbar/m\omega}$, and p in units $\sqrt{\hbar m\omega}$; hence the system of units, which is arbitrary, has been chosen in such a way that $\hbar = m\omega = 1$. Any result expressed in this system can be adjusted to the original (cgs) system by multiplication with suitable powers of $\hbar$ and $m\omega$, chosen in such a way that a dimensionally correct equation is obtained.

Because of this symmetry, it is clear that if $\psi(x)$ is an eigenfunction, then the function $\psi(-x)$ is also an eigenfunction. The eigenfunctions, however, have been shown to be nondegenerate. Hence, these two functions must be linearly dependent, i.e., there must exist a number C such that the expression

$$\psi(-x) = C\psi(x) \tag{5-159}$$

is an identity in x. If the transformation $x \to -x$ is made once more in Eq. (5-159), the result is

$$\psi(x) = C\psi(-x) = C^2\psi(x),$$

whence $C = \pm 1$, and

$$\psi(-x) = \pm\psi(x). \tag{5-160}$$

Every eigenfunction for a bound state in a symmetric field $[u(x) = u(-x)]$ is therefore either an even or an odd function of x. This fact, which has already been noted in connection with Eqs. (5-109) and (5-110), is expressed by the statement that $\psi(x)$ has a definite *parity*. If $\psi(-x) = \psi(x)$, the parity of ψ is *even;* if $\psi(x) = -\psi(-x)$, it is *odd*.

If the system under study is invariant to the transformation $x \to -x$, then conclusions as to the energy levels, etc., cannot be influenced by the choice of which of the two directions along x is to be positive. In other words, if no feature of the environment of the particle, as expressed in the function $V(x)$, specifies a particular direction, then the eigenfunctions of nondegenerate states have a definite parity.

That these simple considerations are not trivial is apparent from the fact that the result $\langle x^{2m+1} \rangle = 0$ can be deduced immediately from parity considerations, without reference to the explicit form of ψ. In more complex situations, the concept of parity is of fundamental importance for the classification of quantum states.

5-13 The Wentzel-Kramers-Brillouin approximation. Only a few problems in quantum mechanics can be solved exactly, and approximation methods are therefore of great practical importance. We shall conclude this chapter on one-dimensional problems with a discussion of an approximate treatment, due to Wentzel, Kramers, and Brillouin.[1] This approach, commonly known as the *WKB method*, is also called the *classical approximation*, since it deals with situations in which $\hbar$ is small compared

[1] G. Wentzel, *Z. Physik* **38**, 518 (1926); H. A. Kramers, *Z. Physik* **39**, 828 (1926); L. Brillouin, *Compt. rend.* **183**, 24 (1926) and *J. phys. et radium* **7**, 353 (1926); R. E. Langer, *Phys. Rev.* **51**, 669 (1937).

to the action. The method leads to a quantization rule which is essentially the same as that of Wilson and Sommerfeld (Section 1–12).

The one-dimensional Schrödinger equation,

$$\frac{d^2\psi}{dx^2} + \frac{2m}{\hbar^2}[E - V(x)]\psi = 0,$$

can be written in the form

$$\frac{d^2\psi}{dx^2} + \frac{p^2}{\hbar^2}\psi = 0, \tag{5-161}$$

where p is the classical momentum at the point x:

$$p = \sqrt{2m[E - V(x)]}. \tag{5-162}$$

If the energy is high enough so that the wave length $\lambda = h/p$ is very short in the classical region, compared to the extent of this region, and if the potential function changes smoothly, then the "index of refraction" for the waves varies slowly. In the discussion of geometrical optics in Section 4–2, it has been shown that, under these circumstances, the wave function can be approximated by

$$\psi(x) = \phi(x)\exp\left[\pm\frac{i}{\hbar}\int^x p(x)\,dx\right], \tag{5-163}$$

where $\phi(x)$ is a slowly varying function [Eq. (4–12)]. This is the basis for the WKB method.

By straightforward substitution of the approximate solution (5–163) into the Schrödinger equation (5–161), the differential equation for the function $\phi(x)$ is obtained:

$$\frac{\hbar}{ip}\frac{d^2\phi}{dx^2} \pm \left(2\frac{d\phi}{dx} + \frac{1}{p}\frac{dp}{dx}\phi\right) = 0. \tag{5-164}$$

It is assumed that $\hbar/p$ is small, compared to the other dimensions of the problem, and that ϕ varies slowly. Hence, we neglect the first term in Eq. (5–164) and obtain

$$\frac{2}{\phi}\frac{d\phi}{dx} + \frac{1}{p}\frac{dp}{dx} = \frac{d}{dx}\ln(\phi^2 p) = 0, \tag{5-165}$$

which yields

$$\phi = Kp^{-1/2} \quad (K = \text{a constant}). \tag{5-166}$$

The approximate wave function is therefore

$$\psi_{\text{WKB}} = Kp^{-1/2} \exp\left(\pm \frac{i}{\hbar} \int^x p \, dx\right).$$ (5–167)

The classical approximation is expected to hold in regions where the fractional change in p in one wavelength is small, that is, where

$$\left|\frac{p'\lambda}{p}\right| = \left|\frac{\hbar p'}{p^2}\right| \ll 1.$$ (5–168)

The WKB approximation is valid under similar conditions: ψ_{WKB} satisfies the differential equation

$$\frac{d^2\psi}{dx^2} + \left[\frac{p^2}{\hbar^2} - Q\right]\psi = 0,$$ (5–169)

where

$$Q = \frac{3}{4}\left(\frac{p'}{p}\right)^2 - \frac{p''}{2p},$$ (5–170)

and Eq. (5–169) is an approximation to Eq. (5–161) if

$$|Q| \ll \frac{p^2}{\hbar^2},$$

or

$$\left|\frac{\hbar p'}{p^2}\right| \sqrt{\tfrac{3}{4} - \tfrac{1}{2}(pp''/p'^2)} \ll 1.$$ (5–171)

In nearly all practical cases, this condition is equivalent to (5–168). The condition will, in general, be fulfilled for problems where the mass is large, the energy high, and the potential smooth. However, it is clear that the WKB solutions cannot be valid near a classical turning point, where the momentum is zero.

We shall now consider the problem of finding the wave function for a particle in a given potential well. Let $V(x)$ have the form shown in Fig. 5–21. In region 1, the wave function decreases exponentially for $x \rightarrow -\infty$, and since p is imaginary $[V(x) > E]$, ψ is approximated by

$$\psi_1 = K_1|p|^{-1/2} \exp\left(\frac{1}{\hbar} \int_{x_1}^x |p| \, dx\right).$$ (5–172)

In region 2, ψ is oscillatory:

$$\psi_2 = K_2 p^{-1/2} \exp\left(\frac{i}{\hbar} \int^x p \, dx\right) + K_2' p^{-1/2} \exp\left(-\frac{i}{\hbar} \int^x p \, dx\right).$$ (5–173)

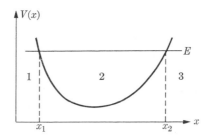

FIG. 5-21. Potential well for discussion of the WKB approximation.

In region 3, the wave function decreases exponentially for $x \to \infty$:

$$\psi_3 = K_3 |p|^{-1/2} \exp \left(-\frac{1}{\hbar} \int_{x_2}^{x} |p|\, dx \right). \tag{5-174}$$

The regions of validity for these forms of the wave function are separated by the classical turning points, near which the approximation fails. However, since ψ_1, ψ_2, and ψ_3 are all approximations to the same function ψ, the constants K_1, K_2, K_2', and K_3 cannot all be arbitrary. In order to evaluate the constants and to connect the approximate solutions in the three regions, we assume that the potential energy function is approximately linear in the neighborhood of x_1 and x_2. Thus, at x_1, we write

$$V(x) \approx E - A(x - x_1), \tag{5-175}$$

and at x_2,

$$V(x) \approx E + B(x - x_2). \tag{5-176}$$

In the neighborhood of x_1, the Schrödinger equation (5-161) then becomes

$$\frac{d^2 \psi}{dx^2} + \frac{2mA}{\hbar^2}(x - x_1)\psi = 0, \tag{5-177}$$

and near x_2,

$$\frac{d^2 \psi}{dx^2} - \frac{2mB}{\hbar^2}(x - x_2)\psi = 0. \tag{5-178}$$

In Eq. (5-177), we now change the variable to

$$z = -\left(\frac{2mA}{\hbar^2} \right)^{1/3}(x - x_1), \tag{5-179}$$

and obtain

$$\frac{d^2 \psi}{dz^2} - z\psi = 0. \tag{5-180}$$

Similarly, the substitution

$$z = \left(\frac{2mB}{\hbar^2}\right)^{1/3} (x - x_2) \tag{5-181}$$

reduces Eq. (5–178) to the same form (5–180).

The solutions of the differential equation (5–180) are the *Airy functions*.[1] We require a function which vanishes asymptotically for large positive z ($z > 0$ corresponds to $x < x_1$ and $x > x_2$). Such a function is

$$Ai(z) = \frac{1}{\pi} \int_0^{\infty} \cos\left(\frac{s^3}{3} + sz\right) ds, \tag{5-182}$$

which, for large $|z|$, has the asymptotic forms

$$Ai(z) \sim \frac{1}{2\sqrt{\pi}\, z^{1/4}} \exp\left(-\tfrac{2}{3} z^{3/2}\right) \quad (z > 0), \tag{5-183}$$

$$Ai(z) \sim \frac{1}{\sqrt{\pi}\,(-z)^{1/4}} \sin\left[\tfrac{2}{3}(-z)^{3/2} + \frac{\pi}{4}\right] \quad (z < 0). \tag{5-184}$$

[See Fig. 5–22 for a graph of $Ai(z)$].

If the energy E is large enough, the regions of validity of the linear approximations (5–175) and (5–176) contain many wavelengths. The

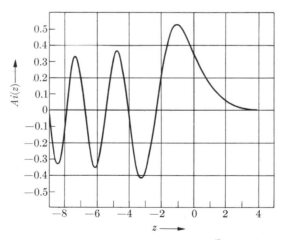

FIG. 5–22. The Airy function $Ai(z) = (1/\pi) \int_0^{\infty} \cos (s^3/3 + sz)\, ds$.

[1] H. and B. S. Jeffreys, *Methods of Mathematical Physics.* Cambridge: Cambridge University Press, 1956, 3rd ed., Section 17.07. J. C. P. Miller, *The Airy Integral* (British Association for the Advancement of Science, Mathematical Tables, Part-Volume B). Cambridge: Cambridge University Press, 1946.

function $Ai(z)$, which passes smoothly through the turning point, provides the required connections among the approximate forms (5–172), (5–173), and (5–174).

In the neighborhood of x_1, we have

$$p^2 \approx 2mA(x - x_1) = -(2mA\hbar)^{2/3}z,$$

and

$$\frac{1}{\hbar}\int_{x_1}^{x} |p|\, dx = \left(\frac{2mA}{\hbar^2}\right)^{1/3} \int_{x_1}^{x} \sqrt{z}\, dx = -\int_{0}^{z} \sqrt{z}\, dz = -\tfrac{2}{3}z^{3/2}. \quad (5\text{–}185)$$

Similarly,

$$\frac{1}{\hbar}\int_{x_1}^{x} p\, dx = \left(\frac{2mA}{\hbar^2}\right)^{1/3} \int_{x_1}^{x} \sqrt{-z}\, dx = -\int_{0}^{z} \sqrt{-z}\, dz = \tfrac{2}{3}(-z)^{3/2},$$

and comparison with Eqs. (5–183) and (5–184) shows that the function approximated to the left of x_1 by

$$\psi_1 \approx |p|^{-1/2} \exp\left(\frac{1}{\hbar}\int_{x_1}^{x} |p|\, dx\right) \qquad (x < x_1) \qquad (5\text{–}186)$$

has, on the right, the approximation

$$\psi \approx 2p^{-1/2} \sin\left(\frac{1}{\hbar}\int_{x_1}^{x} p\, dx + \frac{\pi}{4}\right) \qquad (x > x_1). \qquad (5\text{–}187)$$

A similar analysis in the neighborhood of point x_2 shows that the function approximated to the right of x_2 by

$$\psi_3 = |p|^{-1/2} \exp\left(-\frac{1}{\hbar}\int_{x_2}^{x} |p|\, dx\right) \qquad (x > x_2), \qquad (5\text{–}188)$$

is approximated in region 2 by

$$\psi \approx 2p^{-1/2} \sin\left(\frac{1}{\hbar}\int_{x}^{x_2} p\, dx + \frac{\pi}{4}\right) \qquad (x < x_2). \qquad (5\text{–}189)$$

The functions (5–187) and (5–189) are the continuations, into the classical region, of the functions (5–186) and (5–188), respectively, which have the proper behavior at $x = \pm\infty$. Now if ψ_1 and ψ_3 are approximations to the same eigenfunction ψ, they must be the same except perhaps for a constant multiplier:

$$\sin\left(\frac{1}{\hbar}\int_{x_1}^{x} p\, dx + \frac{\pi}{4}\right) = C \sin\left(\frac{1}{\hbar}\int_{x}^{x_2} p\, dx + \frac{\pi}{4}\right). \qquad (5\text{–}190)$$

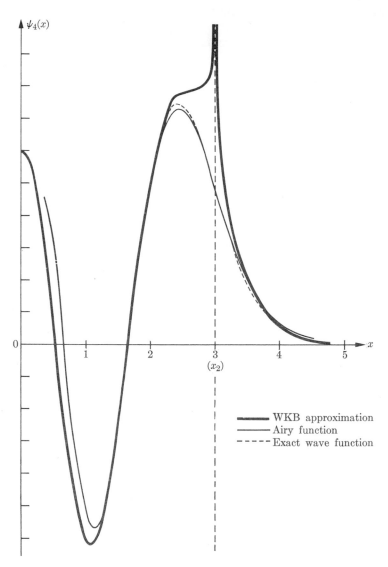

FIG. 5–23. WKB approximation to the harmonic-oscillator wave function in the state $n = 4$. To the accuracy of the graph, the WKB wave function (heavy line) coincides with the exact wave function (broken line) in the interior of the well. Near the classical turning point $x_2 = 3$, the WKB approximation breaks down. The Airy function (light line) coincides with the exact wave function at x_2 and connects the WKB approximations in the classical and non-classical regions. At small and large x, the Airy function deviates from the exact wave function.

Setting $\int_{x_1}^{x} = \int_{x_1}^{x_2} - \int_{x}^{x_2}$, we require that the expression

$$\sin\left(\frac{1}{\hbar}\int_{x_1}^{x_2} p\,dx - \frac{1}{\hbar}\int_{x}^{x_2} p\,dx + \frac{\pi}{4}\right) = C\sin\left(\frac{1}{\hbar}\int_{x}^{x_2} p\,dx + \frac{\pi}{4}\right),$$

be an identity in x. This condition is satisfied only if

$$\frac{1}{\hbar}\int_{x_1}^{x_2} p\,dx = (n + \tfrac{1}{2})\pi \qquad (n \text{ an integer}); \qquad (5\text{–}191)$$

the constant C is then equal to $(-1)^n$.

The (unnormalized) WKB approximation to the bound-state wave function is therefore

$$\psi_{\text{WKB}} = \begin{cases} (-)^n |p|^{-1/2} \exp\left(-\frac{1}{\hbar}\int_{x}^{x_1} |p|\,dx\right) & (x < x_1), \\[2ex] (-)^n 2p^{-1/2} \sin\left(\frac{1}{\hbar}\int_{x_1}^{x} p\,dx + \frac{\pi}{4}\right) & (x_1 < x < x_2), \\[2ex] |p|^{-1/2} \exp\left(-\frac{1}{\hbar}\int_{x_2}^{x} |p|\,dx\right) & (x_2 < x). \end{cases} \qquad (5\text{–}192)$$

(Note that the approximate wave function for the nth bound state has $n + 1$ zeros.)

The WKB approximation to the state ψ for the harmonic oscillator is compared to the correct wave function in Fig. 5–23.

The condition (5–191) can be written

$$\oint p\,dx = (n + \tfrac{1}{2})h. \qquad (5\text{–}193)$$

The symbol $\oint$ denotes the integral taken over a complete cycle of the classical motion, i.e., the area included by the path of the representative point in the p–x plane. This is the Wilson-Sommerfeld condition [Eq. (1–43)], except that n is replaced by $n + 1/2$. Since the classical approximation is reliable only when n is large, this modification is not of great significance.

5–14 Penetration of a potential barrier; WKB approximation. The penetration of a square potential barrier has been discussed in Section 5–2. For a barrier of more complicated shape, the Schrödinger equation cannot usually be solved exactly, and the WKB approximation is often suited for the problem. The wave function is oscillatory outside the barrier and has exponential character in the nonclassical region. In the approximate wave functions (5–192), the exponentially increasing solution in the nonclassical region was discarded because it violates the boundary conditions for ψ at $\pm\infty$. In the present case, however, the nonclassical region is of finite width, and both exponential solutions must be included. We require therefore a second connection formula.

We use the second solution of the differential equation (5–180), which is the Airy function

$$Bi(z) = \frac{1}{\pi} \int_0^\infty \left[e^{-sz-(1/3)s^3} + \sin\left(\frac{s^3}{3} + sz\right) \right] ds, \qquad (5\text{--}194)$$

with the asymptotic forms

$$Bi(z) \sim \frac{1}{\sqrt{\pi}\, z^{1/4}} \exp\left(\tfrac{2}{3}z^{3/2}\right) \qquad (z > 0), \qquad (5\text{--}195)$$

$$Bi(z) \sim \frac{1}{\sqrt{\pi}\,(-z)^{1/4}} \cos\left[\tfrac{2}{3}(-z)^{3/2} + \frac{\pi}{4}\right] \qquad (z < 0). \qquad (5\text{--}196)$$

An argument which follows the same lines as that of the preceding section leads to the connection formula linking an increasing exponential solution in region 1 to an oscillatory solution in region 2 (Fig. 5–24):

$$\psi_{\text{WKB}} = \begin{cases} |p|^{-1/2} \exp\left(\dfrac{1}{\hbar}\displaystyle\int_x^{x_1} |p|\, dx\right) & (x < x_1), \\[4mm] p^{-1/2} \cos\left(\dfrac{1}{\hbar}\displaystyle\int_{x_1}^x p\, dx + \dfrac{\pi}{4}\right) & (x > x_1). \end{cases} \qquad (5\text{--}197)$$

A potential barrier of arbitrary shape is indicated in Fig. 5–25. We assume that a beam of particles is incident from the left. In region 3, the wave function for the transmitted particles is of the form

$$\psi_3 = A p^{-1/2} \exp i\left(\frac{1}{\hbar}\int_{x_2}^x p\, dx + \frac{\pi}{4}\right) \qquad (x > x_2), \qquad (5\text{--}198)$$

where the phase factor $e^{i\pi/4}$ has been included to facilitate the application of Eq. (5–197). In terms of trigonometric functions, ψ_3 can be written

$$\psi_3 = A p^{-1/2} \left[\cos\left(\frac{1}{\hbar}\int_{x_2}^x p\, dx + \frac{\pi}{4}\right) + i\sin\left(\frac{1}{\hbar}\int_{x_2}^x p\, dx + \frac{\pi}{4}\right) \right]. \qquad (5\text{--}199)$$

The connecting wave function of exponential type in region 2 is obtained by comparison with Eqs. (5–197) and (5–192):

$$\psi_2 = A|p|^{-1/2} \left[\exp\left(\frac{1}{\hbar}\int_x^{x_2} |p|\, dx\right) + \frac{i}{2}\exp\left(-\frac{1}{\hbar}\int_x^{x_2} |p|\, dx\right) \right]. \qquad (5\text{--}200)$$

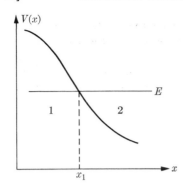

FIG. 5–24. Potential near the classical turning point x_1 at the edge of a barrier.

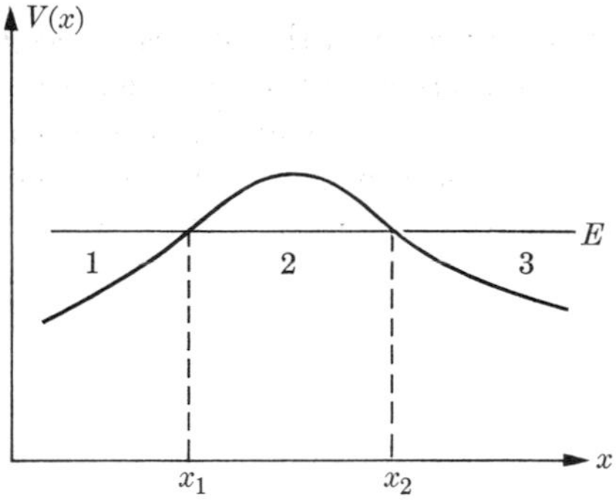

FIG. 5–25. Potential barrier.

In order to find the appropriate wave function in region 1, we rewrite the integrals in the last expression, using (see Fig. 5–25)

$$\int_x^{x_2} |p| \, dx = \int_{x_1}^{x_2} |p| \, dx - \int_{x_1}^x |p| \, dx,$$

and introducing the definition

$$T = \exp\left(-\frac{1}{\hbar} \int_{x_1}^{x_2} \sqrt{2m(V - E)} \, dx\right), \qquad (5\text{–}201)$$

so that Eq. (5–200) becomes

$$\psi_2 = A|p|^{-1/2}\left[T^{-1} \exp\left(-\frac{1}{\hbar} \int_{x_1}^x |p| \, dx\right) + \frac{i}{2} T \exp\left(\frac{1}{\hbar} \int_{x_1}^x |p| \, dx\right)\right]. \qquad (5\text{–}202)$$

By comparison with Eqs. (5–192) and (5–197), the connecting oscillatory wave function in region 1 is now seen to be

$$\psi_1 = Ap^{-1/2}\left[2T^{-1} \sin\left(\frac{1}{\hbar} \int_x^{x_1} p \, dx + \frac{\pi}{4}\right) + \frac{i}{2} T \cos\left(\frac{1}{\hbar} \int_x^{x_1} p \, dx + \frac{\pi}{4}\right)\right]. \qquad (5\text{–}203)$$

It is convenient to rewrite this expression in terms of exponentials:

$$\psi_1 = \frac{A}{ip^{1/2}} \left\{ (T^{-1} - \tfrac{1}{4}T) \exp\left[i\left(\frac{1}{\hbar} \int_x^{x_1} p \, dx + \frac{\pi}{4}\right)\right] \right.$$

$$\left. - (T^{-1} + \tfrac{1}{4}T) \exp\left[-i\left(\frac{1}{\hbar} \int_x^{x_1} p \, dx + \frac{\pi}{4}\right)\right] \right\}. \qquad (5\text{–}204)$$

The first term in the braces is recognized as a wave moving to the left, and hence represents the reflected wave, while the second term represents the incoming wave, which moves to the right.

The constant A can be adjusted for unit incoming current, so that the absolute magnitude of the amplitude of the incoming wave is $v^{-1/2}$. Then

$$A = \frac{\sqrt{m}}{T^{-1} + \frac{1}{4}T}. \qquad (5\text{-}205)$$

With this value for A, the amplitude of the reflected wave has the magnitude

$$v^{-1/2} \frac{1 - T^2/4}{1 + T^2/4}. \qquad (5\text{-}206)$$

The transmitted wave [Eq. (5–198)] has the amplitude

$$Ap^{-1/2} = v^{-1/2} \frac{T}{1 + T^2/4}. \qquad (5\text{-}207)$$

The reflection coefficient R is defined as the ratio of reflected to incident wave amplitudes:

$$|R| = \frac{1 - T^2/4}{1 + T^2/4}. \qquad (5\text{-}208)$$

The square of the reflection coefficient is equal to the fraction of the incident current that is reflected.

The transmission coefficient is the ratio of transmitted to incident wave amplitudes:

$$|\text{Trans. coeff.}| = \frac{T}{1 + T^2/4}. \qquad (5\text{-}209)$$

It is consistent with the error of the WKB approximation to neglect powers of T higher than the first, so that

$$|\text{Trans. coeff.}| \approx T = \exp\left\{-\frac{1}{\hbar} \int_{x_1}^{x_2} \sqrt{2m[V(x) - E]}\, dx\right\} \qquad (T \ll 1).$$
$$(5\text{-}210)$$

To the same approximation,

$$|R|^2 \approx 1 - T^2. \qquad (5\text{-}211)$$

As an example of the application of Eq. (5–210) for the transmission coefficient, let us consider the cold emission of electrons from a metal. In the absence of an external electric field, the electrons are bound by a po-

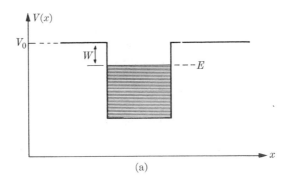

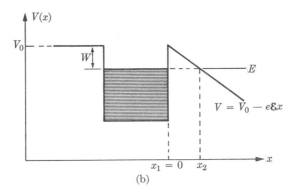

FIG. 5–26.   Potential for electrons in a metal. (a) No external field. (b) With external field $\mathcal{E}$.

tential, as shown in Fig. 5–26(a). The lower levels in the well are filled, according to the Pauli exclusion principle (Chapter 12). The work function W is the energy required to remove an electron from the highest occupied state.

When an external electric field $\mathcal{E}$ is applied to the metal, the potential at the surface takes the form indicated in Fig. 5–26(b). Now the potential barrier has a finite width, and electrons are able to escape. The variation of cold emission with work function and applied field is easily obtained from Eq. (5–210). We set $x_1 = 0$, and find x_2 as follows [cf. Fig. 5–26(b)]:

$$V_0 - e\mathcal{E}x_2 = V_0 - W$$

$$x_2 = \frac{W}{e\mathcal{E}}.$$

Also,

$$V - E = V_0 - e\mathcal{E}x - E = W - e\mathcal{E}x.$$

The transmission probability therefore is

$$T^2 = \exp\left(-\frac{2}{\hbar}\int_{x_1}^{x_2}\sqrt{2m(V-E)}\,dx\right)$$

$$= \exp\left(-\frac{2}{\hbar}\int_0^{W/e\mathcal{E}}\sqrt{2m(W-e\,\mathcal{E}x)}\,dx\right)$$

$$= \exp\left(-\frac{4}{3}\frac{\sqrt{2m}}{\hbar}\frac{W^{3/2}}{e\mathcal{E}}\right). \tag{5-212}$$

This expression is in qualitative agreement with experiment.

REFERENCES

BOHM, DAVID, *Quantum Theory*. New York: Prentice-Hall, Inc., 1951. Sections 11 and 13 deal with square potentials and the harmonic oscillator.

FLÜGGE, SIEGFRIED and MARSCHALL, HANS, *Rechenmethoden der Quantentheorie*. 2nd ed., Berlin: Springer-Verlag, 1952. A useful collection of solved problems, including several that involve square potentials.

KRAMERS, H. A., *Die Grundlagen der Quantentheorie*. Leipzig: Akademische Verlagsgesellschaft m.b.H., 1938.

MACCOLL, L. A., "Note on the Transmission and Reflection of Wave Packets by Potential Barriers," *Phys. Rev.* **40**, 1932, p. 621. An instructive discussion of this problem, which is an extension of the material in Section 5-2.

SCHIFF, L. I., *Quantum Mechanics*. 2nd ed., New York: McGraw-Hill Book Co., Inc., 1955. Sections 8 and 9 cover boundary and continuity conditions as well as square potentials.

PROBLEMS

5-1. Discuss the function ψ_2 [Eq. (5-15)] for the case $E > V_0$.

5-2. Calculate the probability current for the wave function (5-27) and show that it is continuous at each boundary of the potential barrier. Construct a solution of this problem which is an even function of $(x - a/2)$ and draw a graph of $|\psi|^2$ for $E = (1/2)V_0$. What is the amplitude of ψ at $x = a/2$? Study the dependence of $|\psi(a/2)|^2/|\psi(\infty)|^2$ on E.

5-3. Derive relations analogous to (5-31) and (5-33) for the quantity $|R|^2$ as a function of E, and prove that $|R|^2 + |T|^2 = 1$.

5-4. Consider a step potential barrier, as shown in Fig. 5-27. Calculate the transmission coefficient $|T|^2$ and the reflection coefficient $|R|^2$ as functions of the parameter d. What values of d give maximum and minimum transmission?

5-5. Carry out the details of the calculation of ψ for a particle of positive total energy in the potential well (5-34) and draw a graph of the transmission coefficient as a function of E.

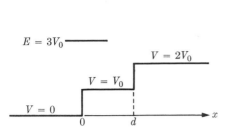

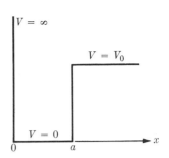

FIG. 5–27. The step potential of Problem 5–4.

FIG. 5–28. The potential well of Problem 5–6.

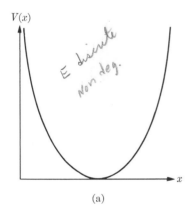

(a)

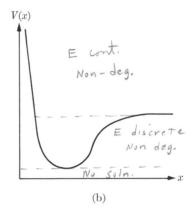

(b)

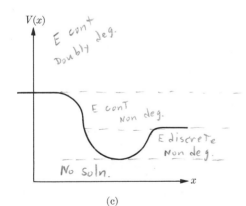

(c)

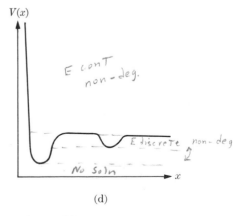

(d)

FIG. 5–29. Potential energy curves for Problem 5–10.

5–6. Find the lowest energy state and the wave function for that state in a potential well, as sketched in Fig. 5–28. What is the least depth of the well for which a bound state exists?

5–7. Derive Eqs. (5–37) and (5–38).

5–8. Obtain Eq. (5–40) from Eq. (5–37) and the definitions (5–39).

5–9. Verify Eq. (5–42) and normalize ψ.

5–10. Discuss the character of the energy spectrum for each of the potential energy curves in Fig. 5–29.

5–11. Prove that a complex function $\psi(x)$ and its complex conjugate $\psi^*(x)$ are linearly dependent if and only if

$$\psi(x) = e^{i\gamma}\phi(x),$$

where γ is a real constant and ϕ is a real function of x.

5–12. Prove that ψ_1 and ψ_2, as given in Eqs. (5–14) and (5–15), are linearly independent.

5–13. Prove that a solution of exponential type can have at most one stationary point.

5–14. If $u_2 > u_1$ in the infinite interval $(-\infty, x_2)$, and ψ_1 and ψ_2 are solutions of the Schrödinger equations for these potential energy functions such that $\psi_2(-\infty) = \psi_2(x_2) = \psi_2'(-\infty) = 0$, prove that ψ_1 has a zero in $(-\infty, x_2)$. (It is implied that ψ_2 has no finite zero smaller than x_2.)

5–15. Prove that the least interval between zeros of ψ is not smaller than $\pi/\sqrt{\lambda - m}$, where $m < u$ is a lower bound for $u(x)$. (Hint: A zero of the function $\sin \sqrt{\lambda - m}\,(x - c)$ can be made to occur at any desired point by proper choice of the arbitrary constant c.)

5–16. Derive Eq. (5–77).

5–17. Show that the functions $v \sin(\phi - \phi_1)$ and $v \sin(\phi - \phi_2)$ [Eq. (5–81)] approach zero as $|x| \to \infty$. (Hint: Write $v \sin(\phi - \phi_1) = [\sin(\phi - \phi_1)]/(1/v)$ and apply l'Hôpital's rule.)

5–18. Prove that $v'(x) \to \infty$ as $|x| \to \infty$. (Hint: Show from Eq. (5–77) that $v'(x) = v'(x_2) + \int_{x_2}^{x} [(u - \lambda)v + 1/v^3]\,dx$.)

5–19. Show that if $\lambda < u(x)$ in the interval (x_2, ∞), and ψ is a nonzero solution of Eq. (5–47) in this interval such that $\psi(\infty) = 0$, then

$$|\psi(x)| < (\text{constant})e^{-Kx},$$

where K is a suitably chosen constant. Hence prove that the integral $\int_{-\infty}^{\infty} |\psi|^2\,dx$ exists for a stationary bound state

5–20. Prove the statement following Eq. (5–84): "The finite zeros of ψ_1 are all included between the largest and smallest finite zeros of ψ_2."

5–21. Differentiate Eq. (5–47) with respect to λ and prove the formula

$$\frac{d}{dx}\left[\psi\frac{\partial\psi}{\partial\lambda} - \frac{\partial\psi'}{\partial\lambda}\psi\right] = \psi^2.$$

Rewrite this relation in terms of v and ϕ, and deduce the expression

$$\left[\left(v' \frac{\partial v}{\partial \lambda} - v \frac{\partial v'}{\partial \lambda} \right) \sin^2 (\phi - \theta) + \frac{2}{v} \frac{\partial v}{\partial \lambda} \sin (\phi - \theta) \cos (\phi - \theta) + \frac{\partial \phi}{\partial \lambda} \right]_{x=x_1}$$

$$= \int_0^{x_1} v^2(x') \sin^2 [\phi(x') - \theta] \, dx'.$$

Hence show that

$$\frac{\partial \phi}{\partial \lambda} = \int_0^x v^2(x') \sin^2 [\phi(x') - \phi(x)] \, dx',$$

and prove the statement that $\phi_2 - \phi_1$ increases steadily with λ. Prove also that

$$\lim_{\lambda \to -\infty} \phi_1 = -\infty, \qquad \lim_{\lambda \to \infty} \phi_2 = \infty.$$

(More complete discussions of the dependence on λ are given in Rojansky, *loc. cit.*, and Milne, *loc. cit.*)

5–22. Prove that the members of an orthogonal set of functions are linearly independent.

5–23. Prove the general theorem following Eq. (5–89). [Hint: A proof can be constructed by studying the behavior of the largest zero of the function $\psi = v \sin (\phi - \phi_1)$ as λ varies continuously between λ_1 and λ_2. Cf. Problem 5–21. An alternative proof can be based upon Eq. (5–64). Cf. R. G. Sachs, *Nuclear Theory*. Reading, Mass.: Addison-Wesley Publishing Co., Inc., 1953, Appendix 1.]

5–24. Substitute the asymptotic solution (5–98) into the left-hand member of Eq. (5–96) and show that the equation is satisfied formally in the limit $|x| \to \infty$.

5–25. Prove that if $\phi_0(x)$ is an even function, and $\phi_1(x)$ an odd function, i.e., if

$$\phi_0(-x) = \phi_0(x), \qquad \phi_1(-x) = -\phi_1(x),$$

then ϕ_0 and ϕ_1 are linearly independent.

5–26. Verify that if the solution $\alpha = 1$ of Eq. (5–102) is chosen, the series obtained is a linear combination of ϕ_0 and ϕ_1.

5–27. Derive Eqs. (5–121) and (5–122), and verify that the recurrence relations are satisfied by the functions (5–115). Continue the list of $H_n(x)$ to $n = 7$.

5–28. Reduce Eq. (5–127) to the form (5–128). (Hint: The calculation depends upon the rule that if $f = f(x + y)$, then $\partial f/\partial x = \partial f/\partial y$.)

5–29. Prove Eq. (5–130) for $m < n$ by substituting the expression (5–128) for H_n and integrating by parts m times.

5–30. Derive Eqs. (5–138), (5–139), and (5–140).

5–31. Prove the integral formula

$$H_n(x) = \frac{(2i)^n}{\sqrt{\pi}} \int_{-\infty}^{\infty} z^n e^{-(z+ix)^2} \, dz.$$

[Hint: The formula obviously yields $H_0 = 1$, which is correct. Differentiate and use mathematical induction based upon Eq. (5–122).] The student who is familiar with the *method of steepest descent* will find it instructive to derive the asymptotic formula (5–144) from this integral.

5–32. Derive the formulae

$$x^{2m} = \sum_{k=0}^{m} \frac{(2m)!}{2^{2m}(2k)!(m-k)!} H_{2k}(x),$$

$$H_n^2(x) = \sum_{k=0}^{n} \frac{2^{n-k}(n!)^2}{(n-k)!(k!)^2} H_{2k}(x),$$

and use these results to show that

$$\langle x^{2m} \rangle = \frac{(2m)!}{2^{2m}m!} \sum_k 2^k \binom{m}{k} \binom{n}{k},$$

where $\binom{n}{k} = n!/k!(n-k)!$ is the binomial coefficient. In particular,

$$\langle x^2 \rangle = n + \tfrac{1}{2}, \qquad \langle x^4 \rangle = \tfrac{3}{2}(n^2 + n + \tfrac{1}{2}), \qquad \text{etc.}$$

5–33. Show that the wave function in momentum space,

$$a_n(p) = \frac{1}{\sqrt{2\pi}} \int_{-\infty}^{\infty} \psi_n(x) e^{-ipx} \, dx,$$

corresponding to the state (5–137) is

$$a_n(p) = \frac{1}{(i)^n \sqrt{2^n n! \sqrt{\pi}}} e^{-(1/2)p^2} H_n(p).$$

(Hint: Use the generating function to calculate the integral.)

5–34. The potential function (5–34) is symmetric in the point $x = a/2$. Prove that the eigenfunction (5–42) satisfies

$$\psi[-(x - a/2)] = \pm\psi(x - a/2) \qquad \text{(cf. Problem 5–2).}$$

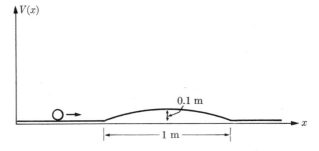

FIG. 5–30.   Ball on track (Problem 5–38).

5–35. Derive a classical approximation for the probability distribution $P(x)$ for the rectangular potential well and show that it is given asymptotically (in the average sense of Section 5–11) by the wave function (5–42).

5–36. Normalize the wave function (5–42), and verify, by explicit calculation, the orthogonality relation (5–88) in this case.

5–37. Show that in the limit of large barrier width a, Eq. (5–33) gives the same penetration probability for a square potential barrier as Eq. (5–210).

5–38. A 1-kgm ball rolls slowly on a level track, then encounters a hump of sinusoidal cross section, 1 m long and 0.1 m high (Fig. 5–30). If the kinetic energy of the ball is neglected, what is the probability that it can overcome the hump?

CHAPTER 6

OPERATORS AND EIGENFUNCTIONS

6–1 Linear operators. Mathematically, quantum mechanics is a *linear* theory. According to the principle of superposition, the wave functions representing various states of a physical system can be combined additively, and the resulting functions represent new states. In this way, a mathematically complex function can be written as a linear combination of simpler functions: E.g., a wave packet can be represented by the Fourier integral as a linear combination of monochromatic components.

The wave functions belonging to a given physical system form a class of functions which describe every possible state of the system. It is the object of the theory to discover a characterization of these functions and to formulate rules by means of which observable properties of the system can be deduced. These rules prescribe the mathematical operations that are to be performed upon the wave functions to yield results which can be interpreted experimentally. Thus, Eq. (2–120) is a rule by which the average momentum of a particle in the state ψ can be calculated. The operator concept is fundamental in the formulation of such rules.

The result obtained from performing a given mathematical operation upon a wave function ψ is conveniently symbolized by writing

$$\phi = A_{\text{op}}\psi.$$

This equation means: The function ϕ is the result of applying to ψ the operation denoted by A_{op}. If, for example, $A_{\text{op}} = x$, then ϕ is the result of multiplying ψ by the independent variable x; or if $A_{\text{op}} = d/dx$, ϕ is the derivative of ψ with respect to x. The subscript $_{\text{op}}$ is used for the present to emphasize that A_{op} is the symbol for a mathematical operation and is not to be interpreted as a multiplicative factor of the ordinary kind. For illustration, the formula

$$\frac{d}{dx}(fg) = f\frac{dg}{dx} + \frac{df}{dx}g$$

can be written

$$A_{\text{op}}(fg) = f(A_{\text{op}}g) + (A_{\text{op}}f)g,$$

in which $A_{\text{op}} = d/dx$. This is, of course, quite different from the rule

$$A(fg) = (Af)g = f(Ag),$$

which holds for ordinary multiplication by the number A.

158

The operator A_{op} is a *linear operator* if it satisfies the rules

$$A_{op}(\psi_1 + \psi_2) = A_{op}\psi_1 + A_{op}\psi_2, \qquad (6\text{–}1)$$

$$A_{op}(c\psi_1) = c(A_{op}\psi_1), \qquad (6\text{–}2)$$

where ψ_1 and ψ_2 are given functions, and c is a complex constant. The examples cited above are linear, as can easily be verified; however, the operation "form the square of ψ" is not. Two linear operators of fundamental importance are the *null* or *zero* operator, defined by $0_{op}\psi = 0$, which "annihilates" the function to which it is applied, and the *identity* or *unit operator* 1_{op}, $1_{op}\psi = \psi$, which produces no change in its operand.

An algebra of linear operators is constructed by defining the terms "sum," "product," "power," etc. The *sum* of two operators is

$$C_{op} = A_{op} + B_{op}$$

if, for every function ψ,

$$C_{op}\psi = A_{op}\psi + B_{op}\psi.$$

Similarly, if

$$A_{op}(B_{op}\psi) = C_{op}\psi,$$

then C_{op} is the *product* of A_{op} and B_{op}. It is essential to note that $A_{op}B_{op}$ and $B_{op}A_{op}$ may not be equal, that is, A_{op} and B_{op} may not commute. For example,

$$x\left(\frac{d}{dx}\psi\right) = x\frac{d\psi}{dx},$$

but

$$\frac{d}{dx}(x\psi) = \psi + x\frac{d\psi}{dx},$$

whence, if

$$A_{op} = \frac{d}{dx}, \qquad B_{op} = x, \qquad (6\text{–}3)$$

$$A_{op}(B_{op}\psi) = 1_{op}\psi + B_{op}(A_{op}\psi).$$

This equation is true for every function ψ and can be expressed as an operator equation:

$$A_{op}B_{op} = 1_{op} + B_{op}A_{op}, \qquad (6\text{–}4)$$

where A_{op} and B_{op} are the operators defined by Eqs. (6–3). If it is agreed that the indicated operations are to be performed from right to left, the parentheses are unnecessary and may be omitted. It is understood that each operation is to be performed on the entire quantity standing to the right of the operator symbol.

The *square* of an operator A_{op} is

$$A_{op}^2 = A_{op}A_{op},$$

and, similarly,

$$A_{op}^3 = A_{op}^2 A_{op} = A_{op}A_{op}^2.$$

By combining the processes of addition and multiplication, a *function* of an operator can be formed. For example, the differential equation

$$\frac{d^2y}{dx^2} + k^2 y = 0$$

is the same as

$$(A_{op}^2 + k^2 \cdot 1_{op})y = 0,$$

where $A_{op} = d/dx$. An algebraic function of a linear operator is itself a linear operator (Problem 6–3).

If two operators A_{op} and B_{op} are related by the equations

$$A_{op}B_{op} = B_{op}A_{op} = 1_{op},$$

then they are *reciprocal* to each other, and we write

$$A_{op}^{-1} = B_{op}, \qquad B_{op}^{-1} = A_{op}.$$

An operator for which a reciprocal exists is *nonsingular*. A nonsingular operation can be inverted, i.e., if $\phi = A_{op}\psi$, and A_{op} has a reciprocal, then ψ can be reconstructed by means of A_{op}^{-1}:

$$\psi = A_{op}^{-1}\phi.$$

In this case, inversion is possible for every function ψ. However, if there is some nonzero ψ for which

$$A_{op}\psi = 0,$$

then, obviously, A_{op} has no reciprocal, i.e., A_{op} is *singular*.

The question whether a given operator is nonsingular can be a difficult one since the answer depends upon the class of functions to which the operation is applied. For example, the operator d/dx is singular with respect to the class of differentiable functions of x, which includes the constant c, because

$$\frac{d}{dx}c = 0,$$

and the number c cannot be reconstructed. If, however, the class is restricted to functions which are integrable in a proper sense (Problem 6–5), the function c is not a member of this class, and d/dx is nonsingular.[1]

6–2 Eigenfunctions and eigenvalues. If A_{op} is a given operator, the function

$$\phi = A_{op}\psi$$

which results from applying the operator A_{op} to ψ will, in general, be linearly independent of ψ. It may happen, however, that for some function ψ

$$A_{op}\psi = \alpha\psi, \tag{6–5}$$

where α is a complex number. In this case, if ψ is a member of the class of physically meaningful functions, it is an *eigenfunction* of the operator A_{op}. Except for multiplication by the number α, ψ is not changed by the operation A_{op}, and is in this sense *invariant*. The number α is called the *eigenvalue* of A_{op} associated with the eigenfunction ψ. It is to be noted that eigenfunctions are selected from a special class of functions. In a bound-state problem, for example, all wave functions are required to be continuous, to have continuous derivatives, and to vanish at infinity in such a way as to have an integrable square. In the continuum states, discussed in Sections 5–1 and 5–2, ψ is not allowed to become infinite at a large distance.

6–3 The operator formalism in quantum mechanics. A comparison of the expressions (2–114) and (2–121),

$$\langle x \rangle = \frac{\int \psi^* x \psi \, dx}{\int \psi^* \psi \, dx}, \qquad \langle p \rangle = \frac{\int \psi^*(\hbar/i)(d/dx)\psi \, dx}{\int \psi^* \psi \, dx},$$

shows that the expectation of the momentum of a particle in the state ψ can be computed in the same way as the expectation of the coordinate, if the operator $(\hbar/i)(d/dx)$ is substituted in place of x.

Furthermore, the Schrödinger equation for a state of energy E,

$$-\frac{\hbar^2}{2m}\frac{d^2\psi}{dx^2} + V\psi = E\psi,$$

---

[1] A linear operator A_{op} defines the *mapping* of a given class of functions ψ onto the class of functions $A_{op}\psi$. In quantum mechanics, it is assumed that this is the same class, i.e., that the wave functions of a physical system are the elements of a linear vector space which is mapped onto itself (or a part of itself) by the linear operator A_{op}.

is obtained from the classical Hamiltonian expression

$$\frac{p^2}{2m} + V = E$$

by making the replacement

$$p \rightarrow p_{\text{op}} = \frac{\hbar}{i} \frac{d}{dx} \tag{6-6}$$

and forming the equation

$$H_{\text{op}}\psi = E\psi, \tag{6-7}$$

in which H_{op} is the Hamiltonian operator

$$H_{\text{op}} = \frac{p_{\text{op}}^2}{2m} + V. \tag{6-8}$$

These observations suggest that the quantum-mechanical theory is linked to the classical description of the system by suitable correspondences between the classical dynamical variables and quantum-mechanical operators. This idea is made definite by the following fundamental assumption:

Let a mechanical system be described in terms of momenta and coordinates, according to classical Hamiltonian mechanics. Then *to every classically defined function $F(x, p)$ of momentum and coordinate there corresponds a quantum-mechanical operator*

$$F_{\text{op}} = F\left(x, \frac{\hbar}{i} \frac{d}{dx}\right), \tag{6-9}$$

and every individual measurement of the quantity F will yield an eigenvalue of F_{op}. The expectation of F in a series of measurements performed on an ensemble of systems in identical states ψ is given by

$$\langle F \rangle = \frac{\int \psi^* F_{\text{op}} \psi \, dx}{\int \psi^* \psi \, dx}. \tag{6-10}$$

This assumption, which was first formulated by Dirac, is a generalization of the correspondence principle. It is a prescription by means of which the observable properties of a system can be deduced from its wave function. The correspondence described by Eqs. (6-9) and (6-10) has been stated for a one-dimensional system only. It will be generalized to a three-dimensional system in Section 6-10, and systems containing more than one particle will be considered in Chapter 7.

The special significance of an eigenfunction of the operator F_{op}, i.e., of a function satisfying

$$F_{\mathrm{op}}\psi = \lambda\psi, \tag{6–11}$$

is made evident by substitution in Eq. (6–10):

$$\langle F \rangle = \frac{\int \psi^*\lambda\psi\, dx}{\int \psi^*\psi\, dx} = \lambda. \tag{6–12}$$

The expectation of an operator F_{op} in a state represented by an eigenfunction of F_{op} is the corresponding eigenvalue. If the physical quantity corresponding to F_{op} can be measured, it must be real; hence, the eigenvalues of a quantum-mechanical operator corresponding to an observable are real numbers. Operators which admit only real eigenvalues are therefore of special importance in the theory. We shall return to this point in Section 6–7.

It is clear that if λ is an eigenvalue of F_{op} in the state ψ, then λ^2 is an eigenvalue of F_{op}^2, for it follows from Eq. (6–11) and the linearity of the operator F_{op} that

$$F_{\mathrm{op}}^2\psi = F_{\mathrm{op}}(\lambda\psi) = \lambda(F_{\mathrm{op}}\psi) = \lambda^2\psi. \tag{6–13}$$

The quantity $\langle F \rangle$ is the expectation, or average, of F, which would be obtained as a result of many measurements of F performed on identical systems in the same state ψ (cf. Section 2–10). The results of individual measurements will, in general, be statistically distributed with respect to the average. The width ΔF of this distribution is defined, as in Section 3–2, by

$$(\Delta F)^2 = \langle (F - \langle F \rangle)^2 \rangle = \langle F^2 \rangle - \langle F \rangle^2.$$

The quantity ΔF measures the mean square difference between the individual results and the average. If ψ is an eigenfunction of F_{op}, then by Eqs. (6–12) and (6–13),

$$\Delta F = 0.$$

By its definition, ΔF vanishes only if every individual measurement yields the result $\langle F \rangle$. In other words, the result of a measurement of F is certainly λ if the state of the system is described by the eigenfunction belonging to λ.

6–4 The operator $(\hbar/i)(d/dx)$. The eigenvalue equation for the momentum operator is

$$p_{\mathrm{op}}\psi = \frac{\hbar}{i}\frac{d\psi}{dx} = \lambda\psi. \tag{6–14}$$

Equation (6–14) can be solved immediately and yields

$$\psi = ce^{(i/\hbar)\lambda x}. \qquad (6\text{–}15)$$

Now, if ψ is to be a physically admissible wave function, the eigenvalue λ must be purely real, for otherwise ψ would be divergent either at $x = \infty$ or at $x = -\infty$. The eigenvalues of the operator p_{op} are the quantities

$$\lambda = p,$$

where p is any real number. Consequently, this operator has a continuous, infinite spectrum of eigenvalues, and the corresponding eigenfunctions, $e^{(i/\hbar)px}$, are the plane-wave states of a free particle of momentum p. The normalized eigenfunctions are (Section 3–5)

$$\psi = \frac{1}{\sqrt{2\pi\hbar}} e^{(i/\hbar)px}. \qquad (6\text{–}16)$$

According to Fourier's theorem, these functions are a complete set, in terms of which every state of the system can be expressed by superposition.

The Schrödinger equation,

$$H_{\text{op}}\psi = E\psi,$$

can be interpreted similarly as an eigenvalue equation for the Hamiltonian operator H_{op}. The eigenfunctions of H_{op} are states for which the total energy of the system has a definite value. Again, these functions are a complete set with respect to the possible states of the system.

6–5 Orthogonal systems. The Schrödinger equation for a one-dimensional bound system leads to an infinite set of orthonormal functions, ψ_n, satisfying

$$\int \psi_m^*(x)\psi_n(x)\, dx = \delta_{mn}. \qquad (6\text{–}17)$$

The terms "orthonormal" and "orthogonal" are adapted from the language of vector analysis. Equations (6–17) are analogous to the familiar relations

$$\hat{e}_m \cdot \hat{e}_n = \delta_{mn}, \qquad (6\text{–}18)$$

in which the three-dimensional unit vectors $\hat{e}_i$ are mutually orthogonal. The scalar product [symbolized by the dot in Eq. (6–18)] is, by definition,

$$\sum_{\alpha=1}^{3} e_m(\alpha)e_n(\alpha) = \delta_{mn}, \qquad (6\text{–}19)$$

where $e_m(\alpha)$ is the component of the vector $\hat{e}_m$ in the direction of the coordinate axis designated by the index α ($\alpha = 1, 2, 3$). In Chapter 9, we shall see that a very useful generalization of Eq. (6–19) can be made, in which the components of the vectors are complex numbers. In this generalization, the scalar product of vectors $\mathbf{e}_m$ and $\mathbf{e}_n$ is, by definition,

$$(\mathbf{e}_m, \mathbf{e}_n) = \sum_{\alpha=1}^{N} e_m^*(\alpha)e_n(\alpha), \tag{6–20}$$

in which N is the dimension. The familiar vectors describing position, velocity, etc., are of course all three-dimensional, but there is nothing to prevent the construction of an algebraic theory of vectors in a space of an arbitrary number of dimensions. However, the "dot" notation will be reserved for the scalar product of real three-dimensional vectors. The complex conjugate of the component $e_m(\alpha)$ appears in the preceding definition to insure that the scalar product of a vector with itself, namely,

$$(\mathbf{e}_m, \mathbf{e}_m) = \sum_{\alpha} |e_m(\alpha)|^2,$$

be non-negative.

The analogy between Eq. (6–17) and the equation

$$\sum_{\alpha=1}^{N} e_m^*(\alpha)e_n(\alpha) = \delta_{mn} \tag{6–21}$$

is complete if the functions ψ_m and ψ_n are made to correspond to the vectors $\mathbf{e}_m$ and $\mathbf{e}_n$, respectively, while the integration variable x is considered to be the analogue of the index α. In other words, the function ψ can be regarded as a vector in a space of infinitely many dimensions whose axes are labeled by the index x. The "component of ψ in the direction x" is then the value $\psi(x)$ of the given function at the point x. Since x is a continuous index, the sum over α in Eq. (6–21) is replaced, in the analogue (6–17), by integration over x. The ideas suggested by these observations will be more fully developed in Section 9–8; they are sketched here to familiarize the reader with the terminology of the subject.

By analogy with Eq. (6–20), the scalar product of two functions ψ and ϕ is defined as

$$(\phi, \psi) = (\psi, \phi)^* = \int \phi^*\psi \, dx. \tag{6–22}$$

The parenthesis notation will indicate, in general, that the product $\phi^*\psi$ is to be integrated over the entire space in which the functions are defined.[1]

---

[1] Note that $(c\phi, \psi) = c^*(\phi, \psi)$, and $(\phi, c\psi) = c(\phi, \psi)$, where c is a complex constant.

Thus, for functions of three spatial variables (x, y, z),

$$(\phi, \psi) = \int_{-\infty}^{\infty} \int_{-\infty}^{\infty} \int_{-\infty}^{\infty} \phi^*(x, y, z)\psi(x, y, z)\, dx\, dy\, dz = \int \phi^*(\mathbf{r})\psi(\mathbf{r})\, d\mathbf{r},$$

or, if one deals with the momentum functions $a(\mathbf{p})$ and $b(\mathbf{p})$,

$$(a, b) = \int_{-\infty}^{\infty} \int_{-\infty}^{\infty} \int_{-\infty}^{\infty} a^*(p_x, p_y, p_z)b(p_x, p_y, p_z)\, dp_x\, dp_y\, dp_z$$

$$= \int a^*(\mathbf{p})b(\mathbf{p})\, d\mathbf{p}.$$

In this notation, the orthogonality relations are

$$(\psi_m, \psi_n) = \delta_{mn} \tag{6–23}$$

6–6 Expansion in eigenfunctions. We shall frequently have to deal with a linear combination of the members of an orthonormal set,

$$\sum_n f_n \psi_n(x), \tag{6–24}$$

where the f_n are a given set of complex numbers. In general, (6–24) is an infinite series and, if it is convergent, it defines a function of x:

$$f(x) = \sum_n f_n \psi_n(x). \tag{6–25}$$

In many cases, the members of a large class of functions can be expressed as sums of this form, and the functions ψ_n are said to be a *complete set* with respect to the members of such a class. The ψ_n are obviously complete with respect to the class of functions defined by the sets of numbers f_n for which the series (6–25) is convergent. In physical applications, it is often necessary to assume that all functions of physical interest can be expressed linearly in terms of a given orthonormal set. Proofs of completeness exist in special cases (for example, boundary-value problems of the Sturm-Liouville type), so that only general requirements (e.g., Dirichlet conditions) need to be satisfied for the expansions to be possible. When completeness is assumed on physical grounds without mathematical proof, the results are tentative to this extent.

If a function $f(x)$ can be expanded in the series (6–25), then the coefficients f_n are

$$f_n = (\psi_n, f), \tag{6–26}$$

since, due to the orthogonality relations (6–23),

$$(\psi_n, f) = \left(\psi_n, \sum_{n'} f_{n'} \psi_{n'}\right) = \sum_{n'} f_{n'}(\psi_n, \psi_{n'}) = \sum_{n'} f_{n'}\, \delta_{nn'} = f_n.$$

The expansion is convergent in the mean to $f(x)$ (Problem 6–6). One also has

$$(f, f) = \left(\sum_n f_n \psi_n, \sum_{n'} f_{n'} \psi_{n'} \right) = \sum_{n,n'} f_n^* f_{n'} (\psi_n, \psi_{n'})$$

$$= \sum_{n,n'} f_n^* f_{n'} \, \delta_{nn'} = \sum_n |f_n|^2, \quad (6\text{–}27)$$

or, more generally,

$$(f, g) = \sum_{n,n'} f_n^* g_{n'} (\psi_n, \psi_{n'}) = \sum_n f_n^* g_n, \qquad (6\text{–}28)$$

in which the numbers g_n are the expansion coefficients of the function $g(x)$. Equation (6–27) is called the *completeness relation* for the set ψ_n.

When the functions ψ_n are eigenfunctions of a quantum-mechanical operator representing an observable quantity in the sense of Section 6–3, the expansion (6–25) has a special significance, which we shall now investigate. Let ψ be any given state, and assume that the eigenfunctions ψ_n of the operator F_{op} are a complete set with respect to the states of the system, that is,

$$F_{\text{op}} \psi_n = \lambda_n \psi_n, \qquad (6\text{–}29)$$

where the numbers λ_n are the eigenvalues of F_{op}, and that ψ can be written

$$\psi = \sum_n a_n \psi_n; \qquad (6\text{–}30)$$

$$a_n = (\psi_n, \psi). \qquad (6\text{–}31)$$

If a measurement of the dynamical variable F is made on the system in the state ψ, the expectation of F is[1]

$$\langle F \rangle = (\psi, F_{\text{op}} \psi) = \left(\sum_n a_n \psi_n, F_{\text{op}} \sum_{n'} a_{n'} \psi_{n'} \right)$$

$$= \left(\sum_n a_n \psi_n, \sum_{n'} a_{n'} \lambda_{n'} \psi_{n'} \right)$$

$$= \sum_{n,n'} \lambda_{n'} a_n^* a_{n'} (\psi_n, \psi_{n'}) = \sum_n \lambda_n |a_n|^2. \quad (6\text{–}32)$$

Similarly, the expectation of F^2 is

$$\langle F^2 \rangle = (\psi, F_{\text{op}}^2 \psi) = \sum_n \lambda_n^2 |a_n|^2, \qquad (6\text{–}33)$$

---

[1] ψ is assumed to be normalized, so that $(\psi, \psi) = \sum |a_n|^2 = 1$.

and, in general,

$$\langle F^k \rangle = (\psi, F^k_{\mathrm{op}}\psi) = \sum_n \lambda_n^k |a_n|^2 \tag{6-34}$$

for any positive integer k.

According to the assumption of Section 6–3, however, the results of individual measurements of F are the eigenvalues λ_n, which have a statistical distribution depending upon ψ. If $P(\lambda_n)$ denotes the probability that an individual measurement yields the value λ_n, we have, by definition,

$$\langle F^k \rangle = \sum_n \lambda_n^k P(\lambda_n). \tag{6-35}$$

This relation is the same as Eq. (6–34) if we assume that

$$P(\lambda_n) = |a_n|^2, \tag{6-36}$$

i.e., that *the absolute square of the coefficient a_n in the expansion of ψ in eigenfunctions of F_{op} is the probability that a measurement of F will yield the eigenvalue λ_n.*

It is reasonable that the identification (6–36) is implied by Eqs. (6–34) and (6–35). Indeed, in statistical theory it is shown that, in a general sense, the distribution function for a statistical variable is uniquely determined by the expectations of all positive integral powers of the variable.[1] It is obviously true that the expectation of any polynomial in F is obtained correctly if the assumption (6–36) is made. Also, it is generally true that the set of all eigenfunctions of an operator which represents a physical observable is a complete set, so that the expansion (6–30) is valid.

The formal structure under discussion is well exemplified by the quantum description of a free particle (developed in Sections 2–8, 2–9, and 2–10). As we have seen, the function

$$\psi_p = \frac{1}{\sqrt{2\pi\hbar}} e^{(i/\hbar)px} \tag{6-37}$$

is an eigenfunction of the operator p_{op} belonging to the eigenvalue p. The Fourier transform

$$\psi(x) = \int a(p)\psi_p \, dp \tag{6-38}$$

is a representation of the state ψ as a linear superposition of the states ψ_p; in this case, it is the expansion (6–30). Note that the eigenvalues of p_{op}

---

[1] This is the "problem of moments" in statistics. See J. V. Uspensky, *Introduction to Mathematical Probability*. New York: McGraw-Hill Book Co., Inc., 1937, Appendix II.

form a continuum, so that the sum in Eq. (6–30) is replaced by an integral. The orthogonality relations for the functions ψ_p are (cf. Section 3–5)

$$(\psi_p, \psi_{p'}) = \frac{1}{2\pi\hbar} \int \psi_p^*(x)\psi_{p'}(x) \, dx = \delta(p - p');$$

because of the continuity of the eigenvalues, the δ_{mn} of Eq. (6–23) is replaced by the Dirac δ-function. In this case, formula (6–31) for the expansion coefficients becomes

$$a(p) = (\psi_p, \psi) = \frac{1}{\sqrt{2\pi\hbar}} \int \psi(x)e^{-(i/\hbar)px} \, dx,$$

which is the Fourier inversion formula [cf. Eq. (2–38)]. The interpretation of $|a(p)|^2$ as a probability distribution for the momentum has already been discussed in Section 2–10, and the relation (2–124), that is,

$$\langle p \rangle = (\psi, p_{op}\psi) = \int p|a(p)|^2 \, dp,$$

is the specialization of Eq. (6–32), in which λ_n is replaced by the eigenvalue p, and the sum changed to an integral.

6–7 Hermitian operators. An operator[1] F representing an observable quantity must, for every state ψ, yield an expectation value

$$\langle F \rangle = (\psi, F\psi), \tag{6–39}$$

which is a real number. Consequently, F must satisfy the condition

$$(\psi, F\psi) = (\psi, F\psi)^* = (F\psi, \psi) \tag{6–40}$$

for every function ψ to which it may be applied. A linear operator which obeys rule (6–40) is called a *Hermitian* operator. The rule is sufficient to insure that the eigenvalues of F are real, for if ψ is an eigenfunction of F belonging to the eigenvalue λ, then

$$(\psi, F\psi) = (\psi, \lambda\psi) = \lambda(\psi, \psi);$$

according to Eqs. (6–40), this expression is also equal to

$$(F\psi, \psi) = (\lambda\psi, \psi) = \lambda^*(\psi, \psi),$$

---

[1] The subscript $_{op}$ will henceforth be omitted unless required to avoid misunderstanding.

and, since $(\psi, \psi) \neq 0$,

$$\lambda = \lambda^*,$$

i.e., λ is real.

More generally, a Hermitian operator satisfies the relation (Problem 6–7)

$$(\psi, F\phi) = (F\psi, \phi), \tag{6–41}$$

that is, the operator can be applied to either factor in the scalar product (6–41).

Equation (2–119) proves that the operator p_{op} is Hermitian, and Eq. (3–15) is a special case of Eq. (6–41), namely,

$$(\psi, p^2\psi) = (p\psi, p\psi).$$

It is clear from these examples that the Hermitian property is associated in an essential way with the boundary conditions imposed upon the functions ψ and is not to be regarded as a property of the operator symbol itself (independent of the class of functions to which it is applied.)

It will now be shown that *two eigenfunctions of a Hermitian operator, belonging to different eigenvalues, are orthogonal.* Let A be a Hermitian operator, and ψ_1 and ψ_2 two eigenfunctions of A such that

$$A\psi_1 = \alpha_1\psi_1, \qquad A\psi_2 = \alpha_2\psi_2.$$

Forming the scalar product of the first of these relations with ψ_2, we obtain

$$(\psi_2, A\psi_1) = \alpha_1(\psi_2, \psi_1). \tag{6–42}$$

But A is Hermitian, and Eq. (6–41) yields

$$(\psi_2, A\psi_1) = (A\psi_2, \psi_1) = \alpha_2^*(\psi_2, \psi_1) = \alpha_2(\psi_2, \psi_1), \tag{6–43}$$

where the last equality holds because α_2 is real. Comparison of Eqs. (6–42) and (6–43) shows that

$$(\alpha_2 - \alpha_1)(\psi_2, \psi_1) = 0, \tag{6–44}$$

and since it has been assumed that $\alpha_2 \neq \alpha_1$,

$$(\psi_2, \psi_1) = 0.$$

This theorem guarantees that the orthogonality relations (6–23) are satisfied for a set of normalized eigenfunctions belonging to different eigenvalues α_n. Normalization of the ψ_n does not, of course, affect the validity of Eq. (6–29), which is homogeneous in ψ_n.

We have said that the set of all eigenfunctions of a given Hermitian operator is, in general, a complete set. There is no assurance, however, that each eigenfunction belongs to an eigenvalue which is different from all the rest. It is frequently true that an eigenvalue α belongs to each of two or more linearly independent eigenfunctions. When this happens, α is said to be degenerate. The degree of degeneracy is the number of linearly independent eigenfunctions which can be constructed. For example, in Section 5–1 we have found that two linearly independent functions [Eqs. (5–14) and (5–15)] could be formed, each belonging to the same total energy. Hence in this example, the energy is a doubly degenerate eigenvalue of H.

If ψ_1 and ψ_2 are two linearly independent eigenfunctions of A, belonging to the same eigenvalue α, then every linear combination of these functions is also an eigenfunction, for the equations

$$A\psi_1 = \alpha\psi_1, \qquad A\psi_2 = \alpha\psi_2$$

imply, since A is linear, that

$$A(c_1\psi_1 + c_2\psi_2) = c_1 A\psi_1 + c_2 A\psi_2 = \alpha(c_1\psi_1 + c_2\psi_2),$$

where c_1 and c_2 are arbitrary constants. By selecting these constants properly, we can always construct orthogonal eigenfunctions. Thus, the functions

$$\psi^{(1)} = c_1\psi_1 + c_2\psi_2, \tag{6–45}$$

$$\psi^{(2)} = c_3\psi_1 + c_4\psi_2, \tag{6–46}$$

are orthogonal provided

$$0 = (\psi^{(1)}, \psi^{(2)}) = c_1^* c_3(\psi_1, \psi_1) + c_1^* c_4(\psi_1, \psi_2) + c_2^* c_3(\psi_2, \psi_1)$$
$$+ c_2^* c_4(\psi_2, \psi_2); \tag{6–47}$$

this equation can be satisfied (in fact, in infinitely many ways) by proper choice of the constants c. By this device, the orthogonality relations (6–23) can be preserved for the eigenfunctions belonging to degenerate eigenvalues. However, this procedure for orthogonalization of degenerate states is arbitrary, and we shall see in the next section that another, more satisfactory, solution of this problem exists in cases of physical interest.

6–8 Simultaneous eigenfunctions; commutators. If A and B are linear operators, and ψ is a function satisfying both the equations

$$A\psi = \alpha\psi, \qquad B\psi = \beta\psi, \tag{6–48}$$

then ψ is a *simultaneous* eigenfunction of A and B, belonging to the eigenvalues α and β, respectively. For example, the wave function (6–37),

$$\psi_p = \frac{1}{\sqrt{2\pi\hbar}} e^{(i/\hbar)px},$$

satisfies simultaneously the equations

$$p_{\text{op}}\psi = p\psi \quad \text{and} \quad H_{\text{op}}\psi = E\psi,$$

where $H_{\text{op}} = p_{\text{op}}^2/2m$ is the Hamiltonian operator for a free particle. Equations (6–48) imply that

$$BA\psi = B(\alpha\psi) = \alpha B\psi = \alpha\beta\psi,$$

and

$$AB\psi = A(\beta\psi) = \beta A\psi = \beta\alpha\psi,$$

whence, by subtraction,

$$(AB - BA)\psi = 0. \tag{6–49}$$

This equation shows that ψ is also an eigenfunction of the operator $(AB - BA)$, belonging to the eigenvalue zero. The condition (6–49) is necessary in order that ψ be a simultaneous eigenfunction of A and B.

The operator in Eq. (6–49) is called the *commutator* of A and B and is written, for brevity,

$$[A, B] = AB - BA. \tag{6–50}$$

The commutator $[A, B]$ satisfies the following identities:

$$[A, B] = -[B, A], \tag{6–51}$$

$$[A, BC] = [A, B]C + B[A, C], \tag{6–52}$$

$$[AB, C] = [A, C]B + A[B, C], \tag{6–53}$$

$$[A,[B, C]] + [B,[C, A]] + [C,[A, B]] = 0, \tag{6–54}$$

which are readily deduced from the definition (6–50) (Problem 6–10).

Two operators satisfying the equation

$$[A, B] = 0 \tag{6–55}$$

are said to *commute*. This equation means that Eq. (6–49) is true for every ψ which is a member of the class of functions under consideration.

The eigenfunctions of commuting operators can always be constructed in such a way that they are simultaneous eigenfunctions. Thus, suppose that ψ

is an eigenfunction of A, that is,

$$A\psi = \alpha\psi,$$

and that A and B commute. By multiplying this equation on the left with B, we obtain

$$BA\psi = AB\psi = B\alpha\psi = \alpha B\psi,$$

or,

$$A(B\psi) = \alpha(B\psi);$$

this shows (if we assume $B\psi \neq 0$) that the function $B\psi$ is an eigenfunction of A belonging to the same eigenvalue α. Now two fundamentally different cases present themselves: (1) If the eigenvalue α is nondegenerate, i.e., if there is only one state ψ corresponding to α, then the function $B\psi$ can differ from ψ only by a constant multiplier, that is,

$$B\psi = \beta\psi, \qquad \beta \text{ constant.} \tag{6–56}$$

Hence, if ψ is a nondegenerate eigenfunction of A, then Eq. (6–49) is both a necessary and a sufficient condition for ψ to be a simultaneous eigenfunction of A and B. (2) The state ψ may be degenerate. In this case, the argument leading to Eq. (6–56) cannot be made. However, we have seen in the last section that every linear combination of the degenerate eigenfunctions of A is also an eigenfunction of A, and it is possible to choose linear combinations in such a way that Eq. (6–56) is satisfied. Suppose, for example, that α is doubly degenerate, having linearly independent eigenfunctions ψ_1 and ψ_2 which may, for convenience, be assumed to be orthonormal. We can attempt to determine the constants c_1 and c_2 in the linear combination

$$\psi = c_1\psi_1 + c_2\psi_2 \tag{6–57}$$

in such a way that Eq. (6–56), i.e.,

$$c_1 B\psi_1 + c_2 B\psi_2 = \beta(c_1\psi_1 + c_2\psi_2)$$

is satisfied. Taking the scalar product of this equation with ψ_1 and ψ_2, respectively, we obtain

$$c_1(B_{11} - \beta) + c_2 B_{12} = 0,$$
$$c_1 B_{21} + c_2(B_{22} - \beta) = 0, \tag{6–58}$$

where $(\psi_1, B\psi_1) = B_{11}$, $(\psi_1, B\psi_2) = B_{12}$, etc. These simultaneous linear equations determine c_1 and c_2. A nontrivial solution exists provided

$$\begin{vmatrix} B_{11} - \beta & B_{12} \\ B_{21} & B_{22} - \beta \end{vmatrix} = 0. \tag{6–59}$$

This equation, which is quadratic in β, has two roots, say β_1 and β_2; for each of these roots constants satisfying Eqs. (6–58) can be found (Problem 6–13). If these roots are not the same, the eigenfunctions of B constructed in this way are orthogonal (Section 6–7)[1] and therefore linearly independent, so that the degeneracy has been resolved by means of the commuting operator B. If the roots of the determinantal equation (6–59) are equal, then the functions ψ_1 and ψ_2 are also simultaneous eigenfunctions of B (Problem 6–15). In this case, of course, nothing further is learned about the nature of the degeneracy.

This procedure can be generalized immediately to cases in which the degree of degeneracy is larger than two provided it is finite. The equation corresponding to Eq. (6–59) for an N-fold degenerate state is of Nth degree in β and has N roots. If these are all distinct, the degeneracy is completely resolved. If they are not, then it may be possible to discover a third operator C which commutes with both A and B; by means of C the remaining degeneracy is further resolved, and so on. Ultimately, we can expect to obtain a *complete set of commuting operators* which determines a complete set of orthonormal functions, in terms of which every state of the system is expressible in the form (6–30).

These considerations lead to a rather complete picture of the quantum-mechanical description of a physical system. Given a physical system, the Hamiltonian function is determined classically, and the Schrödinger equation is constructed by means of the operator correspondence (6–6). One then has an eigenvalue problem to determine the eigenfunctions and eigenvalues of the Hamiltonian, i.e., the possible states of definite energy. By assumption, these states are a complete set permitting, in principle, the description of every state by superposition. However, the energy states (or some of them) may be degenerate. If this is the case, one seeks other operators which commute with the Hamiltonian and resolve the degeneracies, so that ultimately every eigenfunction is uniquely specified by the list of the eigenvalues to which it belongs. It is assumed that such a set of commuting Hermitian operators always exists and is defined by the physical properties of the system, and furthermore, that physically meaningful descriptions of all states can always be made by means of an expansion in terms of the eigenfunctions of a complete set of commuting operators.[2] The justification of these assumptions poses difficult and complex mathematical problems,[3] especially in the case of operators having eigenvalues

---

[1] Throughout this discussion, A and B are assumed to be Hermitian.

[2] P. A. M. Dirac, *The Principles of Quantum Mechanics*. 3rd ed. Oxford: The Clarendon Press, 1947. V. Rojansky, *Introductory Quantum Mechanics*. New York: Prentice-Hall, Inc., 1946, p. 260.

[3] J. von Neumann, *Mathematical Foundations of Quantum Mechanics*. Princeton: Princeton University Press, 1955.

which form a continuum (e.g., the operator p_{op}). The reader who is concerned with the theoretical background of the subject will devote future study to these questions.

It has been shown in Chapter 5 that the quantum states of a one-dimensional system are at most doubly degenerate. Therefore, all degeneracy can be resolved if a single operator can be found that commutes with the Hamiltonian and has different eigenvalues in the two degenerate states. In cases of symmetry, i.e., for potential functions which satisfy

$$V(x) = V(-x), \tag{6-60}$$

it has also been demonstrated that every eigenfunction of the energy can be written as a linear combination of an even and an odd function of x. This fact, an example of the foregoing general theory, can be explained in terms of the *reflection operation*, which we shall now describe by means of a linear operator.

6–9 The parity operator. Let the operator Π be defined, for functions of the variable x, by

$$\Pi\psi(x) = \psi(-x). \tag{6-61}$$

The operator Π is linear, for

$$\Pi[\psi_1(x) + \psi_2(x)] = \psi_1(-x) + \psi_2(-x) = \Pi\psi_1(x) + \Pi\psi_2(x)$$

and

$$\Pi[c\psi(x)] = c\psi(-x) = c\Pi\psi(x),$$

for every function ψ. Moreover, Π is Hermitian, for it follows from the definition (6–22) of the scalar product that

$$(\Pi\psi, \phi) = \int_{-\infty}^{\infty} \psi^*(-x)\phi(x)\,dx = \int_{-\infty}^{\infty} \psi^*(x')\phi(-x')\,dx',$$

in which the second integral is obtained from the first by the substitution $x' = -x$. Since the value of the integral is unaffected by renaming the variable of integration, we have

$$(\Pi\psi, \phi) = \int_{-\infty}^{\infty} \psi^*(x)\phi(-x)\,dx = (\psi, \Pi\phi),$$

and Eq. (6–41) is satisfied by Π. The eigenvalues of Π are therefore real. Indeed, the eigenvalue equation $\Pi\psi = \lambda\psi$ can be multiplied by Π, yielding

$$\Pi^2\psi = \lambda\Pi\psi = \lambda^2\psi = \psi.$$

The last equality follows from the fact that the original function is restored when the transformation $x \to -x$ is performed twice. Consequently, the eigenvalues of Π satisfy the equation $\lambda^2 = 1$, whence $\lambda = \pm 1$. The eigenfunctions belonging to the eigenvalue $\lambda = +1$ of the parity operator are the even functions ψ_e, which satisfy

$$\psi_e(x) = \psi_e(-x). \qquad (6\text{-}62)$$

The odd functions ψ_o belong to the eigenvalue $\lambda = -1$ and satisfy the relation

$$\psi_o(x) = -\psi_o(-x). \qquad (6\text{-}63)$$

The eigenfunctions of Π are a complete set with respect to the class of functions of x. This is shown by the identity

$$\psi(x) = \tfrac{1}{2}[\psi(x) + \psi(-x)] + \tfrac{1}{2}[\psi(x) - \psi(-x)], \qquad (6\text{-}64)$$

which expresses $\psi(x)$ as a linear combination of an even function of x, $[\psi(x) + \psi(-x)]$, and an odd function of x, $[\psi(x) - \psi(-x)]$.

Finally, if the potential-energy function satisfies $V(x) = V(-x)$ [Eq. (6-60)], then Π commutes with the Hamiltonian. Thus, writing

$$H = \frac{p^2}{2m} + V(x),$$

we have

$$H\psi(x) = -\frac{\hbar^2}{2m}\frac{d^2\psi(x)}{dx^2} + V(x)\psi(x),$$

and

$$\Pi H\psi(x) = -\frac{\hbar^2}{2m}\,\Pi\frac{d^2\psi(x)}{dx^2} + \Pi V(x)\psi(x)$$

$$= -\frac{\hbar^2}{2m}\frac{d^2\psi(-x)}{dx^2} + V(-x)\psi(-x)$$

$$= -\frac{\hbar^2}{2m}\frac{d^2\psi(-x)}{dx^2} + V(x)\psi(-x) = H\Pi\psi(x),$$

whence, since this equation is true for arbitrary ψ,

$$H\Pi - \Pi H = [H, \Pi] = 0. \qquad (6\text{-}65)$$

The consequences of this result have already been pointed out in Chapter 5: Nondegenerate energy states are automatically either even or odd in

x and hence are eigenfunctions of Π; degenerate states, on the other hand, can be resolved into an even and an odd eigenfunction of Π [Eq. (6–64)].

It is apparent that Eq. (6–65) expresses the *symmetry* of the physical system: The description of the system is not changed by the substitution $x \rightarrow -x$. It happens very frequently that a geometrical property of this kind lies at the root of a degeneracy.

As a further example, consider the equation

$$[H, p] = 0 \qquad \left(H = \frac{p^2}{2m} \right), \qquad (6\text{--}66)$$

which is true for a free particle. This equation states that the position of the origin of x does not affect the theoretical description of the system. In other words, if $\psi(x)$ is an eigenfunction of H belonging to the energy eigenvalue E, i.e., if

$$H\psi(x) = E\psi(x), \qquad (6\text{--}67)$$

then, by symmetry, the function $\psi(x + \epsilon)$ is also an eigenfunction belonging to the same energy:

$$H\psi(x + \epsilon) = E\psi(x + \epsilon). \qquad (6\text{--}68)$$

Here, ϵ is any real number. Now if ϵ is small, we have

$$\psi(x + \epsilon) = \psi(x) + \epsilon \frac{d\psi}{dx} = \psi(x) + \epsilon \frac{i}{\hbar} p\psi(x),$$

and Eq. (6–68) becomes

$$H\psi(x) + \epsilon \frac{i}{\hbar} Hp\psi(x) = E\psi(x) + \epsilon \frac{i}{\hbar} pE\psi(x).$$

Hence from Eq. (6–67),

$$\epsilon \frac{i}{\hbar} Hp\psi(x) = \epsilon \frac{i}{\hbar} pH\psi(x),$$

or, if we cancel a common factor and transpose,

$$(Hp - pH)\psi(x) = 0. \qquad (6\text{--}69)$$

However, since the eigenfunctions of H are a complete set of functions for the free particle, Eq. (6–69) is true for every state ψ, whence

$$[H, p] = 0. \qquad (6\text{--}70)$$

Hence the eigenfunctions of p [Eq. (6–37)] must also be eigenfunctions of H; this is a result of the invariance of the system to the symmetry transformation $x \rightarrow x + \epsilon$.

6–10 The fundamental commutation rule. The connection between the classical and quantum theories has been established by means of the correspondence relations

$$x \rightarrow x_{op}, \qquad p \rightarrow p_{op} = \frac{\hbar}{i} \frac{d}{dx} \qquad (6\text{–}71)$$

for the basic quantities in terms of which observables are expressed. The operators x_{op} and p_{op} satisfy the *commutation rule*

$$[x, p] = i\hbar, \qquad (6\text{–}72)$$

which, upon introduction of the factor $i\hbar$, follows immediately from Eq. (6–4).

The relation (6–72) is fundamental; it may be considered the basis for most of the formal structure of quantum theory. This point has been emphasized by Dirac. The commutation rule for x and p is, in a sense, a precise formulation of the correspondence principle. This is most clearly seen in the Hamiltonian formulation of classical mechanics. The *Poisson bracket*

$$\{f, g\} = \frac{\partial f}{\partial x} \frac{\partial g}{\partial p} - \frac{\partial f}{\partial p} \frac{\partial g}{\partial x}, \qquad (6\text{–}73)$$

in which $f(x, p)$ and $g(x, p)$ are functions of the coordinate x and momentum p,[1] plays a fundamental role in this theory. For example, the equations of motion for a particle whose Hamiltonian function is

$$H = \frac{p^2}{2m} + V(x), \qquad (6\text{–}74)$$

are (Problem 6–16)

$$\dot{x} = \{x, H\}, \qquad \dot{p} = \{p, H\}. \qquad (6\text{–}75)$$

Furthermore, it is evident that

$$\{x, p\} = 1. \qquad (6\text{–}76)$$

---

[1] The discussion is restricted to a one-particle system. For generalization to many-particle systems, see Chapter 7.

It can be shown[1] that the identities (6–51)–(6–54) are satisfied if the commutator symbols in these equations are replaced by the corresponding Poisson brackets, and the symbols A, B, C are interpreted as functions of the classical variables x and p.

The parallelism between Poisson brackets and quantum-mechanical commutators allows an elegant expression of the correspondence principle, first formulated by Dirac. *The quantum-mechanical operators f_{op} and g_{op} which, in quantum theory, replace the classically defined functions f and g, must always be such that the commutator of f_{op} and g_{op} corresponds to the Poisson bracket of f and g according to*

$$i\hbar\{f, g\} \rightarrow [f_{op}, g_{op}]. \qquad (6\text{–}77)$$

For example: By this assumption, the relation (6–72),

$$[x, p] = i\hbar,$$

is a consequence of Eq. (6–76),

$$\{x, p\} = 1.$$

Every classical equation in Poisson brackets has its quantum counterpart in the operator formalism. Thus the operator equation $[H_{op}, p_{op}] = 0$, which is true for a free particle (Problem 6–11), implies that

$$\{H, p\} = 0, \qquad (6\text{–}78)$$

where $H = p^2/2m$ is the Hamiltonian function for a free particle. Note that, as a consequence, $\dot{p} = 0$ [Eq. (6–75)], that is, the momentum of a free particle is a constant of the motion in the classical theory.

The generalization of these rules to three dimensions is immediate. If the coordinates of the particle are denoted by x_1, x_2, x_3 and the momenta by p_1, p_2, p_3, then the Poisson bracket is

$$\{f, g\} = \sum_{i=1}^{3} \left(\frac{\partial f}{\partial x_i} \frac{\partial g}{\partial p_i} - \frac{\partial f}{\partial p_i} \frac{\partial g}{\partial x_i} \right). \qquad (6\text{–}79)$$

The functions f and g depend upon the six independent variables x_i, p_i, and the partial derivative signs refer to this set of variables. For example,

---

[1] H. Goldstein, *Classical Mechanics*. Reading, Mass.: Addison-Wesley Publishing Co., Inc., 1950, Sections 8–5 and 8–6.

$\partial f/\partial x_1$ means: Differentiate f with respect to x_1, treating $x_2, x_3,$ and $p_1, p_2,$ p_3 as constants.

In this notation, the Hamiltonian function for a particle of potential energy $V(x, y, z) = V(x_1, x_2, x_3)$ is

$$H = \sum_i \frac{p_i^2}{2m} + V(x_1, x_2, x_3), \qquad (6\text{--}80)$$

and the equations of motion, in Poisson brackets, are

$$\dot{x}_i = \{x_i, H\}, \qquad \dot{p}_i = \{p_i, H\}. \qquad (6\text{--}81)$$

By means of the identity (6–79), the *fundamental Poisson bracket relations*

$$\{x_i, p_j\} = \delta_{ij} \qquad (6\text{--}82)$$

are easily obtained. Rule (6–77) now leads to the commutation rules for the quantum operators x_i, p_i:

$$[x_i, p_j] = i\hbar\, \delta_{ij}. \qquad (6\text{--}83)$$

The wave function for a three-dimensional system is a function of the three position coordinates (x_1, x_2, x_3), and Eq. (6–83) is seen to be consistent with

$$p_i \rightarrow \frac{\hbar}{i} \frac{\partial}{\partial x_i}; \qquad (6\text{--}84)$$

this can also be inferred from the work of Section 2–10. The operator $\partial/\partial x_i$ in this relation is, of course, to be applied to the wave function ψ, and $\partial\psi/\partial x_1$ therefore means: Differentiate ψ with respect to x_1, treating x_2 and x_3 as constants.

According to Eq. (6–83), noncommuting coordinate and momentum operators occur in *conjugate pairs*. Thus, x_1 and p_1 satisfy

$$[x_1, p_1] = i\hbar, \qquad (6\text{--}85)$$

i.e., they do not commute; but for x_1 and p_2 we have

$$[x_1, p_2] = 0, \qquad (6\text{--}86)$$

so that the order of these two operators can be changed without affecting the result. In the next section, we shall show that two quantities represented by noncommuting operators always are the complementary variables in a *principle of uncertainty*. They cannot be measured simultaneously with arbitrary precision. Therefore, the classical concept of a pair of canonically conjugate variables, related by[1]

$$p_i = \frac{\partial L}{\partial \dot{x}_i},$$

where L is the Lagrangian function, acquires fundamental significance in quantum mechanics. The limitations upon the theoretical definition of state, due to the wave-like properties of the system, are expressed directly by the commutation rules in terms of conjugate variables.

6-11 Equations of motion. According to the definition of quantum operators by the correspondence (6–77), these operators do not change with the time, but are functions of x_{op} and p_{op} alone. The time dependence of the states is contained in the wave functions. Hence, the time derivative of a classical variable, such as $\dot{p} = dp/dt$ in Eq. (6–75), does not have an obvious quantum-mechanical interpretation. Nevertheless, such an interpretation is possible if *the symbol $\dot{A}$ is defined to mean an operator whose expectation in any state ψ is the time derivative of the expectation of the operator A.*

The time dependence of the state ψ is governed by the Schrödinger equation (Section 4–3):

$$H\psi = i\hbar \frac{\partial \psi}{\partial t}.$$

The expectation of an operator A is

$$\langle A \rangle = (\psi, A\psi).$$

Consequently, the expectation changes at the rate

$$\frac{d\langle A \rangle}{dt} = \left(\frac{\partial \psi}{\partial t}, A\psi \right) + \left(\psi, A \frac{\partial \psi}{\partial t} \right) = \frac{i}{\hbar} \{ (H\psi, A\psi) - (\psi, AH\psi) \}.$$

Since H is a Hermitian operator, this equation may be written

$$\frac{d\langle A \rangle}{dt} = \frac{i}{\hbar} (\psi, (HA - AH)\psi),$$

whence it follows from the above definition that

$$\dot{A} = \frac{i}{\hbar} [H, A], \qquad\qquad (6\text{–}87)$$

---

[1] H. Goldstein, *Classical Mechanics.* Reading, Mass.: Addison-Wesley Publishing Co., Inc., 1950, p. 48.

i.e., the time derivative of A is the commutator of H and A, multiplied by $i/\hbar$.

It is clear that the classical equations of motion (6–75) are the classical counterparts of Eq. (6–87) obtained by means of the correspondence (6–77). The equation (6–87) is called the equation of motion for A. The steps leading to this equation are a generalization of the derivation of Ehrenfest's relations, discussed in Section 4–6.

It has already been seen that operators which commute with the Hamiltonian function are of special importance because their eigenfunctions are simultaneously eigenfunctions of H. Equation (6–87) shows further that the time derivative of such an operator is zero, i.e., the operator represents a constant of the motion. The momentum operator p_{op}, for example, commutes with the Hamiltonian for a free particle, and the momentum is therefore independent of time. In general, the probability distribution function $P(\lambda)$ for the eigenvalues of an operator which commutes with H is independent of time for every state. Thus, the momentum distribution for a one-particle system is $|a(p)|^2$; it is an essential feature of the wave-packet formalism that, for a free particle, this quantity does not contain t (Sections 2–6 and 2–8). If the particle is not free, then $[H, p_{\text{op}}] \neq 0$, and the momentum distribution changes in time. This is expressed by Ehrenfest's relations (Section 4–6), which are the quantum equivalent of Newton's laws of motion.

It has been shown in Section 3–7, however, that the probability distribution in x is time-dependent, and this fact agrees with the commutation relation

$$\frac{i}{\hbar} [H, x] = \frac{p}{m} \neq 0. \tag{6–88}$$

6–12 Commutation rules and the uncertainty principle. We have explained in Section 6–8 that commuting Hermitian operators have simultaneous eigenfunctions. This means that the quantities corresponding to two operators can be arbitrarily well defined in the same state if and only if these operators commute. If the commutator of A and B is not zero, i.e., if

$$[A, B] = iC, \tag{6–89}$$

where the Hermitian operator C is not the zero operator, then an eigenfunction of A is not, in general,[1] an eigenfunction of B, but rather a linear combination of several eigenfunctions. The results of the measurements performed on B will then define a probability distribution for which the width ΔB is finite. For example, the operators p and x do not commute,

---

[1] It may happen that, for a particular state ψ, $C\psi = 0$, but this is exceptional.

and it has been shown that these quantities are subject to the uncertainty relation

$$\Delta x \, \Delta p \geq \frac{\hbar}{2}.$$

We shall now demonstrate that this situation is general:[1] The commutation rule (6–89) implies that, for any state ψ,

$$\Delta A \, \Delta B \geq \tfrac{1}{2} \langle C \rangle. \tag{6–90}$$

The quantity $\langle C \rangle = (\psi, C\psi)$ is the expectation of C in the (normalized) state ψ. To prove this, we introduce the operator

$$D = A + \alpha B + i\beta B,$$

in which α and β are arbitrary real numbers. The scalar product $(D\psi, D\psi)$ is the norm of the function $D\psi$; hence it is not negative. Using the Hermitian property of A and B, we can write this statement (Problem 6–21) in the form

$$(D\psi, D\psi) = \langle A^2 \rangle + (\alpha^2 + \beta^2)\langle B^2 \rangle + \alpha \langle C' \rangle + \beta \langle C \rangle \geq 0, \tag{6–91}$$

in which $C' = AB + BA$ (C' is called the *anticommutator* of A and B). If $B\psi \neq 0$, this expression can be rearranged as follows:

$$\langle A^2 \rangle + \langle B^2 \rangle \left(\alpha + \frac{1}{2} \frac{\langle C' \rangle}{\langle B^2 \rangle} \right)^2$$
$$+ \langle B^2 \rangle \left(\beta + \frac{1}{2} \frac{\langle C \rangle}{\langle B^2 \rangle} \right)^2 - \frac{1}{4} \frac{\langle C' \rangle^2}{\langle B^2 \rangle} - \frac{1}{4} \frac{\langle C \rangle^2}{\langle B^2 \rangle} \geq 0. \tag{6–92}$$

This inequality holds for every value of α and β; in particular, we can choose these numbers in such a way that the quantities in parentheses are both zero, and the inequality becomes

$$\langle A^2 \rangle \langle B^2 \rangle \geq \tfrac{1}{4}(\langle C \rangle^2 + \langle C' \rangle^2) \geq \tfrac{1}{4}\langle C \rangle^2. \tag{6–93}$$

The widths of the distributions in A and B are defined by

$$(\Delta A)^2 = \langle A^2 \rangle - \langle A \rangle^2, \qquad (\Delta B)^2 = \langle B^2 \rangle - \langle B \rangle^2,$$

whence, in the special case $\langle A \rangle = \langle B \rangle = 0$, (6–92) reduces immediately to the result (6–90), which was to be proved.

---

[1] A. Gamba, *Nuovo Cimento* **7**, 378 (1950).

If $\langle A \rangle$ and $\langle B \rangle$ are not zero, then the operators

$$A' = A - \langle A \rangle \cdot 1, \qquad B' = B - \langle B \rangle \cdot 1$$

also satisfy the commutation rule (6–89), that is,

$$[A', B'] = iC.$$

The expectations of A' and B' are obviously zero. Hence, the above reasoning can be applied to these operators. If this is done, and the operators A and B are restored, the result is (Problem 6–22)

$$(\Delta A)^2 (\Delta B)^2 \geq \left(\frac{C}{2} \right)^2 + \left(\frac{\langle C' \rangle}{2} - \langle A \rangle \langle B \rangle \right)^2. \qquad (6\text{–}94)$$

This inequality, from which (6–90) follows, is more restrictive than (6–90).[1] Incidentally, note that ΔB can be finite for an eigenstate of A if and only if both terms in the right-hand member of (6–94) are zero for the state ψ. Hence, precise definition of A implies, in general, that B is not even approximately well defined since the corresponding probability distribution has an infinite width.

6–13 Remarks on the correspondence principle. The reader may have been led to believe that the role of the correspondence principle consists merely in providing a connection with classical theory and in suggesting the direction to be taken in formulating quantum mechanics, so that, in the end, the principle will be discarded when the new theory is completely developed. However, this is not true, and the correspondence principle is continually required. Any description of a physical system is ultimately classical because all experiments are performed in the macroscopic world. The correspondence principle supplies the link by means of which the quantum-mechanical description is formulated from a classical model. Specifically, the form of the Hamiltonian operator is dictated by the classical model of the system. For example, the operator equivalent to the classical Hamiltonian (6–74) is the starting point for the quantum theory of a one-dimensional system. To be sure, modifications of the description may be required which are of purely quantum-mechanical origin and have no classical analogue (e.g., the electron spin). Nevertheless, the construction of the Hamiltonian function always requires a reference to the classical variables, and the correspondence principle is therefore always an essential part of the theory.

---

[1] The significance of the last term in (6–94) is discussed in: D. Bohm, *Quantum Theory*. New York: Prentice-Hall, Inc., 1951, p. 205.

An important attempt, initiated by Heisenberg,[1] has been made to arrive at a formulation of quantum mechanics in which only observable quantities enter the theoretical description, such as the cross sections for scattering and other processes and the half-lives and energies of quantum states. This "S-matrix" theory has been of great value in clarifying theoretical conclusions in a language which is independent of the concept of wave function. It is interesting, however, that there seems to be no simple way of incorporating the correspondence principle into this theory. Therefore, Heisenberg's original goal to replace the conventional structure of quantum mechanics has not been reached.

REFERENCES

KRAMERS, H. A., Die Grundlagen der Quantentheorie. Leipzig: Akademische Verlagsgesellschaft m. b. H., *Leipzig*, 1938.

ROJANSKY, VLADIMIR, *Introductory Quantum Mechanics*. New York: Prentice-Hall, Inc., 1946. Chapter I contains a clear discussion of operator algebra.

PROBLEMS

6–1. Show that x and d/dx are linear operators.

6–2. Prove that the equation

$$A_{op}\psi(x) = \int_a^b G(x, x')\psi(x')\,dx'$$

defines a linear operation. $G(x, x')$ is a *given* function of x, x', the same for all $\psi(x)$.

6–3. Show that an operator formed by addition and multiplication of a set of given linear operators is a linear operator.

6–4. Prove that an operator cannot have more than one reciprocal.

6–5. Let the operation A_{op} be defined (cf. Problem 6–2) by

$$A_{op}\psi(x) = \int_{-\infty}^{\infty} U(x - x')\psi(x')\,dx,$$

where the function $U(x)$ is the *unit step function:*

$$U(x) = \begin{cases} 0, & x < 0 \\ 1, & x > 0. \end{cases}$$

If the class of functions ψ is the class for which ψ and $d\psi/dx$ are integrable for $|x| \to \infty$, show that

$$A_{op} = \left(\frac{d}{dx}\right)^{-1}.$$

[1] W. Heisenberg, Z. *Physik* **120,** 513 and 673 (1943); Z. *Naturforschung* **1,** 608 (1946).

6–6. Show that the completeness relation (6–27) is a necessary and sufficient condition for the expansion (6–25) to be convergent in the mean to the function $f(x)$ (cf. Section 2–4).

6–7. Prove Eq. (6–41). [Hint: Apply rule (6–40) to the functions $\psi + \phi$ and $\psi + i\phi$ and compare.]

6–8. Show that the operator

$$H = \frac{p^2}{2m} + V(x)$$

is Hermitian with respect to the class of integrable, twice differentiable functions of x. [Cf. Problem 3–2, Chapter 3, and Eq. (3–15).]

6–9. Construct an explicit solution of Eq. (6–47) and generalize the result for the case of more than two degenerate eigenfunctions. Look up the Schmidt orthogonalization process in a mathematics text.[1]

6–10. Prove the identities (6–51) to (6–54).

6–11. Show that $[H_{op}, p_{op}] = 0$, where $H_{op} = p_{op}^2/2m$.

6–12. Prove that the operator $i[A, B]$ is Hermitian if A and B are Hermitian.

6–13. Solve the quadratic equation (6–59) and construct the eigenfunctions of β_1 and β_2 explicitly. Show that they are orthogonal and normalize them.

6–14. Prove that the roots of Eq. (6–59) are real if B is a Hermitian operator.

6–15. If B is Hermitian, show that the conditions

$$B_{11} = B_{22}, \qquad B_{12} = 0$$

are necessary and sufficient for the roots of Eq. (6–59) to be equal. Hence, prove that, in this case, ψ_1 and ψ_2 are simultaneous eigenfunctions of B as well as of A.

6–16. Verify by means of the definition (6–73) that the equations (6–75) are the equations of motion: $\dot{x} = p/m$, $\dot{p} = -\partial V/\partial x$. Also prove that

$$\dot{f} = \{f, H\},$$

where $f = f(x, p)$ is any function of x and p.

6–17. By mathematical induction on n, show that

$$[x^n, p] = i\hbar n x^{n-1},$$

where n is any positive integer. Hence, prove that

$$[f(x), p] = i\hbar \frac{\partial f}{\partial x},$$

where $f(x)$ is any polynomial in x.

---

[1] R. Courant and D. Hilbert, *Methods of Mathematical Physics*. New York: Interscience Publishers, Inc., 1953, Vol. I, p. 4 and p. 50. E. T. Whittaker and G. N. Watson, *A Course of Modern Analysis*. 4th ed. Cambridge: Cambridge University Press, 1927, Section 11–6.

6–18. Prove that

$$[x, p^n] = i\hbar n p^{n-1}$$

and, similarly to Problem 6–17, that

$$[x, f(p)] = i\hbar \frac{\partial f}{\partial p},$$

where f is a polynomial in p. These results can be extended to much larger classes of functions and are assumed to be generally true. They are very useful in the evaluation of commutators.

6–19. From the equation of motion (6–87) and the result of Problem 6–17, deduce

$$\dot{p} = -\frac{\partial V}{\partial x}, \qquad \dot{x} = \frac{p}{m},$$

where $H = p^2/2m + V$ is the Hamiltonian operator for a one-dimensional system (cf. Section 4–6).

6–20. Show that the commutation rule (6–72) is satisfied by the operators

$$p_{op} = p, \qquad x_{op} = -\frac{\hbar}{i}\frac{\partial}{\partial p},$$

in which it is implied that the operands are the wave functions $a(p)$ in momentum space (Section 2–10).

6–21. Derive the inequality (6–91) and show that it is equivalent to (6–92).

6–22. Verify the inequality (6–94) by the method indicated in the text.

6–23. Show that $[A, 1/B] = -(1/B)[A, B](1/B)$.

CHAPTER 7

SPHERICALLY SYMMETRIC SYSTEMS

7–1 The Schrödinger equation for spherically symmetric potentials. The potential energy of a particle which moves in a central, spherically symmetric field of force depends only upon the distance r between the particle and the center of force. The Schrödinger equation for the energy states of such a system is therefore

$$\nabla^2 \psi + \frac{2m}{\hbar^2}[E - V(r)]\psi = 0. \tag{7-1}$$

Spherical polar coordinates (Fig. 7–1),

$$x = r \sin \theta \cos \phi,$$
$$y = r \sin \theta \sin \phi,$$
$$z = r \cos \theta, \tag{7-2}$$

are appropriate to the symmetry of the problem, since the potential function $V(r)$ is independent of the angular variables θ and ϕ. The Schrödinger equation (7–1), expressed in these coordinates, is

$$\frac{1}{r^2}\frac{\partial}{\partial r}\left(r^2 \frac{\partial \psi}{\partial r}\right) + \frac{1}{r^2}\left[\frac{1}{\sin\theta}\frac{\partial}{\partial\theta}\left(\sin\theta \frac{\partial\psi}{\partial\theta}\right) + \frac{1}{\sin^2\theta}\frac{\partial^2\psi}{\partial\phi^2}\right]$$
$$+ \frac{2m}{\hbar^2}[E - V(r)]\psi = 0. \tag{7-3}$$

The energy states of the system are determined by those solutions of this equation which are continuous, have continuous derivatives in r, θ, and ϕ, and are, for bound states, quadratically integrable.

Solutions of Eq. (7–3) can be constructed by the method of separation of variables. To apply this method, we attempt to find a solution of the form

$$\psi = R(r)Y(\theta, \phi), \tag{7-4}$$

in which $R(r)$ is independent of the angles, and $Y(\theta, \phi)$ is independent

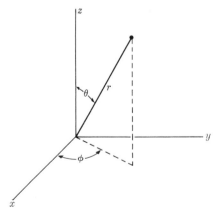

FIG. 7–1. Spherical polar coordinate system.

of r. Substituting Eq. (7–4) into Eq. (7–3) and rearranging, we obtain

$$\frac{1}{R}\left[\frac{d}{dr}\left(r^2\frac{dR}{dr}\right) + \frac{2m}{\hbar^2}\,[E - V(r)]r^2 R\right]$$

$$= -\frac{1}{Y}\left[\frac{1}{\sin\theta}\frac{\partial}{\partial\theta}\left(\sin\theta\frac{\partial Y}{\partial\theta}\right) + \frac{1}{\sin^2\theta}\frac{\partial^2 Y}{\partial\phi^2}\right]. \qquad (7\text{–}5)$$

In this equation, the left-hand member depends, by hypothesis, only upon the variable r, while the right-hand member is independent of r. Consequently, the equation can be satisfied identically only if each member is a constant, C. The energy is determined by the equation for the *radial wave function* $R(r)$,

$$\frac{1}{r^2}\frac{d}{dr}\left(r^2\frac{dR}{dr}\right) + \left\{\frac{2m}{\hbar^2}\,[E - V(r)] - \frac{C}{r^2}\right\}R = 0, \qquad (7\text{–}6)$$

whereas the angular part of the solution, $Y(\theta, \phi)$, satisfies

$$\frac{1}{\sin\theta}\frac{\partial}{\partial\theta}\left(\sin\theta\frac{\partial Y}{\partial\theta}\right) + \frac{1}{\sin^2\theta}\frac{\partial^2 Y}{\partial\phi^2} = -CY. \qquad (7\text{–}7)$$

Equation (7–7) is independent of the energy E and of the potential energy $V(r)$. Therefore, the angular dependence of the wave functions is determined solely by the property of spherical symmetry, and admissible solutions of Eq. (7–7) are valid for every spherically symmetric system regardless of the special form of the potential function. We shall first give attention to the solutions of the angular equation and return to the radial equation in the discussion of specific examples.

Equation (7–7) for the functions Y can be separated again by the substitution

$$Y(\theta, \phi) = P(\theta)\Phi(\phi). \qquad (7\text{–}8)$$

We obtain

$$\frac{1}{P}\left[\sin\theta\frac{d}{d\theta}\left(\sin\theta\frac{dP}{d\theta}\right)\right] + C\sin^2\theta = -\frac{1}{\Phi}\frac{d^2\Phi}{d\phi^2} = m^2, \qquad (7\text{–}9)$$

in which the separation constant is written as m^2. The second of Eqs. (7–9) is

$$\frac{d^2\Phi}{d\phi^2} + m^2\Phi = 0, \qquad (7\text{–}10)$$

which has the solutions

$$\Phi = e^{\pm im\phi}. \qquad (7\text{–}11)$$

By the substitution $\mu = \cos\theta$, the first of Eqs. (7–9) is reduced to

$$\frac{d}{d\mu}\left[(1 - \mu^2)\frac{dP}{d\mu}\right] + \left(C - \frac{m^2}{1 - \mu^2}\right)P = 0, \qquad (7\text{–}12)$$

which is the differential equation defining the *associated Legendre functions*.[1]

It is possible to solve this equation by the method of series, that is, by a procedure similar to that used for the harmonic oscillator (Section 5–9). This is done in detail in mathematical works devoted to the subject of spherical harmonics, and it is found that bounded, differentiable solutions of Eq. (7–12) exist if and only if the constant C is

$$C = l(l + 1), \qquad (7\text{–}13)$$

where l is a non-negative integer, and m has one of the integer values $-l$, $-l + 1, \ldots, l - 1, l$. These are therefore the only admissible values of C and m if ψ is to be the wave function for a physical system. Note that the functions defined by Eq. (7–11) are therefore *single-valued* functions of ϕ. In other subjects, such as electrostatics, the functions $Y(\theta, \phi)$ must represent physical quantities known in advance to be single-valued, so that the requirement that m be an integer follows immediately from Eq. (7–11). The wave function ψ, however, is interpreted physically through the product $\psi^*\psi$ which is independent of m (provided m is real); hence the condition of single-valuedness of $\psi^*\psi$ cannot be applied directly. The requirement that m be an integer arises, rather, from the condition of boundedness of ψ and $\nabla\psi$, for if m is not an integer, the solutions of Eq. (7–12) are irregular at the poles $\mu = \pm 1$.[2]

7–2 Spherical harmonics.   We shall now construct the functions $Y(\theta, \phi)$ by a method (due to Kramers)[3] which is more direct than the method of series. We have already remarked that the angular functions Y are independent of E and $V(r)$; therefore, no generality is lost in this

---

[1] E. T. Whittaker and G. N. Watson, *A Course of Modern Analysis*. 4th ed., Cambridge: Cambridge University Press, 1927, Section 15.5 ff.

[2] W. Pauli, *Die Allgemeinen Prinzipien der Wellenmechanik*.   Ann Arbor: J. W. Edwards, Publisher, 1947, p. 126.

[3] H. A. Kramers, *Quantum Mechanics*. Amsterdam: North-Holland Publishing Co., 1957, §45, p. 168. H. C. Brinkman, *Applications of Spinor Invariants in Atomic Physics*. Amsterdam: North-Holland Publishing Co., 1956. Cf. also R. Courant and D. Hilbert, *Methods of Mathematical Physics*. New York: Interscience Publishers, Inc., 1953, Vol. I, Appendix to Chapter VII.

part of the problem if Eq. (7–1) is replaced by Laplace's equation

$$\nabla^2\psi = 0. \tag{7–14}$$

Solutions of this equation are called harmonic functions. A single-valued harmonic function which is continuous in a neighborhood of the origin can be approximated arbitrarily well by a polynomial in x, y, z. If such a polynomial is to have the form of Eq. (7–4), it must be homogeneous in these variables, i.e., it must have the form

$$\psi_l = \sum_{p+q+r=l} a_{pqr}x^p y^q z^r, \tag{7–15}$$

in which the numbers a_{pqr} must be chosen so that Eq. (7–14) is satisfied. The integer l is, of course, the degree of the polynomial. The number of terms in the sum (7–15) is $(\frac{1}{2})(l+1)(l+2)$: For a given value of p, the index q can have the $l - p + 1$ values $0, 1, \ldots, l - p$, while p can have any of the $l+1$ values $0, 1, 2, \ldots, l$; thus the total number of combinations,

$$(l+1) + (l) + (l-1) + \cdots + 1 = \tfrac{1}{2}(l+1)(l+2),$$

is the number of linearly independent polynomials of degree l. Now the Laplacian of the function (7–15) is a polynomial of degree $l - 2$; hence, the requirement that Laplace's equation be satisfied imposes $(\frac{1}{2})(l-1)l$ conditions on the coefficients a_{pqr}. There remain, therefore,

$$\tfrac{1}{2}(l+1)(l+2) - \tfrac{1}{2}(l-1)l = 2l+1$$

linearly independent harmonic polynomials of degree l. Explicitly, these may be taken to be

$$
\begin{aligned}
l &= 0: \quad 1; \\
l &= 1: \quad x, y, z; \\
l &= 2: \quad xy, yz, zx, x^2 - y^2, 2z^2 - x^2 - y^2; \\
l &= 3: \quad x(x^2 - 3z^2), x(x^2 - 3y^2), y(y^2 - 3x^2), y(y^2 - 3z^2), \\
&\qquad\quad z(z^2 - 3x^2), z(z^2 - 3y^2), xyz;
\end{aligned}
\tag{7–16}
$$

and so on.

It is evident that each term in the polynomial (7–15) is proportional to r^l. Hence, ψ_l can be written

$$\psi_l = r^l Y_l(\theta, \phi), \tag{7–17}$$

in which $Y_l(\theta, \phi)$ is a *spherical harmonic of order l*. The separation constant

C is now easily evaluated: Eqs. (7–14) and (7–5) yield

$$\nabla^2\psi_l = \frac{1}{r^2}\frac{\partial}{\partial r}\left(r^2\frac{\partial\psi_l}{\partial r}\right) - \frac{C}{r^2}\psi_l = \frac{1}{r^2}l(l+1)r^lY_l - \frac{C}{r^2}\psi_l$$

$$= [l(l+1) - C]\frac{1}{r^2}\psi_l = 0, \quad (7\text{–}18)$$

whence Eq. (7–13) follows.

Explicit forms for the $2l+1$ linearly independent functions of the form (7–17) can be constructed by means of the special polynomial

$$T^l = (ax + by + cz)^l, \quad (7\text{–}19)$$

which is harmonic if

$$\nabla^2 T^l = l(l-1)(a^2 + b^2 + c^2)(ax + by + cz)^{l-2} = 0, \quad (7\text{–}20)$$

i.e., provided

$$a^2 + b^2 + c^2 = 0. \quad (7\text{–}21)$$

The coefficients a, b, c, which cannot all be real, are not independent. The condition (7–21) is automatically satisfied, however, if parameters ξ, η are introduced such that

$$a = \frac{1}{2i}(\xi^2 + \eta^2), \quad b = \tfrac{1}{2}(\xi^2 - \eta^2), \quad c = -\xi\eta, \quad (7\text{–}22)$$

for then

$$a^2 + b^2 + c^2 = \tfrac{1}{4}(\xi^4 - 2\xi^2\eta^2 + \eta^4 - \xi^4 - 2\xi^2\eta^2 - \eta^4) + \xi^2\eta^2 = 0.$$

In terms of ξ and η, T^l becomes

$$T^l = \left[\frac{1}{2i}(\xi^2 + \eta^2)x + \tfrac{1}{2}(\xi^2 - \eta^2)y - \xi\eta z\right]^l$$

$$= \left(\frac{1}{2i}\right)^l [(x + iy)\xi^2 - 2i\xi\eta z + (x - iy)\eta^2]^l,$$

or, since $x \pm iy = r\sin\theta\, e^{\pm i\phi}$, $z = r\cos\theta$,

$$T^l = \left(\frac{r}{2i}\right)^l [\sin\theta\, e^{i\phi}\xi^2 - 2i\cos\theta\,\xi\eta + \sin\theta\, e^{-i\phi}\eta^2]^l. \quad (7\text{–}23)$$

This expression is a homogeneous polynomial of degree $2l$ in the parameters ξ and η; therefore, it can be written in the form

$$T^l = \sum_{m=-l}^{l} \xi^{l+m}\eta^{l-m}Q_l^m(x, y, z), \quad (7\text{–}24)$$

in which the harmonic polynomials $Q_l^m(x, y, z)$ can be found by expanding the expression (7–23). This expansion is facilitated if the quantities

$$\mu = \cos\theta, \qquad \sigma = \sin\theta\, e^{i\phi}\,\frac{\xi}{\eta}, \qquad \tau = \sin\theta\, e^{-i\phi}\,\frac{\eta}{\xi}$$

are introduced temporarily. Thus, in terms of μ and σ,

$$T^l = \left(\frac{r}{2i}\right)^l \left(\frac{\eta^2 e^{-i\phi}}{\sin\theta}\right)^l [\sigma^2 - 2i\mu\sigma + 1 - \mu^2]^l$$

$$= \left(\frac{r}{2i}\right)^l \left(\frac{\eta^2 e^{-i\phi}}{\sin\theta}\right)^l [1 + (\sigma - i\mu)^2]^l. \quad (7\text{–}25)$$

The last factor in this expression can be written out by means of Taylor's theorem as follows:

$$F(\sigma, \mu) = [1 + (\sigma - i\mu)^2]^l =$$

$$\sum_{m=-l}^{l} \frac{\sigma^{l+m}}{(l+m)!} \left[\frac{\partial^{l+m}}{\partial\sigma^{l+m}} [1 + (\sigma - i\mu)^2]^l\right]_{\sigma=0}.$$

The function $F(\sigma, \mu)$ contains σ and μ only in the combination $\sigma - i\mu$; hence (cf. Problem 5–28, Chapter 5),

$$\frac{\partial^{l+m}}{\partial\sigma^{l+m}} F(\sigma, \mu) = (i)^{l+m} \frac{\partial^{l+m}}{\partial\mu^{l+m}} F(\sigma, \mu).$$

Substituting in Eq. (7–25) and setting $\sigma = 0$, we obtain

$$T^l = \left(\frac{r}{2i}\right)^l \left(\frac{\eta^2 e^{-i\phi}}{\sin\theta}\right)^l \sum_{m=-l}^{l} \frac{\sigma^{l+m}}{(l+m)!} (i)^{l+m} \frac{d^{l+m}}{d\mu^{l+m}} (1 - \mu^2)^l.$$

Finally, restoring the variables ξ, η and rearranging, we have

$$T^l = \left(\frac{r}{2i}\right)^l \sum_{m=-l}^{l} \xi^{l+m}\eta^{l-m} e^{im\phi} \sin^m\theta \frac{(i)^{l+m}}{(l+m)!} \frac{d^{l+m}}{d\mu^{l+m}} (1 - \mu^2)^l. \quad (7\text{–}26)$$

Thus the function Q_l^m of Eq. (7–24) is

$$Q_l^m = \left(\frac{r}{2i}\right)^l e^{im\phi} \sin^m\theta \frac{(i)^{l+m}}{(l+m)!} \frac{d^{l+m}}{d\mu^{l+m}} (1 - \mu^2)^l. \quad (7\text{–}27)$$

The *associated Legendre function* $P_l^m(\mu)$ is defined by the equation[1]

---

[1] This is Ferrer's definition of P_l^m. Cf. E. T. Whittaker and G. N. Watson, *A Course of Modern Analysis*. 4th ed., Cambridge: Cambridge University Press, 1927, Section 15–5. P. M. Morse and H. Feshbach, *Methods of Theoretical Physics*. New York: McGraw-Hill Book Co., Inc., 1953, p. 601.

$$P_l^m(\mu) = \frac{(1-\mu^2)^{\frac{m}{2}}}{2^l l!} \frac{d^{l+m}}{d\mu^{l+m}}(\mu^2-1)^l, \tag{7-28}$$

in terms of which the function Q_l^m is

$$Q_l^m = (-r)^l(i)^m \frac{l!}{(l+m)!} e^{im\phi}P_l^m(\mu). \tag{7-29}$$

The $2l+1$ functions Q_l^m are linearly independent and form an explicit representation of the spherical harmonics of degree l.

An alternative expression for Q_l^m can be obtained from Eq. (7–23) by factoring out the quantity $(\xi^2 e^{i\phi}/\sin\theta)^l$ rather than $(\eta^2 e^{-i\phi}/\sin\theta)^l$, as was done above. The same manipulation of the coefficient in the Taylor's series expansion of the function $[1+(\tau-i\mu)^2]^l$ then leads to (Problem 7–5)

$$Q_l^m = \left(\frac{r}{2i}\right)^l e^{im\phi} \sin^{-m}\theta \frac{(i)^{l-m}}{(l-m)!} \frac{d^{l-m}}{d\mu^{l-m}}(1-\mu^2)^l. \tag{7-30}$$

By comparison with Eq. (7–29) it follows that

$$P_l^m(\mu) = (-)^m \frac{(l+m)!}{(l-m)!} \frac{(1-\mu^2)^{-m/2}}{2^l l!} \frac{d^{l-m}}{d\mu^{l-m}}(\mu^2-1)^l, \tag{7-31}$$

a result that is not easily obtained directly from the definition (7–28). If m is replaced by $-m$ in Eq. (7–31), the result is, by Eq. (7–28),

$$P_l^{-m}(\mu) = (-)^m \frac{(l-m)!}{(l+m)!} P_l^m(\mu). \tag{7-32}$$

This expression will be referred to later [cf. Eq. (7–45)].

The polynomials Q_l^m can be written

$$Q_l^m = \text{constant} \times r^l Y_l^m(\theta, \phi), \tag{7-33}$$

in which the spherical harmonics

$$Y_l^m(\theta, \phi) = A_l^m e^{im\phi} P_l^m(\cos\theta) \tag{7-34}$$

constitute a complete orthonormal set provided the constants A_l^m are suitably chosen, as we shall now show.

The *scalar product* of two functions Y_1 and Y_2 of the variables θ and ϕ is, by definition,

$$(Y_1, Y_2) = \int Y_1^* Y_2 \, d\Omega = \int_0^{2\pi}\int_0^\pi Y_1^* Y_2 \sin\theta \, d\theta \, d\phi,$$

where $d\Omega = \sin\theta \, d\theta \, d\phi$ is the element of solid angle, i.e., the element of area on the unit sphere. In terms of $\mu = \cos\theta$, this can also be written

$$(Y_1, Y_2) = \int_0^{2\pi} \int_{-1}^1 Y_1^* Y_2 \, d\mu \, d\phi.$$

Now we shall show that two spherical harmonics Y_l and $Y_{l'}$ of degrees l and l' are orthogonal, i.e.,

$$(Y_{l'}, Y_l) = 0 \qquad (l \neq l'). \tag{7–35}$$

The proof is based on the fact that the corresponding harmonic polynomials,

$$\psi_l = r^l Y_l, \qquad \psi_{l'} = r^{l'} Y_{l'} \tag{7–36}$$

satisfy Laplace's equation; hence

$$\psi_{l'}^* \nabla^2 \psi_l - \psi_l \nabla^2 \psi_{l'}^* = 0.$$

If Green's second identity is applied in the volume enclosed by the unit sphere, one obtains

$$\int_0^1 \int_\Omega [\psi_{l'}^* \nabla^2 \psi_l - \psi_l \nabla^2 \psi_{l'}^*] r^2 \, dr \, d\Omega = \int_\Omega \left[\psi_{l'}^* \frac{\partial \psi_l}{\partial r} - \psi_l \frac{\partial \psi_{l'}^*}{\partial r} \right]_{r=1} d\Omega = 0. \tag{7–37}$$

The derivatives in the second form can be evaluated immediately from Eq. (7–36), and Eq. (7 37) becomes

$$(l - l') \int_\Omega Y_{l'}^* Y_l \, d\Omega = 0,$$

which is equivalent to Eq. (7–35).

The functions Y_l^m contain ϕ only in the factor $e^{im\phi}$; hence if $l = l'$, we have

$$(Y_l^{m'}, Y_l^m) = A_l^{m'*} A_l^m \int_0^{2\pi} \int_{-1}^1 e^{i(m-m')\phi} P_l^m(\mu) P_l^{m'}(\mu) \, d\mu \, d\phi.$$

The integration over ϕ can be carried out immediately, and if $m' \neq m$, one obtains

$$(Y_l^{m'}, Y_l^m) = 0 \qquad (m \neq m'). \tag{7–38}$$

The functions Y_l^m are therefore orthogonal with respect to each of the indices l and m. They are also normalized, provided the constants A_l^m are chosen to satisfy

$$(Y_l^m, Y_l^m) = |A_l^m|^2 2\pi \int_{-1}^1 [P_l^m(\mu)]^2 \, d\mu = 1. \tag{7–39}$$

The integral in this expression can be evaluated most easily by substituting the expression (7–28) for one of the P_l^m and (7–31) for the other; thus

$$\int_{-1}^{1} [P_l^m(\mu)]^2 \, d\mu = (-)^m \frac{(l+m)!}{(l-m)!} \frac{1}{2^{2l}(l!)^2}$$

$$\times \int_{-1}^{1} \left[\frac{d^{l+m}}{d\mu^{l+m}} (\mu^2 - 1)^l \right] \left[\frac{d^{l-m}}{d\mu^{l-m}} (\mu^2 - 1)^l \right] d\mu. \quad (7\text{–}40)$$

The function $(\mu^2 - 1)^l$ has a zero of order l at each of the points $\mu = \pm 1$, whence, if (7–40) is integrated by parts $l - m$ times, the integrated part will vanish each time. The result is

$$\int_{-1}^{1} [P_l^m(\mu)]^2 \, d\mu = (-)^m \frac{(l+m)!}{(l-m)!} \frac{(-)^{l-m}}{2^{2l}(l!)^2}$$

$$\times \int_{-1}^{1} \left[\frac{d^{2l}}{d\mu^{2l}} (\mu^2 - 1)^l \right] (\mu^2 - 1)^l \, d\mu$$

$$= \frac{(l+m)!}{(l-m)!} \frac{(2l)!}{2^{2l}(l!)^2} \int_{-1}^{1} (1 - \mu^2)^l \, d\mu. \quad (7\text{–}41)$$

The integral in this equation is (Problem 7–6)

$$\int_{-1}^{1} (1 - \mu^2)^l \, d\mu = 2^{2l+1} \frac{(l!)^2}{(2l+1)!},$$

whence

$$\int_{-1}^{1} [P_l^m(\mu)]^2 \, d\mu = \frac{2}{2l+1} \frac{(l+m)!}{(l-m)!}, \quad (7\text{–}42)$$

and Eq. (7–39) becomes

$$|A_l^m|^2 \cdot 2\pi \cdot \frac{2}{2l+1} \frac{(l+m)!}{(l-m)!} = 1. \quad (7\text{–}43)$$

The orthonormal functions Y_l^m are therefore

$$Y_l^m(\theta, \phi) = (-)^m \sqrt{\frac{2l+1}{4\pi} \frac{(l-m)!}{(l+m)!}} \, e^{im\phi} P_l^m (\cos \theta), \quad (7\text{–}44)$$

in which the phase factor $(-)^m$ has been chosen to agree with that most commonly used in the literature.[1] The function Y_l^{-m} is related to Y_l^m

---

[1] E. U. Condon and G. H. Shortley, *The Theory of Atomic Spectra*. Cambridge: Cambridge University Press, 1953, Chapter III, Section 4. J. M. Blatt and V. F. Weisskopf, *Theoretical Nuclear Physics*. New York: John Wiley and Sons, 1952, Appendix A, Section 2.

through the identity (7–32):

$$Y_l^{-m} = (-)^m Y_l^{m*}. \tag{7-45}$$

The functions P_l^m can be computed from the *Legendre polynomials*,

$$P_l = P_l^0 = \frac{1}{2^l l!} \frac{d^l}{d\mu^l} (\mu^2 - 1)^l, \tag{7-46}$$

by means of the relation

$$P_l^m = \sin^m \theta \frac{d^m}{d\mu^m} P_l. \tag{7-47}$$

In this way, one can construct the table

$$Y_0^0 = \frac{1}{\sqrt{4\pi}}$$

$$Y_1^0 = \sqrt{\frac{3}{4\pi}} \cos \theta, \qquad\qquad Y_1^1 = -\sqrt{\frac{3}{8\pi}} e^{i\phi} \sin \theta,$$

$$Y_2^0 = \sqrt{\frac{5}{16\pi}} (3\cos^2 \theta - 1), \qquad Y_2^1 = -\sqrt{\frac{15}{8\pi}} e^{i\phi} \sin \theta \cos \theta, \tag{7-48}$$

$$Y_2^2 = \sqrt{\frac{15}{32\pi}} e^{2i\phi} \sin^2 \theta,$$

. . .

The spherical harmonics $Y_l^m(\theta, \phi)$ are a complete orthonormal set of functions, i.e.,

$$(Y_{l'}^{m'}, Y_l^m) = \delta_{ll'} \delta_{mm'}, \tag{7-49}$$

and any function $f(\theta, \phi)$ which is continuous and has continuous first and second derivatives can be expanded in the form

$$f(\theta, \phi) = \sum_{l=0}^{\infty} \sum_{m=-l}^{l} f_l^m Y_l^m(\theta, \phi). \tag{7-50}$$

The coefficients f_l^m are given by

$$f_l^m = (Y_l^m, f), \tag{7-51}$$

and the value of $f(\theta, \phi)$ at the pole $(\theta = 0)$ is (Problem 7–11)

$$f(0, \phi) = \sum_{l=0}^{\infty} f_l^0 Y_l^0(0, \phi) = \sum_{l=0}^{\infty} \sqrt{\frac{2l + 1}{4\pi}} f_l^0. \tag{7-52}$$

In the particular case that $f(\theta, \phi)$ is a spherical harmonic Y_l, we have $f_{l'}^m = (Y_{l'}^m, Y_l)$, which is zero unless $l' = l$. Consequently, Eq. (7–52)

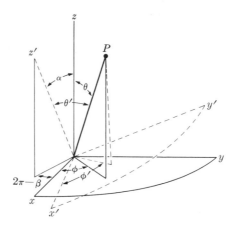

FIG. 7–2. Coordinate systems for the proof of the addition theorem.

becomes (cf. Problem 7–10)

$$Y_l(0, \phi) = \sqrt{\frac{2l+1}{4\pi}} \, (Y_l^0, \, Y_l) = \frac{2l+1}{4\pi} \, (P_l \, (\cos \theta), \, Y_l). \qquad (7\text{--}53)$$

This relation can be used to prove an important formula, called the *addition theorem for spherical harmonics*, which establishes the relation between spherical harmonics referred to two differently oriented systems of axes.

Let the point P on the unit sphere be defined by the coordinates θ and ϕ with respect to the axes x, y, z and by θ' and ϕ' with respect to x', y', z' (Fig. 7–2). The rectangular coordinates of P in the two coordinate systems are related linearly by means of the table of direction cosines of the angles between the primed and unprimed axes.[1] Consequently, a homogeneous polynomial of degree l in x, y, z becomes, after transformation to the primed coordinates, a homogeneous polynomial of degree l in x', y', z'. Also, the Laplace equation

$$\nabla^2 \psi = \frac{\partial^2 \psi}{\partial x^2} + \frac{\partial^2 \psi}{\partial y^2} + \frac{\partial^2 \psi}{\partial z^2} = 0 \qquad (7\text{--}54)$$

is invariant to this change of variables (Problem 7–12), i.e., it becomes

$$\nabla'^2 \psi = \frac{\partial^2 \psi}{\partial x'^2} + \frac{\partial^2 \psi}{\partial y'^2} + \frac{\partial^2 \psi}{\partial z'^2} = 0. \qquad (7\text{--}55)$$

---

[1] G. Joos, *Theoretical Physics.* New York: G. E. Stechert and Co., 1934, Chapter I.

It follows that a spherical harmonic of degree l in x', y', z' is also a spherical harmonic of degree l (although one of different form[1]) in x, y, z.

The meaning of these considerations is clear: The physical system is spherically symmetric, and the direction chosen for the coordinate axes is of no significance for the mathematical description of the system. The angular dependence of the wave functions has been shown to be given by the functions $Y_l^m(\theta, \phi)$, which are a complete set of $2l + 1$ harmonics of degree l. Consequently, it must be possible to express the spherical harmonics of the same degree, $Y_l^m(\theta', \phi')$, with respect to any other system of axes as linear combinations of the $Y_l^m(\theta, \phi)$. This *invariance to rotation of the coordinate system* characterizes the symmetry. It is the fundamental property upon which the entire theory of spherical harmonics depends.

In particular, the function $P_l(\cos \theta')$ is a spherical harmonic of degree l, and therefore

$$P_l(\cos \theta') = \sum_{m=-l}^{l} a_m Y_l^m(\theta, \phi), \qquad (7\text{–}56)$$

in which the coefficients a_m are

$$a_m = \left(Y_l^m(\theta, \phi), P_l(\cos \theta')\right) = \left(P_l(\cos \theta'), Y_l^{m*}(\theta, \phi)\right). \quad (7\text{–}57)$$

The last equation follows from the fact that the Legendre polynomial is a real function. Substituting θ', ϕ' for θ, ϕ in Eq. (7–53), we obtain

$$\left(P_l(\cos \theta'), Y_l^{m*}(\theta, \phi)\right) = \frac{4\pi}{2l + 1} Y_l^{m*}(\theta, \phi)|_{\theta'=0}. \qquad (7\text{–}58)$$

Now at $\theta' = 0$, the angles θ and ϕ are equal to the angles α and β which are the polar coordinates of the z'-axis with respect to the x, y, z-system (Fig. 7–2). Hence,

$$a_m = \frac{4\pi}{2l + 1} Y_l^{m*}(\alpha, \beta), \qquad (7\text{–}59)$$

and

$$P_l(\cos \theta') = \frac{4\pi}{2l + 1} \sum_{m=-l}^{l} Y_l^{m*}(\alpha, \beta) Y_l^m(\theta, \phi). \qquad (7\text{–}60)$$

The relation between θ' and the angles θ, ϕ is determined from the geometry of Fig. 7–2:

$$\cos \theta' = \cos \alpha \cos \theta + \sin \alpha \sin \theta \cos(\phi - \beta). \qquad (7\text{–}61)$$

---

[1] The two polynomials have different coefficients a_{pqr}.

The angles θ and ϕ are determined by the direction of the unit vector $\hat{\mathbf{r}}$ pointing from the origin toward P, and it is often convenient to use the notation

$$Y_l^m(\theta, \phi) = Y_l^m(\hat{\mathbf{r}}). \tag{7-62}$$

Thus if $\hat{\mathbf{r}}'$ is a vector in the direction of the z'-axis, we have

$$\cos \theta' = \hat{\mathbf{r}}' \cdot \hat{\mathbf{r}},$$

and Eq. (7-60) becomes

$$P_l(\hat{\mathbf{r}}' \cdot \hat{\mathbf{r}}) = \frac{4\pi}{2l+1} \sum_{m=-l}^{l} Y_l^{m*}(\hat{\mathbf{r}}') Y_l^m(\hat{\mathbf{r}}). \tag{7-63}$$

Equation (7-60) or Eq. (7-63) is the addition theorem.

7-3 Degeneracy; angular momentum. The energies of the stationary states of a spherically symmetric system are those values of E for which the radial wave equation (7-6) has solutions which are admissible as wave functions. This equation is equivalent to

$$\frac{d^2u}{dr^2} + \left\{\frac{2m}{\hbar^2} [E - V(r)] - \frac{l(l+1)}{r^2}\right\} u = 0, \tag{7-64}$$

where $rR = u$. Except for a change in the range of the independent variable ($r \geq 0$), this equation is of the same form as the one-dimensional Schrödinger equation (5-3) provided the function

$$`V' = V(r) - \frac{\hbar^2}{2m} \frac{l(l+1)}{r^2} \tag{7-65}$$

is regarded as an "equivalent one-dimensional potential-energy function." The solutions of Eq. (7-64) are similar in character to those of Eq. (5-3), in that there is, in general, a partly continuous and partly discrete spectrum of allowed values of E. It is clear that the energy eigenvalues depend, in general, upon the quantum number l, but that they are independent of m. Therefore, each of the functions RY_l^m ($m = -l, \ldots, l$) is a solution of the Schrödinger equation (7-1) for the same value of E, and since these functions are linearly independent, *the states of energy E are $(2l+1)$-fold degenerate.* The reason for this degeneracy is the rotational symmetry of the system.

In order to see this more clearly, let $\psi(r, \theta, \phi)$ be an eigenfunction belonging to the energy E, i.e.,

$$H\psi(r, \theta, \phi) = E\psi(r, \theta, \phi). \tag{7-66}$$

Since the angle ϕ can be measured with respect to any direction perpendicular to the z-axis, the function $\psi(r, \theta, \phi + \epsilon)$ must also be an eigenfunction for the same value of E, i.e.,

$$H\psi(r, \theta, \phi + \epsilon) = E\psi(r, \theta, \phi + \epsilon), \qquad (7\text{–}67)$$

where ϵ is an arbitrary angle. Now if ϵ is small, we have

$$\psi(r, \theta, \phi + \epsilon) = \psi(r, \theta, \phi) + \epsilon \frac{\partial}{\partial \phi} \psi(r, \theta, \phi);$$

hence, introducing the operator $D = \partial/\partial\phi$, Eq. (7–67) becomes

$$H\psi(r, \theta, \phi) + \epsilon H D\psi(r, \theta, \phi) = E\psi(r, \theta, \phi) + \epsilon DE\psi(r, \theta, \phi).$$

By Eq. (7–66), this reduces to

$$(HD - DH)\psi(r, \theta, \phi) = 0. \qquad (7\text{–}68)$$

This equation is true for all eigenfunctions of the Hamiltonian, and since these are, by hypothesis, a complete set of states of the system, we have

$$[H, D] = 0, \qquad (7\text{–}69)$$

i.e., the operator $D = \partial/\partial\phi$ commutes with the Hamiltonian (cf. Section 6–9). The relation (7–69), which can be verified by writing out the Schrödinger equation in the form (7–3) (Problem 7–18), states in quantum-mechanical terms that H describes a system invariant to rotation about the z-axis.

According to the general theory of Chapter 6, the operators H and D must have simultaneous eigenfunctions. The functions Y_l^m contain the angle ϕ only in the factor $e^{im\phi}$, whence

$$De^{im\phi} = ime^{im\phi}.$$

The eigenvalues of D are purely imaginary, and hence D is not Hermitian. The operator

$$L_z = \frac{\hbar}{i} D = \frac{\hbar}{i} \frac{\partial}{\partial \phi}, \qquad (7\text{–}70)$$

however, satisfies the eigenvalue equation

$$L_z e^{im\phi} = m\hbar e^{im\phi} \qquad (7\text{–}71)$$

and is Hermitian.   Explicitly, if ψ is any single-valued function of ϕ, we have

$$(\psi, L_z\psi) = \int_0^{2\pi} \psi^* \frac{\hbar}{i} \frac{\partial\psi}{\partial\phi}\, d\phi = \int_0^{2\pi} \left(-\frac{\hbar}{i} \frac{\partial\psi^*}{\partial\phi}\right) \psi\, d\phi = (L_z\psi, \psi),$$

where the second integral is obtained from the first by integration by parts.

The physical interpretation of the operator L_z follows immediately from the rules (6–83).   Thus, we have

$$D\psi = \frac{\partial\psi}{\partial\phi} = \frac{\partial\psi}{\partial x}\frac{\partial x}{\partial\phi} + \frac{\partial\psi}{\partial y}\frac{\partial y}{\partial\phi} = x\frac{\partial\psi}{\partial y} - y\frac{\partial\psi}{\partial x},$$

whence

$$L_z = x\left(\frac{\hbar}{i}\frac{\partial}{\partial y}\right) - y\left(\frac{\hbar}{i}\frac{\partial}{\partial x}\right) = xp_y - yp_x. \qquad (7\text{--}72)$$

The operator L_z therefore corresponds to the z-component of the angular momentum of the particle.   The classical angular-momentum vector is defined to be

$$\mathbf{L} = \mathbf{r} \times \mathbf{p}.$$

The corresponding vector operator

$$\mathbf{L}_{\mathrm{op}} = \mathbf{r}_{\mathrm{op}} \times \mathbf{p}_{\mathrm{op}} \qquad (7\text{--}73)$$

has the components L_z [Eq. (7–72)] and

$$L_x = yp_z - zp_y, \qquad L_y = zp_x - xp_z. \qquad (7\text{--}74)$$

Note that $[x, p_y] = 0$, etc., so that the order of the factors in the terms of Eq. (7–72) and Eq. (7–74) can be changed if desired:

$$\mathbf{L} = \mathbf{r} \times \mathbf{p} = -\mathbf{p} \times \mathbf{r}. \qquad (7\text{--}75)$$

The negative sign in this equation results, of course, from the definition of the vector product.

Equation (7–69), which is equivalent to

$$[H, L_z] = 0, \qquad (7\text{--}76)$$

expresses the fact that the z-component of $\mathbf{L}$ is a constant of the motion for any spherically symmetric system.   Hence, the law of conservation of angular momentum holds in quantum as well as in classical mechanics, as it must, since it is a direct consequence of the geometrical symmetry of the system.   Furthermore, since

$$[H, L_x] = [H, L_y] = 0, \qquad (7\text{--}77)$$

or, in general,

$$[H, \mathbf{L}] = 0, \tag{7–78}$$

an eigenfunction of H can be simultaneously an eigenfunction of L_x, or L_y, or L_z, or of any linear combination of these operators. The function RY_l^m is now seen to be a simultaneous eigenfunction of H and L_z; the $(2l + 1)$-fold degeneracy has been resolved by means of the commuting operator L_z, according to the general procedure described in Section 6–8.

The rotational degeneracy of the states does not occur, of course, for a system which is not spherically symmetric. If forces are introduced which destroy the spherical symmetry, then the energy will depend, in general, upon m as well as l. Such forces are represented by additional terms in the Hamiltonian, which do not commute with $\mathbf{L}$, so that Eq. (7–78) is no longer true. The introduction of such forces *removes* the degeneracy. We shall show in Section 10–5 that, for a charged particle, the degeneracy is completely removed by the application of an external magnetic field, which defines a special direction in space.

The functions RY_l^m are eigenfunctions of the z-component of angular momentum and represent states which are *quantized with respect to the z-axis.* We have seen that any direction in space can be chosen as the axis of quantization since every component of $\mathbf{L}$ commutes with H. However, *the components of* $\mathbf{L}$ *are not commuting operators;* hence it is usually impossible to form a simultaneous eigenfunction of two different components of $\mathbf{L}$. It is a matter of algebra to show (Problem 7–19) that the operators L_x, L_y, L_z satisfy the commutation rules

$$[L_x, L_y] = i\hbar L_z, \qquad [L_y, L_z] = i\hbar L_x, \qquad [L_z, L_x] = i\hbar L_y, \quad (7–79)$$

or, in vector form (Problem 7–20),

$$\mathbf{L} \times \mathbf{L} = i\hbar\mathbf{L}. \tag{7–80}$$

The equations (7–79) are the fundamental relations among the components of any angular-momentum vector. They express, in precise form, that successive rotations of the coordinate frame about axes in two different directions are not commutable operations.[1] We shall show in Chapter 9 that a complete characterization of the angular-momentum operators can be obtained from the relations (7–79). Some of the consequences of these relations, which result from the definitions (7–72), (7–74) and the commutation rules (6–83), are listed in Problems 7–21 and 7–22.

---

[1] Cf., e.g., L. Page, *Introduction to Theoretical Physics.* 2nd ed. New York: D. Van Nostrand Co., Inc., 1935, p. 101. G. Joos, *Theoretical Physics.* New York: G. E. Stechert and Co., 1934, p. 132.

The square of the angular-momentum operator $\mathbf{L}$ is

$$\mathbf{L}^2 = \mathbf{L} \cdot \mathbf{L} = L_x^2 + L_y^2 + L_z^2. \tag{7-81}$$

The operator $\mathbf{L}^2$ *commutes with every component of* $\mathbf{L}$, i.e.,

$$[\mathbf{L}^2, L_x] = [\mathbf{L}^2, L_y] = [\mathbf{L}^2, L_z] = 0, \tag{7-82}$$

for it follows from Eqs. (7–81) and (7–79) that, for example,

$$
\begin{aligned}
[\mathbf{L}^2, L_x] &= [L_x^2, L_x] + [L_y^2, L_x] + [L_z^2, L_x] \\
&= L_y[L_y, L_x] + [L_y, L_x]L_y + L_z[L_z, L_x] + [L_z, L_x]L_z \\
&= i\hbar(-L_yL_z - L_zL_y + L_zL_y + L_yL_z) = 0.
\end{aligned}
$$

Furthermore, because of the relations (7–76) and (7–77), $\mathbf{L}^2$ commutes with the Hamiltonian for a spherically symmetric system:

$$[H, \mathbf{L}^2] = 0. \tag{7-83}$$

The operators H, L_z and $\mathbf{L}^2$ are commuting operators, and the energy states of our problem can therefore be written as simultaneous eigenfunctions of these operators (cf. Section 6–8). We have already seen that the functions $RY_l^m(\theta, \phi)$ are eigenfunctions of H and L_z, and it will now be shown that

$$\mathbf{L}^2 Y_l^m(\theta, \phi) = l(l + 1)\hbar^2 Y_l^m(\theta, \phi), \tag{7-84}$$

i.e., the spherical harmonic of degree l is an eigenfunction of the square of the total angular momentum, belonging to the eigenvalue $l(l + 1)\hbar^2$.

The spherical harmonics are functions of the angles θ and ϕ, and the operator $\mathbf{L}$ has been defined in terms of the rectangular coordinates x, y, z. A change of variables is therefore required, which can be carried out by straightforward substitution [Eqs. (7–2); Problem 7–25]. Alternatively, using vector methods, we can proceed directly from the definition

$$\mathbf{L} = \frac{\hbar}{i} \mathbf{r} \times \mathbf{\nabla} \tag{7-85}$$

to arrive at

$$
\begin{aligned}
\mathbf{L}^2\psi &= -\hbar^2 \mathbf{r} \times \mathbf{\nabla} \cdot (\mathbf{r} \times \mathbf{\nabla}\psi) = -\hbar^2 \mathbf{r} \cdot \mathbf{\nabla} \times (\mathbf{r} \times \mathbf{\nabla}\psi) \\
&= -\hbar^2 \mathbf{r} \cdot [\mathbf{r}\nabla^2\psi + (\mathbf{\nabla}\psi \cdot \mathbf{\nabla})\mathbf{r} - (\mathbf{\nabla} \cdot \mathbf{r})\mathbf{\nabla}\psi - (\mathbf{r} \cdot \mathbf{\nabla})\mathbf{\nabla}\psi]. \tag{7-86}
\end{aligned}
$$

Now if $\mathbf{A}$ is any vector and ϕ any scalar function, we have, by the definition of the gradient operator,

$$(\mathbf{A} \cdot \mathbf{\nabla})\mathbf{r} = \mathbf{A}, \qquad (\mathbf{r} \cdot \mathbf{\nabla})\mathbf{A} = r\frac{\partial \mathbf{A}}{\partial r}, \qquad (\mathbf{r} \cdot \mathbf{\nabla})\phi = r\frac{\partial \phi}{\partial r},$$

and

$$\mathbf{\nabla} \cdot \mathbf{r} = 3.$$

Application of these rules in Eq. (7–86) yields

$$L^2\psi = -\hbar^2 r^2 \nabla^2 \psi - \hbar^2 \left[r \frac{\partial \psi}{\partial r} - 3r \frac{\partial \psi}{\partial r} - \mathbf{r} \cdot \left(r \frac{\partial}{\partial r} \mathbf{\nabla}\psi \right) \right]$$

$$= -\hbar^2 r^2 \nabla^2 \psi - \hbar^2 \left[-2r \frac{\partial \psi}{\partial r} - \frac{\partial}{\partial r} (r\mathbf{r} \cdot \mathbf{\nabla}\psi) + \mathbf{\nabla}\psi \cdot \frac{\partial}{\partial r} (r\mathbf{r}) \right]$$

$$= -\hbar^2 r^2 \nabla^2 \psi + \hbar^2 \frac{\partial}{\partial r} \left(r^2 \frac{\partial \psi}{\partial r} \right),$$

or

$$\frac{1}{\hbar^2 r^2} L^2\psi = -\nabla^2 \psi + \frac{1}{r^2} \frac{\partial}{\partial r} \left(r^2 \frac{\partial \psi}{\partial r} \right).$$

Now, if $\nabla^2\psi$ is expressed in terms of spherical polar coordinates, the preceding equation becomes

$$L^2\psi = -\hbar^2 \left[\frac{1}{\sin \theta} \frac{\partial}{\partial \theta} \left(\sin \theta \frac{\partial \psi}{\partial \theta} \right) + \frac{1}{\sin^2 \theta} \frac{\partial^2 \psi}{\partial \phi^2} \right], \qquad (7\text{–}87)$$

i.e., the operator $-L^2/\hbar^2 r^2$ is just the "angular part" of the Laplacian operator. The relation (7–84) is therefore equivalent to Eq. (7–7), in which $C = l(l + 1)$ [Eq. (7–18)].

The results of this section are summarized in the statement that the spherical harmonics $Y_l^m(\theta, \phi)$ are simultaneous eigenfunctions of the operators L_z and L^2 belonging to the eigenvalues $m\hbar$ and $l(l + 1)\hbar^2$, respectively. Since these functions are a complete set with respect to functions of θ and ϕ, L_z and L^2 are a complete set of commuting operators for this class of functions. The Hamiltonian H, representing a spherically symmetric system, commutes with L_z and L^2, and these three operators form a complete set with respect to the quantum states of the system.

7–4 The three-dimensional harmonic oscillator. A particle attracted toward a fixed point by a force proportional to the distance from the point has the potential energy

$$V(r) = \tfrac{1}{2}kr^2 = \tfrac{1}{2}k(x^2 + y^2 + z^2),$$

which is spherically symmetric. The Schrödinger equation for this system is

$$-\frac{\hbar^2}{2m} \nabla^2 \psi + \tfrac{1}{2}kr^2\psi = E\psi,$$

or, writing $E = \frac{1}{2}\hbar\omega \cdot \lambda$ and measuring r in the unit $(\hbar/m\omega)^{1/2}$, where $\omega^2 = k/m$ (cf. Section 5–9),

$$\nabla^2\psi + (\lambda - r^2)\psi = 0. \tag{7–88}$$

In rectangular coordinates, this equation is immediately separable. The substitution

$$\psi = X(x)Y(y)Z(z) \tag{7–89}$$

results in separate equations for X, Y, and Z, that is,

$$\frac{d^2X}{dx^2} + (\lambda_x - x^2)X = 0, \tag{7–90}$$

$$\frac{d^2Y}{dy^2} + (\lambda_y - y^2)Y = 0, \tag{7–91}$$

$$\frac{d^2Z}{dz^2} + (\lambda_z - z^2)Z = 0, \tag{7–92}$$

where the constants λ_x, λ_y, λ_z are related by

$$\lambda_x + \lambda_y + \lambda_z = \lambda. \tag{7–93}$$

Each of Eqs. (7–90), (7–91), and (7–92) is the differential equation for a one-dimensional harmonic oscillator (Section 5–9), for which the normalized wave functions are given by Eq. (5–137). Integrable solutions of the form (7–89) are obtained provided the constants λ_x, λ_y, λ_z are

$$\lambda_x = 2n_x + 1, \qquad \lambda_y = 2n_y + 1, \qquad \lambda_z = 2n_z + 1, \tag{7–94}$$

in which n_x, n_y, and n_z are non-negative integers. The corresponding normalized wave function is

$$\psi_{n_x,n_y,n_z} = \frac{1}{\sqrt{2^n n_x!n_y!n_z!\pi^{3/2}}}\, e^{-r^2/2}H_{n_x}(x)H_{n_y}(y)H_{n_z}(z), \tag{7–95}$$

where

$$n = n_x + n_y + n_z. \tag{7–96}$$

According to Eq. (7–93), the energy is

$$E_n = (n_x + n_y + n_z + \tfrac{3}{2})\hbar\omega = (n + \tfrac{3}{2})\hbar\omega \tag{7–97}$$

and depends only upon the integer n. For a given value of n, the wave functions (7–95) are a set of linearly independent functions, each of which corresponds to one of the sets of integers n_x, n_y, n_z which satisfy Eq. (7–96). The states of energy E_n are therefore degenerate, and the degree of the degeneracy is $(\frac{1}{2})(n+1)(n+2)$, which is the number of ways in which Eq. (7–96) can be satisfied by non-negative integers n_x, n_y, n_z (cf. Section 1–16).

In spherical polar coordinates, which are adapted to the symmetry of the harmonic-oscillator potential, solutions exist of the form

$$\psi_l^m = R(r)\,Y_l^m(\theta,\phi), \tag{7–98}$$

in which the differential equation for the radial function $R(r)$ [cf. Eq. (7–6)] is

$$\frac{1}{r^2}\frac{d}{dr}\left(r^2\frac{dR}{dr}\right) + \left(\lambda - r^2 - \frac{l(l+1)}{r^2}\right)R = 0. \tag{7–99}$$

Since, by Eq. (7–95), every solution contains the factor $e^{-r^2/2}$, we write

$$R = e^{-r^2/2}\phi, \tag{7–100}$$

by which substitution Eq. (7–99) reduces (Problem 7–26) to

$$\phi'' + \left(\frac{2}{r} - 2r\right)\phi' + \left(\lambda - 3 - \frac{l(l+1)}{r^2}\right)\phi = 0. \tag{7–101}$$

In analogy with the earlier treatment of Eq. (5–100), this equation can be solved by the series method. If a series in r beginning with a term in r^α is substituted into Eq. (7–101), then the coefficient of $r^{\alpha-2}$ is seen to be

$$\alpha(\alpha-1) + 2\alpha - l(l+1) = \alpha(\alpha+1) - l(l+1)$$

and this must vanish identically. This condition is satisfied if $\alpha = l$ or $\alpha = -(l+1)$. The latter value, however, leads to a series beginning with $r^{-(l+1)}$, which is infinite at $r = 0$. Consequently, only $\alpha = l$ is allowed, and the series can be written

$$\phi = \sum_{k=l}^{\infty} a_k r^k \qquad (a_l \text{ arbitrary}). \tag{7–102}$$

A recurrence relation for the coefficients a_k is obtained by substituting Eq. (7–102) into Eq. (7–101) and requiring the coefficient of each power of r in the resulting series to vanish identically. The result of this substitution is

$$\sum_{k=l}^{\infty} [k(k+1)a_k r^k - 2k a_k r^{k+2} + (\lambda-3)a_k r^{k+2} - l(l+1)a_k r^k] = 0. \tag{7–103}$$

The coefficient of r^l in this equation is

$$l(l+1)a_l - l(l+1)a_l$$

which is, of course, identically zero. The coefficient of r^{l+1} is

$$(l+1)(l+2)a_{l+1} - l(l+1)a_{l+1} = 2(l+1)a_{l+1} = 0.$$

This is satisfied only if

$$a_{l+1} = 0. \tag{7-104}$$

In general, the coefficient of r^k is

$$[k(k+1) - l(l+1)]a_k + [-2(k-2) + (\lambda - 3)]a_{k-2} = 0, \tag{7-105}$$

which gives a recurrence formula for the remaining coefficients, namely,

$$a_k = \frac{2(k-2) - (\lambda - 3)}{(k-l)(k+l+1)}\, a_{k-2} \qquad (k > l+1). \tag{7-106}$$

Because of Eq. (7-104), all terms in the series (7-102) in which the exponent of r is (l + an odd integer) are absent; hence the series is of the form

$$\phi = r^l \times (\text{power series in } r^2).$$

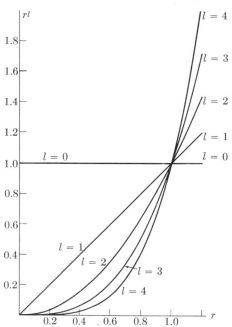

FIG. 7-3. The function r^l for various values of l.

In general, the leading term in the power series for an admissible solution of Eq. (7–6) is r^l, for unless $V(r)$ becomes infinite at $r = 0$ more rapidly than the function $1/r^2$, the term $l(l + 1)/r^2$ is dominant for small r. The equation obtained by neglecting the quantity $(2m/\hbar^2)[E - V(r)]$, that is,

$$\frac{1}{r^2} \frac{d}{dr}\left(r^2 \frac{dR}{dr}\right) - \frac{l(l + 1)}{r^2} R = 0 \qquad (7\text{–}107)$$

therefore holds approximately near $r = 0$. The solutions of (7–107) are:

$$R = r^l \quad \text{and} \quad R = r^{-(l+1)}, \qquad (7\text{–}108)$$

the second of which is inadmissible as a wave function. The functions r^l for $l = 0$,[1] 1, 2, 3, and 4 are shown in Fig. 7–3. We see that for large l the quantity $|\psi_l^m|^2$ is small in the immediate neighborhood of the origin. This corresponds to the classical situation, in which orbits of large angular momentum are remote from the center of force. Equation (7–6) can be rewritten in the form

$$\frac{1}{r^2} \frac{d}{dr}\left(r^2 \frac{dR}{dr}\right) + \frac{2m}{\hbar^2}\left\{E - \left[V(r) + \frac{l(l + 1)\hbar^2}{2mr^2}\right]\right\} R = 0. \quad (7\text{–}109)$$

The term $l(l + 1)\hbar^2/2mr^2$ represents the effect of the centrifugal force in the equivalent one-dimensional problem; it is often called the *centrifugal barrier potential*.

According to Eq. (7–106), the ratio of consecutive terms in the series (7–102) is

$$\frac{2(k - 2) - (\lambda - 3)}{(k - l)(k + l + 1)} r^2, \qquad (7\text{–}110)$$

which approaches the value $2r^2/k$ as $k \to \infty$. Consequently, according to the argument of Section 5–9, ϕ behaves asymptotically like e^{r^2} and does not yield an integrable solution unless a_k is zero for some value of k. This occurs if and only if

$$\lambda - 3 = 2n, \qquad (7\text{–}111)$$

where $n = $ (l plus an even integer). When this relation holds, the function ϕ reduces to a polynomial. Thus, we have the quantum condition

$$E_n = (n + \tfrac{3}{2})\hbar\omega, \qquad (7\text{–}112)$$

---

[1] The centrifugal barrier is absent for $l = 0$; and in this case ψ is, in general, finite and not zero at $r = 0$.

as before [Eq. (7–97)]. The eigenfunctions are

$$\psi_{n,l,m} = e^{-r^2/2}\phi_n^l(r)\,Y_l^m(\theta,\phi) \qquad (n \geq l), \qquad (7\text{–}113)$$

where n and l have the same parity and $\phi_n^l(r)$ is a polynomial in r of degree n, in which the lowest term in r is r^l. These eigenfunctions can be classified according to n and l as indicated in Table 7–1: The entries in the body of the table are the numbers $2l+1$ of linearly independent states belonging to the corresponding values of l. The total number of states for a given value of l is easily found by inspection. Thus, there is one state for $n = 0$, there are 3 states for $n = 1$, $1 + 5 = 6$ for $n = 2$, $3 + 7 = 10$ for $n = 3$, etc. In general, there are $(\frac{1}{2})(n+1)(n+2)$ states for each value of n, in agreement with the conclusion reached earlier in this section. The functions (7–95) and (7–113) are, of course, merely different representations of the degenerate states belonging to a given E_n, and the total number of such states must be independent of the system of coordinates in terms of which they are expressed. Each of the functions (7–95) is a linear combination of the $(\frac{1}{2})(n+1)(n+2)$ functions (7–113) which have the same n (Problem 7–36).

The polynomials $\phi_n^l(r)$ are solutions of the equation

$$r^2\phi'' + 2(r - r^3)\phi' + [2nr^2 - l(l+1)]\phi = 0; \qquad (7\text{–}114)$$

they are of the form

$$\phi_n^l = r^l v, \qquad (7\text{–}115)$$

TABLE 7–1

DEGENERACY OF HARMONIC-OSCILLATOR EIGENFUNCTIONS
CHARACTERIZED BY THE QUANTUM NUMBERS n AND l

l / n	0	1	2	3	4	5
5		3		7		11
4	1		5		9	
3		3		7		
2	1		5			
1		3				
0	1					

in which v is a polynomial of degree $k = (\frac{1}{2})(n - l)$ in the variable

$$t = r^2. \tag{7–116}$$

In terms of v and t, the differential equation (7–114) reduces (Problem 7–28) to

$$t \frac{d^2v}{dt^2} + (l + \tfrac{3}{2} - t) \frac{dv}{dt} + kv = 0. \tag{7–117}$$

This is a special case of the differential equation satisfied by the *Laguerre polynomials*. We shall present a brief description of these functions and find the normalization constant for the wave functions (7–113).

7–5 Laguerre polynomials. The differential equation

$$t \frac{d^2v}{dt^2} + (\alpha + 1 - t) \frac{dv}{dt} + kv = 0, \tag{7–118}$$

in which k is an integer and $\alpha > -1$, can be shown to have a solution (Problem 7–29)

$$v = L_k^\alpha(t), \tag{7–119}$$

which is a polynomial of degree k. No solution of Eq. (7–118) which is linearly independent of (7–119) is analytic at $t = 0$, i.e., there is only one polynomial solution (Problem 7–30). Consequently, the polynomial (7–119) is uniquely defined by the differential equation except for a constant multiplier which is conventionally chosen so that the coefficient of x^k in $L_k^\alpha(t)$ is $(-)^k/k!$. With this convention, (7–119) is the Laguerre polynomial[1]

$$L_k^\alpha(t) = \sum_{\nu=0}^{k} \binom{k+\alpha}{k-\nu} \frac{(-t)^\nu}{\nu!}. \tag{7–120}$$

The symbol $\binom{p}{q}$ is used to denote the quantity[2]

$$\binom{p}{q} = \frac{\Gamma(p+1)}{\Gamma(q+1)\Gamma(p-q+1)} = \frac{p(p-1)\ldots(p-q+2)(p-q+1)}{q!}$$

$$(q \text{ an integer} \geq 0). \tag{7–121}$$

---

[1] The term "Laguerre polynomial" is sometimes restricted to mean the polynomial $L_k(t) = L_k^0(t)$, while the function (7–120) for $\alpha \neq 0$ is called an "associated Laguerre polynomial." The definition of $L_k^\alpha(t)$ used here is that of G. Szegö, *Orthogonal Polynomials*. New York: American Mathematical Society Colloquium Publications, 1939, Vol. XXIII, p. 96. Cf. also A. Erdélyi *et al.*, *Higher Transcendental Functions*. New York: McGraw-Hill Book Co., Inc., 1953, Vol. II, Section 10–12.

[2] $\Gamma(p + 1) = \int_0^\infty x^p e^{-x} \, dx = p\Gamma(p)$; $\Gamma(1) = 1$; $\Gamma(\frac{1}{2}) = \sqrt{\pi}$. If p is an integer, $\Gamma(p + 1) = p!$.

The first three Laguerre polynomials are (Problem 7–31):

$$L_0^\alpha = 1, \qquad L_1^\alpha = -t + \alpha + 1,$$

$$L_2^\alpha = \tfrac{1}{2}[t^2 - 2(\alpha + 2)t + (\alpha + 1)(\alpha + 2)]. \qquad (7\text{–}122)$$

Differentiation of Eq. (7–118) yields

$$t\,\frac{d^2}{dt^2}\left(\frac{dv}{dt}\right) + (\alpha + 2 - t)\,\frac{d}{dt}\left(\frac{dv}{dt}\right) + (k - 1)\left(\frac{dv}{dt}\right) = 0. \quad (7\text{–}123)$$

This differential equation, however, is obtained from Eq. (7–118) by the substitutions

$$v \to \frac{dv}{dt}, \qquad \alpha \to \alpha + 1, \qquad k \to k - 1, \qquad (7\text{–}124)$$

and it follows that the polynomials $(d/dt)L_k^\alpha$ and $L_{k-1}^{\alpha+1}$ are proportional:

$$\frac{d}{dt}\,L_k^\alpha = C L_{k-1}^{\alpha+1}. \qquad (7\text{–}125)$$

By comparison of the terms in t^{k-1} in this equation, one has

$$\frac{(-)^k}{k!}\,k = C\,\frac{(-)^{k-1}}{(k - 1)!}, \qquad (7\text{–}126)$$

whence $C = -1$, and we have the recurrence formula

$$\frac{d}{dt}\,L_k^\alpha = -L_{k-1}^{\alpha+1}. \qquad (7\text{–}127)$$

Also the polynomial

$$t\,\frac{d}{dt}\,L_k^\alpha - k L_k^\alpha = -\frac{(k + \alpha)}{(k - 1)!}\,(-t)^{k-1} + \cdots \qquad (7\text{–}128)$$

satisfies the differential equation obtained from Eq. (7–118) by the substitution $k \to k - 1$ (Problem 7–32) and is therefore proportional to

$$L_{k-1}^\alpha = \frac{(-t)^{k-1}}{(k - 1)!} + \cdots.$$

The constant of proportionality is $-(k + \alpha)$; hence

$$t\,\frac{d}{dt}\,L_k^\alpha - k L_k^\alpha = -(k + \alpha)L_{k-1}^\alpha. \qquad (7\text{–}129)$$

By means of this relation and Eq. (7–118), it can also be shown that

$$t \frac{d}{dt} L_k^\alpha + (k + \alpha + 1 - t)L_k^\alpha = (k + 1)L_{k+1}^\alpha, \qquad (7\text{–}130)$$

which can be used to construct the polynomials L_k^α successively, beginning with $L_0^\alpha = 1$. Finally, by combining Eqs. (7–129) and (7–130), we have the recurrence formula

$$(k + 1)L_{k+1}^\alpha - (2k + \alpha + 1 - t)L_k^\alpha + (k + \alpha)L_{k-1}^\alpha = 0. \quad (7\text{–}131)$$

Using the above formulae, the generating function

$$g(t, h) = \sum_{k=0}^\infty L_k^\alpha(t)h^k \qquad (7\text{–}132)$$

can be shown to satisfy the equations

$$(1 - h)\frac{\partial g}{\partial t} + hg = 0, \qquad (1 - h)\frac{\partial g}{\partial h} = t\frac{\partial g}{\partial t} + (\alpha + 1 - t)g. \quad (7\text{–}133)$$

From these equations and the fact that

$$g(t, 0) = L_0^\alpha = 1, \qquad (7\text{–}134)$$

one obtains (Problem 7–33)

$$g(t, h) = \frac{1}{(1 - h)^{\alpha+1}} e^{-th/(1-h)}. \qquad (7\text{–}135)$$

The function

$$w_k = e^{-t/2} t^{(\alpha+1)/2} L_k^\alpha(t) \qquad (7\text{–}136)$$

is found, by substitution in Eq. (7–118), to satisfy the differential equation

$$t \frac{d^2 w_k}{dt^2} + \left(k + \frac{\alpha + 1}{2} - \frac{t}{4} + \frac{1 - \alpha^2}{4t}\right) w_k = 0, \qquad (7\text{–}137)$$

which does not contain a term in dw_k/dt. If $k \neq k'$, it follows that (Problem 7–34)

$$w_{k'} \frac{d^2 w_k}{dt^2} - w_k \frac{d^2 w_{k'}}{dt^2} + \frac{k - k'}{t} w_k w_{k'} = 0, \qquad (7\text{–}138)$$

or

$$\frac{d}{dt}\left(w_{k'} \frac{dw_k}{dt} - w_k \frac{dw_{k'}}{dt}\right) = \frac{k' - k}{t} w_k w_{k'}. \qquad (7\text{–}139)$$

If this equation is integrated between the limits $0, \infty$, the left-hand member vanishes $(\alpha > -1)$, and we have

$$(k' - k) \int_0^\infty w_k w_{k'} \frac{dt}{t} = 0, \tag{7-140}$$

which is to say that

$$\int_0^\infty e^{-t} t^\alpha L_k^\alpha(t) L_{k'}^\alpha(t) \, dt = 0 \qquad (k \neq k'). \tag{7-141}$$

The polynomials $L_k^\alpha(t)$ are orthogonal in $(0, \infty)$ with respect to the weight function $e^{-t} t^\alpha$.

The integral (7–141) for $k = k'$ can be evaluated with the help of the generating function. Thus, by Eqs. (7–132) and (7–141), we obtain

$$\int_0^\infty e^{-t} t^\alpha g^2(t, h) \, dt = \sum_{k=0}^\infty \int_0^\infty e^{-t} t^\alpha [L_k^\alpha(t)]^2 \, dt \cdot h^{2k}, \tag{7-142}$$

and substitution of Eq. (7–135) yields[1]

$$\frac{1}{(1 - h)^{2\alpha+2}} \int_0^\infty t^\alpha e^{-t(1+h)/(1-h)} \, dt = \frac{\Gamma(\alpha + 1)}{(1 - h^2)^{\alpha+1}}$$

$$= \sum_{k=0}^\infty \Gamma(\alpha + 1) \binom{k + \alpha}{k} h^{2k}. \tag{7-143}$$

Comparing the series (7–142) and (7–143), we have

$$\int_0^\infty e^{-t} t^\alpha [L_k^\alpha(t)]^2 \, dt = \Gamma(\alpha + 1) \binom{k + \alpha}{k}. \tag{7-144}$$

Therefore, the *Laguerre functions*

$$\Lambda_k^\alpha(t) = \left[\Gamma(\alpha + 1) \binom{k + \alpha}{k} \right]^{-1/2} e^{-t/2} t^{\alpha/2} L_k^\alpha(t) \tag{7-145}$$

are orthonormal in $(0, \infty)$, that is,

$$\int_0^\infty \Lambda_k^\alpha(t) \Lambda_{k'}^\alpha(t) \, dt = \delta_{kk'}. \tag{7-146}$$

---

[1] $\dfrac{1}{(1 - x)^n} = \displaystyle\sum_{k=0}^\infty \binom{n + k - 1}{k} x^k.$

According to Eqs. (7–113), (7–115), (7–116), and (7–117), the harmonic oscillator wave functions are

$$\psi_{n,l,m} = Ne^{-r^2/2}r^l L_k^{l+1/2}(r^2)Y_l^m(\theta, \phi), \tag{7-147}$$

in which N is a normalization factor determined by

$$(\psi_{n,l,m}, \psi_{n,l,m}) = 1 = |N|^2 \int_0^\infty e^{-t}t^{l+1/2}[L_k^\alpha(t)]^2 \frac{dt}{2}.$$

This integral has been evaluated above, i.e.,

$$1 = \frac{|N|^2}{2}\Gamma(l+\tfrac{3}{2})\binom{k+l+\tfrac{1}{2}}{k}$$

$$= \frac{|N|^2}{2k!}\frac{[2(k+l)+1]}{2}\cdot\frac{[2(k+l)-1]}{2}\cdots\tfrac{3}{2}\cdot\tfrac{1}{2}\sqrt{\pi}, \tag{7-148}$$

whence

$$\psi_{n,l,m} = \sqrt{\frac{2}{r}}\Lambda_k^{l+1/2}(r^2)Y_l^m(\theta, \phi) \quad [k=\tfrac{1}{2}(n-l)]. \tag{7-149}$$

TABLE 7–2

THE LAGUERRE POLYNOMIALS $L_k^{l+1/2}(t)$

$n=0,$	$l=0,$	$k=0$	$L_0^{1/2}=1$
$n=1,$	$l=1,$	$k=0$	$L_0^{3/2}=1$
$n=2,$	$l=0,$	$k=1$	$L_1^{1/2}=\tfrac{3}{2}-t$
	$l=2,$	$k=0$	$L_0^{5/2}=1$
$n=3,$	$l=1,$	$k=1$	$L_1^{3/2}=\tfrac{5}{2}-t$
	$l=3,$	$k-0$	$L_0^{7/2}=1$
$n=4,$	$l=0,$	$k=2$	$L_2^{1/2}=\tfrac{15}{8}-\tfrac{5}{2}t+\tfrac{1}{2}t^2$
	$l=2,$	$k=1$	$L_1^{5/2}=\tfrac{7}{2}-t$
	$l=4,$	$k=0$	$L_0^{9/2}=1$
$n=5,$	$l=1,$	$k=2$	$L_2^{3/2}=\tfrac{35}{8}-\tfrac{7}{2}t+\tfrac{1}{2}t^2$
	$l=3,$	$k=1$	$L_1^{7/2}=\tfrac{9}{2}-t$
	$l=5,$	$k=0$	$L_0^{11/2}=1$

The functions $L_k^{l+1/2}(t)$ for $n = 0, 1, 2, 3, 4, 5$ are given in Table 7–2, and the normalized radial functions

$$\sqrt{\frac{2}{r}}\, \Lambda_k^{l+1/2}\,(r^2)$$

are shown graphically in Fig. 7–4, in which states belonging to the same energy are grouped together.

The $(2l + 1)$-fold degeneracy of the states of given l is fundamental and arises from the symmetry. However, in this problem it is seen that states with different values of l may also have the same energy, so that the total degeneracy is greater than $2l + 1$. For example, the states with $n = 2$ have angular momentum 0 and 2. This degeneracy is associated with the special form of the potential-energy function $\frac{1}{2}kr^2$ and is called *accidental*. We shall see in Section 11–3 that the accidental degeneracy is removed by the addition of a spherically symmetric perturbation.

7–6 Many-particle systems. The next system to be considered is the hydrogen atom. This atom is a two-particle system, consisting of a proton and an electron, attracted by the Coulomb force. It is well known (Section 1–11) that, classically, a two-particle problem can be reduced to two equivalent one-particle problems, in which the motion of the center of mass and the relative motion are treated separately. This reduction can also be accomplished in the quantum-mechanical problem. Before proceeding, however, we must extend the definition and interpretation of the wave function to systems containing more than one particle. (Cf. Section 4–1).

The appropriate generalization is indicated by the Born interpretation of the single-particle wave function, i.e.: The quantity $|\psi(\mathbf{r})|^2\,d\mathbf{r}$ is proportional to the probability that the particle will be found in $d\mathbf{r}$ at the point $\mathbf{r}$. Consider two *independent* one-particle systems (denoted by 1 and 2), each consisting of a single particle moving in its own field of force. The wave functions of the two systems are $\psi_1(\mathbf{r}_1)$ and $\psi_2(\mathbf{r}_2)$, and the corresponding Hamiltonian operators are

$$H_1 = -\frac{\hbar^2}{2m_1}\,\nabla_1^2 + V_1(\mathbf{r}_1) \quad \text{and} \quad H_2 = -\frac{\hbar^2}{2m_2}\,\nabla_2^2 + V_2(\mathbf{r}_2).$$

If the functions $\psi_1(\mathbf{r}_1)$ and $\psi_2(\mathbf{r}_2)$ are eigenfunctions of the energy,

$$H_1\psi_1 = E_1\psi_1, \qquad H_2\psi_2 = E_2\psi_2, \tag{7–150}$$

then the function

$$\Psi(\mathbf{r}_1, \mathbf{r}_2) = \psi_1(\mathbf{r}_1)\psi_2(\mathbf{r}_2) \tag{7–151}$$

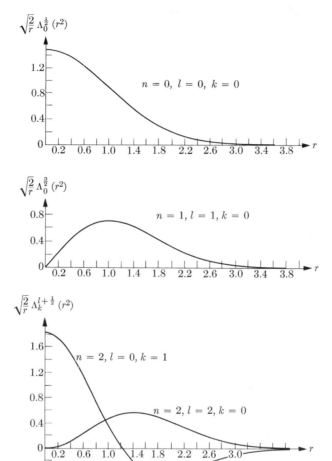

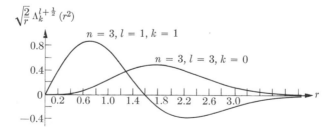

FIG. 7–4. The normalized radial functions $\sqrt{2/r}\ \Lambda_k^{l+1/2}\ (r^2)$ for the three-dimensional harmonic oscillator.

satisfies the equation

$$H\Psi = E\Psi, \tag{7-152}$$

in which

$$H = H_1 + H_2 \quad \text{and} \quad E = E_1 + E_2. \tag{7-153}$$

The interpretation of Eq. (7–152) is clear: The wave function (7–151) describes the particles 1 and 2 considered together as a single system. It is an eigenfunction of the total Hamiltonian which belongs to the energy E, i.e., to the total energy of the combined system. Furthermore, since the two systems are assumed to be independent, the probability that particles 1 and 2 are at the same instant in $d\mathbf{r}_1$ and $d\mathbf{r}_2$, respectively, is the product

$$|\psi_1(\mathbf{r}_1)|^2 \, d\mathbf{r}_1 \cdot |\psi_2(\mathbf{r}_2)|^2 \, d\mathbf{r}_2 = |\Psi(\mathbf{r}_1, \mathbf{r}_2)|^2 \, d\mathbf{r}_1 \, d\mathbf{r}_2. \tag{7-154}$$

These observations can be generalized as follows: *The quantum-mechanical description of a system of N particles is given in terms of a wave function*

$$\Psi(\mathbf{r}_1, \mathbf{r}_2, \ldots, \mathbf{r}_N, t), \tag{7-155}$$

which is a function of the $3N$ coordinates of the particles and of the time. The behavior of the wave function is governed by the Schrödinger equation

$$H\Psi + \frac{\hbar}{i} \frac{\partial \Psi}{\partial t} = 0, \tag{7-156}$$

where the Hamiltonian operator

$$H = H\left(\frac{\hbar}{i}\, \nabla_1, \frac{\hbar}{i}\, \nabla_2, \ldots, \frac{\hbar}{i}\, \nabla_N; \mathbf{r}_1, \mathbf{r}_2, \ldots, \mathbf{r}_N\right) \tag{7-157}$$

is obtained from the classical Hamiltonian function

$$H(\mathbf{p}_1, \mathbf{p}_2, \ldots, \mathbf{p}_N; \mathbf{r}_1, \mathbf{r}_2, \ldots, \mathbf{r}_N)$$

by the correspondence

$$\mathbf{p}_i \to \mathbf{p}_{i_{\mathrm{op}}} = \frac{\hbar}{i}\, \nabla_i, \qquad \mathbf{r}_i \to \mathbf{r}_{i_{\mathrm{op}}} = \mathbf{r}_i.$$

The quantity

$$|\Psi|^2 \, d\mathbf{r}_1, d\mathbf{r}_2, \ldots, d\mathbf{r}_N$$

is interpreted as the probability for particle 1 to be found in $d\mathbf{r}_1$, particle 2 in $d\mathbf{r}_2$, etc., at the instant t. The energy states of the system are described by wave functions of the form

$$\Psi(\mathbf{r}_i, t) = \Psi(\mathbf{r}_i) e^{-(i/\hbar)Et}, \tag{7-158}$$

where

$$H\Psi = E\Psi. \qquad (7\text{–}159)$$

Several remarks about these generalizations must be made. First, the function Ψ depends upon the $3N$ coordinates of all the particles. $|\Psi|^2$ is therefore a probability density in the $3N$-dimensional *configuration space* of the system. The "waves" represented by Ψ are waves in this configuration space, and are not to be regarded as having a substantial reality. The interpretation of $|\Psi|^2$ in terms of a particle density (see Section 2–9) is therefore restricted to one-particle noninteracting systems. A clear understanding of this fact, which was first recognized by Born, is essential if confusion in the interpretation of Ψ is to be avoided. Second, the meaning of the operators p_i has been extended: they are now regarded as operating on functions of $3N$ variables, rather than of 3 variables, as before. The partial derivative sign in $p_{x_1}\Psi = (\hbar/i)(\partial\Psi/\partial x_1)$ now means: Calculate the derivative of Ψ with respect to x_1, keeping the variables y_1, z_1, and $x_2, y_2, z_2, x_3, y_3, z_3, \ldots, x_N, y_N, z_N$ constant. The commutation rules for momentum and coordinates are therefore

$$[p_{x_1}, x_1] = \frac{\hbar}{i}, \qquad [p_{x_1}, y_1] = 0, \qquad [p_{x_1}, x_2] = 0, \qquad \text{etc.} \quad (7\text{–}160)$$

In other words, p_{x_1} commutes with all the variables x_i, y_i, z_i except x_1. Third, the Hamiltonian function (7–157) is not, in general, separable in the form (7–153), but contains parts which depend jointly on the positions (or velocities) of two or more particles. These parts of the Hamiltonian describe the interactions among the particles. A separation like (7–153) is, in general, possible only if the system is composed of two noninteracting parts. Finally, all previous discussions of wave packets, probability current, expectation, etc., are now to be generalized to apply to the $3N$-dimensional configuration space. The formula (6–10) for the expectation, that is,

$$\langle F \rangle = \frac{(\Psi, F\Psi)}{(\Psi, \Psi)}, \qquad (7\text{–}161)$$

is retained, the scalar product having now the extended meaning

$$(\Psi, \Phi) = \int \Psi^*(\mathbf{r}_1, \mathbf{r}_2, \ldots, \mathbf{r}_N)\Phi(\mathbf{r}_1, \mathbf{r}_2, \ldots, \mathbf{r}_N)\, d\mathbf{r}_1, d\mathbf{r}_2, \ldots, d\mathbf{r}_N,$$
$$(7\text{–}162)$$

in which the integration is over the entire configuration space of the system.

It is clear that the quantum mechanics of many-particle systems is mathematically very complex. No general procedure for obtaining exact results is known, and a large part of the methods and techniques of the subject is associated with attempts to reduce complex systems to combinations of simpler independent ones. This procedure is illustrated in the problem of the hydrogen atom, where it can be carried out exactly.

7–7 The hydrogen atom. The Hamiltonian function for the hydrogen atom is

$$H = \frac{p_1^2}{2m_1} + \frac{p_2^2}{2m_2} - \frac{e^2}{|\mathbf{r}_1 - \mathbf{r}_2|}, \qquad (7\text{–}163)$$

in which the subscripts 1 and 2 denote the electron and the proton, respectively. The wave function $\Psi(\mathbf{r}_1, \mathbf{r}_2)$ depends upon six variables and is a solution of the Schrödinger equation

$$H\Psi = -\frac{\hbar^2}{2m_1}\nabla_1^2\Psi - \frac{\hbar^2}{2m_2}\nabla_2^2\Psi - \frac{e^2}{|\mathbf{r}_1 - \mathbf{r}_2|}\Psi = E_t\Psi, \quad (7\text{–}164)$$

where E_t is the total energy.

The separation of the problem is accomplished by changing the variables to $\mathbf{R}$ and $\mathbf{r}$, where

$$\mathbf{R} = [X, Y, Z] = \frac{m_1\mathbf{r}_1 + m_2\mathbf{r}_2}{m_1 + m_2} \qquad (7\text{–}165)$$

is the position vector of the center of mass, and

$$\mathbf{r} = [x, y, z] = \mathbf{r}_1 - \mathbf{r}_2 \qquad (7\text{–}166)$$

is the separation between electron and proton. Solving for $\mathbf{r}_1$ and $\mathbf{r}_2$ in terms of $\mathbf{R}$ and $\mathbf{r}$, we have

$$\mathbf{r}_1 = \mathbf{R} + \frac{m}{m_1}\mathbf{r}, \qquad \mathbf{r}_2 = \mathbf{R} - \frac{m}{m_2}\mathbf{r}, \qquad (7\text{–}167)$$

in which

$$m = \frac{m_1 m_2}{m_1 + m_2}$$

is the reduced mass. For the partial derivatives of Ψ, we find

$$\frac{\partial\Psi}{\partial x_1} = \frac{m_1}{m_1 + m_2}\frac{\partial\Psi}{\partial X} + \frac{\partial\Psi}{\partial x}, \qquad \frac{\partial\Psi}{\partial x_2} = \frac{m_2}{m_1 + m_2}\frac{\partial\Psi}{\partial X} - \frac{\partial\Psi}{\partial x},$$

or

$$\nabla_1 = \frac{m}{m_2}\nabla_R + \nabla, \qquad \nabla_2 = \frac{m}{m_1}\nabla_R - \nabla, \qquad (7\text{–}168)$$

where

$$\nabla_R = \left[\frac{\partial}{\partial X}, \frac{\partial}{\partial Y}, \frac{\partial}{\partial Z}\right], \qquad \text{and} \qquad \nabla = \left[\frac{\partial}{\partial x}, \frac{\partial}{\partial y}, \frac{\partial}{\partial z}\right].$$

By these substitutions, Eq. (7–164) becomes ($r = |\mathbf{r}|$)

$$-\frac{\hbar^2}{2(m_1 + m_2)} \, \nabla_R^2 \Psi - \frac{\hbar^2}{2m} \, \nabla^2 \Psi - \frac{e^2}{r} \, \Psi = E_t \Psi, \qquad (7\text{–}169)$$

where Ψ is now to be regarded as a function of $\mathbf{R}$ and $\mathbf{r}$. Equation (7–169) is separable; it has a solution in the form of a product function:

$$\Psi(\mathbf{R}, \mathbf{r}) = \phi(\mathbf{R})\psi(\mathbf{r}). \qquad (7\text{–}170)$$

By the usual procedure, one finds that the functions ϕ and ψ satisfy the equations

$$-\frac{\hbar^2}{2(m_1 + m_2)} \, \nabla_R^2 \phi(\mathbf{R}) = E_c \phi(\mathbf{R}), \qquad (7\text{–}171)$$

$$-\frac{\hbar^2}{2m} \, \nabla^2 \psi(\mathbf{r}) - \frac{e^2}{r} \, \psi(\mathbf{r}) = E\psi(\mathbf{r}), \qquad (7\text{–}172)$$

where $E_c + E = E_t$.

The equation for $\phi(\mathbf{R})$ has the general solution

$$\phi(\mathbf{R}) = \text{constant} \times e^{(i/\hbar)\mathbf{P}\cdot\mathbf{R}}, \qquad (7\text{–}173)$$

where $\mathbf{P}$ is a constant vector of arbitrary direction; the magnitude of $\mathbf{P}$ is

$$|\mathbf{P}| = \sqrt{2(m_1 + m_2)E_c}. \qquad (7\text{–}174)$$

The function $\phi(\mathbf{R})$ represents the motion of the center of mass of the system as that of a single particle of mass $m_1 + m_2$ and energy $E_c = P^2/2(m_1 + m_2)$. It describes a plane wave, indicating that the motion of the center of mass is that of a free particle. This corresponds to the classical result that the center of mass moves in a straight line with constant speed.

Equation (7–172) for the relative motion is the Schrödinger equation for an equivalent particle having the reduced mass and moving in a fixed central field, $V(r) = -e^2/r$. The energy of relative motion, E, is determined as the eigenvalue of this equivalent problem. In the treatment of Eq. (7–172), it is convenient to introduce "atomic" units, in which the energy is measured in multiples of the ionization energy of hydrogen, $me^4/2\hbar^2$, and the coordinate in terms of the Bohr radius $\hbar^2/me^2$. Expressed in these units, which are equivalent to writing $\hbar = 1$, $e^2 = 2$, $m = \frac{1}{2}$, Eq. (7–172) becomes

$$\nabla^2 \psi + \left(E + \frac{2}{r}\right)\psi = 0. \qquad (7\text{–}175)$$

The potential-energy function $V(r) = 2/r$ is spherically symmetric, and the eigenfunctions are therefore $(2l + 1)$-fold degenerate, having the form

$$\psi = R(r) Y_l^m(\theta, \phi). \qquad (7\text{--}176)$$

The radial equation (7–64) is

$$\frac{d^2u}{dr^2} + \left[E + \frac{2}{r} - \frac{l(l + 1)}{r^2} \right] u = 0 \qquad (u = rR). \qquad (7\text{--}177)$$

Since R must be finite at $r = 0$ and such that the integral

$$\int_0^\infty R^2 r^2 \, dr = \int_0^\infty u^2 \, dr$$

exists, the boundary conditions on the solutions of Eq. (7–177) are

$$u = 0 \quad \text{at} \quad r = 0 \quad \text{and} \quad u \to 0 \quad \text{as} \quad r \to \infty. \qquad (7\text{--}178)$$

From the discussion of Section 5–4, we expect to find a continuum of solutions for $E > 0$, corresponding to the classical hyperbolic orbits, and a discrete spectrum of energy states for $E < 0$, corresponding to the elliptic orbits of the Bohr theory. Attention will be given to the bound states of negative energy. We proceed by the polynomial method, which is now familiar.

If r is large, Eq. (7–177) is approximated by

$$u'' - (-E)u = 0 \qquad \left[u' = \frac{du}{dr}, u'' = \frac{d^2u}{dr^2} \right], \qquad (7\text{--}179)$$

which has the solutions $u = \exp(\pm\sqrt{-E}\, r)$, and we are led to make the substitution

$$u = w \exp(-\sqrt{-E}\, r). \qquad (7\text{--}180)$$

The differential equation for w is

$$w'' - 2\sqrt{-E}\, w' + \left(\frac{2}{r} - \frac{l(l + 1)}{r^2} \right) w = 0. \qquad (7\text{--}181)$$

Note that this substitution results in an equation which will yield a two-term recurrence relation for the coefficients in the series solution. Equation (7–177) would lead to a three-term recurrence relation.

Equation (7–181) can be solved formally by means of a power series in r, beginning with a term in r^α. The exponent α is determined by the equation

$$\alpha(\alpha - 1) - l(l + 1) = 0, \qquad (7\text{–}182)$$

which is obtained as the coefficient of $r^{\alpha-2}$ if r^{α} is substituted for w in Eq. (7–181). This gives

$$\alpha = l + 1 \quad \text{or} \quad \alpha = -l, \qquad (7\text{–}183)$$

and since l is not negative, the first of the conditions (7–178) is satisfied only by the root $\alpha = l + 1$. Hence the power series is

$$w = \sum_{k=l+1}^{\infty} a_k r^k, \qquad (7\text{–}184)$$

in which a_{l+1} is an arbitrary number and the remaining a_k are determined by the recurrence formula

$$a_{k+1} = 2\, \frac{k\sqrt{-E} - 1}{k(k + 1) - l(l + 1)}\, a_k. \qquad (7\text{–}185)$$

As $k \to \infty$, the ratio a_{k+1}/a_k approaches $2\sqrt{-E}/k$; thus w behaves, for large r, like the function $\exp(2\sqrt{-E}\, r)$. Therefore u is not bounded as $r \to \infty$ unless the series (7–184) is broken off at some value n of the index k. This happens, and w is a polynomial, if

$$E = E_n = -\frac{1}{n^2} \qquad (n > l), \qquad (7\text{–}186)$$

or, in conventional units,

$$E_n = -\frac{me^4}{2n^2\hbar^2} \qquad (n = l + 1, l + 2, \ldots). \qquad (7\text{–}187)$$

The energy levels for hydrogen are therefore exactly the same as those obtained from the Bohr theory described in Chapter 1. The principal quantum number n determines the energy, and for each level E_n, l can have any non-negative value smaller than n. The energy diagram is shown in Fig. 7–5, in which the fundamental degeneracy $(2l + 1)$ of each (n, l) state is indicated by the small numeral. The coincidence of the energies of states belonging to different l is accidental and is removed by any perturbation whose r-dependence is not $1/r$. The total degeneracy of the states of energy E_n is

$$1 + 3 + 5 + \cdots + (2n - 1) = n^2. \qquad (7\text{–}188)$$

The radial eigenfunctions $u = u_{n,l}$ are of the form

$$u_{n,l} = e^{-r/n} r^{l+1} v, \qquad (7\text{–}189)$$

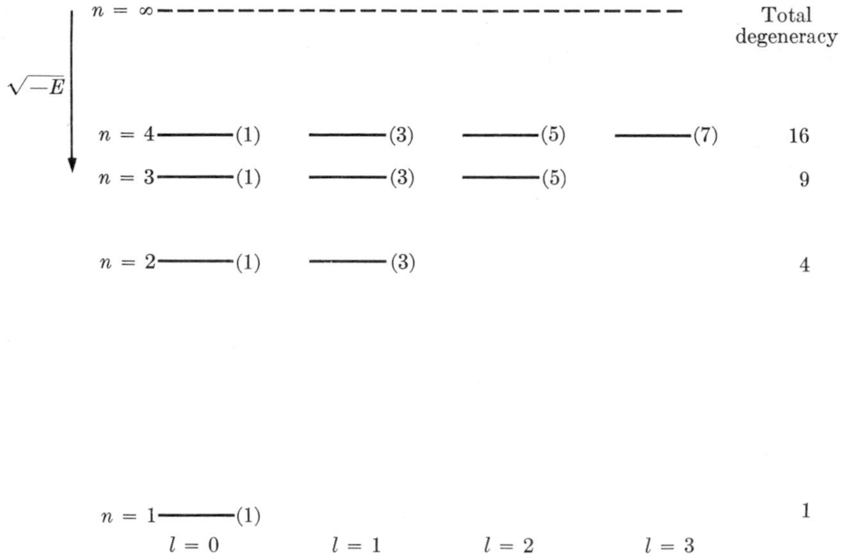

$n = \infty$ ———————————————————————— Total
degeneracy

$\sqrt{-E}$

$n = 4$ ———(1)   ———(3)   ———(5)   ———(7)   16

$n = 3$ ———(1)   ———(3)   ———(5)   9

$n = 2$ ———(1)   ———(3)   4

$n = 1$ ———(1)   1

$l = 0$ $l = 1$ $l = 2$ $l = 3$

FIG. 7–5. Energy level diagram for the hydrogen atom. The $(2l+1)$-fold degeneracy of each state is indicated in parentheses, and the total degeneracy for all states with a given principal quantum number n has been entered on the right.

where v is a polynomial solution of the differential equation

$$t \frac{d^2v}{dt^2} + (2l + 2 - t) \frac{dv}{dt} + (n - l - 1)v = 0, \qquad (7\text{–}190)$$

where $t = 2r/n$. This equation defines the Laguerre polynomial $L_{n-l-1}^{2l+1}(t)$, whence

$$u_{n,l} = N e^{-r/n} r^{l+1} L_{n-l-1}^{2l+1}\left(\frac{2r}{n}\right), \qquad (7\text{–}191)$$

in which the normalization constant N is determined by

$$\int_0^\infty |u|^2 \, dr = |N|^2 \int_0^\infty e^{-2r/n} r^{2l+2} \left[L_{n-l-1}^{2l+1}\left(\frac{2r}{n}\right) \right]^2 dr = 1, \quad (7\text{–}192)$$

or

$$\left(\frac{n}{2}\right)^{2l+3} |N|^2 \int_0^\infty e^{-t} t^{\alpha+1} [L_k^\alpha(t)]^2 \, dt = 1$$

$$(\alpha = 2l + 1, \qquad k = n - l - 1). \qquad (7\text{–}193)$$

This normalization integral differs from (7–144) in that the integrand contains the factor $t^{\alpha+1}$ rather than t^{α}. However, by the recurrence relation (7–131), we have

$$tL_k^{\alpha} = (2k + \alpha + 1)L_k^{\alpha} - (k + 1)L_{k+1}^{\alpha} - (k + \alpha)L_{k-1}^{\alpha}; \quad (7\text{–}194)$$

multiplying by $e^{-t}t^{\alpha}L_k^{\alpha}(t)$ and integrating, we obtain from Eq. (7–141)

$$\int_0^{\infty} e^{-t}t^{\alpha+1}[L_k^{\alpha}]^2 \, dt = (2k + \alpha + 1) \int_0^{\infty} e^{-t}t^{\alpha}[L_k^{\alpha}]^2 \, dt$$

$$= (2k + \alpha + 1)\Gamma(\alpha + 1)\binom{k + \alpha}{k}. \quad (7\text{–}195)$$

Combining Eqs. (7–195), (7–193), and (7–191), and using Eq. (7–145), we have

$$u_{n,l} = \sqrt{\frac{2r}{n^3}} \, \Lambda_{n-l-1}^{2l+1}\left(\frac{2r}{n}\right), \quad (7\text{–}196)$$

and the orthonormal wave functions for the bound states of hydrogen are

$$\psi_{n,l,m} = \frac{1}{r} \, u_{n,l}(r) \, Y_l^m(\theta, \phi). \quad (7\text{–}197)$$

The functions $u_{n,l}$ for $n = 1, 2, 3$ are[1]

$$u_{10} = 2re^{-r}$$

$$u_{20} = \frac{1}{\sqrt{8}} \, e^{-r/2} r(2 - r), \qquad u_{21} = \frac{1}{\sqrt{24}} \, e^{-r/2} r^2$$

$$u_{30} = \frac{2}{81\sqrt{3}} \, e^{-r/3} r(27 - 18r + 2r^2), \qquad (7\text{–}198)$$

$$u_{31} = \frac{4}{81\sqrt{6}} \, e^{-r/3} r^2(6 - r), \qquad u_{32} = \frac{4}{81\sqrt{30}} \, e^{-r/3} r^3.$$

The functions $u_{n,l}^2$ are shown in Fig. 7–6.

The differential equation (7–177) for the radial functions $u_{n,l}$, i.e.,

$$\frac{d^2u}{dr^2} + \left[E + \frac{2}{r} - \frac{l(l + 1)}{r^2}\right] u = 0, \qquad (7\text{–}199)$$

is, for $l \neq 0$, exactly of the form (5–47), in which the equivalent potential

---

[1] E. U. Condon and G. H. Shortley, *The Theory of Atomic Spectra*. Cambridge: Cambridge University Press, 1953, p. 117.

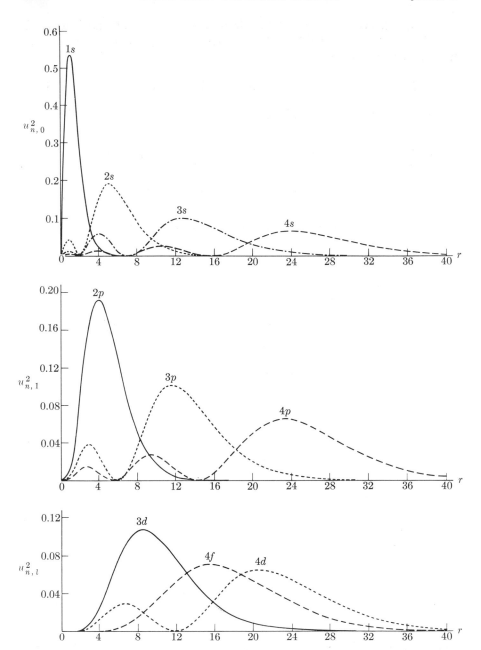

FIG. 7-6.   The radial probability distribution function $u_{n,l}^2$ for several values of the quantum numbers n, l.   (From E. U. Condon and G. H. Shortley, *The Theory of Atomic Spectra*, Cambridge University Press, Cambridge, 1953, by permission.)

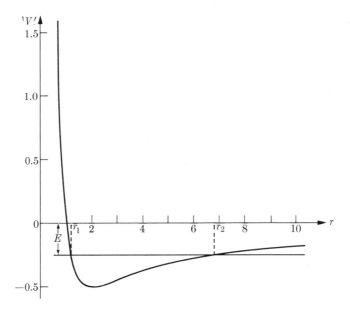

Fig. 7-7. Equivalent potential $'V'$ for the hydrogen atom [Eq. (7-200)] in the state $n = 2, l = 1$.

energy (Fig. 7-7)

$$'V' = -\frac{2}{r} + \frac{l(l+1)}{r^2} \tag{7-200}$$

is infinite at $r = 0$. Hence, the restriction $r \geq 0$ (which was, of course, not imposed upon x in Chapter 5) does not matter since the admissible solutions of Eq. (7-199) for $r < 0$ would vanish identically in any case. The general theory of Sections 5-4 ff. is therefore directly applicable. The eigenfunction of lowest energy corresponds, as we have seen, to $n = l + 1$. This function has zeros only at $r = 0$ and $r = \infty$ [cf. Eq. (7-198)]; the total number of distinct zeros is, in general, $n - l + 1$. The classical turning points are the values of r for which

$$E - \,'V' = -\frac{1}{n^2} + \frac{2}{r} - \frac{l(l+1)}{r^2} = 0, \tag{7-201}$$

giving

$$r_1, r_2 = n[n \mp \sqrt{n^2 - l(l+1)}\,]. \tag{7-202}$$

These points correspond closely to the extrema of the Bohr orbit. The classical energy equation (cf. Section 1-13) is

$$E = \frac{1}{2m}\left(p_r^2 + \frac{p_\phi^2}{r^2}\right) - \frac{e^2}{r}, \tag{7-203}$$

and the maximum and minimum values of r occur when the radial momentum p_r vanishes. Since $p_\phi = \hbar k$ and $E = -me^4/2n^2\hbar^2$, Eq. (7–203), in units $\hbar^2/me^2$ and with $p_r = 0$, is

$$-\frac{1}{n^2} = \frac{k^2}{r^2} - \frac{2}{r}. \tag{7–204}$$

This result becomes the same as Eq. (7–201) if the quantity k^2 is replaced by $l(l + 1)$, showing once more the approximate validity of the Bohr theory.

Because of the exponential character of the wave function outside the classical region, the quantity $|u|^2$ has its largest values within the region of the Bohr orbit. The case $l = n - 1$, which corresponds to the circular orbit, has just one maximum in this region, and as n becomes large, the position of this maximum can be shown to approach the corresponding classical radius. This is an example of the correspondence principle. Further similarities to the Bohr orbits are made evident by the calculation of certain expectation values. For example,

$$\left\langle \frac{1}{r} \right\rangle = \int_0^\infty u_{n,l}^2 \, \frac{dr}{r} = \frac{2}{n^3} \int_0^\infty \left[\Lambda_{n-l-1}^{2l+1} \left(\frac{2r}{n} \right) \right]^2 dr = \frac{1}{n^2}. \tag{7–205}$$

The time average of this quantity in the Bohr-Sommerfeld theory is

$$\overline{\left(\frac{1}{r} \right)} = \frac{1}{P} \int_0^P \frac{1}{r} \, dt = \frac{m}{p_\phi P} \frac{p_\phi^2}{me^2} \int_0^{2\pi} \frac{d\phi}{1 - \epsilon \cos \phi} = \frac{1}{n^2 a_0}, \tag{7–206}$$

in which the integration variable has been changed from t to ϕ by means of

$$p_\phi = mr^2\dot{\phi} = \text{constant}, \tag{7–207}$$

and the expression (1–45) has been substituted. The quantities

$$a_0 = \frac{\hbar^2}{me^2} \quad \text{and} \quad P = \frac{p_\phi^3}{me^4} \frac{2\pi}{(1 - \epsilon^2)^{3/2}} \tag{7–208}$$

are the Bohr radius and the period of the classical motion, respectively. The expectation of the kinetic energy, $T = p^2/2m$, can be obtained directly from Eq. (7–205) and the Schrödinger equation

$$\frac{p^2}{2m} \psi - \frac{e^2}{r} \psi = E\psi. \tag{7–209}$$

Thus,

$$\langle T \rangle = (\psi, T\psi) = \langle E \rangle + \left\langle \frac{2}{r} \right\rangle = E + \frac{2}{n^2} = \frac{1}{n^2} = -\tfrac{1}{2}\langle V \rangle, \tag{7–210}$$

which is referred to as the *virial theorem* (cf. Section 1–9).[1] Further examples of this kind are described in Problem 7–40.

REFERENCES

CONDON, E. U. and SHORTLEY, G. H., *The Theory of Atomic Spectra*. Cambridge: Cambridge University Press, 1953. The definitions of the spherical harmonics (Chapter III, §4) have become generally accepted as standard. Chapter V contains a complete discussion of hydrogen and its relation to the general theory of one-electron systems.

KRAMERS, H. A., *The Foundations of Quantum Theory*. Amsterdam: North-Holland Publishing Company, 1957. Chapter III contains a thorough discussion of the quantum mechanics of many-particle systems. The central-field problem and the hydrogen atom are the subject matter of §46.

ROJANSKY, V., *Introductory Quantum Mechanics*. New York: Prentice-Hall, Inc., 1946. Chapter XII presents tables of the spherical harmonics and a discussion of their theory, based upon the eigenvalue problems for the operators L_z and $\mathbf{L}^2$. The appendix contains a useful collection of formulae relating to the functions $Y_l^m(\theta, \phi)$.

SCHIFF, L. I., *Quantum Mechanics*. 2nd ed., New York: McGraw-Hill Book Co., Inc., 1955. The subject matter of the present chapter is treated in Sections 14, 15, and 16. Section 15 contains a discussion of the three-dimensional square well potential. The solution of the hydrogen problem in parabolic coordinates is given in Section 16.

---

[1] H. Goldstein, *Classical Mechanics*. Reading, Mass.: Addison-Wesley Publishing Co., Inc., 1950, p. 71.

PROBLEMS

7–1. Construct series solutions for the function P [Eq. (7–12)] when $m = 0$. Prove that the functions defined by these series are either divergent or have divergent derivatives at $\mu = 1$ unless $C = l(l + 1)$, and that the only admissible solution is the Legendre polynomial $P_l(\mu)$.

7–2. Show that if P is a solution of Legendre's equation

$$\frac{d}{d\mu}\left[(1 - \mu^2)\frac{dP}{d\mu}\right] + l(l + 1)P = 0,$$

then the function $w = (\mu^2 - 1)^{m/2}(d^m P/d\mu^m)$ is a solution of the associated Legendre equation

$$\frac{d}{d\mu}\left[(1 - \mu^2)\frac{dw}{d\mu}\right] + \left[l(l + 1) - \frac{m^2}{1 - \mu^2}\right]w = 0, \qquad m = \text{an integer.}$$

7–3. From the result of Problem 1, show that

$$P_l(\mu) = \sum_{k=0}^{[l/2]} (-)^k \frac{(2l - 2k)!}{2^l k!(l - k)!(l - 2k)!} \mu^{l-2k},$$

where $[l/2]$ means $l/2$ or $(l - 1)/2$, whichever is an integer.

7–4. Verify that the functions (7–16) are spherical harmonics, and extend the table to $l = 5$.

7–5. Derive the formula (7–30) for Q_l^m.

7–6. Prove the formula following Eq. (7–41).

7–7. Continue the list of Eqs. (7–48) to $l = 4$.

7–8. Show that the polynomials (7–16) are

$$1 = \sqrt{4\pi}\, Y_0^0,$$

$$x = -r\sqrt{\frac{2\pi}{3}}\,(Y_1^1 - Y_1^{-1}), \qquad y = -\frac{r}{i}\sqrt{\frac{2\pi}{3}}\,(Y_1^1 + Y_1^{-1}),$$

$$z = r\sqrt{\frac{4\pi}{3}}\, Y_1^0;$$

$$xy = \frac{r^2}{i}\sqrt{\frac{2\pi}{15}}\,(Y_2^2 - Y_2^{-2}), \qquad yz = -\frac{r^2}{i}\sqrt{\frac{2\pi}{15}}\,(Y_2^1 + Y_2^{-1}),$$

$$zx = -r^2\sqrt{\frac{2\pi}{15}}\,(Y_2^1 - Y_2^{-1});$$

$$x^2 - z^2 = -r^2 \sqrt{\frac{8\pi}{15}} (Y_2^2 + Y_2^{-2}), \quad 2z^2 - x^2 - y^2 = r^2 \sqrt{\frac{16\pi}{5}} Y_2^0,$$

$$y^2 - z^2 = -r^2 \sqrt{\frac{2\pi}{15}} (Y_2^2 + \sqrt{6} Y_2^0 + Y_2^{-2}),$$

and so on.

Normalize these functions and extend the list to $l = 4$.

7–9. Show that the function (7–23) satisfies

$$[T^l(\xi, \eta)]^* = (-)^l T^l(\eta, -\xi),$$

in which ξ and η are regarded as real variables. Hence prove, by comparing the expansions [Eq. (7–24)] for the two members of this identity, that

$$Q_l^{m*} = (-)^m Q_l^{-m}.$$

This provides an alternative derivation of Eq. (7–45).

7–10. Show that

$$Y_l^{-l} = \frac{1}{2^l l!} \sqrt{\frac{(2l+1)!}{4\pi}} \sin^l \theta \, e^{-il\phi},$$

and

$$Y_l^0 = \sqrt{\frac{2l+1}{4\pi}} P_l (\cos \theta).$$

7–11. Deduce that

$$T^l = \left(\frac{r}{2i}\right)^l [\sin \theta \, e^{i\phi} \xi^2 - 2i \cos \theta \, \xi\eta + \sin \theta \, e^{-i\phi} \eta^2]^l$$

$$= (-r)^l l! \sqrt{\frac{4\pi}{2l+1}} \sum_{m=-l}^{l} \frac{(-i)^m}{\sqrt{(l-m)!(l+m)!}} Y_l^m(\theta, \phi) \xi^{l+m} \eta^{l-m},$$

and hence show that

$$Y_l^m(0, \phi) = \begin{cases} \sqrt{\dfrac{2l+1}{4\pi}}, & m = 0, \\ 0, & m \neq 0. \end{cases}$$

Note that this quantity is independent of ϕ, i.e., that the value of Y_l^m at the pole is independent of the direction of approach to $\theta = 0$. This follows from the fact that Y_l^m is a continuous function.

7–12. Prove that $\nabla^2 \psi = \nabla'^2 \psi$, where ∇^2 and ∇'^2 are the operators defined in Eqs. (7–54) and (7–55).

7–13. Show that

$$\sum_{m=-l}^{l} |Y_l^m(\theta, \phi)|^2 = \frac{2l+1}{4\pi},$$

and deduce that $|P_l^m(\mu)|^2 \leq [(l+m)!]/[(l-m)!], \ (-1 < \mu < 1)$.

7–14. Derive the recurrence relations

$$lP_{l-1} = l\mu P_l + (1 - \mu^2)P'_l,$$

$$(l+1)P_{l+1} = (l+1)\mu P_l - (1 - \mu^2)P'_l,$$

$$(l+1)P_{l+1} - (2l+1)\mu P_l + lP_{l-1} = 0,$$

where $P_l = P_l(\mu)$ is the Legendre polynomial of degree l, and $P'_l = dP_l/d\mu$. [Hint: The last of these expressions follows from $\int P_l^2 \, d\mu = 2/(2l+1)$ and the fact that the $P_{l'}(\mu)$, $l' = 0, 1, 2, \ldots, l$ are a complete set of polynomials of degree l.]

7–15. Prove that the generating function

$$g(\mu, h) = \sum_{l=0}^{\infty} P_l(\mu)h^l \qquad (h < 1)$$

is

$$g(\mu, h) = (1 - 2\mu h + h^2)^{-1/2}.$$

Hence show that

$$P_l(1) = 1, \qquad P_l(-1) = (-1)^l.$$

7–16. Demonstrate that

$$Y_l(\pi - \theta, \phi + \pi) = (-)^l Y_l(\theta, \phi),$$

where Y_l is a spherical harmonic of degree l. (Hint: $r^l Y_l$ is a homogeneous polynomial of degree l in x, y, z.)

7–17. Show that Eq. (7–64) is equivalent to Eq. (7–6).

7–18. Prove the commutation rule (7–69) by writing Eq. (7–68) in spherical polar coordinates.

7–19. Derive Eq. (7–79) from the definition (7–73) and the commutation rules (6–83).

7–20. Prove that Eq. (7–80) is equivalent to Eq. (7–79).

7–21. Derive the commutation rules

$$[L_z, x] = i\hbar y, \qquad [L_z, p_x] = i\hbar p_y,$$

$$[L_z, y] = -i\hbar x, \qquad [L_z, p_y] = -i\hbar p_x,$$

$$[L_z, z] = 0, \qquad [L_z, p_z] = 0,$$

etc. \qquad etc.,

and show that $\mathbf{L} \cdot \mathbf{p} = \mathbf{L} \cdot \mathbf{r} = 0$.

7–22. Prove that

$$\mathbf{L} \times \mathbf{r} - i\hbar\mathbf{r} = i\hbar\mathbf{r} - \mathbf{r} \times \mathbf{L} \qquad (=\mathbf{K}),$$

and show that this operator is Hermitian. Show also that

$$[\mathbf{L}^2, \mathbf{r}] = -2i\hbar\mathbf{K}.$$

7–23. Show that

$$[L_x^2, L_y^2] = [L_y^2, L_z^2] = [L_z^2, L_x^2].$$

7–24. Show that

$$[L_x, r^2] = [L_y, r^2] = [L_z, r^2] = 0,$$

where $r^2 = x^2 + y^2 + z^2$.

7–25. By substituting (7–2) in the definition of **L**, show that

$$L_x \pm iL_y = \hbar e^{\pm i\phi}\left(\pm\frac{\partial}{\partial\theta} + i\cot\theta\,\frac{\partial}{\partial\phi}\right),$$

and derive

$$(L_x \pm iL_y)Y_l^m = \hbar\sqrt{(l \mp m)(l \pm m + 1)}\; Y_l^{m-1}.$$

7–26. Derive Eq. (7–101).

7–27. Substitute Eq. (7–102) into Eq. (7–101) and deduce the recurrence formula (7–106).

7–28. Substituting (7–115) and (7–116) into Eq. (7–114), derive Eq. (7–117).

7–29. Solve Eq. (7–118) in series and prove the statements made in the text in connection with Eq. (7–119). Also, derive Eq. (7–120).

7–30. Show that if u and w are both solutions of Eq. (7–118), then

$$e^{-t}t^{\alpha+1}(u'w - uw') = \text{constant}.$$

Hence prove that if u and w are both polynomials in t, they are linearly dependent, i.e., the constant in the above equation is zero.

7–31. Verify the equations (7–122) and extend the list to $k = 4$. Check that each of these polynomials satisfies Eq. (7–118).

7–32. Derive the relation (7–129) by the method indicated in the text and prove Eqs. (7–130) and (7–131). Also, show that

$$\frac{d}{dt}L_k^\alpha = \frac{d}{dt}L_{k-1}^\alpha - L_{k-1}^\alpha.$$

7–33. Derive the relations (7–133) and (7–135). (The equation of Problem 7–32 is useful in obtaining the first of these.)

7–34. Make the substitution (7–136), obtain Eq. (7–137), and complete the steps leading to Eq. (7–144).

7–35. Show that

$$e^{-t}t^\alpha L_k^\alpha(t) = \frac{1}{k!}\left(\frac{d}{dt}\right)^k (e^{-t}t^{k+\alpha}),$$

and construct an alternative proof of the orthogonality relations.

7–36. Prove, by writing out the functions (7–95) in polar coordinates, that

$$[0, 0, 0] = (0, 0, 0),$$

$$[1, 0, 0] = -\frac{1}{\sqrt{2}}\{(1, 1, 1) - (1, 1, -1)\}, \quad (cont.)$$

$$[0, 1, 0] = \frac{i}{\sqrt{2}} \{(1, 1, 1) + (1, 1, -1)\},$$

$$[0, 0, 1] = (1, 1, 0),$$

etc.,

where the symbols $[n_x, n_y, n_z]$ and (n, l, m) represent the functions (7–95) and (7–149), respectively. Work out the corresponding relations for the six functions $(2, l, m)$. Note that the functions on the right in these equations are orthonormal. A linear transformation of this kind, which preserves the orthonormal property, is called *unitary*. Solve the above equations for the functions (n, l, m) in terms of $[n_x, n_y, n_z]$.

7–37. A two-dimensional harmonic oscillator is a particle which moves in the x, y-plane, the potential energy being

$$V(r) = \tfrac{1}{2}kr^2, \qquad r^2 = x^2 + y^2.$$

Set up the Schrödinger equation for this problem and solve it in rectangular coordinates and in polar coordinates (r, ϕ) $[x = r \cos \phi, y = r \sin \phi]$. Show that the energy levels are given by

$$E_n = (n + 1)\hbar\omega, \qquad n = 0, 1, 2, \ldots,$$

and that each E_n is $(n + 1)$-fold degenerate. Obtain the wave functions

$$\psi_{n,m} = \frac{1}{\sqrt{\pi}} \Lambda_k^m(\rho^2)e^{\pm im\phi} \qquad \left(\rho = r\sqrt{\frac{m\omega}{\hbar}}\right),$$

where $n = m + 2k$, $k = 0, 1, 2, \ldots$. These functions are orthonormal, i.e.,

$$(\psi_{n',m'}, \psi_{n,m}) = \int_0^{2\pi} \int_0^\infty \psi_{n',m'}^* \psi_{n,m} \rho \, d\rho \, d\phi = \delta_{nn'} \delta_{mm'}.$$

Finally, construct the unitary transformation from the functions $[n_x, n_y] = \psi_{n_x,n_y}$ to the functions $(n, m) = \psi_{n,m}$.

7–38. Show that the Laguerre and Hermite polynomials are related by

$$H_{2m}(x) = (-)^m 2^{2m} m! L_m^{-1/2}(x^2),$$

$$H_{2m+1}(x) = (-)^m 2^{2m+1} m! x L_m^{1/2}(x^2).$$

7–39. Show that the function $\sqrt{2r}\,\Lambda_k^{l+1/2}(r^2)$ has exactly $k + 2$ zeros in the interval $(0, \infty)$, two of which are $r = 0$ and $r = \infty$. [Hint: The function $L_k^\alpha(t)$ is a polynomial of degree k which has k distinct zeros in $0 < t < \infty$. This follows from Eq. (7–141).] Note that $k = \tfrac{1}{2}(n - l)$ so that the number of zeros increases with n for fixed l and decreases with l for fixed n. These properties of the radial zeros of ψ are common to all spherically symmetric, bound systems.

7-40. Compute the expectations of r, $1/r^2$, and $1/r^3$ for the state (7–196) of hydrogen and compare the results with the corresponding time averages in the Bohr theory.[1]

Answer:

$$\langle r \rangle = n^2\left[1 + \tfrac{1}{2}\left(1 - \frac{l(l+1)}{n^2}\right)\right]a_0, \qquad \bar{r} = n^2\left[1 + \tfrac{1}{2}\left(1 - \frac{k^2}{n^2}\right)\right]a_0,$$

$$\left\langle \frac{1}{r^2} \right\rangle = \frac{1}{n^3(l+\tfrac{1}{2})}\frac{1}{a_0^2}, \qquad\qquad \overline{\left(\frac{1}{r^2}\right)} = \frac{1}{n^3 k}\frac{1}{a_0^2},$$

$$\left\langle \frac{1}{r^3} \right\rangle = \frac{1}{n^3 l(l+\tfrac{1}{2})(l+1)}\frac{1}{a_0^3}, \qquad \overline{\left(\frac{1}{r^3}\right)} = \frac{1}{n^3 k^3}\frac{1}{a_0^3}.$$

[Hint: The integrals required can be evaluated by means of the recurrence relation, Eq. (7–131).][2] Note that the correspondence $k^2 \to l(l+1)$ is not always exact.

7-41. Derive the formula[3]

$$(\psi_{1,0,0}, z\psi_{n,1,0}) = \frac{2^4 n^{7/2}}{\sqrt{3}}(n-1)^{n-5/2}(n+1)^{-n-5/2}.$$

The square of this number is proportional to the intensity of the $np \to 1s$ line in the Lyman series (cf. Chapter 11).

7-42. Show that[4]

$$\cos\theta\, Y_l^m = \sqrt{\frac{(l+m+1)(l-m+1)}{(2l+1)(2l+3)}}\, Y_{l+1}^m + \sqrt{\frac{(l+m)(l-m)}{(2l-1)(2l+1)}}\, Y_{l-1}^m,$$

and prove that

$$(\psi_{n',l',m'}, z\psi_{n,l,m}) = 0$$

unless $m' = m$, $l' = l \pm 1$.

---

[1] E. U. Condon and G. H. Shortley, *The Theory of Atomic Spectra*. Cambridge: Cambridge University Press, 1953, p. 117.

[2] I. Waller, *Z. Physik* **38,** 635 (1926). J. H. Van Vleck, *Proc. Roy. Soc.* (London) **A 143,** 679 (1933).

[3] E. U. Condon and G. H. Shortley, *The Theory of Atomic Spectra*. Cambridge: Cambridge University Press, 1953, pp. 133 ff.

[4] V. Rojansky, *Introductory Quantum Mechanics*. New York: Prentice-Hall, Inc., 1946. The appendix contains an excellent summary of the properties of Y_l^m.

7–43. The discussion following Eq. (7–199) was restricted to $l \neq 0$ because, if $l = 0$, 'V' is negatively infinite at $r = 0$. Show that the conclusions concerning the zeros of $u_{n,l}$ are nevertheless correct for $l = 0$. Note that the quantum number k of the Bohr theory has the values $1, 2, \ldots, n$ while $l = 0$, $1, \ldots$ so that the restriction $k \neq 0$ (Section 1–13) appears automatically in quantum mechanics, without any *ad hoc* hypothesis about the state of zero angular momentum.

CHAPTER 8

THEORY OF SCATTERING

8–1 General remarks. The most direct information about the nature of forces between particles is obtained from the study of collisions. In a scattering experiment, a beam of particles is directed at a target containing the scattering material, and the energy and angular distribution of the deflected beam or of the recoil particles are measured. The experimental data can be interpreted in terms of a model of the microscopic details of a single collision, and conclusions concerning the interaction between the scattered particle and the scatterer can be drawn. In this chapter, we shall develop the quantum-mechanical theory of scattering processes. The results of this theory also find application in the study of matter in bulk. For example, the behavior of a gas is determined by the properties of individual collisions between the molecules, and of the molecules with the walls of the containing vessel.

In our consideration of scattering, we shall assume that the interaction between the scattered particle and the scatterer can be represented by a potential-energy function $V(\mathbf{r})$,[1] where $\mathbf{r}$ is the vector joining the scattered particle and the center of force ($r = 0$). In the case of a two-particle collision, this means that we work within a frame of reference in which the center of mass of the system is at rest. The vector $\mathbf{r}$ thus represents the relative coordinate of the particles. The mass of the scattered particle, denoted by m, is to be replaced, where appropriate, by the reduced mass[2] (cf. Section 1–11). If the mass of the scatterer is large compared to that of the scattered particle (e.g., in the collision between an electron and an atom), the scatterer can be assumed to remain at rest during the entire collision. In this case, no distinction needs to be made between the laboratory and center-of-mass coordinates. Our discussion will be restricted to elastic scattering, in which the kinetic energy of the system is not changed by the collision.

In general, the potential energy $V(\mathbf{r})$ decreases in magnitude as the distance $r = |\mathbf{r}|$ from the scattering center becomes large. It is convenient to choose the arbitrary constant in the definition of $V(\mathbf{r})$ such that $V = 0$

---

[1] Cf. Section 4–1.

[2] The transformation from laboratory to center-of-mass coordinates is treated in detail in books on experimental physics. See, e.g., P. Morrison in E. Segrè, Ed., *Experimental Nuclear Physics*. New York: John Wiley and Sons, Inc., 1953, Vol. II, Part VI, Section 1.

at $r = \infty$. The total energy of the particle is, therefore, $E = \frac{1}{2}mv_\infty^2 = (\hbar^2/2m)k^2$, where v_∞ is the velocity at $r = \infty$, and k is the corresponding wave number. Since $E > V(\mathbf{r})$ for large $|\mathbf{r}|$, we are now concerned with the stationary states in the continuous part of the energy spectrum (cf. Section 5–4). If $V(\mathbf{r})$ decreases to zero sufficiently rapidly[1] as $r \to \infty$, the particle can be considered essentially free when r is large. The asymptotic wave function is therefore, in general, a linear superposition of the free-particle wave functions $\exp(i\mathbf{k} \cdot \mathbf{r})$ (cf. Section 2–8).

Since we are concerned with a central force, spherical polar coordinates are adapted to the theory of scattering, and we shall therefore devote the next section to the description of a free particle in terms of these coordinates.

8–2 Wave functions for a free particle in spherical polar coordinates.

The Schrödinger equation for the stationary states of a free particle of energy E is [Eq. (4–22)]

$$\nabla^2\psi + \frac{2m}{\hbar^2}E\psi = 0, \tag{8-1}$$

or, expressed in spherical polar coordinates,

$$\frac{\partial^2}{\partial r^2}(r\psi) - \frac{\mathbf{L}^2}{\hbar^2 r^2}(r\psi) + k^2(r\psi) = 0, \tag{8-2}$$

in which $k = \sqrt{2mE}/\hbar^2$, and $\mathbf{L}^2$ is the square of the angular-momentum operator [Eq. (7–87)]. Since the system is spherically symmetric, Eq. (8–2) has a complete set of solutions of the form

$$\psi = \frac{u(r)}{r}Y_l^m, \tag{8-3}$$

in which the function $u(r)$ is a solution of

$$\frac{d^2u}{dr^2} + \left[k^2 - \frac{l(l+1)}{r^2}\right]u = 0 \tag{8-4}$$

[cf. Eq. (7–64)].

The boundary conditions to be imposed upon solutions of Eq. (8–4) require some thought. If we were to deal only with a free particle, not subject to any forces, then, for ψ to be finite at $r = 0$, we should require

$$u(0) = 0. \tag{8-5}$$

---

[1] Specifically, this description is possible provided $\lim_{r\to\infty} rV(\mathbf{r}) = 0$.

However, in the theory of scattering, we are interested in solutions only in the asymptotic region, where the potential energy is negligible compared to the total energy, and solutions which do not satisfy Eq. (8–5) are quite acceptable; indeed, they are necessary in order that the asymptotic free-particle functions can be joined smoothly to the solutions appropriate to the region of interaction. From the discussion in Section 7–4, we know that Eq. (8–4) has solutions behaving like r^{l+1} and r^{-l} for small r. For the scattering theory, the latter, irregular solutions are just as important as the regular ones.

If r is large, Eq. (8–4) becomes, approximately,

$$\frac{d^2u}{dr^2} + k^2u = 0, \tag{8–6}$$

which has the solutions $e^{\pm ikr}$. The corresponding wave functions $e^{\pm ikr}/r$ describe spherical waves propagated outward from and inward toward the scattering center (cf. Problem 4–4, Chapter 4). These functions cannot be normalized and are to be interpreted in the relative sense discussed in Section 2–9.

A complete set of solutions of Eq. (8–4) will now be developed in detail. For simplicity, we introduce the variable $\rho = kr$ and write

$$\frac{d^2u}{d\rho^2} + \left[1 - \frac{l(l+1)}{\rho^2}\right]u = 0. \tag{8–7}$$

As already noted, there are regular and irregular solutions [denoted by $f_l(\rho)$ and $g_l(\rho)$],[1] where

$$f_l \approx C_l\rho^{l+1} \qquad \text{and} \qquad g_l \approx D_l\rho^{-l}, \tag{8–8}$$

respectively. The definition of the arbitrary constant multipliers C_l and D_l is temporarily deferred. The sign "$\approx$" will be used in this discussion to mean "approximately equal when ρ is small." Thus, the expressions (8–8) are the leading terms in power series in ρ, which can be completed by the method of series (Problem 8–1).

Asymptotically ($\rho \to \infty$), linearly independent solutions of Eq. (8–7) are $e^{\pm i\rho}$; therefore f_l and g_l, as defined by Eqs. (8–8), are asymptotic to linear combinations of these two functions. Now except for the constant multiplier C_l, the regular function f_l is uniquely defined by the first of Eqs. (8–8), and we shall choose this factor so that f_l is asymptotic to a

---

[1] The terms "regular" and "irregular" are more properly applied to the functions f_l/ρ and g_l/ρ, (g_l is finite for $l = 0$), but the above terminology has become customary.

wave of unit amplitude:

$$f_l \sim \sin(\rho + \beta_l). \tag{8-9}$$

The constant β_l is determined by this choice for the asymptotic form of f_l. It will be evaluated below. The sign "$\sim$" in Eq. (8-9) means "asymptotically equal when ρ is large."

The further properties of the functions f_l and g_l can be deduced by the *method of factorization*.[1]   We note first that the equation (8-7) can be written in either of two "factorized" forms (Problem 8-2), that is,

$$\left[\frac{d}{d\rho} + \frac{l+1}{\rho}\right]\left[\frac{d}{d\rho} - \frac{l+1}{\rho}\right]u = -u, \tag{8-10}$$

$$\left[\frac{d}{d\rho} - \frac{l}{\rho}\right]\left[\frac{d}{d\rho} + \frac{l}{\rho}\right]u = -u. \tag{8-11}$$

We now define the functions u_{l-1} by

$$u_{l-1} = \left[\frac{d}{d\rho} + \frac{l}{\rho}\right]f_l \qquad (l = 1, 2, \ldots), \tag{8-12}$$

so that Eq. (8-11) for the function f_l can be written

$$\left[\frac{d}{d\rho} - \frac{l}{\rho}\right]u_{l-1} = -f_l. \tag{8-13}$$

Replacing l by $l + 1$, we obtain

$$\left[\frac{d}{d\rho} - \frac{l+1}{\rho}\right]u_l = -f_{l+1}. \tag{8-14}$$

If the operator $[d/d\rho + (l + 1)/\rho]$ is now applied to the two members of Eq. (8-14), we have

$$\left[\frac{d}{d\rho} + \frac{l+1}{\rho}\right]\left[\frac{d}{d\rho} - \frac{l+1}{\rho}\right]u_l = -\left[\frac{d}{d\rho} + \frac{l+1}{\rho}\right]f_{l+1} = -u_l, \quad (8-15)$$

in which the second equation follows from Eq. (8-12) upon replacing l by $l + 1$. Comparison of Eqs. (8-15) and (8-10) shows that the function u_l *satisfies the same differential equation as* f_l. Moreover, it is regular, since Eqs. (8-8) yield

$$u_l \approx \left[\frac{d}{d\rho} + \frac{l+1}{\rho}\right]C_{l+1}\rho^{l+2} = (2l + 3)C_{l+1}\rho^{l+1}. \tag{8-16}$$

---

[1] L. Infeld and T. E. Hull, *Revs. Modern Phys.* **23,** 25 (1951).

Equation (8–7) has only one regular solution, and u_l and f_l must therefore be proportional. The constant of proportionality can be found by comparison of Eqs. (8–16) and (8–8); one finds

$$u_l = \frac{(2l + 3)C_{l+1}}{C_l} f_l. \qquad (8\text{–}17)$$

Thus we have derived the recurrence formula

$$f_{l+1} = -\frac{(2l + 3)C_{l+1}}{C_l}\left\{\frac{d}{d\rho} - \frac{l + 1}{\rho}\right\} f_l, \qquad (8\text{–}18)$$

which follows from Eq. (8–14) upon substitution of (8–17). By means of Eqs. (8–18) and (8–9), the regular solutions f_l of Eq. (8–7) can be constructed consecutively, beginning with the function

$$f_0 = \sin \rho \qquad (C_0 = 1), \qquad (8\text{–}19)$$

which is the unique regular solution of Eq. (8–7) for $l = 0$. The constants C_l and β_l are automatically evaluated by this process.

Thus, the regular solution for $l = 1$ is

$$f_1 = -\frac{3C_1}{C_0}\left\{\cos \rho - \frac{\sin \rho}{\rho}\right\}; \qquad (8\text{–}20)$$

the asymptotic form is therefore

$$f_1 = \frac{3C_1}{C_0}\sin\left(\rho - \frac{\pi}{2}\right), \qquad (8\text{–}21)$$

which agrees with Eq. (8–9) provided that

$$C_1 = \tfrac{1}{3}C_0 = \tfrac{1}{3}, \qquad \beta_1 = -\frac{\pi}{2}, \qquad (8\text{–}22)$$

whence

$$f_1 = \frac{\sin \rho}{\rho} - \cos \rho. \qquad (8\text{–}23)$$

The function f_2 can now be calculated similarly; we obtain (Problem 8–3)

$$\left.\begin{aligned}
C_2 &= \tfrac{1}{5}C_1 = \tfrac{1}{5}\cdot\tfrac{1}{3}, \qquad \beta_2 = -2\frac{\pi}{2}, \\
f_2 &= \frac{3\sin \rho}{\rho^2} - \frac{3\cos \rho}{\rho} - \sin \rho.
\end{aligned}\right\} \qquad (8\text{ }24)$$

In general, we have

$$C_l = \frac{1}{2l + 1}\cdot\frac{1}{2l - 1}\cdots\frac{1}{5}\cdot\frac{1}{3} = \frac{2^l l!}{(2l + 1)!}; \qquad \beta_l = -\frac{l\pi}{2}. \qquad (8\text{–}25)$$

In summary, f_l is the regular solution of Eq. (8–7). Asymptotically, it has the form

$$f_l \sim \sin\left(\rho - \frac{l\pi}{2}\right),\qquad (8\text{–}26)$$

and for small ρ,

$$f_l \approx C_l \rho^{l+1},\qquad C_l = \frac{2^l l!}{(2l+1)!}.\qquad (8\text{–}27)$$

With this value for C_l, the recurrence relations (8–18) and (8–12) are

$$\left\{\frac{d}{d\rho} - \frac{l+1}{\rho}\right\} f_l = -f_{l+1},\qquad \left\{\frac{d}{d\rho} + \frac{l}{\rho}\right\} f_l = f_{l-1}.\qquad (8\text{–}28)$$

In contrast to the regular solution, the leading term in the series solution for the irregular function g_l [Eqs. (8–8)] does not determine the function uniquely, because this term is unaffected by adding to g_l the function Af_l, where A is any arbitrarily chosen constant. A unique specification is obtained, however, by the requirement that g_l have the asymptotic form

$$g_l \sim \cos\left(\rho - \frac{l\pi}{2}\right).\qquad (8\text{–}29)$$

This amounts to making a definite choice of the constant A. The function g_l, defined by Eq. (8–29) and the differential equation (8–7), is obviously unique and certainly irregular. This follows from the fact that the Wronskian determinant (Section 5–5),

$$W = \frac{df_l}{d\rho} g_l - f_l \frac{dg_l}{d\rho},\qquad (8\text{–}30)$$

is independent of ρ and has, according to Eqs. (8–26) and (8–29), the value

$$W = \cos^2\left(\rho - \frac{l\pi}{2}\right) + \sin^2\left(\rho - \frac{l\pi}{2}\right) = 1,\qquad (8\text{–}31)$$

i.e., g_l is linearly independent of f_l. The constant D_l in Eqs. (8–8) is now determined by the equation

$$W = 1 = (l+1)C_l \rho^l D_l \rho^{-l} + C_l \rho^{l+1} l D_l \rho^{-l-1} = (2l+1)C_l D_l,\qquad (8\text{–}32)$$

that is,

$$D_l = \frac{1}{(2l+1)C_l}.\qquad (8\text{–}33)$$

We summarize the properties of g_l as follows: The function g_l is the solution of Eq. (8–7) which is asymptotically

$$g_l \sim \cos\left(\rho - l\frac{\pi}{2}\right). \tag{8-34}$$

It is irregular at $\rho = 0$, according to

$$g_l \approx \frac{1}{(2l + 1)C_l}\rho^{-l}. \tag{8-35}$$

It satisfies the Wronskian relation

$$\frac{df_l}{d\rho}g_l - f_l\frac{dg_l}{d\rho} = 1, \tag{8-36}$$

and the same recurrence relations as the function f_l, that is,

$$\left\{\frac{d}{d\rho} - \frac{l+1}{\rho}\right\}g_l = -g_{l+1} \tag{8-37}$$

and

$$\left\{\frac{d}{d\rho} + \frac{l}{\rho}\right\}g_l = g_{l-1} \tag{8-38}$$

(cf. Problem 8–4). In particular, we have

$$g_0 = \cos\rho, \qquad g_1 = \frac{\cos\rho}{\rho} + \sin\rho, \qquad g_2 = \frac{3\cos\rho}{\rho^2} + \frac{3\sin\rho}{\rho} - \cos\rho. \tag{8-39}$$

Since the recurrence relations (8–37) and (8–38) are linear in f_l or g_l, it is clear that they are also satisfied by any linear combination of these functions in which the coefficients are independent of l. Thus, the complex functions

$$y_l = f_l + ig_l \tag{8-40}$$

can be constructed from

$$y_0 = ie^{-i\rho} \tag{8-41}$$

by means of Eq. (8–37) which, in this case, is

$$y_{l+1} = -\left\{\frac{d}{d\rho} - \frac{l+1}{\rho}\right\}y_l. \tag{8-42}$$

Inspection of Eqs. (8–41) and (8–42) shows that y_l has the form

$$y_l = e^{-i\rho}Q_l(\rho), \tag{8-43}$$

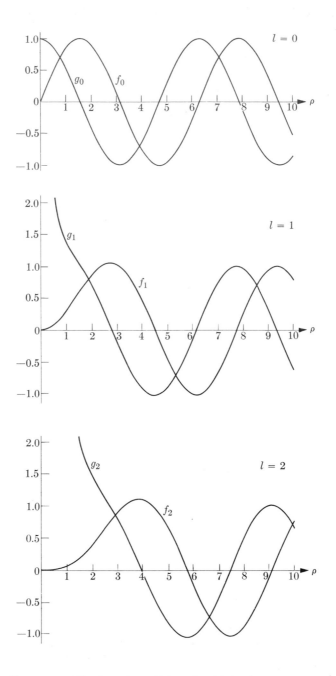

FIG. 8–1.   The functions $f_l(\rho)$ and $g_l(\rho)$ for $l = 0, 1, 2$.

in which $Q_l(\rho)$ is a polynomial in the variable $1/\rho$. In terms of the variable $x = (2i\rho)^{-1}$, the differential equation for Q_l is found from Eq. (8–7) to be

$$x^2 \frac{d^2Q}{dx^2} + 2x \frac{dQ}{dx} - l(l+1)Q = 0, \tag{8–44}$$

and by working out the polynomial solution of this equation by the method of series, we have (Problem 8–6)

$$y_l = ie^{-i(\rho - l\pi/2)} \sum_{k=0}^{l} \frac{1}{k!} \frac{(l+k)!}{(l-k)!} \frac{1}{(2i\rho)^k}, \tag{8–45}$$

which expresses the solution of Eq. (8–7) explicitly in finite form.

The functions y_l and $y_l^* = f_l - ig_l$ have the asymptotic forms[1]

$$y_l \sim ie^{-i(\rho - l\pi/2)}, \qquad y_l^* \sim -ie^{i(\rho - l\pi/2)}, \tag{8–46}$$

and therefore represent incoming and outgoing spherical waves, respectively.

The functions f_l and g_l can be shown to be related to Bessel functions of order $l + \frac{1}{2}$, according to[2]

$$f_l(\rho) = \sqrt{\frac{\pi\rho}{2}} J_{l+1/2}(\rho), \qquad g_l(\rho) = (-)^l \sqrt{\frac{\pi\rho}{2}} J_{-(l+1/2)}(\rho). \tag{8–47}$$

Extensive tables of these functions exist, the most complete being those for the *spherical Bessel functions*,[3]

$$j_l(\rho) = \frac{f_l(\rho)}{\rho}, \qquad n_l(\rho) = -\frac{g_l(\rho)}{\rho}. \tag{8–48}$$

Graphs of the functions f_l and g_l for $l = 0, 1, 2$ are shown in Fig. 8–1.

8–3 Expansion of a plane wave in spherical harmonics. The functions $[f_l(\rho)/\rho]Y_l^m(\theta\phi)$ are a complete set of regular solutions of the Schrödinger equation for a free particle of energy E. Hence, any regular solution of this equation can be expressed as a linear superposition of these functions,

---

[1] The functions f_l and g_l are real functions of ρ.

[2] G. N. Watson, *A Treatise on the Theory of Bessel Functions.* 2nd ed. Cambridge: Cambridge University Press, 1945, Sections 3.4 and 3.41.

[3] *Tables of Spherical Bessel Functions.* Mathematical Tables Project, National Bureau of Standards. New York: Columbia University Press, 1947. See also G. N. Watson, *op. cit.,* Table V, p. 740.

and in particular, it must be possible to express the plane wave $\exp(i\mathbf{k}\cdot\mathbf{r})$ in the form

$$\exp(i\mathbf{k}\cdot\mathbf{r}) = \sum_{l,m} c_{lm} Y_l^m(\hat{\mathbf{r}}) \frac{f_l(\rho)}{\rho}, \qquad (8\text{–}49)$$

in which $\hat{\mathbf{r}}$ is a unit vector in the direction (θ, ϕ) (i.e., $\hat{\mathbf{r}} = \mathbf{r}/r$), and $\mathbf{k}$ is the propagation vector (i.e., $|\mathbf{k}| = \sqrt{2mE}/\hbar$). The expansion coefficients c_{lm} will now be evaluated.

For simplicity, we first consider the plane wave

$$e^{ikz} = e^{i\rho\cos\theta}, \qquad (8\text{–}50)$$

for which the direction of propagation is that of the positive z-axis, i.e., the axis of quantization for the angular momentum. The function (8–50) is independent of the angle ϕ, and hence, in this special case, only the functions Y_l^0 appear in Eq. (8–49). Since Y_l^0 is proportional to the Legendre polynomial of degree l in $\mu = \cos\theta$, we can write

$$e^{i\rho\mu} = \sum_{l=0}^{\infty} a_l P_l(\mu) \frac{f_l(\rho)}{\rho}. \qquad (8\text{–}51)$$

The orthogonality relations for the functions $P_l(\mu)$, namely

$$\int_{-1}^{1} P_l(\mu) P_{l'}(\mu)\, d\mu = \frac{2}{2l+1}\, \delta_{ll'}, \qquad (8\text{–}52)$$

can be used, in the usual way, to obtain

$$\int_{-1}^{1} e^{i\rho\mu} P_l(\mu)\, d\mu = \frac{2}{2l+1}\, a_l \frac{f_l(\rho)}{\rho}. \qquad (8\text{–}53)$$

This relation is an identity in ρ, and the number a_l can be determined if the two members of the equation can be evaluated for any particular value of ρ. By integration by parts, the integral in Eq. (8–53) can be transformed to

$$\int_{-1}^{1} e^{i\rho\mu} P_l(\mu)\, d\mu$$
$$= \frac{2i^l}{\rho} \sin\left(\rho - \frac{l\pi}{2}\right) - \frac{1}{i\rho} \int_{-1}^{1} e^{i\rho\mu} P_l'(\mu)\, d\mu. \qquad (8\text{–}54)$$

One can show without difficulty that if ρ is large, the second term in the right-hand member of this equation is negligible compared to the first.

Therefore, the asymptotic form of Eq. (8-53) is

$$\frac{2}{2l+1} a_l \frac{\sin (\rho - l\pi/2)}{\rho} = \frac{2i^l}{\rho} \sin \left(\rho - \frac{l\pi}{2}\right), \tag{8-55}$$

whence

$$a_l = i^l(2l+1). \tag{8-56}$$

Thus, the required expansion is

$$e^{ikz} = e^{i\rho\mu} = \sum_{l=0}^{\infty} i^l(2l+1)P_l(\mu)\frac{f_l(\rho)}{\rho}; \tag{8-57}$$

this is *Bauer's formula*.[1]
  Incidentally, Eq. (8-53) yields the interesting formula

$$\frac{f_l(\rho)}{\rho} = \frac{1}{2i^l} \int_{-1}^{1} e^{i\rho\mu} P_l(\mu)\, d\mu, \tag{8-58}$$

in which the function $f_l(\rho)/\rho$ is exhibited as a Fourier transform of $P_l(\mu)$.
  In the general case of **k** having an arbitrary direction, the formula (8-57) becomes

$$\exp (i\mathbf{k} \cdot \mathbf{r}) = \sum_{l=0}^{\infty} i^l(2l+1)P_l (\cos \theta')\frac{f_l(\rho)}{\rho}, \tag{8-59}$$

where θ' is the angle between **r** and **k**. Hence, by the addition theorem [Eq. (7-63)], we obtain

$$\exp (i\mathbf{k} \cdot \mathbf{r}) = 4\pi \sum_{l,m} i^l Y_l^{m*}(\hat{\mathbf{k}}) Y_l^m(\hat{\mathbf{r}})\frac{f_l(\rho)}{\rho}, \tag{8-60}$$

which is the general expansion (8-49) referred to at the beginning of this section.

8-4 Scattering by a short-range central field. The wave function for a particle of energy E which moves in a central, spherically symmetric field of force is a solution of the Schrödinger equation

$$\nabla^2\psi + \frac{2m}{\hbar^2} [E - V(r)]\psi = 0, \tag{8-61}$$

---

[1] G. N. Watson, *A Treatise on the Theory of Bessel Functions.* 2nd ed. Cambridge: Cambridge University Press, 1945, Section 4.82.

in which $V(r)$ is the potential-energy function. In the scattering problem, we are interested in states of positive energy, and we assume[1] that
(Section 8–1)

$$\lim_{r \to \infty} r V(r) = 0. \tag{8–62}$$

A potential-energy function satisfying this restriction is said to define a
short-range force. The conditions to be satisfied by ψ are that it must be
regular at $r = 0$ and continuous, together with its derivative, at every
point. Furthermore, the form of ψ at a large distance from the scattering
center must be such as to describe an incident plane wave of energy E
and an outgoing spherical wave representing the scattered particle. If
the direction of incidence is along the positive z-axis, we require

$$\psi \sim e^{ikz} + f(\theta) \frac{e^{ikr}}{r}. \tag{8–63}$$

The first term, e^{ikz}, represents a beam of particles (Section 2–9) of unit
density, incident upon the scattering center in the z-direction. The incident
particle current (cf. Section 4–5) is

$$S_i = \frac{\hbar}{m} \operatorname{Im} e^{-ikz} \frac{d}{dz} (e^{ikz}) = \frac{\hbar k}{m} = v_\infty, \tag{8–64}$$

where v_∞ is the velocity corresponding to the energy E. The second term
in Eq. (8–63) represents an outgoing wave corresponding to the particles
scattered from the beam. The outward current of scattered particles is,
asymptotically,

$$S_r \sim \frac{\hbar}{m} \operatorname{Im} f^*(\theta) \frac{e^{-ikr}}{r} \frac{\partial}{\partial r} \left(f(\theta) \frac{e^{ikr}}{r} \right) \sim \frac{v_\infty}{r^2} |f(\theta)|^2, \tag{8–65}$$

in which the factor $1/r^2$ describes the diminution in particle density
according to the inverse square law.[2] The function $f(\theta)$ is the *scattering*

---

[1] Because of the restriction (8–62), our analysis does not apply to the case of
Coulomb scattering (however, see Section 8–14). A complete analysis of scatter-
ing in the Coulomb field is given in: N. F. Mott and H. S. W. Massey, *The
Theory of Atomic Collisions*. 2nd ed. Oxford: Clarendon Press, 1949, Chapter III.

[2] We are justified in calculating the particle current separately for the two
parts of ψ since, in an actual experiment, the incident and scattered particles are
collimated, so that Eq. (8–63) is only an approximate formulation of the physical
situation. Cf. L. I. Schiff, *Quantum Mechanics*. 2nd ed. New York: McGraw-
Hill Book Co., Inc., 1955, p. 101.

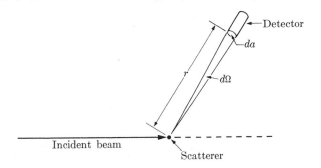

FIG. 8–2.   Diagram of scattering experiment.

amplitude; it describes the angular distribution of the scattered wave.[1] Our problem is to determine this function when $V(r)$ is given. In an experiment, the observed quantities are the particle currents (8–64) and (8–65). If the experimental situation is such that the density of particles in the incident beam is N per cm^3, then the observed incident particle current is $NS_i = Nv_\infty$. Suppose that the scattered particles are intercepted by a detector of area da, at distance r (Fig. 8–2). The solid angle subtended by the detector at the scattering center is $d\Omega = da/r^2$, and the rate at which particles enter the detector is

$$NS_r\, da \,=\, Nr^2 S_r\, d\Omega. \tag{8–66}$$

Experimental data are commonly expressed in terms of the *differential scattering cross section,* $\sigma(\theta)$, which is defined by

$$\sigma(\theta)\, d\Omega \,=\, \frac{\text{number of particles scattered into } d\Omega \text{ at angle } \theta}{\text{number of particles incident per } cm^2} \tag{8–67}$$

or,

$$\sigma(\theta)\, d\Omega \,=\, \frac{Nr^2 S_r\, d\Omega}{NS_i}, \tag{8–68}$$

whence, by Eqs. (8–64) and (8–65),

$$\sigma(\theta) \,=\, |f(\theta)|^2. \tag{8–69}$$

The *total scattering cross section* is

$$\sigma \,=\, \int_\Omega \sigma(\theta)\, d\Omega \,=\, 2\pi \int_{-1}^{1} |f(\theta)|^2\, d\mu. \tag{8–70}$$

---

[1] Since the force is assumed to be spherically symmetric, the scattering amplitude is independent of the azimuthal angle ϕ.

The quantity σ is the total number of particles scattered out of the beam, divided by the number of particles incident per square centimeter.

The function $f(\theta)$ is, by Eq. (8–69), directly related to the experimentally observable cross section. It will now be shown how $f(\theta)$ can be determined from knowledge of the potential-energy function $V(r)$.

8–5 Method of partial waves. The functions

$$\frac{F_l(\rho)}{\rho} Y_l^m(\hat{\mathbf{r}}),\tag{8–71}$$

in which $F_l(\rho)$ is the regular solution of the radial wave equation

$$\frac{d^2F_l}{d\rho^2} + \left\{ 1 - \frac{V(r)}{E} - \frac{l(l+1)}{\rho^2} \right\} F_l = 0,\tag{8–72}$$

are a complete set of regular solutions of the Schrödinger equation (8–61). The functions (8–71) are, therefore, a set of "partial waves," in terms of which the general solution of Eq. (8–61) can be written

$$\psi = \sum_{l,m} c_{lm} \frac{F_l(\rho)}{\rho} Y_l^m(\hat{\mathbf{r}}).\tag{8–73}$$

If, as in Eq. (8–63), the polar axis is chosen parallel to the direction of incidence, the wave function is independent of ϕ, and Eq. (8–73) can be specialized to the form [cf. Eqs. (8–49) and (8–51)]

$$\psi = \sum_{l} b_l P_l(\mu) \frac{F_l(\rho)}{\rho}.\tag{8–74}$$

Now provided (cf. Section 7–4) that[1]

$$\lim_{r \to 0} r^2 V(r) = 0,\tag{8–75}$$

Eq. (8–72) has solutions which behave, near the origin, like ρ^{l+1} and ρ^{-l}; the former of these solutions is the regular function F_l which, except for a constant multiplier, is uniquely defined by the differential equation (8–72). Also, since $V(r)$ is a short-range potential [Eq. (8–62)], this solution is asymptotic, for large ρ, to a linear combination of the functions $e^{\pm i\rho}$.

---

[1] The condition (8–75) is somewhat more restrictive than necessary. However, if $V(r)$ becomes infinite at the origin more rapidly than the function $1/r^2$, the scattering problem has no solution. [N. F. Mott and H. S. W. Massey, *The Theory of Atomic Collisions*. 2nd ed. Oxford: Clarendon Press, 1949, Chapter II, Section 6. Note: In Eq. (46), the term $-\beta/v^2$ should be added to the expression enclosed by braces.]

As in Section 8–2, we choose the constant factor so that F_l is asymptotic to a wave of unit amplitude:

$$F_l \sim \sin\left(\rho - \frac{l\pi}{2} + \delta_l\right). \tag{8-76}$$

The regular function F_l is uniquely defined by Eq. (8–76) and the differential equation (8–72). The quantity δ_l is the *phase shift* of the lth partial wave. It measures the amount by which the regular function (8–76) is displaced in ρ, relative to the free-particle function f_l [Eq. (8–26)]. If $V(r) = 0$, the functions $F_l(\rho)$ and $f_l(\rho)$ are identical. The entire effect of the scattering force is therefore described in terms of the (infinite) set of phase shifts.

With this specification of F_l, the constants b_l in Eq. (8–74) can be evaluated. Thus, by substituting the expansions (8–57) and (8–74) in the asymptotic expression (8–63) and cancelling a common factor, one finds

$$\sum_l b_l P_l(\mu) \sin\left(\rho - \frac{l\pi}{2} + \delta_l\right)$$

$$= \sum_l i^l (2l + 1) P_l(\mu) \sin\left(\rho - \frac{l\pi}{2}\right) + kf(\theta)e^{i\rho}, \quad (8-77)$$

and this equation must be an identity in ρ. If we write the trigonometric functions in terms of $e^{\pm i\rho}$ and rearrange, this equation becomes

$$e^{i\rho}\left[\sum_l b_l P_l(\mu)i^{-l}e^{i\delta_l} - \sum_l (2l+1)P_l(\mu) - 2ikf(\theta)\right]$$

$$-e^{-i\rho}\left[\sum_l b_l P_l(\mu)i^l e^{-i\delta_l} - \sum_l i^{2l}(2l+1)P_l(\mu)\right] = 0. \quad (8-78)$$

Since $e^{i\rho}$ and $e^{-i\rho}$ are linearly independent, each of the quantities in square brackets must vanish. From the coefficient of $e^{-i\rho}$, one obtains

$$b_l = i^l(2l+1)e^{i\delta_l}. \tag{8-79}$$

The constants b_l are therefore determined by the condition that the scattered-wave part of the solution shall contain outgoing waves only, that is, a term in $e^{-i\rho}/\rho$ is not allowed in the scattered-wave part of Eq. (8–63). This is sometimes called the *radiation condition*. Inserting Eq. (8–79) into Eq. (8–78) and solving for $f(\theta)$, one has

$$f(\theta) = \frac{1}{k}\sum_{l=0}^{\infty}(2l+1)e^{i\delta_l}\sin\delta_l P_l(\cos\theta). \tag{8-80}$$

Similarly, the wave function is

$$\psi = \sum_{l=0}^{\infty} i^l(2l + 1)P_l (\cos \theta)e^{i\delta_l}\frac{F_l(\rho)}{\rho}. \tag{8-81}$$

An intuitive understanding of the radiation condition can be obtained from another point of view by writing the asymptotic expression for ψ as

$$\psi \sim \frac{1}{2i} \sum_{l=0}^{\infty} i^l(2l + 1)P_l (\cos \theta) \left\{ e^{2i\delta_l}\frac{e^{i(\rho-l\pi/2)}}{\rho} - \frac{e^{-i(\rho-l\pi/2)}}{\rho} \right\}, \tag{8-82}$$

which follows from Eq. (8–76) when the sine function is expressed in terms of $e^{i\rho}$ and $e^{-i\rho}$. In Eq. (8–82), the phase shift occurs only in the coefficient of the outgoing-wave part of the asymptotic wave function. The part representing incoming waves is identical with the corresponding part of the asymptotic expansion of the incident wave e^{ikz}. We can say, therefore, that the incoming spherical wave components of the incident plane wave are transformed, within the scattering region, into outgoing spherical waves which, because of the interaction $V(r)$, are changed in relative phase by amounts $2\delta_l$. This picture of the scattering process (which can be made precise in terms of spherical wave packets) is helpful in understanding the origin of Eq. (8–79). Separating the expression (8–80) into real and imaginary parts, we have

$$f(\theta) = \frac{1}{2k} \sum_l (2l + 1) \sin 2\delta_l P_l (\cos \theta)$$

$$+ \frac{i}{2k} \sum_l (2l + 1)(1 - \cos 2\delta_l)P_l (\cos \theta), \tag{8-83}$$

whence the differential cross section [Eq. (8–69)] is

$$\sigma(\theta) = \frac{1}{4k^2} \left\{ \sum_l (2l + 1) \sin 2\delta_l P_l (\cos \theta) \right\}^2$$

$$+ \frac{1}{4k^2} \left\{ \sum_l (2l + 1)(1 - \cos 2\delta_l)P_l (\cos \theta) \right\}^2 ; \tag{8-84}$$

and for the total cross section [Eq. (8–70)], one obtains by Eq. (8–52)

$$\sigma = \frac{4\pi}{k^2} \sum_l (2l + 1) \sin^2 \delta_l. \tag{8-85}$$

Equations (8–80), (8–81), (8–84), and (8–85) provide a formal solution of the scattering problem. When the potential-energy function $V(r)$ is given, the scattering cross section at any energy can, in principle, be found by solving the differential equations (8–72) for each value of l and evaluating the phase shifts δ_l by means of Eq. (8–76). Unless $V(r)$ has a special form, for which solutions of the radial equation can be obtained analytically, the solution must be found numerically or by some other approximation procedure. In one practical method,[1] the solution is started at $\rho = 0$ by means of the approximation ρ^{l+1}, and is continued by numerical integration to a value of ρ sufficiently large that $V(r)$ is negligible and the function is sinusoidal. The quantity δ_l is then obtained as the distance between the zeros of (8–76) and the corresponding zeros of (8–26). However, this process is practically useful only if the phase shifts decrease rapidly with l, so that the series (8–80) converges quickly. Fortunately, this situation occurs in important cases of physical interest, and the method of partial waves is of great value in the interpretation of experimental scattering data. It has been particularly important in the study of nuclear interactions.

One of the most important applications of the theory of scattering arises in the attempt to deduce the nature of the interaction between two particles from experimental scattering data. This is the inverse of the problem just described, in that the function $\sigma(\theta)$ is given by experiment, and it is desired to find the corresponding potential-energy function $V(r)$. Thus, one can obtain the phase shifts δ_l by fitting the cross-section formula (8–84) to the experimental data. This can be done, for example, by the method of least squares. Provided that a satisfactory fit is achieved by means of a small number of terms in the partial-wave expansion, the experimental data are thus expressed in terms of a set of experimental phase shifts. This procedure can be repeated for several values of the incident energy E, so that the dependence of the phase shifts on the energy becomes experimentally known. One can then attempt to discover a potential-energy function $V(r)$ which will reproduce this dependence. To the extent that this problem can be solved successfully, the interaction can be represented in the assumed form, namely, in terms of a static force dependent only upon the particle separation.

In practice, one usually proceeds by investigating several likely forms of the function $V(r)$, containing parameters which can be adjusted to yield phase shifts agreeing with experiment. For example, a potential-energy function in the form of a "square well" (Section 5–3) has been devised, which gives reasonably good agreement with experimental data on the scattering of neutrons by protons or of protons by protons.[2]

---

[1] Cf. E. U. Condon and G. H. Shortley, *The Theory of Atomic Spectra*. Cambridge: Cambridge University Press, 1953, Chapter XIV, Section 4.

[2] R. G. Sachs, *Nuclear Theory*. Reading, Mass.: Addison-Wesley Publishing Co., Inc., 1953. G. Breit, E. U. Condon, and R. D. Present, *Phys. Rev.* **50,** 825 (1936).

From a theoretical point of view, this inverse problem raises interesting questions: Does a potential-energy function corresponding to a given set of phase shifts exist, and if so, is it unique? These questions have received much attention recently, and the interested reader will find it instructive to consult the references cited.[1] In general, it has been demonstrated that the phase shifts and energies of bound states for *two* values of the angular momentum are sufficient to determine $V(r)$ uniquely.

8–6 Dependence of δ_l on l and E. The connection between the phase shift δ_l and the potential-energy function is given implicitly by the formula

$$\sin \delta_l = - \int_0^\infty \frac{V}{E} F_l(\rho) f_l(\rho)\, d\rho. \tag{8–86}$$

To derive this relation, multiply the differential equations

$$\frac{d^2 f_l}{d\rho^2} + \left\{ 1 - \frac{l(l+1)}{\rho^2} \right\} f_l = 0, \tag{8–87}$$

$$\frac{d^2 F_l}{d\rho^2} + \left\{ 1 - \frac{V}{E} - \frac{l(l+1)}{\rho^2} \right\} F_l = 0 \tag{8–88}$$

by F_l and f_l, respectively, and subtract, obtaining

$$\frac{d}{d\rho} \left\{ f_l' F_l - f_l F_l' \right\} + \frac{V}{E} F_l f_l = 0. \tag{8–89}$$

Now as $\rho \to \infty$, the quantity $f_l' F_l - f_l F_l'$ approaches the limit

$$\cos \left(\rho - \frac{l\pi}{2} \right) \sin \left(\rho - \frac{l\pi}{2} + \delta_l \right)$$

$$- \sin \left(\rho - \frac{l\pi}{2} \right) \cos \left(\rho - \frac{l\pi}{2} + \delta_l \right) = \sin \delta_l. \tag{8–90}$$

Hence, integration of Eq. (8–89) between the limits 0 and ∞ yields Eq. (8–86).[2]

The potential-energy function appears explicitly as a factor in the integrand in Eq. (8–86) and also implicitly in the function $F_l(\rho)$, which is

[1] R. Jost and W. Kohn, *Phys. Rev.* **88**, 382 (1952). V. Bargmann, *Revs. Modern Phys.* **21**, 488 (1949). R. G. Newton, *Journal of Mathematical Physics* **1**, 319 (1960).
[2] In Eq. (8–86), δ_l is determined only to within an arbitrary multiple of 2π. We adopt the convention that this constant is to be chosen such that $\delta_l = 0$ at $E = 0$. Cf. Eq. (8–96).

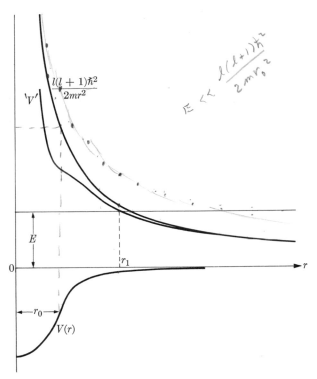

FIG. 8–3.   The potential function $V(r)$ and the potential function $`V'$ for the equivalent one-dimensional problem.

the regular solution of Eq. (8–88). However, we can show that if l is sufficiently large, the dependence of F_l upon $V(r)$ can be neglected, to good approximation, and this function can be replaced, within the integral, by $f_l(\rho)$. To see this qualitatively, consider the function

$$`V' = V + \frac{l(l+1)\hbar^2}{2mr^2},\qquad (8\text{–}91)$$

which is the potential-energy function for the equivalent one-dimensional problem (Section 7–3). This quantity is large and positive near the origin, and decreases with increasing r, becoming equal to the energy E at the point r_1, the classical turning point. It is clear that for a given energy, r_1 can be made as large as desired by choosing l sufficiently large (Fig. 8–3). To the left of the point r_1, the function F_l is of exponential type, and since it is initially proportional to r^{l+1}, it is small in the region in which $V(r)$ is appreciable. To the right of the point r_1, F_l is sinusoidal in character, and since $V(r)$ is essentially zero in this region, it behaves exactly like the free-particle function f_l. In other words, for small r, the centrifugal barrier potential is dominant, and the effect of $V(r)$ is negligible; for large r, $V(r)$ is practically zero. Hence, if the intermediate region in the neighborhood of r_1 is outside the range of $V(r)$, the effect of the force is small everywhere. Therefore, for large l,

$$F_l(\rho) \approx f_l(\rho), \tag{8-92}$$

or

$$\lim_{l \to \infty} \delta_l = 0. \tag{8-93}$$

Explicitly, this approximation is valid provided that $r_1 > r_0$, where r_0 is the "range" of $V(r)$, i.e., we require $V(r_1) \ll l(l+1)\hbar^2/2mr_1^2$. Under these circumstances, $l(l+1)\hbar^2/2mr_1^2 = E$, whence we have the criterion $l(l+1) \gg 2mEr_0^2/\hbar^2$, or

$$l \gg kr_0. \tag{8-94}$$

This effect of the centrifugal barrier corresponds, in the classical picture, to the fact that an orbit of large angular momentum is so remote from the force center that the motion is essentially unaffected by the short-range interaction.

Whenever the criterion (8-94) is satisfied, the phase shift δ_l is small, and Eq. (8-86) becomes, approximately,

$$\delta_l \approx - \int_0^\infty \frac{V}{E} [f_l(\rho)]^2 \, d\rho. \tag{8-95}$$

The inequality (8-94) is satisfied by any nonzero value of the angular momentum if the energy is sufficiently small. Hence, if the energy is near zero, only the s ($l = 0$) phase shift is appreciable, and the higher angular momenta only become important successively as the energy is increased. The initial dependence of the phase shifts on the energy can be obtained from Eq. (8-95). If the energy is very low, then the approximation

$$f_l \sim C_l \rho^{l+1}$$

is valid throughout the entire region in which $V(r)$ is appreciable, so that

$$\delta_l \approx - \int_0^\infty \frac{V}{E} C_l^2 \rho^{2l+2} \, d\rho = -C_l^2 \frac{k^{2l+3}}{E} \int_0^\infty V(r) r^{2l+2} \, dr. \tag{8-96}$$

Since k^2 is proportional to E, δ_l is proportional to $E^{l+1/2}$ for small E. Furthermore, it should be noted that the phase shifts are positive for an attractive interaction [$V(r) < 0$] and negative for a repulsive interaction [$V(r) > 0$].

8–7 Low-energy scattering. The above discussion shows that the partial-wave method is useful, in practice, in the study of low-energy scattering, where at most two or three values of l are important. If the energy is so low that only the s-state ($l = 0$) is involved, the theory is especially simple. In this case, all the phase shifts except δ_0 are negligible, and the formula (8–80) for the scattering amplitude becomes

$$f(\theta) = \frac{1}{k} e^{i\delta_0} \sin \delta_0, \qquad (8\text{–}97)$$

and the total cross section is

$$\sigma = \frac{4\pi}{k^2} \sin^2 \delta_0. \qquad (8\text{–}98)$$

The scattering amplitude, and therefore the differential cross section, is independent of θ. This means that the scattering is *isotropic*. To understand this fact, we only have to remember that we are dealing with very low energy, so that the wavelength associated with the incident particles is very long, compared to the range of force. The direction of incidence is "detected" by the scatterer by means of the difference in phase of the incident wave between two points separated by a distance of order r_0. If this distance is a small fraction of a wavelength, the difference of phase is inappreciable, and the direction of incidence does not matter for the scattering. In general, an obstacle which is small compared to the wavelength scatters isotropically.

We have also seen that the scattering is the result of the modification of the incident wave produced by the potential energy inside the range of force. In the case of s-waves, the behavior of the wave function in this region is governed by the term $E - V(r)$ in the Schrödinger equation. Consequently, if E is small compared to $V(r)$, i.e., to the depth of the potential well, the scattering is practically independent of E. The limiting value of $-f(\theta)$, as E approaches zero, is called the *scattering length, a:*

$$\lim_{E \to 0} f(\theta) = -a. \qquad (8\text{–}99)$$

In terms of a, the zero-energy cross section is

$$\sigma_0 = 4\pi a^2. \qquad (8\text{–}100)$$

If the interaction is weak and the phase shift small, then it follows from Eqs. (8–99) and (8–97) that

$$\lim_{k \to 0} \frac{\delta_0}{k} = -a, \qquad (8\text{–}101)$$

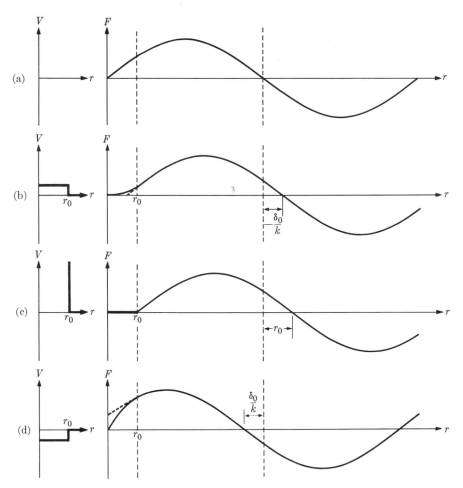

FIG. 8–4. The s-wave function for low-energy scattering. (a) Free-particle wave function (no scatterer). (b) Positive potential energy at $r < r_0$ (repulsive force). (c) Positive infinite potential energy at $r < r_0$ (impenetrable sphere). (d) Negative potential energy at $r < r_0$ (attractive force).

i.e., the phase shift for small energy is given approximately by

$$\delta_0 = -ka. \qquad (8\text{--}102)$$

The behavior of the s-wave function for low-energy scattering from various potentials is illustrated in Fig. 8–4. For convenience, the potential energy is assumed to be zero outside the sphere $r = r_0$. The "free-particle" function $\sin kr$, whose wavelength is long compared to r_0, is shown in

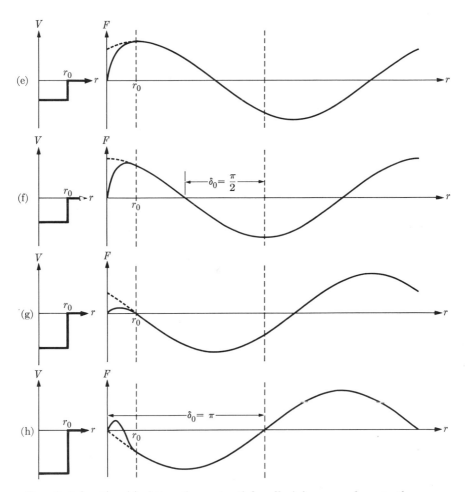

FIG. 8–4 (*cont.*). (e) Attractive potential well giving zero slope to the wave function at $r = r_0$. (f) The wave function has negative slope at the edge of the well (a bound state appears). (g) A zero of the wave function appears inside the well. (h) Phase shift of π results in zero s cross section.

Fig. 8–4(a). If the potential energy is positive [corresponding to a repulsive force, cf. Fig. 8–4(b)], then the interior wave function is of exponential type, and the slope of the wave function F at the boundary is increased relative to that in Fig. 8–4(a). Consequently, the zeros of the exterior wave function are moved out, and the phase shift is therefore negative. In the limit $V \to \infty$, the interior wave function is zero, and the exterior function is zero at the boundary of the well [Fig. 8–4(c)]. The phase shift is now $\delta_0 = -kr_0$, and the scattering cross section for such an "impenetrable sphere" is

$$\sigma = \frac{4\pi}{k^2} \sin^2 kr_0 \approx 4\pi r_0^2. \tag{8-103}$$

This is four times the "geometrical" cross section.

The behavior of the wave function for an attractive potential is quite different. The interior wave function is sinusoidal in this case, and the phase shift is always positive. In Fig. 8–4(d), the interaction is weakly attractive, the slope of the wave function at the boundary of the well is decreased relative to the free-particle function, and the exterior wave function is displaced to the left. In Fig. 8–4(e), the depth of the potential well is such that the wave function F has zero slope at $r = r_0$, and since $\lambda \gg r_0$, the phase shift is very nearly $\pi/2$. The cross section is now $4\pi/k^2$ and varies inversely with the energy; in the limit of zero energy, it is theoretically infinite. This effect is believed to account for the anomalously large cross section for the scattering of thermal neutrons by protons.[1]

The large cross section at $\delta_0 = \pi/2$ is associated with the appearance of a bound state of the system at zero energy. If the energy is small compared to V, then the interior wave function is nearly independent of E and would not be sensibly different if E were slightly negative. In Fig. 8–4(f), the slope of F at $r = r_0$ is negative, so that it is possible for the interior wave function to be joined smoothly to a decreasing exponential function in the exterior region. Since this occurs whenever the phase shift is an odd multiple of $\pi/2$, the number of bound states within the well is n, where $(n - \frac{1}{2})\pi < \delta_0 < (n + \frac{1}{2})\pi$. It should be noted that in this case Eq. (8–101) becomes

$$\lim_{k \to 0} \frac{\delta_0 - n\pi}{k} = -a. \tag{8-104}$$

This equation is valid whenever the indicated limit is finite. The scattering length is, of course, infinite for $\delta_0 = \pi/2, 3\pi/2$, etc.

It is apparent from Fig. 8–4 that the phase shift grows steadily as the strength of the potential increases. In parts (g) and (h) of Fig. 8–4, a zero of F appears inside the well for the first time, and the phase shift passes through the value π. The cross section is zero when the phase shift is π [Fig. 8–4(h)]: the s-state makes no contribution to the scattering. Since the effect of higher angular-momentum states is small at low energy, abnormally low total scattering can result. These properties of the scattering by an attractive potential will now be illustrated by an example.

---

[1] R. G. Sachs, *Nuclear Theory.* Reading, Mass.: Addison-Wesley Publishing Company, Inc., 1953, Sections 4–5 and 4–6.

8–8 Scattering by an attractive square potential well. An attractive potential-energy function of the form

$$V(r) = \begin{cases} -V_0 & (r < r_0), \\ 0 & (r > r_0), \end{cases} \tag{8–105}$$

represents a square potential well (Fig. 8–5).

The radial function $F_0(kr) = F$ for the s-state obeys the differential equations

$$\frac{d^2F}{dr^2} + \frac{2m}{\hbar^2}(E + V_0)F = 0 \qquad (r < r_0),$$

$$\frac{d^2F}{dr^2} + \frac{2m}{\hbar^2}EF = 0 \qquad (r > r_0), \tag{8–106}$$

and the conditions

$$F = 0 \text{ at } r = 0, \qquad F \text{ and } dF/dr \text{ continuous at } r = r_0.$$

The solution for $r < r_0$ is

$$F = A \sin \kappa r \qquad (r < r_0), \tag{8–107}$$

where

$$\kappa = \frac{1}{\hbar}\sqrt{2m(E + V_0)}. \tag{8–108}$$

Outside the well, F is a linear combination of the functions $f_0 = \sin kr$ and $g_0 = \cos kr$, and according to Eq. (8–78), we write

$$F = \sin (kr + \delta_0) \qquad (r > r_0). \tag{8–109}$$

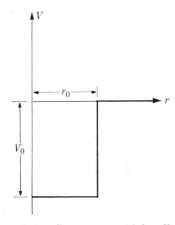

FIG. 8–5.   Square potential well.

The continuity condition at $r = r_0$ is expressed, independently of the normalization of F, by the requirement that the logarithmic derivative, $(1/F)(dF/dr)$, shall be continuous, i.e.,

$$\kappa \cot \kappa r_0 = k \cot (kr_0 + \delta_0). \qquad (8\text{--}110)$$

In the limit $k \to 0$, we have $\delta_0 \approx n\pi - ka$, and Eq. (8–110) is reduced to

$$K \cot Kr_0 = (r_0 - a)^{-1}, \qquad (8\text{--}111)$$

or

$$a = r_0 \left(1 - \frac{\tan Kr_0}{Kr_0} \right), \qquad (8\text{--}112)$$

where $K = (1/\hbar)\sqrt{2mV_0}$ is the value of κ at $E = 0$. Figure 8–6 shows a graph of the cross section derived from Eq. (8–112), which can be considered to represent the zero-energy cross section as a function of the well depth V_0 for a fixed range. If the well is very shallow, the cross section is nearly zero. It becomes infinite at $Kr_0 = (n + \frac{1}{2})\pi$ and is zero at intermediate points, corresponding to zero scattering length. As we have seen, this behavior is characteristic of any short-range attractive force.

The phase shift is given as a function of the wave number k by Eq. (8–110), which can be rewritten in the form

$$\delta_0 = n\pi - kr_0 + \tan^{-1}\left(\frac{k}{\kappa} \tan \kappa r \right). \qquad (8\text{--}113)$$

The number n is, as before, the number of bound states of the system, and the inverse tangent is an angle in the range $(0, \pi/2)$. This relation enables us to determine the quantity $\sin \delta_0$, and Eq. (8–98) yields the s cross section. A little trigonometric manipulation gives the result

$$\sin \delta_0 = \frac{(-)^n \sin kr_0}{\sqrt{k^2 + \kappa^2 \cot^2 \kappa r_0}} (k \cot kr_0 - \kappa \cot \kappa r_0). \qquad (8\text{--}114)$$

In practice, one is usually interested in the scattering for small energy, and this rather cumbersome formula can be simplified by expanding it in a power series. Instead of working directly with the expression (8–114), however, it is somewhat easier to expand the function

$$L_0 = k \cot \delta_0, \qquad (8\text{--}115)$$

where L_0 is the logarithmic derivative of the asymptotic wave function (8–109), evaluated at $r = 0$. It is evident from Eq. (8–110) that L_0 is an even function of k and therefore can be expanded in powers of k^2

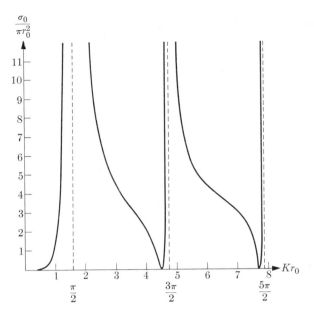

FIG. 8–6.   Zero-energy cross section (in units πr_0^2) for an attractive potential well, as a function of well depth $[Kr_0 = (r_0/\hbar)\sqrt{2mV_0}]$.

(cf. Problem 8–9). Explicitly, one finds that L_0 satisfies the equation

$$L_0\left(\cos kr_0 - L\frac{\sin kr_0}{k}\right) = k\sin kr_0 + L\cos kr_0, \qquad (8\text{–}116)$$

where $L = \kappa\cot\kappa r_0$ is the logarithmic derivative of F at the boundary of the well. Each of the quantities in Eq. (8–116) is an even function of k, and can be easily expanded in powers of k^2. The result of the calculation is

$$k\cot\delta_0 = -\frac{1}{a} + \frac{1}{2}r_0\left\{1 - \frac{1}{(Kr_0)^2}\left(\frac{r_0}{a}\right) - \frac{1}{3}\left(\frac{r_0}{a}\right)^2\right\}k^2 + O(k^4).$$
$$(8\text{–}117)$$

It is instructive to examine the behavior of this function in the neighborhood of the point $Kr_0 = \pi/2$, at which the zero-energy cross section is infinite (Fig. 8–6). The quantity r_0/a is small in this region, and it is apparent that the energy dependence of the cross section is almost entirely determined by the range of the force, r_0. The coefficient of $k^2/2$ in Eq. (8–117), namely

$$r_\text{eff} = r_0\left\{1 - \frac{1}{(Kr_0)^2}\left(\frac{r_0}{a}\right) - \frac{1}{3}\left(\frac{r_0}{a}\right)^2\right\}, \qquad (8\text{–}118)$$

is called the *effective range* for the square well. The scattering length and the effective range determine the low-energy cross section, which is given by

$$\sigma = \frac{4\kappa}{k^2 + [\frac{1}{2}r_{\text{eff}}k^2 - 1/a]^2},$$ (8–119)

and these quantities can therefore be determined from experiment.

It can be shown that a formula of the form (8–119) holds for any attractive short-range force, so that the above treatment can be generalized for potential wells which do not have the simple shape assumed in this example. In the case of "resonant" scattering, when there is a bound state near zero energy, the effective range is related in a simple way to the wave function for the bound state. By exploiting these facts, it is possible to formulate the low-energy scattering theory in a way that is largely independent of detailed assumptions concerning the potential function. Instead, the experimental results are associated directly with the wave function of the system. This point of view has been extensively developed in discussions of neutron-proton scattering and in the theory of the deuteron.[1]

8–9 Theoretical connection between scattering and bound states; S-matrix. The asymptotic wave function (8–82) can be made the basis of an interesting conjecture about the connection between the continuum states of the system and the discrete spectrum of bound states. The s-wave part of the asymptotic wave function is

$$\psi_0 \sim e^{2i\delta_0(k)} \frac{e^{ikr}}{r} - \frac{e^{-ikr}}{r},$$ (8–120)

in which the dependence of the phase shift on the energy has been emphasized by writing $\delta_0 = \delta_0(k)$. The wave number k is a real variable. However, by the process of analytic continuation, the mathematical function $\delta_0(k)$ can be regarded as a function of a complex variable k. Suppose that this is done, and that one finds a negative imaginary value of k, i.e.,

$$k' = -i|k'|,$$ (8–121)

such that

$$S(k') = e^{2i\delta_0(k')} = 0.$$ (8–122)

[1] G. Breit, *Revs. Modern Phys.* **23**, 238 (1951). H. A. Bethe, *Phys. Rev.* **76**, 38 (1949).

This means, of course, that at the point k',

$$\delta(k') = i\infty.$$

Provided that the asymptotic behavior of the wave function is still given by (8–120), we should then have

$$\psi_0 \sim -\frac{e^{-|k'|r}}{r} \, ; \tag{8–123}$$

this is exactly the condition which must be satisfied by a function representing a bound state of energy

$$E = -\frac{\hbar^2|k'|^2}{2m} \, . \tag{8–124}$$

In other words, one is led to the conjecture that those negative imaginary values of k which are zeros of the function $S(k) = e^{2i\delta(k)}$ correspond to bound states of energy E, given by Eq. (8–124). This conjecture, which was made first by H. A. Kramers,[1] is easily confirmed in the example given in the preceding section. According to Eq. (8–110), we have

$$\kappa \cot \kappa r_0 = k \cot (kr_0 + \delta_0) = ik\, \frac{e^{i(kr_0+\delta_0)} + e^{-i(kr_0+\delta_0)}}{e^{i(kr_0+\delta_0)} - e^{-i(kr_0+\delta_0)}}, \tag{8–125}$$

or

$$\kappa \cot \kappa r_0 = ik\, \frac{S(k) + e^{-2ikr_0}}{S(k) - e^{-2ikr_0}} \, . \tag{8–126}$$

Solving for $S(k)$, we obtain

$$S(k) = \frac{1 + (ik/\kappa)\tan \kappa r_0}{1 - (ik/\kappa)\tan \kappa r_0} \, . \tag{8–127}$$

Hence, writing $k = -i|k'|$ and requiring $S(-i|k'|) = 0$, we have the equation

$$1 + \frac{|k'|}{\sqrt{K^2 - |k'|^2}} \tan \sqrt{K^2 - |k'|^2}\, r_0 = 0 \tag{8–128}$$

as the condition for a bound state of energy E given by Eq. (8–124). The

---

[1] See C. Møller, *Kgl. Danske Videnskab. Selskab, Mat.-fys. Medd.* **24**, No. 19 (1946), and V. Bargmann, *Revs. Modern Phys.* **21**, 488 (1949).

correctness of this result is verified directly by writing the solution of
Eq. (7–64) as

$$
u = \begin{cases}
A \sin \dfrac{\sqrt{2m(E + V_0)}}{\hbar}\, r & (r < r_0), \\[2ex]
\exp\left(-\dfrac{\sqrt{-2mEr}}{\hbar}\right) & (r > r_0),
\end{cases}
\tag{8–129}
$$

and by applying the usual conditions of continuity at $r = r_0$ (Problem
8–13).

To achieve a general verification of the foregoing conjectures, it is
necessary to examine carefully the analytic properties of the solutions of the
Schrödinger equation, and detailed study shows that very considerable re-
finements are required to provide a sound basis for these ideas.[1] Nevertheless, it is interesting that the quantum properties of a given potential
function can, in a sense, be summarized in the function $S(k)$, which no
longer contains explicit reference to the wave functions of the system. This
function is the $(l = 0)$ element of the S-matrix of Heisenberg, mentioned
in Section 6–13.

8–10 General formulation of scattering theory. In practice, the analysis
of scattering by the partial-wave method is limited to scattering at relatively low energy. Furthermore, the expansion of the wave function in
spherical harmonics is generally useful only for spherically symmetric
systems. We shall now present a more general analysis, which is not, in
principle, subject to these limitations.

The Schrödinger equation for the scattering problem can be written

$$
\nabla^2 \psi + k^2 \psi = \frac{2m}{\hbar^2}\, V(\mathbf{r})\psi,
\tag{8–130}
$$

in which the potential-energy function $V(\mathbf{r})$ is no longer required to be
spherically symmetric. A solution is sought which satisfies the condition

$$
\psi \sim \exp\,(i\mathbf{k} \cdot \mathbf{r}) + f(\mathbf{k}, \mathbf{k}')\,\frac{e^{ikr}}{r}
\tag{8–131}
$$

when $\mathbf{r}$ is so large that the particle is beyond the range of the force. The
scattering amplitude is written as $f(\mathbf{k}, \mathbf{k}')$, where $\mathbf{k}$ and $\mathbf{k}'$ are propagation
vectors of magnitude k. The vector $\mathbf{k}$ points in the direction of incidence,
and the vector $\mathbf{k}'$ in that of $\mathbf{r}$ (direction of observation).

---

[1] R. Jost, *Helv. Phys. Acta* **20,** 256 (1947); R. G. Newton, *Journal of Mathematical Physics* **1,** 319 (1960).

It is convenient to separate the incident wave, $\exp(i\mathbf{k} \cdot \mathbf{r})$, from the wave function, and to write

$$\psi = \exp(i\mathbf{k} \cdot \mathbf{r}) + \psi_s. \tag{8-132}$$

The function ψ_s, the "scattered wave," is asymptotic to the second term of Eq. (8–131). Since $\exp(i\mathbf{k} \cdot \mathbf{r})$ is a solution of the homogeneous wave equation, i.e.,

$$(\nabla^2 + k^2)\exp(i\mathbf{k} \cdot \mathbf{r}) = 0, \tag{8-133}$$

the Schrödinger equation (8–130) now becomes

$$(\nabla^2 + k^2)\psi_s = \frac{2m}{\hbar^2} V(\mathbf{r})\psi. \tag{8-134}$$

A formal solution of this equation can be constructed by applying the principle of superposition. Let us write, temporarily,

$$\frac{2m}{\hbar^2} V(\mathbf{r})\psi = -4\pi\rho(\mathbf{r}), \tag{8-135}$$

so that Eq. (8–134) assumes the form

$$(\nabla^2 + k^2)\psi_s = -4\pi\rho(\mathbf{r}). \tag{8-136}$$

The quantity $\rho(\mathbf{r})$ can be regarded as a source density for divergent spherical waves. It is clear that if ψ_{s1} and ψ_{s2} are solutions of Eq. (8–136), belonging to the density functions $\rho_1(\mathbf{r})$ and $\rho_2(\mathbf{r})$ and satisfying

$$\psi_{s1} \sim f_1 \frac{e^{ikr}}{r}, \qquad \psi_{s2} \sim f_2 \frac{e^{ikr}}{r}, \tag{8-137}$$

then the function $\psi_s = \psi_{s1} + \psi_{s2}$ is a solution belonging to $\rho(\mathbf{r}) = \rho_1(\mathbf{r}) + \rho_2(\mathbf{r})$ such that

$$\psi_s \sim f \frac{e^{ikr}}{r}, \tag{8-138}$$

where $f = f_1 + f_2$. By means of this principle of superposition, a solution of Eq. (8–136) can be built up by addition of solutions for simple point sources of unit strength. The identity

$$\rho(\mathbf{r}) = \int \delta(\mathbf{r} - \mathbf{r}')\rho(\mathbf{r}') \, d\mathbf{r}' \tag{8-139}$$

represents the arbitrary density $\rho(\mathbf{r})$ as a sum of point sources $\delta(\mathbf{r} - \mathbf{r}')$ at the points $\mathbf{r}'$. Therefore, if a solution $G(\mathbf{r}, \mathbf{r}')$ can be found for the equation

$$(\nabla^2 + k^2)G(\mathbf{r}, \mathbf{r}') = -4\pi \, \delta(\mathbf{r} - \mathbf{r}') \tag{8-140}$$

such that $G(\mathbf{r}, \mathbf{r}')$ is asymptotic to a function of r of the form (8–138), then the solution of the scattering problem for the density $\rho(\mathbf{r})$ can be calculated from

$$\psi_s = \int G(\mathbf{r}, \mathbf{r}')\rho(\mathbf{r}')\, d\mathbf{r}'. \tag{8–141}$$

The function $G(\mathbf{r}, \mathbf{r}')$ is called *Green's function*.[1]

8–11 Green's function. It will now be shown that the function

$$G(\mathbf{r}, \mathbf{r}') = \frac{\exp\,(ik|\mathbf{r} - \mathbf{r}'|)}{|\mathbf{r} - \mathbf{r}'|} \tag{8–142}$$

is the solution of the scattering problem for a source of unit intensity at the point $\mathbf{r}'$. To prove this, it must be shown that the equation (8–140) is satisfied, and that the solution has the proper asymptotic form. For simplicity, we first change the origin of coordinates to the point $\mathbf{r}'$, so that Eq. (8–140) becomes

$$(\nabla^2 + k^2)G = -4\pi\,\delta(\mathbf{r}), \tag{8–143}$$

where

$$G = \frac{e^{ikr}}{r}, \tag{8–144}$$

and r is the radial distance in the new coordinates. Now one can verify (Problem 8–14) that if $r \neq 0$,

$$(\nabla^2 + k^2)\frac{e^{ikr}}{r} = 0. \tag{8–145}$$

Equation (8–143) is, therefore, satisfied in every region which does not contain the source point. To prove that the singularity at $\mathbf{r} = 0$ is properly represented by G, it must be established that the function

$$\Delta(\mathbf{r}) = -\frac{1}{4\pi}(\nabla^2 + k^2)\frac{e^{ikr}}{r} \tag{8–146}$$

is, in fact, a delta function. In other words, we must prove that

$$\int_\tau \Delta(\mathbf{r})F(\mathbf{r})\, d\mathbf{r} = F(0), \tag{8–147}$$

where $F(\mathbf{r})$ is any smooth function of $\mathbf{r}$ which has the value $F(0)$ at the origin. The region of integration τ in Eq. (8–147) is any finite volume which

---

[1] R. Courant and D. Hilbert, *Methods of Mathematical Physics.* New York: Interscience Publishers, Inc., 1953, Vol. I, Chapter 5, Section 14.

contains the origin. We choose a small sphere of radius a and consider the identity

$$\int_{\tau} [(\nabla^2 G + k^2 G)F - (\nabla^2 F + k^2 F)G] \, d\mathbf{r}$$

$$= \int_{\Sigma} \left[\frac{\partial G}{\partial r} F - \frac{\partial F}{\partial r} G \right] da, \qquad (8\text{–}148)$$

where Σ is the surface of the sphere τ. If $F(\mathbf{r})$ is sufficiently regular within τ, it can be assumed that positive numbers M and N can be found such that

$$|\nabla^2 F + k^2 F| < M \qquad \text{and} \qquad \left| \frac{\partial F}{\partial r} \right| < N \qquad (r \le a), \quad (8\text{–}149)$$

i.e., that these functions are bounded in τ. It follows that

$$\left| \int_{\tau} (\nabla^2 F + k^2 F)G \, d\mathbf{r} \right| \le M \int_{\tau} |G| \, d\mathbf{r} = M \cdot 4\pi \int_0^a \frac{1}{r} r^2 \, dr = 2\pi M a^2,$$

$$(8\text{–}150)$$

and

$$\left| \int_{\Sigma} \frac{\partial F}{\partial r} G \, da \right| \le N \int_S |G| \, da = 4\pi N a. \qquad (8\text{–}151)$$

Therefore, if we take the limit of Eq. (8–148) as $a \to 0$, we find

$$\lim_{a \to 0} \int_{\tau} (\nabla^2 G + k^2 G)F \, d\mathbf{r} = \lim_{a \to 0} \int_{\Sigma} \frac{\partial G}{\partial r} F \, da$$

$$= \lim_{a \to 0} 4\pi a^2 F(0) \left\{ -\frac{e^{ika}}{a^2} + ik \frac{e^{ika}}{a} \right\} = -4\pi F(0). \qquad (8\text{–}152)$$

Because of the relation (8–145), the only contribution to the integral (8–147) must come from the singularity of $\Delta(\mathbf{r})$ at $\mathbf{r} = 0$, and by Eq. (8–152), this is $F(0)$. Hence, Eq. (8–147) is proved, and

$$\Delta(\mathbf{r}) = \delta(\mathbf{r}). \qquad (8\text{–}153)$$

Returning to the original coordinate system by the substitution $\mathbf{r} \to \mathbf{r} - \mathbf{r}'$, we obtain Eq. (8–140).

The asymptotic form of $G(\mathbf{r}, \mathbf{r}')$ is easily found by referring to Fig. 8–7. If $|\mathbf{r}|$ is large compared to $|\mathbf{r}'|$, then it is clear that

$$|\mathbf{r} - \mathbf{r}'| \sim |\mathbf{r}| - \frac{\mathbf{r}' \cdot \mathbf{r}}{|\mathbf{r}|}. \qquad (8\text{–}154)$$

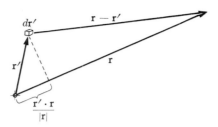

F IG . 8–7.   The vectors **r** and **r**′.

The error involved in this approximation can be made arbitrarily small by choosing |**r**| sufficiently large. Writing

$$k|\mathbf{r} - \mathbf{r}'| = kr - \mathbf{k}' \cdot \mathbf{r}', \qquad (8\text{--}155)$$

where $\mathbf{k}' = k(\mathbf{r}/|\mathbf{r}|)$, we therefore have

$$G(\mathbf{r}, \mathbf{r}') \sim e^{-i\mathbf{k}'\cdot\mathbf{r}'} \frac{e^{ikr}}{r}, \qquad (8\text{--}156)$$

and the first factor depends only upon the orientation of **k**′ relative to the fixed vector **r**′, i.e., it is of the form $f(\theta, \phi)$. This completes the demonstration that the expression (8–142) is Green's function.

8–12 Integral equation for the wave function. If the expression for the density function [Eq. (8–135)] is now restored in Eq. (8–141), the scattering problem is formulated in terms of the integral equation

$$\psi(\mathbf{r}) = \exp{(i\mathbf{k} \cdot \mathbf{r})} - \frac{1}{4\pi} \cdot \frac{2m}{\hbar^2} \int \frac{\exp{(ik|\mathbf{r} - \mathbf{r}'|)}}{|\mathbf{r} - \mathbf{r}'|} V(\mathbf{r}')\psi(\mathbf{r}')\, d\mathbf{r}'. \quad (8\text{--}157)$$

This equation shows that the scattered wave at the point **r** is composed of spherical waves originating at each point of space **r**′. The amplitude of each contribution is proportional to the product $V(\mathbf{r}')\psi(\mathbf{r}')$, i.e., it is proportional jointly to the strength of the interaction and the amplitude of the wave function at **r**′. All these spherical waves are compounded at the point **r** and generate the total scattered wave, which is then added to the incident wave to produce the total wave function at **r**. If the potential-energy function is confined to a limited region of space (this is necessary if the notion of scattering is to make sense), then the asymptotic form of $G(\mathbf{r}, \mathbf{r}')$ can be substituted in Eq. (8–157) for points remote from this region. This yields for the scattering amplitude,

$$f(\mathbf{k}, \mathbf{k}') = - \frac{1}{4\pi} \frac{2m}{\hbar^2} \int \exp{(-i\mathbf{k}' \cdot \mathbf{r}')}\, V(\mathbf{r}')\psi(\mathbf{r}')\, d\mathbf{r}', \quad (8\text{--}158)$$

a general formula by which the cross section can be computed from the wave function.[1]

Of course, Eq. (8–157) is not an explicit solution of the scattering problem, because the function ψ appears under the integral sign and must be known before the integral can be evaluated. However, the formulation of the problem according to Eq. (8–157) provides a single equation which expresses both the content of the differential equation (8–130) *and* the boundary condition (8–131). The theory of integral equations allows one to develop the entire subject on the basis of this formulation. This aspect of the theory will not be considered in this book; however, it is instructive to verify that the formula (8–158) is consistent with the results of the method of partial waves. Equation (8–157) is also a convenient basis for the Born approximation, described in Section (8–14).

8–13 Connection with partial-wave analysis. Let the z-axis be chosen in the direction of incidence. Then the wave function for scattering in a spherically symmetric potential [Eq. (8–81)] is

$$\psi(\mathbf{r}') = \sum_l i^l \sqrt{4\pi(2l+1)}\ Y_l^0(\hat{\mathbf{r}}')e^{i\delta_l}\frac{F_l(\rho')}{\rho'}, \qquad (8\text{–}159)$$

where we have written

$$P_l(\mu') = \sqrt{\frac{4\pi}{2l+1}}\ Y_l^0(\hat{\mathbf{r}}'), \qquad (8\text{–}160)$$

and $\rho' = kr'$. By Eq. (8–60), the function $\exp(-i\mathbf{k}' \cdot \mathbf{r}')$ is

$$\exp(-i\mathbf{k}' \cdot \mathbf{r}') = 4\pi \sum_{l',m'} (-i)^{l'} Y_{l'}^{m'}(\hat{\mathbf{k}}') Y_l^{m*}(\hat{\mathbf{r}}')\frac{f_l(\rho')}{\rho'}. \qquad (8\text{–}161)$$

Also the volume element $d\mathbf{r}'$ in Eq. (8–158) can be written $d\mathbf{r}' = r'^2\,dr'\,d\Omega'$, where $d\Omega'$ is the element of solid angle. Making these substitutions in Eq. (8–158) and carrying out the integration over the angles, one finds, because of the orthonormality of the functions $Y_l^m(\hat{\mathbf{r}}')$, that

$$f(\mathbf{k}, \mathbf{k}') = f(\theta) =$$

$$= -\frac{2m}{\hbar^2 k^2} \sum_l \sqrt{4\pi(2l+1)}\ Y_l^0(\hat{\mathbf{k}}')e^{i\delta_l} \int_0^\infty V(r')f_l(\rho')F_l(\rho')\,dr'; \qquad (8\text{–}162)$$

---

[1] The dependence of $f(\mathbf{k}, \mathbf{k}')$ on $\mathbf{k}$ is contained implicitly in the right-hand member of this equation in the function $\psi(\mathbf{r}')$, which depends upon the direction of incidence.

by Eq. (8–160), this reduces to

$$f(\theta) = \frac{1}{k} \sum_l (2l + 1) P_l (\cos \theta) e^{i\delta_l} \left\{ - \int_0^\infty \frac{V}{E} f_l(\rho) F_l(\rho) \, d\rho \right\}, \quad (8\text{–}163)$$

which, by virtue of Eq. (8–86), is seen to be identical with Eq. (8–80). A similar verification can be made, showing that Eq. (8–157) is satisfied by the expansion (8–81) (Problem 8–18).

8–14 The Born approximation. Equation (8–157) can be solved formally by the process of iteration. Thus, if $\mathbf{r}$ is replaced by $\mathbf{r}'$ in the equation

$$\psi(\mathbf{r}) = \exp(i\mathbf{k} \cdot \mathbf{r}) - \frac{1}{4\pi} \frac{2m}{\hbar^2} \int G(\mathbf{r}, \mathbf{r}'') V(\mathbf{r}'') \psi(\mathbf{r}'') \, d\mathbf{r}'' \quad (8\text{–}164)$$

and the result is substituted for $\psi(\mathbf{r}')$ in the integral, one obtains

$$\psi(\mathbf{r}) = \exp(i\mathbf{k} \cdot \mathbf{r}) - \frac{1}{4\pi} \frac{2m}{\hbar^2} \int G(\mathbf{r}, \mathbf{r}') V(\mathbf{r}') \exp(i\mathbf{k} \cdot \mathbf{r}') \, d\mathbf{r}'$$

$$+ \left(-\frac{1}{4\pi} \frac{2m}{\hbar^2} \right)^2 \iint G(\mathbf{r}, \mathbf{r}') V(\mathbf{r}') G(\mathbf{r}', \mathbf{r}'') V(\mathbf{r}'') \psi(\mathbf{r}'') \, d\mathbf{r}'' \, d\mathbf{r}'. \quad (8\text{–}165)$$

This equation is called the *first iterated form* of Eq. (8–164). The iteration process can be repeated indefinitely, resulting in an infinite series (the *Neumann series*) which, if it converges, can be expected to represent a solution. This series has a rather direct intuitive meaning.[1] The first term in the scattered wave represents single scattering of the incident wave $\exp(i\mathbf{k} \cdot \mathbf{r}')$ by the interaction $V(\mathbf{r}')$ in the volume element $d\mathbf{r}'$. This produces a wave which travels from $\mathbf{r}'$ to the point of observation $\mathbf{r}$, and the total wave arising from single scattering is obtained by integration over the region in which the force is effective. In the second term, the incident wave $\exp(i\mathbf{k} \cdot \mathbf{r}'')$ is scattered at the point $\mathbf{r}''$, $[V(\mathbf{r}'') \exp(i\mathbf{k} \cdot \mathbf{r}'')]$, travels to the point $\mathbf{r}'$, $[G(\mathbf{r}', \mathbf{r}'') V(\mathbf{r}'') \exp(i\mathbf{k} \cdot \mathbf{r}'')]$, where it is again scattered, and then travels from $\mathbf{r}'$ to $\mathbf{r}$. The total effect of all such double scatterings is obtained by integration over $\mathbf{r}''$ and $\mathbf{r}'$. This interpretation is illustrated in Fig. 8–8, and it is seen that, in general, the nth term represents the contribution of waves which have been scattered n times in the region of interaction before traveling to the point $\mathbf{r}$, where their total contribution is observed. The Green's functions in each term can be thought of as "propagators," which carry the waves from their last point of scattering to the next.

If the interaction is weak, so that the scattered wave is not large, it can

---

[1] R. P. Feynman, *Phys. Rev.* **76**, 749 (1949).

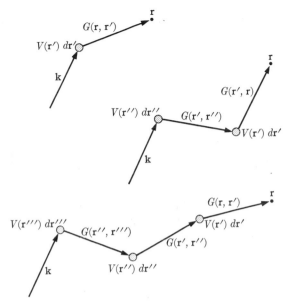

FIG. 8–8.   Illustration of first, second, and third terms in the Neumann series.

be expected that the Neumann series will converge rapidly, and that the first term in the Neumann expansion will provide an approximation to ψ. This is the first *Born approximation*:

$$\psi = \exp\left(i\mathbf{k} \cdot \mathbf{r}\right) - \frac{1}{4\pi} \frac{2m}{\hbar^2} \int G(\mathbf{r}, \mathbf{r}')\, V(\mathbf{r}')\, \exp\left(i\mathbf{k} \cdot \mathbf{r}'\right)\, d\mathbf{r}'. \quad (8\text{--}166)$$

The equation obtained by cutting off the Neumann series at the nth term is called the nth Born approximation. As we have seen, it amounts to neglecting multiply scattered waves which have been scattered more than n times by the interaction.

It is clear from the discussion of Section 8–6 that, for weak interactions, the Born approximation is equivalent to the approximation (8–95), which arose in treating the problem by the method of partial waves. (This will be shown in detail below.) However, this is not the only condition under which Eq. (8–166) can give useful results. Equation (8–165) shows that meaningful results can be obtained whenever the second integral is negligibly small. If the energy is sufficiently high, the wave function $\psi(\mathbf{r})$ is a rapidly oscillating function within the region of interaction, and it can be expected that the integral will be small because of cancellation of contributions from different parts of the region of interaction. Hence, although the interaction may be quite strong, the scattered wave is small, and the Born approximation is valid provided the energy is sufficiently high. In this sense, the Born approximation is complementary to the method of partial waves, which is most effective for low energy.

The asymptotic form (8–156) leads, by (8–131) and (8–166), to the expression

$$f(\mathbf{k}, \mathbf{k}') = -\frac{1}{4\pi} \frac{2m}{\hbar^2} \int \exp\left(-i\mathbf{k}' \cdot \mathbf{r}\right) V(\mathbf{r}) \exp\left(i\mathbf{k} \cdot \mathbf{r}\right) d\mathbf{r} \quad (8\text{--}167)$$

for the scattering amplitude. We shall illustrate the application of this formula by a discussion of the scattering of fast electrons by atoms.

8–15 Atomic scattering of electrons. The scattering of fast electrons by an atom can be represented, to good approximation, as the scattering in a spherically symmetric field $V(r)$, which is the electrostatic potential energy due to the nuclear charge Ze and the charge of the atomic electrons,

$$V(r) = -\frac{Ze^2}{r} + Ze^2 \int \frac{\rho(\mathbf{r}') \, d\mathbf{r}'}{|\mathbf{r} - \mathbf{r}'|}. \quad (8\text{--}168)$$

In this equation, $\mathbf{r}$ is the position of the scattered electron, and $\rho(\mathbf{r}')$ is the particle density of the atomic electrons at the point $\mathbf{r}'$. This expression can be written, for convenience, as

$$V(r) = -Ze^2 \int \frac{\rho_t(\mathbf{r}') \, d\mathbf{r}'}{|\mathbf{r} - \mathbf{r}'|},$$

where $\rho_t(\mathbf{r}') = \delta(\mathbf{r}') - \rho(\mathbf{r}')$ represents the total charge density (nuclear and electronic). Substituting in Eq. (8–167) and writing $\mathbf{K} = \mathbf{k} - \mathbf{k}'$, one obtains

$$f(\mathbf{k}, \mathbf{k}') = \frac{Ze^2}{4\pi} \frac{2m}{\hbar^2} \iint \frac{\rho_t(\mathbf{r}')}{|\mathbf{r} - \mathbf{r}'|} \exp\left(i\mathbf{K} \cdot \mathbf{r}\right) d\mathbf{r}' \, d\mathbf{r}, \quad (8\text{--}169)$$

or

$$f(\mathbf{k}, \mathbf{k}') = \frac{Ze^2}{4\pi} \frac{2m}{\hbar^2} \int \rho_t(\mathbf{r}') \exp\left(i\mathbf{K} \cdot \mathbf{r}'\right) \left\{ \int \frac{\exp\left[i\mathbf{K} \cdot (\mathbf{r} - \mathbf{r}')\right]}{|\mathbf{r} - \mathbf{r}'|} d\mathbf{r} \right\} d\mathbf{r}'.$$

The integral within the braces is evaluated by taking the point $\mathbf{r}'$ as origin in the integration over $\mathbf{r}$. It becomes

$$\int \frac{\exp\left(i\mathbf{K} \cdot \mathbf{r}\right)}{r} \, d\mathbf{r}. \quad (8\text{--}170)$$

This integral is not absolutely convergent, but it can be integrated by inserting a *convergence factor* $e^{-\lambda r}$. Thus, we have

$$\int e^{-\lambda r} \frac{\exp{(i\mathbf{K} \cdot \mathbf{r})}}{r} \, d\mathbf{r} = 2\pi \int_0^\infty e^{-\lambda r} \int_{-1}^1 e^{iK\mu r} \, d\mu r \, dr$$

$$= 2\pi \int_0^\infty e^{-\lambda r} \left\{ \frac{e^{iKr} - e^{-iKr}}{iK} \right\} dr = \frac{4\pi}{K} \operatorname{Im} \int_0^\infty e^{-(\lambda - iK)^r} \, dr$$

$$= \frac{4\pi}{K} \operatorname{Im} \frac{1}{\lambda - iK} = \frac{4\pi}{\lambda^2 + K^2}.$$

In the limit $\lambda \to 0$, this becomes

$$\int \frac{\exp{(i\mathbf{K} \cdot \mathbf{r})}}{r} \, d\mathbf{r} = \frac{4\pi}{K^2}. \tag{8-171}$$

Inserting Eq. (8–171) into Eq. (8–169), one has

$$f(\mathbf{k}, \mathbf{k}') = Ze^2 \cdot \frac{2m}{\hbar^2 K^2} \int [\delta(\mathbf{r}) - \rho(\mathbf{r})] \exp{(i\mathbf{K} \cdot \mathbf{r})} \, d\mathbf{r}$$

$$= Ze^2 \frac{2m}{\hbar^2 K^2} \{1 - F(K)\}, \tag{8-172}$$

where

$$F(K) = \int \rho(\mathbf{r}) \exp{(i\mathbf{K} \cdot \mathbf{r})} \, d\mathbf{r}. \tag{8-173}$$

Now $K = 2k \sin \theta/2$, where θ is the scattering angle, and $\hbar^2 k^2 = 2mE$, whence Eq. (8–172) can be written

$$f(\theta) = \frac{Ze^2}{4E \sin^2 \theta/2} [1 - F(K)]. \tag{8-174}$$

The quantity $F(K)$ is the *atomic scattering factor*, which is shown by Eq. (8–173) to be the Fourier transform of the electronic density. The particle density in the ground state of an atom is spherically symmetric, and the integration over the angles in Eq. (8–173) can be carried out immediately. We obtain

$$F(K) = \frac{4\pi}{K} \int_0^\infty \rho(r) \sin Kr \cdot r \, dr. \tag{8-175}$$

A crude approximation to the charge distribution within an atom is given by the function

$$\rho(r) = \frac{1}{(\sqrt{\pi} \, a)^3} e^{-(r^2/a^2)}, \tag{8-176}$$

where the constant a is the "atomic radius." In this case, Eq. (8–175) gives (Problem 8–21)

$$F(K) = e^{-(Ka)^2/4}. \tag{8-177}$$

The atomic scattering factor represents the shielding effect of the atomic electrons upon the nuclear charge. It is clear that this effect is small unless the quantity Ka is of order unity, i.e., unless

$$ka \sin \frac{\theta}{2} \approx 1. \tag{8-178}$$

If the scattering angle is large, this means that the shielding is ineffective if the electron energy is much greater than the ionization energy for the atom. In this case, Eq. (8-174) gives for the cross section

$$\sigma(\theta) = |f(\theta)|^2 = \left(\frac{Ze^2}{4E}\right)^2 \frac{1}{\sin^4 \theta/2}, \tag{8-179}$$

which is the classical Rutherford formula for scattering by a bare nucleus of charge Ze. At sufficiently small angles, the condition (8-178) is satisfied, however great the energy may be, and the effect of shielding becomes important. Thus, the infinite total cross section, which results from the strong singularity in the Rutherford cross section at $\theta = 0$, is made finite by the shielding effect. However, the total cross section calculated from the Born approximation is incorrect in principle, because the correct scattering amplitude in the forward direction cannot be purely real (Problem 8-20). It is obvious that the quantity $F(K)$, and therefore $f(\theta)$, as calculated from Eq. (8-174), is purely real for any spherically symmetric atom.

General conditions for the validity of the Born approximation cannot be given. In some cases, it is found to be reliable over a wide range of energy and angle; in others, it has no practical range of validity at all.[1] We shall show that the Born approximation is equivalent to the approximation (8-95) in the method of partial waves and therefore applicable whenever the phase shifts δ_l are small for all values of l. For a spherically symmetric potential, the angular part of the integration in Eq. (8-167) can be carried out (Problem 8-24), with the result

$$f(\theta) = -\frac{1}{k} \int_0^\infty \frac{V}{E} \frac{\sin Kr}{Kr} \rho^2 \, d\rho, \tag{8-180}$$

where $\rho = kr$.

The function $(\sin Kr)/Kr$ is easily expanded in partial waves, for we have (Problem 8-25)

$$4\pi \frac{\sin k|\mathbf{r} - \mathbf{r}'|}{k|\mathbf{r} - \mathbf{r}'|} = \int \exp [i\mathbf{k} \cdot (\mathbf{r} - \mathbf{r}')] \, d\Omega_k, \tag{8-181}$$

---

[1] R. Jost and A. Pais, *Phys. Rev.* **82**, 840 (1951). L. I. Schiff, *Quantum Mechanics.* 2nd ed. New York: McGraw-Hill Book Co., Inc., 1955, p. 169.

where $d\Omega_k$ is the element of solid angle in $\mathbf{k}$-space. By Eq. (8–60), the integral is

$$\int \exp\left[i\mathbf{k}\cdot(\mathbf{r}-\mathbf{r}')\right] d\Omega_k = (4\pi)^2 \sum_{l,m}' Y^{m*}(\hat{\mathbf{r}}')Y_l^m(\hat{\mathbf{r}}) \frac{f_l(kr')f_l(kr)}{kr'\cdot kr}, \quad (8\text{–}182)$$

whence, by the addition theorem for spherical harmonics,

$$\frac{\sin k|\mathbf{r}-\mathbf{r}'|}{k|\mathbf{r}-\mathbf{r}'|} = \sum_l (2l+1)P_l(\hat{\mathbf{r}}\cdot\hat{\mathbf{r}}') \frac{f_l(\rho)}{\rho}\frac{f_l(\rho')}{\rho'}. \quad (8\text{–}183)$$

In the special case $|\mathbf{r}'| = |\mathbf{r}|$, Eq. (8–183) reduces to

$$\frac{\sin Kr}{Kr} = \sum_l (2l+1)P_l(\cos\theta)\frac{f_l^2(\rho)}{\rho^2}. \quad (8\text{–}184)$$

Equation (8–180) is therefore

$$f(\theta) = \frac{1}{k}\sum_l (2l+1)P_l(\cos\theta)\left\{-\int_0^\infty \frac{V}{E}f_l^2(\rho)\,d\rho\right\}, \quad (8\text{–}185)$$

which, by (8–95), is the same as

$$f(\theta) = \frac{1}{k}\sum_l (2l+1)\,\delta_l P_l(\cos\theta). \quad (8\text{–}186)$$

The approximation $\delta_l \ll 1$, however, reduces the exact expression (8–80) to precisely this form.

In some applications, it may happen that only the first one or two phase shifts are large, while all others are small. The Born approximation is then useful in calculating the contributions to $f(\theta)$ which come from large l. For instance, if only the s phase shift is large, it can be calculated precisely by other methods, and the effect of higher angular momenta approximated by writing

$$f(\theta) = \frac{1}{k}\sum_{l=0}^\infty (2l+1)e^{i\delta_l}\sin\delta_l P_l(\cos\theta)$$

$$\approx \frac{1}{k}e^{i\delta_0}\sin\delta_0 + \sum_{l=1}^\infty (2l+1)\,\delta_l P_l(\cos\theta)$$

$$= \frac{1}{k}e^{i\delta_0}\sin\delta_0 + f_B(\theta) - \frac{1}{k}\delta_{0B}, \quad (8\text{–}187)$$

where $f_B(\theta)$ and δ_{0B} are the (incorrect) values computed from the Born approximations (8–167) and (8–95).

REFERENCES

BLATT, J. M., and WEISSKOPF, V. F., *Theoretical Nuclear Physics*. New York: John Wiley and Sons, 1952.

MOTT, N. F., and MASSEY, H. S. W., *The Theory of Atomic Collisions*. 2nd ed. Oxford: Clarendon Press, 1949.

SACHS, R. G., *Nuclear Theory*. Reading, Mass.: Addison-Wesley Publishing Co., Inc., 1953.

PROBLEMS

8–1. Evaluate the coefficients in the series solution

$$f_l = \sum_{k=l}^{\infty} a_k \rho^k$$

of Eq. (8–7), taking $a_l = C_l = 2^l l!/(2l+1)!$.

8–2. Carry out the operations in Eqs. (8–10) and (8–11) and verify that each is equivalent to Eq. (8–7).

8–3. Verify Eqs. (8–24), continue the calculation by induction, and thus prove Eq. (8–25).

8–4. Show that the irregular function g_l satisfies the relations (8–37) and (8–38); verify Eqs. (8–39).

8–5. Prove that $f_l g_{l+1} - f_{l+1} g_l = 1$, and show that both f_l and g_l satisfy the relation

$$f_{l+1} - \frac{2l+1}{\rho} f_l + f_{l-1} = 0.$$

8–6. Derive the formula (8–45) by the method indicated in the text.

8–7. Show that

$$\frac{d}{d\rho}\left[\frac{f_l}{\rho^{l+1}}\right] = -\frac{f_{l+1}}{\rho^{l+1}},$$

and deduce[1] that

$$\frac{f_l(\rho)}{\rho} = (-\rho)^l \left\{\frac{1}{\rho}\frac{d}{d\rho}\right\}^l \left(\frac{\sin\rho}{\rho}\right).$$

8–8. Obtain the relation (8–58) by comparing the expansions of the two members of Eq. (8–53) in power series in ρ. [Hint: Expand the term $e^{i\rho\mu}$ in the integral and evaluate the integrals $\int_{-1}^{1}\mu^k P_l(\mu)\,d\mu$ for suitable values of k].

Also, show that the term

$$\frac{1}{-i\rho}\int_{-1}^{1} e^{i\rho\mu} P'_l(\mu)\,d\mu$$

in Eq. (8–54) is of order $1/\rho^2$ for large ρ.

---

[1] G. N. Watson, *A Treatise on the Theory of Bessel Functions*. 2nd ed. Cambridge: Cambridge University Press, 1945, p. 54.

8–9. Show that the regular solution $F_l(kr)$ of Eq. (8–72) satisfies the functional equation

$$F_l(-kr) = (-)^{l+1} F_l(kr).$$

Hence, prove that

$$\delta_l(-k) = -\delta_l(k).$$

[Hint: The differential equation

$$\frac{d^2 F_l}{dr^2} + \left\{ k^2 - \frac{2mV(r)}{\hbar^2} - \frac{l(l+1)}{r^2} \right\} F_l = 0$$

is not changed if k is replaced by $-k$. Thus, if $F_l(kr)$ is a *regular* solution, so is $F_l(-kr)$.]

8–10. In Fig. 8–6, the scattering length changes sign also at the points $Kr_0 = \pi, 2\pi$, etc. Explain why these points do not correspond to bound states at zero energy.

8–11. Derive Eq. (8–114). (Hint: The ambiguity in sign which arises from the use of trigonometric identities can be resolved by the requirement that $\delta_0 \approx n\pi - ka$ for small k.)

8–12. Analyze the p-scattering $(l = 1)$ of a particle by a spherical potential well and construct a graph showing the contribution of p-wave scattering to the total cross section. Also, calculate the angular distribution for $Kr_0 = \pi/2, \pi$, and 2π. What is the ratio of the maximum to the minimum total cross section in each case?

8–13. Derive the condition (8–128) for the bound states directly from (8–129). Show also that the number of bound states in a well of depth V_0 agrees with the number obtained on the basis of the scattering length in Section 8–9.

8–14. Show that, if $r \neq 0$,

$$(\nabla^2 + k^2) \frac{e^{ikr}}{r} = 0.$$

8–15. Supply the details of the calculation leading to Eq. (8–162).

8–16. Show that the function

$$G_l(\rho, \rho') = f_l(\rho_<) g_l(\rho_>)$$

satisfies the differential equation

$$\frac{d^2 G_l}{d\rho^2} + \left\{ 1 - \frac{l(l+1)}{\rho^2} \right\} G_l = -\delta(\rho - \rho').$$

The symbols $\rho_<$ and $\rho_>$ denote, respectively, the larger and the smaller of the quantities ρ and ρ'. [Hint: The function $G_l(\rho, \rho')$ has a discontinuity of slope at $\rho = \rho'$, and the δ-function arises from the representation (A4–7).] This function is Green's function for the regular solution of the differential equation (8–72) for the lth partial wave.

8–17. Derive the integral equation

$$F_l(\rho) = \cos \delta_l f_l(\rho) - \int_0^\infty G_l(\rho, \rho')v(\rho')F_l(\rho')\, d\rho',$$

where $v(\rho') = V(r')/E$. Show that this is a regular solution of Eq. (8–72), and that it leads to the asymptotic form (8–76) for $F_l(\rho)$.

8–18. Using the expansions[1]

$$\frac{\sin k|\mathbf{r} - \mathbf{r}'|}{k|\mathbf{r} - \mathbf{r}'|} = \sum_{l=0}^\infty (2l + 1)P_l(\hat{\mathbf{r}} \cdot \hat{\mathbf{r}}') \frac{f_l(\rho)}{\rho} \frac{f_l(\rho')}{\rho'},$$

$$\frac{\cos k|\mathbf{r} - \mathbf{r}'|}{k|\mathbf{r} - \mathbf{r}'|} = \sum_{l=0}^\infty (2l + 1)P_l(\hat{\mathbf{r}} \cdot \hat{\mathbf{r}}') \frac{f_l(\rho_<)}{\rho_<} \frac{g_l(\rho_>)}{\rho_>},$$

establish that the integral equation (8–157) is satisfied by the solution (8–81) for a spherically symmetric force.

8–19. Let ψ_k and $\psi_{k'}$ be solutions of the Schrödinger equation (8–130), corresponding to incident waves of the same wavelength in the directions $\mathbf{k}$ and $\mathbf{k}'$, respectively, i.e.,

$$\psi_k \sim \exp(i\mathbf{k} \cdot \mathbf{r}) + f(\mathbf{k}, \mathbf{k}'') \frac{e^{ikr}}{r},$$

$$\psi_{k'} \sim \exp(i\mathbf{k}' \cdot \mathbf{r}) + f(\mathbf{k}', \mathbf{k}'') \frac{e^{ikr}}{r},$$

where $\mathbf{k}''$ is the propagation vector in the direction of observation. Prove the *reciprocity theorem*

$$f(\mathbf{k}', \mathbf{k}) = f(-\mathbf{k}, -\mathbf{k}')$$

and explain its physical meaning.

8–20. Prove that

$$\frac{1}{2i}\{f(\mathbf{k}', \mathbf{k}) - f^*(\mathbf{k}, \mathbf{k}')\} = \frac{k}{4\pi} \int f^*(\mathbf{k}'', \mathbf{k}')f(\mathbf{k}'', \mathbf{k})\, d\Omega_{k''},$$

where $d\Omega_{k''}$ is the element of solid angle in the $\mathbf{k}''$ space. Hence show that

$$\sigma(\mathbf{k}) = \frac{4\pi}{k} \operatorname{Im} f(\mathbf{k}, \mathbf{k}),$$

---

[1] G. N. Watson, *A Treatise on the Theory of Bessel Functions*. 2nd ed. Cambridge: Cambridge University Press, 1945, p. 366. P. M. Morse and H. Feshbach, *Methods of Theoretical Physics*. New York: McGraw-Hill Book Co., Inc., 1953, Vol. II, p. 1466.

where $\sigma(\mathbf{k})$ is the total scattering cross section for waves incident in the direction $\mathbf{k}$.[1] The second of these relations is the subject of a very interesting discussion by L. I. Schiff.[2] Note: It follows that the Born approximation (8–167) can never be relied upon to give the correct scattering amplitude in the forward direction. Why?

8–21. Verify Eq. (8–175) for a spherically symmetric particle density, and obtain the scattering factor (8–177).

8–22. The potential energy for scattering of an electron by an atom can be represented approximately as a "shielded" Coulomb field by

$$V(r) = \frac{Ze^2}{r} e^{-r/a},$$

where a is a "shielding radius." Show that, in the Born approximation, the scattering amplitude is

$$f(\theta) = -\frac{2mZe^2}{\hbar^2 K^2} \left\{ 1 - \frac{1}{1 + (Ka)^2} \right\}.$$

8–23. The electronic density in hydrogen is given by

$$\rho(r) = |\psi(1, 0, 0)|^2,$$

where $\psi(1, 0, 0)$ is the ground-state wave function [Eq. 7–197]. Show that the scattering factor for hydrogen is

$$F(K) = \frac{(2/a_0)^4}{\{(2/a_0)^2 + K^2\}^2},$$

where a_0 is the Bohr radius, and that the total cross section is

$$\sigma = \frac{\pi a_0^2}{3} \frac{(12 + 18\epsilon + 7\epsilon^2)}{(1 + \epsilon)^3},$$

where ϵ is the energy in units of the ionization energy of hydrogen.

8–24. Verify Eq. (8–180).

8–25. Supply the details of the calculation leading to Eq. (8–185).

---

[1] See R. Glauber and V. Schomaker, *Phys. Rev.* **89,** 667 (1953).

[2] L. I. Schiff, *Progr. Theoret. Phys.* (Kyoto) **11,** 288 (1954). See also E. Feenberg, *Phys. Rev.* **40,** 40 (1932).

CHAPTER 9

MATRIX MECHANICS

9–1 Introduction. The early development of quantum mechanics followed two very different mathematical lines. The theory of quantum states, developed by Schrödinger, seemed, in a sense, to reinstate the position of continuity as an essential element of nature: it appeared that the observed discreteness of certain atomic phenomena followed from the requirement that the wave function be continuous. Heisenberg, on the other hand, adopted the idea of discrete quantum jumps as the essential feature of atomic phenomena, which had to be described by the mathematical theory. He abandoned the search for an underlying picture and dealt directly with collections of discrete numbers, representing the observed quantum transitions. The algebraic theory of matrices is the mathematical discipline appropriate to this approach.

It soon became evident that the existence of these two apparently very different formulations of quantum theory is not accidental, and that they are indeed alternate expressions of the same underlying mathematical structure. The identity of the Schrödinger and Heisenberg formulations depends essentially upon the linear nature of the theory. In Chapter 6, it has been noted that the Schrödinger wave function ψ can be regarded as a representative of a vector, and that the operators of the theory have the effect of transforming one vector into another. These vectors represent different states of a given physical system, and they can be combined linearly to form representatives of other states, according to the principle of superposition. In the mathematical theory of *linear vector spaces*,[1] it is shown how transformations which preserve linear relations among vectors can be represented by *matrices*. Thus, in the language of vector theory, the operators of the Schrödinger theory appear as matrices, and the wave functions as representatives of vectors.

The application of the theory of transformations in a vector space to the formulation of quantum mechanics has been extensively developed.[2] We shall first explain the basic elements of the mathematical theory of a finite-dimensional vector space and then describe qualitatively those extensions

---

[1] P. R. Halmos, *Finite Dimensional Vector Spaces*. Princeton: Princeton University Press, 1942. R. Courant and D. Hilbert, *Methods of Mathematical Physics*. New York: Interscience Publishers, Inc., 1953.

[2] P. A. M. Dirac, *The Principles of Quantum Mechanics*. 2nd ed. Oxford: The Clarendon Press, 1935. M. Born and P. Jordan, *Elementare Quantenmechanik*. Berlin: Verlag von Julius Springer, 1930.

and generalizations which are required for a complete physical theory. Much of the mathematical work of previous chapters will be repeated or translated into the new language, and it will be seen that there is a gain in clarity and unification. Furthermore, the practical algebraic methods of calculation which will be evolved are essential techniques for the solution of many physical problems.

9–2 Linear vector spaces. A linear vector space is defined as an abstract set of elements, called vectors and denoted by $x, y, z, \ldots$, which satisfy the algebraic axioms set forth in the following paragraph:[1]

For every pair of vectors x, y in the space, there exists a unique vector z, the sum of x and y:

$$z = x + y. \tag{9–1}$$

Vector addition is commutative,

$$x + y = y + x, \tag{9–2}$$

and associative:

$$(x + y) + z = x + (y + z) = x + y + z. \tag{9–3}$$

(These rules are satisfied by the familiar position vectors of points in three-dimensional space.) The origin of the vector space is the *zero vector* 0, which has, by definition, the property

$$x + 0 = x \tag{9–4}$$

for every vector x. Associated with every vector x, there is a unique vector $(-x)$, such that

$$x + (-x) = 0. \tag{9–5}$$

This statement, defining the subtraction of vectors, can be shortened to

$$x - x = 0. \tag{9–6}$$

The elements of a vector space can be multiplied by complex numbers. Thus, for every vector x and every complex number a, there exists a unique vector ax, such that multiplication is distributive with respect to vector addition,

$$a(x + y) = ax + ay, \tag{9–7}$$

---

[1] P. R. Halmos, *Finite Dimensional Vector Spaces*. Princeton: Princeton University Press, 1942. G. Birkhoff and S. MacLane, *A Survey of Modern Algebra*. New York: The Macmillan Co., 1949.

and also with respect to addition of complex numbers,

$$(a + b)\mathbf{x} = a\mathbf{x} + b\mathbf{x}. \tag{9-8}$$

Multiplication is associative, i.e.,

$$a(b\mathbf{x}) = (ab)\mathbf{x}. \tag{9-9}$$

Multiplication by the complex numbers 0 and 1 is defined by

$$0\mathbf{x} = \mathbf{0}, \qquad 1\mathbf{x} = \mathbf{x}. \tag{9-10}$$

The M vectors $\mathbf{x}^1, \mathbf{x}^2, \ldots, \mathbf{x}^M$ are *linearly dependent* if they satisfy a relation of the form

$$c_1\mathbf{x}^1 + c_2\mathbf{x}^2 + \cdots + c_M\mathbf{x}^M = \mathbf{0}, \tag{9-11}$$

where the numbers $c_1, c_2, \ldots, c_M$ are not all zero. If no such numbers c_i exist, the vectors $\mathbf{x}^i$ are *linearly independent*. If any vector $\mathbf{x}$ is a linear combination of M other vectors $\mathbf{x}^i$, so that

$$\mathbf{x} = c_1\mathbf{x}^1 + c_2\mathbf{x}^2 + \cdots + c_M\mathbf{x}^M, \tag{9-12}$$

then the set $\mathbf{x}, \mathbf{x}^1, \ldots, \mathbf{x}^M$ is linearly dependent, since

$$(-1)\mathbf{x} + c_1\mathbf{x}^1 + \cdots + c_M\mathbf{x}^M = \mathbf{0}. \tag{9-13}$$

Conversely, every linearly dependent set contains at least one vector which is a linear combination of the others.

If a given linearly independent set $\{\mathbf{x}^i\}$ of vectors has the property that *every* vector $\mathbf{x}$ in the space is a linear combination of the given vectors,

$$\mathbf{x} = \sum_i c_i\mathbf{x}^i, \tag{9-14}$$

then the set $\{\mathbf{x}^i\}$ is called a *coordinate system* or *basis* for the vector space. The set $\{\mathbf{x}^i\}$ is then said to *span* the space. The representation (9–14) of the vector $\mathbf{x}$ is unique, for if we also have

$$\mathbf{x} = \sum_i c_i'\mathbf{x}^i, \tag{9-15}$$

then, by subtraction,

$$\mathbf{0} = \sum_i (c_i' - c_i)\mathbf{x}^i, \tag{9-16}$$

and since the vectors $\mathbf{x}^i$ are linearly independent,

$$c_i' - c_i = 0.$$

The numbers c_i, which are uniquely defined by Eq. (9–14), are the components of $\mathbf{x}$ relative to the basis $\{\mathbf{x}^i\}$.

The number of vectors $\mathbf{x}^i$ in a basis can be finite or infinite. It can be shown that, if a vector space has a finite basis containing N vectors, every basis in the space has the same number of vectors. The space is then finite-dimensional, of dimension N.

In a finite-dimensional space, there is a one-to-one correspondence between the vectors $\mathbf{x}$ of the space and the ordered sets $[c_1, c_2, \ldots, c_N]$ of complex numbers which are the *expansion coefficients* of $\mathbf{x}$ with respect to the basis $\{\mathbf{x}^i\}$ [Eq. (9–14)]. We denote these ordered sets by x:

$$x = [c_1, c_2, \ldots, c_N]. \tag{9–17}$$

Thus, x is the *representative* of $\mathbf{x}$ with respect to the basis $\{\mathbf{x}^i\}$. If

$$y = [d_1, d_2, \ldots, d_N] \tag{9–18}$$

is a representative of the vector $\mathbf{y}$ with respect to the same basis, then the linear combination $ax + by$ is, by definition,

$$ax + by = [ac_1 + bd_1, ac_2 + bd_2, \ldots, ac_N + bd_N]. \tag{9–19}$$

This is a representative of the vector $a\mathbf{x} + b\mathbf{y}$.

Every relation among vectors has its exact copy in a relation among the representatives of the vectors with respect to a given basis. Because of this isomorphism, it is possible to develop the entire theory in terms of a representation, chosen once for all, without making further reference to the abstractly defined vectors $\mathbf{x}$. However, we shall see that all physically significant properties of the vector spaces of quantum mechanics are independent of the representations or coordinate systems selected for the description of the spaces. In other words, only the intrinsic properties of a space, which are independent of the basis, are physically (or, for that matter, mathematically) important. This fact is so fundamental and useful that it is better to retain the concept of an abstract space and express results directly in terms of vectors. We shall, of course, employ a basis whenever convenient for computational purposes.

With every ordered pair of vectors $\mathbf{x}$ and $\mathbf{y}$, there is associated a complex number $(\mathbf{x}, \mathbf{y})$, which is called the *scalar product* of $\mathbf{x}$ and $\mathbf{y}$ and satisfies the following relations:

$$(\mathbf{x}, \mathbf{y}) = (\mathbf{y}, \mathbf{x})^*, \tag{9–20}$$

$$(\mathbf{x}, a\mathbf{y} + b\mathbf{z}) = a(\mathbf{x}, \mathbf{y}) + b(\mathbf{x}, \mathbf{z}), \tag{9–21}$$

$$(\mathbf{x}, \mathbf{x}) \geq 0, \tag{9–22}$$

$$(\mathbf{x}, \mathbf{x}) = 0 \quad \text{if and only if} \quad \mathbf{x} = 0. \tag{9–23}$$

As usual, the asterisk denotes the complex conjugate. Two vectors whose scalar product is zero are *orthogonal*.

The positive square root of the scalar product of a vector with itself, $\sqrt{(\mathbf{x}, \mathbf{x})}$, is called the *norm* of $\mathbf{x}$. If $\mathbf{x}$ is given, the vector

$$\mathbf{y} = \frac{e^{i\gamma}}{\sqrt{(\mathbf{x}, \mathbf{x})}}\, \mathbf{x} \qquad (9\text{--}24)$$

has unit norm:

$$(\mathbf{y}, \mathbf{y}) = e^{i\gamma} e^{-i\gamma} \frac{1}{(\mathbf{x}, \mathbf{x})}\, (\mathbf{x}, \mathbf{x}) = 1. \qquad (9\text{--}25)$$

A vector of unit norm is said to be *normalized*, and the process of forming $\mathbf{y}$ from $\mathbf{x}$, as in Eq. (9–24), is called *normalization*. Every nonzero vector can be normalized, and the result is unique, except for the real number γ, the phase of the vector.

Since the norm of a vector is never negative, the scalar product satisfies the Schwarz inequality (Appendix 3):

$$|(\mathbf{x}, \mathbf{y})|^2 \leq (\mathbf{x}, \mathbf{x})(\mathbf{y}, \mathbf{y}). \qquad (9\text{--}26)$$

The equality sign holds if and only if $\mathbf{x}$ and $\mathbf{y}$ are linearly dependent.

9–3 Orthonormal systems. A set of M vectors, $\mathbf{x}^1, \mathbf{x}^2, \ldots, \mathbf{x}^M$ is *orthonormal* if

$$(\mathbf{x}^i, \mathbf{x}^j) = \delta_{ij}, \qquad i, j = 1, 2, \ldots, M. \qquad (9\text{--}27)$$

The vectors of an orthonormal set are linearly independent, for if we calculate the coefficients c_i in the equation

$$\sum_{i=1}^{M} c_i \mathbf{x}^i = 0 \qquad (9\text{--}28)$$

by forming the scalar product of this equation with the vector $\mathbf{x}^j$, we obtain, by Eq. (9–27), $c_j = 0$ for every value of j from 1 through M.

If a set of M linearly independent vectors is given, an orthonormal set can be constructed. This can be done explicitly by means of the Schmidt orthogonalization process (Problem 6–9).

A set of N orthonormal vectors $\{\mathbf{e}^1, \mathbf{e}^2, \ldots, \mathbf{e}^N\}$ forms a basis for a vector space of N dimensions; any vector in the space can be expressed as a linear combination of these basic vectors. Thus,

$$\mathbf{x} = \sum_{i=1}^{N} x_i \mathbf{e}^i, \qquad \text{where} \qquad x_i = (\mathbf{e}^i, \mathbf{x}). \qquad (9\text{--}29)$$

The orthonormal set represented by

$$e^1 = [1, 0, 0, \ldots 0],$$
$$e^2 = [0, 1, 0, \ldots 0],$$
$$\vdots$$
$$e^N = [0, 0, 0, \ldots 1] \tag{9–30}$$

is such a basis, and the equations (9–29) show that the numbers x_i in the representative

$$x = [x_1, x_2, \ldots, x_N]$$

are the components of x relative to this basis.

If

$$x = [x_1, x_2, \ldots, x_N] \tag{9–31}$$

and

$$y = [y_1, y_2, \ldots, y_N] \tag{9–32}$$

are representatives of the vectors x and y with respect to the orthonormal basis $\{e^i\}$, then the scalar product of x and y is

$$(x, y) = \left(\sum_i x_i e^i, \sum_j y_j e^j \right) = \sum_{i,j} x_i^* y_j (e^i, e^j)$$
$$= \sum_i x_i^* y_i, \tag{9–33}$$

and the norm of x is

$$(x, x) = \sum_i |x_i|^2. \tag{9–34}$$

The norm of a vector is the sum of the squares of the absolute values of its components with respect to an orthonormal basis. The positive definite character of the norm is explicit in this form.

9–4 Linear transformations. A transformation in a vector space is a correspondence whereby each vector x of the space is uniquely associated with another vector Ax. The operator A is said to *transform* the vector x into the vector y = Ax.

The transformation A is linear if it satisfies

$$A(ax + by) = a(Ax) + b(Ay) \tag{9–35}$$

for any vectors x and y and any complex numbers a and b. The *identity* is the transformation which leaves every vector unchanged, that is,

$$1x = x. \tag{9-36}$$

The zero operator annihilates every vector in the space:

$$0x = 0. \tag{9-37}$$

If **A** and **B** are linear operators, then the transformation **C**, defined by

$$Cx = B(Ax) = (BA)x, \tag{9-38}$$

is also linear. It is called the product of **A** and **B**, i.e.,

$$C = BA. \tag{9-39}$$

It has already been pointed out that linear operators do not necessarily commute, that is, in general, $AB \neq BA$. However, the other laws of ordinary algebra are valid for operators. Thus, the linear combination

$$C = aA + bB \tag{9-40}$$

is defined by the equation

$$Cx = a(Ax) + b(Bx) \tag{9-41}$$

for every vector **x** and any complex numbers a and b.

The existence of an *inverse* of a linear transformation in a vector space depends upon two distinct requirements for the transformation. First, if **A** is to have an inverse, the vector $y = Ax$ must be uniquely defined by **x**. In other words, $x_1 \neq x_2$ must imply $Ax_1 \neq Ax_2$. If this were not the case, then we would have $y = Ax_1 = Ax_2$ for some **y**, and the vector **x** giving rise to **y** would not be uniquely defined. Second, for every vector **y** in the space, there must exist a vector **x** such that $y = Ax$. If this requirement were not satisfied, then the operation designated as the inverse of **A** would not be defined for every vector in the space.

The first of these requirements is satisfied if there is an operator **B** such that, for every $y = Ax$, we have $x = By$. Since **x** is any vector whatever, the equation

$$x = BAx \tag{9-42}$$

implies that

$$BA = 1. \tag{9-43}$$

The second requirement, on the other hand, implies that an **x** can be found for every vector **y**, i.e., the existence of an operator **C** such that $x = Cy$

implies $y = Ax$. In this case, the equation

$$y = ACy, \qquad (9\text{–}44)$$

which is true for every y, asserts the existence of an operator C such that

$$AC = 1. \qquad (9\text{–}45)$$

If operators B and C satisfying both Eqs. (9–43) and (9–45) exist, then A is *nonsingular*, and

$$B = C = A^{-1} \qquad (9\text{–}46)$$

is called the inverse of A. The inverse of a linear operator is itself a linear operator.

In a finite-dimensional vector space, the two conditions for the existence of an inverse are equivalent. To show this, we note that, for a linear operator, the condition

$$x_1 \neq x_2 \qquad \text{implies} \qquad Ax_1 \neq Ax_2, \qquad (9\text{–}47)$$

is equivalent to the condition

$$Ax = 0 \qquad \text{only if} \qquad x = 0. \qquad (9\text{–}48)$$

Consequently, a set $\{x^i\}$ of linearly independent vectors is transformed by A into a linearly independent set $\{y^i\}$, for the equation

$$\sum_i c_i y^i = 0 \qquad (9\text{–}49)$$

implies that

$$\sum_i c_i A x^i = A \sum_i c_i x^i = 0, \qquad (9\text{–}50)$$

whence, by Eq. (9–48),

$$\sum_i c_i x^i = 0, \qquad (9\text{–}51)$$

and $c_i = 0$ if the set $\{x^i\}$ is linearly independent. A basis

$$\{x^i\} \quad (i = 1, 2, \ldots, N)$$

in a space of N dimensions is, therefore, transformed into a basis $\{Ax^i\}$, and every vector y can be expressed in terms of this basis in the form

$$y = \sum_{i=1}^{N} a_i (Ax^i) = A \sum_{i=1}^{N} a_i x^i = Ax. \qquad (9\text{–}52)$$

Conversely, if x exists for every y such that $y = Ax$, then given a basis $\{y^i\}$, there is a set of N vectors $\{x^i\}$, such that

$$y^i = Ax^i, \qquad (9\text{--}53)$$

by hypothesis. The set $\{x^i\}$ is also a basis, for $\sum_i c_i x^i = 0$ implies that $\sum_i c_i Ax^i = 0$ because A is linear, whence $\sum_i c_i y^i = 0$ and $c_i = 0$ since $\{y^i\}$ is a basis. Now suppose $Ax = 0$. Expand x in the basis $\{x^i\}$:

$$x = \sum_i b_i x^i. \qquad (9\text{--}54)$$

Then

$$Ax = \sum b_i Ax^i = \sum b_i y^i = 0.$$

Therefore, $b_i = 0$ and $x = 0$; consequently the hypothesis that x exists for every y such that $y = Ax$ is equivalent to

$$Ax = 0 \qquad \text{only if} \qquad x = 0. \qquad (9\text{--}55)$$

We can summarize by saying that A has a reciprocal (is nonsingular) if and only if $Ax = 0$ implies $x = 0$. The inverse has been defined very carefully, because the distinction between the operators B and C in Eqs. (9–43) and (9–45) is important for an infinite-dimensional space (cf. Problem 9–7).

The definitions of linear combination, multiplication, and inversion of nonsingular operators enables one to define functions of an operator, and to construct an operator algebra of the kind discussed in Chapter 6. We shall now consider the description of linear transformations in terms of an orthonormal representation.

9–5 Matrices. Let $\{e^i\}$ be an orthonormal basis in a vector space of N dimensions, so that every vector x has a representative in the form

$$x = \sum_{j=1}^{N} x_j e^j, \qquad x_j = (e^j, x). \qquad (9\text{--}56)$$

The operator equation $y = Ax$, when expressed in terms of the e^i, is

$$\sum_{i=1}^{N} y_i e^i = A \sum_{j=1}^{N} x_j e^j = \sum_{j=1}^{N} x_j Ae^j, \qquad (9\text{--}57)$$

and, by the orthonormal property of the e^i,

$$y_i = \sum_{j=1}^{N} (e^i, Ae^j) x_j. \qquad (9\text{--}58)$$

The N^2 scalar products $(\mathbf{e}^i, \mathbf{A}\mathbf{e}^j)$ are entirely determined by the operator $\mathbf{A}$ and the basis $\{\mathbf{e}^i\}$. We write

$$(\mathbf{e}^i, \mathbf{A}\mathbf{e}^j) = a_{ij}, \tag{9-59}$$

and the linear transformation becomes

$$y_i = \sum_j a_{ij} x_j \qquad (i = 1, 2, \ldots, N). \tag{9-60}$$

In terms of the representatives [cf. (9–17)],

$$x = [x_1, x_2, \ldots, x_N],$$
$$y = [y_1, y_2, \ldots, y_N], \tag{9-61}$$

this set of N linear algebraic equations can be written in the shorthand form

$$y = Ax, \tag{9-62}$$

where A denotes the array of coefficients a_{ij}:

$$A = (a_{ij}) = \begin{pmatrix} a_{11} & a_{12} & \cdots & a_{1N} \\ a_{21} & a_{22} & \cdots & a_{2N} \\ \vdots & & & \vdots \\ a_{N1} & a_{N2} & \cdots & a_{NN} \end{pmatrix}. \tag{9-63}$$

The numbers a_{ij} are the *elements* of A, which is called "the *matrix A*."

The *product* of the matrix A into the vector x is a new vector y whose components are defined by Eqs. (9–60).

Two matrices A and B are equal if they define the same transformation of the vector space, that is, $A = B$ implies that the equations (9–60) are identical with the equations

$$y_i = \sum_j b_{ij} x_j \qquad (i = 1, 2, \ldots, N), \tag{9-64}$$

where the b_{ij} are the elements of the matrix B. These equations are identical for all vectors x if and only if

$$a_{ij} = b_{ij}. \tag{9-65}$$

The corresponding elements of equal matrices are equal.

The *zero* matrix,

$$0 = \begin{pmatrix} 0 & 0 & \cdots & 0 \\ 0 & 0 & \cdots & 0 \\ \vdots & & & \vdots \\ 0 & 0 & \cdots & 0 \end{pmatrix}, \tag{9-66}$$

reduces every vector x to the zero vector:

$$0x = 0. \tag{9-67}$$

The unit matrix, 1, defined by

$$(1_{ij}) = (\delta_{ij}) = \begin{pmatrix} 1 & 0 & \ldots & 0 \\ 0 & 1 & \ldots & 0 \\ \vdots & & & \vdots \\ 0 & 0 & \ldots & 1 \end{pmatrix}, \tag{9-68}$$

corresponds to the *identity transformation*

$$y_i = \sum_{j=1}^{N} \delta_{ij} x_j = x_i \qquad \text{or} \qquad \mathbf{y} = 1\mathbf{x} = \mathbf{x}. \tag{9-69}$$

Let the vector $\mathbf{y}$ be again subject to a transformation $\mathbf{z} = \mathbf{B}\mathbf{y}$, represented, with respect to the same basis, by

$$z_k = \sum_{i=1}^{N} b_{ki} y_i \qquad \text{or} \qquad z = By, \tag{9-70}$$

with a matrix $B = (b_{ki})$. The relation between z and x is then

$$z = By = BAx = Cx, \tag{9-71}$$

where the matrix $C = BA$ has the elements

$$c_{kj} = \sum_{i=1}^{N} b_{ki} a_{ij}. \tag{9-72}$$

This is the law of *matrix multiplication*. It expresses the effect of performing first the transformation represented by A, and then the transformation represented by B. It is easy to prove from Eq. (9–72) that $(AB)C = A(BC) = ABC$, i.e., matrix multiplication is *associative*. However, the order of the factors in the product BA is not arbitrary. Matrix multiplication is, in general, *not commutative*.

Suppose that the vectors x of the space are subject to the two transformations $y = Ax, z = Bx$. Then the ith component of the sum

$$\alpha y + \beta z = \alpha Ax + \beta Bx \tag{9-73}$$

is

$$(\alpha y + \beta z)_i = \alpha(Ax_i) + \beta(Bx_i) = \alpha \sum_{j=1}^{N} a_{ij} x_j + \beta \sum_{j=1}^{N} b_{ij} x_j, \tag{9-74}$$

in which α and β are given complex numbers. This equation can be re-written in the form

$$(\alpha y + \beta z)_i = \sum_{j=1}^{N} (\alpha a_{ij} + \beta b_{ij}) x_j = (Cx)_i, \qquad (9\text{–}75)$$

whence we have the general law of linear combination of matrices, namely,

$$\alpha A + \beta B = C, \qquad (9\text{–}76)$$

where

$$c_{ij} = \alpha a_{ij} + \beta b_{ij}. \qquad (9\text{–}77)$$

The transformation $y = Ax$, that is,

$$y_i = \sum_{j=1}^{N} a_{ij} x_j \qquad (i = 1, 2, \ldots, N), \qquad (9\text{–}78)$$

can be *inverted*, provided these equations can be solved for the components x_j in terms of the y_i. This is possible for an arbitrary vector y if and only if the determinant of the coefficients a_{ij} is not zero:

$$\det A = \begin{vmatrix} a_{11} & a_{12} & \cdots & a_{1N} \\ a_{21} & a_{22} & \cdots & a_{2N} \\ \vdots & & & \vdots \\ a_{N1} & a_{N2} & \cdots & a_{NN} \end{vmatrix} \neq 0. \qquad (9\text{–}79)$$

If this is true, Cramer's rule[1] yields the solution

$$x_j = \sum_{i=1}^{N} \frac{\alpha_{ij}}{\det A} y_i, \qquad (9\text{–}80)$$

where α_{ij} is the cofactor[2] of the element a_{ij} of A. This is a linear trans-formation, which can be written

$$x = A^{-1} y, \qquad (9\ 81)$$

where the matrix A^{-1} has the elements

$$(A^{-1})_{ji} = \frac{\alpha_{ij}}{\det A}. \qquad (9\text{–}82)$$

It is the representative, in the basis $\{e^i\}$, of the reciprocal of **A**. The

---

[1] G. Birkhoff and S. MacLane, *A Survey of Modern Algebra*. New York: The Macmillan Co., 1949, p. 290.

[2] The cofactor of the element a_{ij} is the quantity $\partial \det A / \partial a_{ij}$.

relation

$$x = A^{-1}y = A^{-1}Ax \tag{9-83}$$

is true for every vector x, which shows that

$$A^{-1}A = 1. \tag{9-84}$$

Furthermore, the equation $y = Ax = AA^{-1}y$ is true for every vector y, whence $AA^{-1} = 1$. A matrix commutes with its reciprocal. The equations

$$A^{-1}A = AA^{-1} = 1 \tag{9-85}$$

are equivalent to the *Laplace developments*

$$\delta_{ij} \det A = \sum_k \alpha_{ik} a_{jk} = \sum_k \alpha_{ki} a_{kj} \tag{9-86}$$

of the determinant of A.

If $\det A = 0$, then the matrix A has no reciprocal, and is *singular*. A singular transformation has no inverse. The law of multiplication of determinants shows that if $C = AB$, then

$$\det C = \det A \cdot \det B, \tag{9-87}$$

whence, if either A or B is singular, C is also singular, and conversely. (Note that $AB = 0$ does not imply that either A or B is the zero matrix.)

The reciprocal of the product AB of nonsingular matrices A and B is

$$(AB)^{-1} = B^{-1}A^{-1}, \tag{9-88}$$

for if we write $(AB)^{-1} = D$, then, by the definition of reciprocal, we obtain

$$1 = (AB)D = ABD, \quad A^{-1} = BD, \quad B^{-1}A^{-1} = D. \tag{9-89}$$

The reciprocal of a product of given matrices is the product of the reciprocals of the given matrices, written in the reverse order.

Other matrices related to $A = (a_{ij})$, which occur frequently in the theory, are the *transpose* of A,

$$\tilde{A} = (a_{ji}), \tag{9-90}$$

which is obtained from A by interchanging its rows and columns; the *complex conjugate* of A,

$$A^* = (a_{ij}^*), \tag{9-91}$$

whose elements are the complex conjugates of the corresponding elements of A; and the *Hermitian conjugate* of A,

$$A^\dagger = (a_{ji}^*), \qquad (9\text{-}92)$$

which is obtained by conjugation and transposition of the elements. The reader will easily verify the following identities:

$$(\widetilde{AB}) = \widetilde{B}\widetilde{A}, \qquad (9\text{-}93)$$

$$\det \widetilde{A} = \det A, \qquad (9\text{-}94)$$

$$(AB)^* = A^*B^*, \qquad (9\text{-}95)$$

$$\det A^* = (\det A)^*, \qquad (9\text{-}96)$$

$$(AB)^\dagger = B^\dagger A^\dagger, \qquad (9\text{-}97)$$

$$\det A^\dagger = (\det A)^*. \qquad (9\text{-}98)$$

Matrices of certain special types have been given descriptive names. Among those of principal interest in quantum mechanics are:

symmetric: $\quad A = \widetilde{A},\qquad$ real: $\qquad A = A^*,$

skew-symmetric: $A = -\widetilde{A},\qquad$ imaginary: $\quad A = -A^*,$

Hermitian: $\quad A = A^\dagger,\qquad$ unitary: $\quad A^{-1} = A^\dagger,$

skew-Hermitian: $A = -A^\dagger,\qquad$ diagonal: $\quad a_{ij} = a_{ii}\,\delta_{ij}.$ $\quad$ (9-99)

Algebraic functions of a given matrix A can be constructed. The integral *powers* of A are, by definition,

$$A^0 = 1, \qquad A^1 = A, \qquad A^2 = AA, \qquad \text{etc.} \qquad (9\text{-}100)$$

Negative powers are defined similarly in terms of A^{-1}, if A is nonsingular. To any polynomial in λ of degree m,

$$f(\lambda) = \sum_{i=0}^{m} c_i \lambda^i = c_m(\lambda - \lambda_1)(\lambda - \lambda_2) \ldots (\lambda - \lambda_m), \quad (9\text{-}101)$$

there corresponds a matrix polynomial

$$f(A) = \sum_{i=0}^{m} c_i A^i = c_m \prod_i (A - \lambda_i \cdot 1), \qquad (9\text{-}102)$$

in which the numbers λ_i are the roots of $f(\lambda)$. Also, one can formally construct transcendental functions of A, for example,

$$e^A = 1 + A + \frac{A^2}{2!} + \cdots \qquad (9\text{-}103)$$

This definition has meaning if each of the elements of the matrix formed by the first n terms of the sum on the right converges to a limit as $n \rightarrow \infty$.[1]

Finally, the notation of matrices can be extended to vectors by writing the vector x as a *column vector* or one-column matrix,

$$
x = \begin{pmatrix} x_1 \\ x_2 \\ \vdots \\ x_N \end{pmatrix},
\tag{9–104}
$$

so that the equation $y = Ax$ appears, in analogy with Eq. (9–72), as

$$
\begin{pmatrix} y_1 \\ y_2 \\ \vdots \\ y_N \end{pmatrix} = \begin{pmatrix} a_{11} & a_{12} & \cdots & a_{1N} \\ a_{21} & a_{22} & \cdots & a_{2N} \\ \vdots & & & \vdots \\ a_{N1} & a_{N2} & \cdots & a_{NN} \end{pmatrix} \begin{pmatrix} x_1 \\ x_2 \\ \vdots \\ x_N \end{pmatrix}
$$

$$
= \begin{pmatrix} a_{11}x_1 + a_{12}x_2 + \cdots + a_{1N}x_N \\ a_{21}x_1 + a_{22}x_2 + \cdots + a_{2N}x_N \\ \vdots & & \vdots \\ a_{N1}x_1 + a_{N2}x_2 + \cdots + a_{NN}x_N \end{pmatrix}.
\tag{9–105}
$$

Furthermore, if we define row matrices by

$$
\tilde{x} = (x_1, x_2, \ldots, x_N), \qquad x^\dagger = (x_1^*, x_2^*, \ldots, x_N^*), \qquad \text{etc.,}
\tag{9–106}
$$

then the equation $y = Ax$ implies $\tilde{y} = \tilde{x}\tilde{A}$, $y^\dagger = x^\dagger A^\dagger$, and so on. With this discussion of definitions and notation, we return to the subject of vector spaces.

9–6 Change of basis. If the vectors $\mathbf{e}^i$ are an orthonormal basis for a space of N dimensions,

$$
(\mathbf{e}^i, \mathbf{e}^j) = \delta_{ij} \qquad (i, j = 1, 2, \ldots, N),
\tag{9–107}
$$

then a new basis can be constructed by forming N linear combinations of these vectors,

$$
\mathbf{e}^{i\prime} = \sum_{j=1}^{N} u_{ij}^* \mathbf{e}^j,
\tag{9–108}
$$

in which the u_{ij} are suitably chosen complex numbers. This new basis is

----

[1] In the case of the exponential function, this is always true.

also orthonormal, provided that

$$(\mathbf{e}^{i\prime}, \mathbf{e}^{j\prime}) = \delta_{ij}. \qquad (9\text{–}109)$$

If Eq. (9–108) is substituted, this condition becomes

$$\sum_{k,l} u_{ik} u_{jl}^*(\mathbf{e}^k, \mathbf{e}^l) = \delta_{ij}, \qquad (9\text{–}110)$$

or, by Eqs. (9–107),

$$\sum_{k} u_{ik} u_{jk}^* = \delta_{ij}. \qquad (9\text{–}111)$$

In other words, the matrix $U = (u_{ij})$ must satisfy

$$UU^\dagger = 1. \qquad (9\text{–}112)$$

It follows that U must be nonsingular, and that

$$U^{-1} = U^\dagger. \qquad (9\text{–}113)$$

It is also apparent that this condition is sufficient to insure that the vectors $\mathbf{e}^{i\prime}$, formed according to Eq. (9–108), are orthonormal. Hence, the necessary and sufficient condition for the vectors $\mathbf{e}^{i\prime}$ to form an orthonormal basis is that the matrix U is unitary. In view of (9–85), the condition (9–113) is often written in the more symmetrical form

$$UU^\dagger = U^\dagger U = 1. \qquad (9\text{–}114)$$

(Note that the second of these equations is a consequence of the first for any space of finite dimension.)

Such a *unitary transformation* has the effect of substituting the basis $\{\mathbf{e}'\}$ for the basis $\{\mathbf{e}\}$. If a vector $\mathbf{x}$ is represented with respect to the basis $\{\mathbf{e}\}$ by

$$\mathbf{x} = \sum_{i} x_i \mathbf{e}^i, \qquad (9\text{–}115)$$

and with respect to the basis $\{\mathbf{e}'\}$ by

$$\mathbf{x} = \sum_{i} x_i' \mathbf{e}^{i\prime}, \qquad (9\text{–}116)$$

then, since these expressions are representatives of one and the same vector,

$$\sum_{j} x_j \mathbf{e}^j = \sum_{i,j} x_i' u_{ij}^* \mathbf{e}^j. \qquad (9\text{–}117)$$

Hence, by the orthogonality of the $\mathbf{e}^j$,

$$x_j = \sum_{i} u_{ij}^* x_i'. \qquad (9\text{–}118)$$

This equation provides the connection between the components x_i and x_i' of the vector with respect to the bases $\{e\}$ and $\{e'\}$. Since U is a unitary matrix, the equations (9–118) can be solved for the x_i' in terms of the x_i. Thus, Eq. (9–111) implies that

$$\sum_j u_{kj} x_j = \sum_{i,j} u_{kj} u_{ij}^* x_i' = \sum_i \delta_{ki} x_i', \qquad (9\text{–}119)$$

or

$$x_k' = \sum_j u_{kj} x_j. \qquad (9\text{–}120)$$

The student should again note carefully that we are adopting the point of view that a vector has an identity independent of the basis with respect to which its components are given. Thus, in the preceding discussion, x and x' represent the same vector. We have emphasized this fact by using the symbol $\mathbf{x}$ to denote a vector, and the symbols x, x', x'', etc., to denote representatives of $\mathbf{x}$ with respect to different bases. Thus x, x', x'' are column matrices, while $\mathbf{x}$ is an abstract symbol denoting a definite directed line in the vector space. With this in mind, we write the foregoing equations in the form

$$x' = Ux, \qquad x = U^\dagger x'. \qquad (9\text{–}121)$$

Consider now a linear transformation with matrix A:

$$y = Ax \qquad \text{or} \qquad y_i = \sum_j a_{ij} x_j. \qquad (9\text{–}122)$$

This is a linear relation between the representatives of $\mathbf{x}$ and $\mathbf{y}$ relative to the basis $\{e\}$. Changing to the basis $\{e'\}$, we obtain, in the new representation,

$$y' = Uy = UAx = UAU^\dagger x', \qquad (9\text{–}123)$$

and this can be written

$$y' = A'x', \qquad (9\text{–}124)$$

in which the matrix A' is

$$A' = UAU^\dagger = UAU^{-1}. \qquad (9\text{–}125)$$

In Eqs. (9–124) and (9–125), A' is the matrix which represents the linear transformation when the components of the vectors are given relative to the basis $\{e'\}$. In other words, a given linear transformation can be expressed as a matrix equation in any representation, and Eq. (9–125) provides the connection between the matrices A and A', which are representatives of $\mathbf{A}$ with respect to different bases.

Matrix equations such as

$$y = Ax, \qquad AB = C, \qquad A + B = C, \qquad \text{etc.,} \qquad (9\text{--}126)$$

become, in a new representation,

$$y' = A'x', \qquad A'B' = C', \qquad A' + B' = C', \qquad \text{etc.,} \qquad (9\text{--}127)$$

i.e., they are of the same *form* in every representation. For example,

$$AB = C$$

implies that

$$UABU^\dagger = UCU^\dagger = UAU^\dagger UBU^\dagger = A'B' = C'. \qquad (9\text{--}128)$$

This is simply an expression of the isomorphism between operators in the abstract vector space and their matrix representatives. The equations (9–126) are representatives of the operator equations

$$\mathbf{y} = \mathbf{Ax}, \qquad \mathbf{AB} = \mathbf{C}, \qquad \mathbf{A} + \mathbf{B} = \mathbf{C}. \qquad (9\text{--}129)$$

Because of this isomorphism, it is not really necessary that the notation should distinguish vectors and matrices from their representatives, and in the following we shall frequently use the forms (9–126), with the understanding that the equations (9–129) are implied. Also, a scalar product will frequently be denoted by (x, y) which is, by definition, the sum

$$(x, y) = \sum_i x_i^* y_i = (\mathbf{x}, \mathbf{y}). \qquad (9\text{--}130)$$

Since the scalar product was originally defined in terms of the abstract vectors $\mathbf{x}$ and $\mathbf{y}$, it is, of course, independent of the representation. This can be explicitly verified by means of the identity

$$(x, Ay) = (A^\dagger x, y), \qquad (9\text{--}131)$$

which holds for any matrix A and its Hermitian conjugate $A^\dagger$ (Problem 9–16). Thus, if x, y and x', y' are different representatives of $\mathbf{x}$ and $\mathbf{y}$ connected by the unitary matrix U, we have

$$(x', y') = (Ux, Uy) = (U^\dagger Ux, y) = (x, y). \qquad (9\text{--}132)$$

9–7 Hermitian operators; diagonalization. A Hermitian operator is an operator which satisfies

$$(\mathbf{x}, \mathbf{Ay}) = (\mathbf{Ax}, \mathbf{y}) \qquad (9\text{--}133)$$

for all vectors **x** and **y**. In terms of a representation, this is

$$(x, Ay) = (Ax, y), \tag{9-134}$$

or

$$\sum_{i,j} a_{ij} x_i^* y_j = \sum_{i,j} a_{ji}^* x_i^* y_j, \tag{9-135}$$

which can only be true for every x and y if

$$a_{ji}^* = a_{ij}. \tag{9-136}$$

Thus, a Hermitian operator is represented by a Hermitian matrix.

The definition (9-136) of a Hermitian matrix is equivalent to the statement $A = A^\dagger$, or

$$(x, Ay) = (Ax, y), \tag{9-137}$$

where x and y are any vectors. The *Hermitian form*

$$(x, Ax) = \sum_{i,j} a_{ij} x_i^* x_j \tag{9-138}$$

is real for a Hermitian matrix A, and this condition is sufficient to insure that A is Hermitian. To prove this last statement, we are given that

$$(x, Ax) = (x, Ax)^* \tag{9-139}$$

or

$$(x, Ax) = (Ax, x) \qquad \text{for every } x. \tag{9-140}$$

Therefore, if x and y are any given vectors, we have

$$((x + y), A(x + y)) = (A(x + y), (x + y)), \tag{9-141}$$

which is immediately reducible to

$$(x, Ay) + (y, Ax) = (Ax, y) + (Ay, x). \tag{9-142}$$

Since this relation holds for every pair of vectors x and y, it holds for the pair x, iy, whence

$$(x, Ay) - (y, Ax) = (Ax, y) - (Ay, x). \tag{9-143}$$

By addition of Eqs. (9-142) and (9-143), we obtain

$$(x, Ay) = (Ax, y). \tag{9-144}$$

Thus A is Hermitian if (x, Ax) is real for every x.

A fundamental theorem, of central importance in quantum mechanics, is related to the problem of finding a basis in which a Hermitian operator is represented by a *diagonal* matrix. To arrive at this theorem, we recall that the scalar product of $\mathbf{x}$ and $\mathbf{y} = \mathbf{A}\mathbf{x}$ is independent of the representation, and real for every $\mathbf{x}$, that is,

$$(\mathbf{x}, \mathbf{A}\mathbf{x}) = (x, Ax) = \sum_{i,j} a_{ij} x_i^* x_j. \qquad (9\text{-}145)$$

In a given representation, (x, Ax) is a Hermitian form in the variables x_i. It will now be convenient to use geometrical terminology, which can be explained by recalling a familiar situation in real three-dimensional space. A *quadratic form* in $[x, y, z] = [x_1, x_2, x_3]$, namely

$$F(\mathbf{r}) = \sum_{i,j} a_{ij} x_i x_j \qquad (9\text{-}146)$$

(where the numbers $a_{ij} = a_{ji}$ are real) represents a family of confocal quadric surfaces whose equations are

$$F(\mathbf{r}) = C = \text{constant.} \qquad (9\text{-}147)$$

The different members of the family are obtained by giving various values to the constant C. If the form $F(\mathbf{r})$ has the same sign for every vector $\mathbf{r}$, the surfaces are ellipsoids. If both signs are allowed for C, the surfaces are hyperboloids of one or two sheets, depending upon particular properties[1] of the symmetric matrix $A = (a_{ij})$. In any case, a system of mutually perpendicular *principal axes* is defined by the equation (9-147), and if the system of coordinates is rotated to coincide with the principal axes, the equation of the surface in the new coordinates (x_1', x_2', x_3') becomes

$$F(\mathbf{r}) = a_1 x_1'^2 + a_2 x_2'^2 + a_3 x_3'^2 = C. \qquad (9\text{-}148)$$

This form contains no cross products, such as $x_1' x_2'$ or $x_3' x_1'$. In other words, the matrix of the quadratic form (9 148) is diagonal; it has the form

$$\begin{pmatrix} a_1 & 0 & 0 \\ 0 & a_2 & 0 \\ 0 & 0 & a_3 \end{pmatrix}. \qquad (9\text{-}149)$$

The directions of the principal axes are easily found from geometrical considerations. The vector $\mathbf{r}$ has a principal direction when it is parallel to

---

[1] R. Courant, *Differential and Integral Calculus.* New York: Nordemann Publishing Co., Inc., 1936, Vol. II, p. 204 ff.

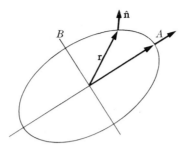

FIG. 9–1. At the points A and B, $\mathbf{r}$ is parallel to the normal $\hat{\mathbf{n}}$.

the normal to the quadric surface (Fig. 9–1). Since the direction cosines of the normal are proportional to the numbers $\partial F/\partial x_i$, the components of $\mathbf{r}$ must satisfy

$$\sum_j a_{ij}x_j = \alpha x_i, \qquad (9\text{–}150)$$

where α is a real number.

In matrix terms, this is the *eigenvalue equation*

$$Ax = \alpha x,$$

where we denote the vector $[x_1, x_2, x_3]$ simply by x. The problem of *transformation to principal axes* is therefore equivalent to this eigenvalue problem.

An analytic specification of the principal directions is obtained by noting that the magnitude of the radius vector to the quadric surface has a local *stationary value* when this vector points along a principal axis. This means that the function

$$\mathbf{r}^2 = \sum_i x_i^2 \qquad (9\text{–}151)$$

is an extremum for the principal direction, the variables x_i being subject to the condition (9–146). Unless the quadratic form is positive definite, the quantity $\mathbf{r}^2$ becomes infinite for certain directions on the surface, and this is an uncomfortable analytic situation which can be avoided by turning the problem around. The principal directions are also specified if one requires that the quadratic form

$$F(\mathbf{r}) = \sum_{i,j} a_{ij}x_i x_j \qquad (9\text{–}152)$$

have a stationary value for values of x_i which are restricted by

$$\sum_i x_i^2 = 1. \qquad (9\text{–}153)$$

In this formulation, the point $\mathbf{r}$ moves on the unit sphere and we wish to determine those directions for which the function $F(\mathbf{r})$ has a local extremum. Geometrically, this occurs at the points at which the unit sphere is tangent to a member of the family (9–147), and these points obviously lie on the principal axes. The function $F(\mathbf{r})$ is now, however, a continuous function of bounded variables, and thus no infinities occur in the problem. This point of view will be generalized to complex N-dimensional space, and we shall prove the following theorem:

For a Hermitian operator $\mathbf{A}$ *in a linear vector space of N dimensions, there exists an orthonormal basis* $\mathbf{u}^1, \mathbf{u}^2, \ldots, \mathbf{u}^N$, *relative to which* $\mathbf{A}$ *is represented by a diagonal matrix,*

$$A' = \begin{pmatrix} \alpha_1 & 0 & \ldots & 0 \\ 0 & \alpha_2 & \ldots & 0 \\ \vdots & & & \vdots \\ 0 & 0 & \ldots & \alpha_N \end{pmatrix} = (\alpha_i\, \delta_{ij}). \qquad (9\text{--}154)$$

The vectors $\mathbf{u}^i$ *and the corresponding real numbers* α_i *are solutions of the eigenvalue equation*

$$\mathbf{A}\mathbf{u} = \alpha\mathbf{u}, \qquad (9\text{--}155)$$

and there are no other eigenvalues.

The vectors $\mathbf{u}^i$ are constructed by means of a stationary value problem, analogous to the one just discussed. In an arbitrarily chosen basis $\{\mathbf{e}^i\}$, let the vector $\mathbf{x}$ and the operator $\mathbf{A}$ be represented by x and A, and let x range over the set of all vectors of unit norm:

$$\sum_i |x_i|^2 = 1. \qquad (9\text{--}156)$$

The Hermitian form

$$(x,\, Ax) = \sum_{i,j} a_{ij} x_i^* x_j \qquad (9\text{--}157)$$

is then a real, continuous function of $2N$ bounded, real variables, namely the real and imaginary parts of the complex numbers x_i. The boundedness follows from Eq. (9–156), which shows that none of these $2N$ variables can exceed unity. It follows from Weierstrass' theorem [1] that (x, Ax) has

---

[1] G. H. Hardy, *A Course of Pure Mathematics*. 9th ed. Cambridge: Cambridge University Press, 1945, Section 103. The discussion of diagonalization on this and the following pages closely follows R. Courant and D. Hilbert, *Methods of Mathematical Physics*. New York: Interscience Publishers, Inc., 1953, Vol. I, Chapter I, §3, Section 1.

a largest value α_1 which it takes on for some vector $x = u^1$. There may be more than one such x. If this happens, we choose one arbitrarily and call it u^1. Now let x be restricted further by the requirement that it be orthogonal to u^1, i.e.,

$$(x, u^1) = 0. \tag{9–158}$$

The function (x, Ax) now takes on a maximum value α_2 for some $x = u^2$, which cannot be larger than α_1 because x is more restricted than before. If there is ambiguity again, we make an arbitrary choice. Proceeding step by step in this way, we construct an orthonormal set of vectors u^i such that the Hermitian form (x, Ax) has its largest value α_m for $x = u^m$:

$$(x, Ax) \leq (u^m, Au^m) = \alpha_m, \tag{9–159}$$

where x is any normalized vector satisfying

$$(x, u^i) = 0 \qquad (i = 1, 2, \ldots, m - 1). \tag{9–160}$$

This process must stop at the Nth step, because the u^i are orthonormal and therefore linearly independent; we have seen that a set of linearly independent vectors in a space of N dimensions can contain at most N members. Incidentally, the numbers α_i are ordered, i.e.:

$$\alpha_1 \geq \alpha_2 \geq \cdots \geq \alpha_N.$$

The vectors u^i and the numbers α_i are the eigenvectors and eigenvalues of the above theorem. To prove this, we construct a unitary matrix U from the components of the u^i,

$$u_{ij} = u_j^{i*}, \tag{9–161}$$

and introduce a change of basis to a new set of orthonormal vectors according to Eq. (9–108), that is,

$$e^{i\prime} = \sum_{j=1}^{N} u_{ij}^* e^j = \sum_{j=1}^{N} u_j^i e^j = u^i. \tag{9–162}$$

Thus, the u^i are chosen as the basic vectors in the new representation. We must now establish four points in order to complete the proof: (1) The matrix which represents $\mathbf{A}$ in the new representation, namely

$$A' = UAU^\dagger, \tag{9–163}$$

has the numbers α_i for its diagonal elements. (2) All other elements of A' are zero. (3) The eigenvalue equation (9–155) is satisfied by each u^i and the corresponding α_i. (4) There are no other eigenvalues.

The first of these points can be seen immediately, for the mth diagonal element of A' is, by Eq. (9–163) and the definition (9–159) of α_m,

$$a'_{mm} = \sum_{k,l} u_{mk} a_{kl} u^*_{ml} = \sum_{k,l} u^{m*}_k a_{kl} u^m_l = (u^m, Au^m) = \alpha_m. \quad (9\text{–}164)$$

The second point follows from the maximum property (9–159) of the vectors u^i, namely, α_m is the largest value of the function

$$(x, Ax) = (x', A'x'), \quad (9\text{–}165)$$

where the x_i are subject to the restrictions (9–160). In the new representation, these restrictions are

$$x'_i = 0 \qquad (i = 1, 2, \ldots, m - 1), \quad (9\text{–}166)$$

that is, the first $m - 1$ components of x' are zero. We can show that a contradiction follows from the assumption that $a_{m,m+k} \neq 0$, where $k > 0$. Let λ and μ be any complex numbers, such that $|\lambda|^2 + |\mu|^2 = 1$, and consider the normalized vector x' whose mth component is λ and whose $(m + k)$th component is μ, all others being zero. This vector satisfies (9–166), and the statement

$$(x', A'x') = \sum_{i,j} x'^*_i a'_{ij} x'_j \leq \alpha_m \quad (9\text{–}167)$$

becomes for this x' (A' is Hermitian):

$$|\lambda|^2 a'_{mm} + 2\,\mathrm{Re}\,\lambda^* \mu a'_{m,m+k} + |\mu|^2 a'_{m+k,m+k} \leq \alpha_m. \quad (9\text{–}168)$$

If we substitute (9–164) for the diagonal elements and rearrange (9–168), the inequality reduces to

$$2\,\mathrm{Re}\,\frac{\lambda^* \mu}{|\mu|^2}\, a'_{m,m+k} \leq \alpha_m - \alpha_{m+k}. \quad (9\text{–}169)$$

Now the number $\lambda^* \mu / |\mu|^2$ assumes every complex value, notwithstanding the restriction $|\lambda|^2 + |\mu|^2 = 1$. Hence (9–169) is a contradiction, unless

$$a'_{m,m+k} = 0, \quad (9\text{–}170)$$

i.e., A' is a diagonal matrix. This is the essential point of the method of diagonalization discussed here. Incidentally, since Eq. (9–170) holds, the inequality (9–169) is the statement that the α_m are ordered, which we already know.

The third point of the proof is obtained if Eq. (9–163) is multiplied by the matrix $U^\dagger$,

$$AU^\dagger = U^\dagger A', \quad (9\text{–}171)$$

and written out in terms of the vectors u^i:

$$\sum_k a_{ik}u_{jk}^* = \sum_k a_{ik}u_k^j = \sum_k u_{ki}^* a_{kj}' = \sum_k u_i^k \delta_{jk}\alpha_j = \alpha_j u_i^j. \qquad (9\text{--}172)$$

The second and last of the above expressions are the ith components of

$$Au^j = \alpha_j u^j. \qquad (9\text{--}173)$$

Finally, the α_j are the only numbers for which the eigenvalue equation

$$Au = \alpha u \qquad (9\text{--}174)$$

has nontrivial solutions. If this equation is written in terms of components in the form

$$\sum_j (a_{ij} - \alpha\,\delta_{ij})u_j = 0, \qquad (9\text{--}175)$$

then it is clear from the fundamental theorem on linear homogeneous equations that the only solution is $u_j = 0$, unless α is a root of the Nth-degree determinantal equation

$$D(\alpha) = \det(a_{ij} - \alpha\,\delta_{ij}) = 0. \qquad (9\text{--}176)$$

In the new representation, however, the matrix $A - \alpha \cdot 1$ becomes the matrix

$$U(A - \alpha \cdot 1)U^\dagger = A' - \alpha \cdot 1, \qquad (9\text{--}177)$$

and the determinants of these two matrices are equal, whence

$$\begin{aligned}
D(\alpha) &= \det(a'_{ij} - \alpha\,\delta_{ij}) \\
&= \det((\alpha_i - \alpha)\,\delta_{ij}) = \prod_{i=1}^{N}(\alpha_i - \alpha), \qquad (9\text{--}178)
\end{aligned}$$

from which it follows by inspection that the only roots of $D(\alpha) = 0$ are the α_i. This completes the proof of the fundamental theorem on page 303.

The function

$$D(\alpha) = \det(a_{ij} - \alpha\,\delta_{ij}) = \prod_{i=1}^{N}(\alpha_i - \alpha) = \sum_{i=1}^{N} c_i\alpha^i \qquad (9\text{--}179)$$

is called the *characteristic function* for the operator $\mathbf{A}$. It is a polynomial of degree N in α. The equation (9–176), whose roots are the characteristic values or eigenvalues of $\mathbf{A}$, is the *characteristic equation* for $\mathbf{A}$. The characteristic function is invariant to unitary transformations, and thus is associated with the operator $\mathbf{A}$ independently of a representation. The invariant coefficients c_i in $D(\alpha)$ are functions of the eigenvalues. In particular, the coefficient c_0 is

$$c_0 = \prod_i \alpha_i = \det A. \qquad (9\text{--}180)$$

The determinant of a matrix is the product of its eigenvalues. It follows from the definition of a singular matrix that a matrix (and hence the corresponding operator) is singular if and only if it has the eigenvalue zero. These statements have been proved here only for Hermitian matrices, but they are generally true.[1]

Another very important function of the eigenvalues of a matrix A is the coefficient of α^{N-1} in Eq. (9–179). This is

$$c_{N-1} = (-)^{N-1} \sum_i \alpha_i, \qquad (9\text{--}181)$$

i.e., the quantity $(-)^{N-1}c_{N-1}$ is the sum of the eigenvalues of A. This quantity, called the *trace* of A, can easily be evaluated in any representation, for we have

$$\operatorname{tr} A = \sum_i \alpha_i = \sum_i a'_{ii} = \sum_{i,k,l} u_{ik}a_{kl}u^*_{il} = \sum_{k=1}^{N} a_{kk}. \qquad (9\text{--}182)$$

The trace of a matrix is the sum of its diagonal elements. Since it is the sum of the eigenvalues of $\mathbf{A}$, the trace is invariant to the representation, i.e., it is intrinsically associated with $\mathbf{A}$. Indeed, the equations (9–182) are an explicit proof of this statement, since they are true for any unitary matrix U. Another important property of the trace is expressed by

$$\operatorname{tr} AB = \operatorname{tr} BA. \qquad (9\text{--}183)$$

In more general terms, the trace of a product of matrices is not changed by cyclic permutation of the factors, e.g.,

$$\operatorname{tr} ABCD = \operatorname{tr} BCDA = \operatorname{tr} CDAB, \qquad \text{etc.} \qquad (9\text{--}184)$$

It follows also from Eq. (9–183) and from

$$\operatorname{tr} (A + B) = \operatorname{tr} A + \operatorname{tr} B \qquad (9\text{--}185)$$

that the trace of a commutator is always zero, i.e.,

$$\operatorname{tr} (AB - BA) = \operatorname{tr} [A, B] = 0. \qquad (9\text{--}186)$$

This theorem, which is true for a finite-dimensional vector space, has an

---

[1] P. R. Halmos, *Finite Dimensional Vector Spaces*. Princeton: Princeton University Press, 1942.

important consequence for the matrix formulation of quantum mechanics (Section 9–13).

As a final remark concerning the characteristic function $D(\alpha)$, we shall prove the *Hamilton-Cayley theorem* for Hermitian operators, which states:

Every Hermitian operator A *satisfies its characteristic equation*

$$D(\mathsf{A}) = 0. \qquad (9\text{--}187)$$

The proof is simple. Since Eq. (9–187) is algebraic in A, the corresponding matrix equation $D(A) = 0$ is invariant to the representation, and it suffices therefore to prove the equation for the representation in which A is diagonal. In this representation, however, the matrix elements of the powers of A' are

$$(A')_{ij} = a'_{ij} = \alpha_i\,\delta_{ij}, \qquad (A'^2)_{ij} = \sum_k \alpha_i\alpha_j\,\delta_{ik}\,\delta_{jk} = \alpha_i^2\,\delta_{ij},$$

$$(A'^3)_{ij} = \alpha_i^3\,\delta_{ij}, \qquad \text{and, in general,} \qquad (A'^k)_{ij} = \alpha_i^k\,\delta_{ij}. \qquad (9\text{--}188)$$

Hence, the (i,j) matrix element of the characteristic polynomial $D(A')$ is

$$(D(A'))_{ij} = \left(\sum_k c_k A'^k\right)_{ij} = \sum_k c_k (A'^k)_{ij} = \left(\sum_k c_k \alpha_i^k\right)\delta_{ij}, \qquad (9\text{--}189)$$

or, since α_i is a root of $D(\alpha) = 0$,

$$(D(A'))_{ij} = D(\alpha_i)\,\delta_{ij} = 0. \qquad (9\text{--}190)$$

Every matrix element of $D(A')$ is zero, i.e., $D(A') = 0$.

The Hamilton-Cayley theorem is of great usefulness in applications. In general, any algebraic equation in the matrix A is satisfied by each of its eigenvalues. For example, the equation

$$A^3 - A = 0 \qquad (9\text{--}191)$$

implies that the eigenvalues of A are among the three numbers $\pm 1, 0$. It does not, of course, imply that these numbers are all necessarily eigenvalues of A. (The equation is obviously satisfied by the unit matrix, which does not have the eigenvalues 0 or -1.)

The Hamilton-Cayley theorem holds for an arbitrary matrix, although it has been proved here for Hermitian matrices only.

9–8 Degenerate matrices. The eigenvectors u^i are determined *uniquely* in the process of diagonalization of A (except for a factor of absolute magnitude unity) whenever the corresponding eigenvalue α_i is "single," i.e., not a repeated root of the characteristic equation. However, if α_i

is a repeated root, then u^i is not uniquely determined, and, in constructing the orthonormal set, we are forced to make an arbitrary choice. Indeed, if α_i is repeated p times, the vectors u^i, u^{i+1}, ..., u^{i+p-1} are determined only to the extent that they must be orthogonal to all other u^j for which $\alpha_j \neq \alpha_i$. Thus any orthonormal set formed from the u^i, ..., u^{i+p-1} will serve equally well to diagonalize A. Hence, the matrix A does not define a unique basis in the space, and is *degenerate*. A nondegenerate matrix has N distinct eigenvalues.

In Section 6–8, it has been shown that a degeneracy can be resolved by introducing new commuting operators which, together with **A**, are a complete set of operators and define a basis uniquely. We translate the arguments of that section into the language of vector spaces in the form of a theorem:

Two Hermitian matrices A and B can be simultaneously diagonalized by the same unitary transformation if and only if they commute, i.e.,

$$[A, B] = AB - BA = 0. \tag{9–192}$$

The proof that the condition is necessary is immediate, for if A and B are brought to diagonal forms, say A' and B', by the same unitary transformation, then

$$A'B' - B'A' = 0. \tag{9–193}$$

This is, however, a matrix equation, and therefore independent of the representation. Hence, $AB - BA = 0$.

The sufficiency of the condition is proved by examining the diagonalization process. We assume that A is diagonalized by the transformation U, and that Eq. (9–192) holds. Then the eigenvalues of A correspond to vectors u^i for which

$$Au^i = \alpha_i u^i. \tag{9–194}$$

Consequently,

$$BAu^i = ABu^i = \alpha_i Bu^i, \tag{9–195}$$

and the vector Bu^i is also an eigenvector of A, belonging to α_i. Now if α_i is not repeated, every eigenvector belonging to α_i is a multiple of u^i, whence

$$Bu^i = \beta_i u^i, \tag{9–196}$$

where the real number β_i is an eigenvalue of B.

The diagonalization process for A therefore diagonalizes B as well, so long as the eigenvalues of A are single. If, however, we reach at some step a set of repeated eigenvalues of A, the corresponding u^i is not uniquely determined. To make it unique, we can impose the additional condition that the Hermitian form (x, Bx) also be a maximum for $x = u^i$.

This will determine an eigenvalue of B, and if it is single, a unique vector u^i. We now proceed to u^{i+1} by maximizing (x, Bx) subject to $(x, u^i) = 0$. In this way, the repeated eigenvalues of A can be exhausted provided that those of B are all single. If both A and B have simultaneous, repeated eigenvalues, an element of arbitrariness remains. The result, however, is that A' and B' are both diagonal.

If the vectors u^i, which are generated by diagonalization of both A and B, are not unique, i.e., if there still exist independent vectors u^i and u^{i+1} for which

$$Au^i = \alpha_i u^i, \quad Au^{i+1} = \alpha_i u^{i+1}, \quad Bu^i = \beta_i u^i, \quad Bu^{i+1} = \beta_i u^{i+1},$$
$$(9\text{–}197)$$

then a third matrix C, which commutes with both A and B, can be used to resolve the degeneracy further. Proceeding in this way, we finally arrive at a set of commuting matrices $A, B, C, \ldots$, which determines uniquely a representation $\{e^{i'}\} = \{u^i\}$. Such a set is a complete set of commuting matrices. Every vector has a unique representation as a linear combination of the eigenvectors u^i.

9–9 Infinite-dimensional spaces. The theory of a vector space of N dimensions can be generalized by going to the limit $N \to \infty$. Each vector in an infinite-dimensional space is represented by an infinite number of components

$$x = [x_1, x_2, x_3, \ldots],$$

and linear transformations in the space are represented by *infinite matrices* with elements

$$a_{ij} \quad (i, j = 1, 2, \ldots).$$

The scalar product of two infinite-dimensional vectors x and y can be defined in analogy with the scalar product of finite-dimensional vectors (Section 9–2) as

$$(x, y) = \lim_{N \to \infty} \sum_{i=1}^{N} x_i^* y_i, \qquad (9\text{–}198)$$

provided this limit exists.

The properties of an infinite-dimensional space can be inferred by generalization of the definitions and theorems of the preceding sections. However, any argument which depends upon the assumption that N is finite must be re-examined to establish that the implied limits exist. In particular, the theorem that a Hermitian operator can be diagonalized depends upon a process which ends after N steps. Hence, this theorem cannot be assumed to hold in the more general case without further investigation.

The theorem that N linearly independent vectors span a space of N dimensions loses its meaning for $N \rightarrow \infty$. A set of infinitely many orthonormal vectors $\{\mathbf{e}^i\}$ is said to be *complete* if every vector $\mathbf{x}$ can be expressed as a convergent sum

$$\mathbf{x} = \sum_{i=1}^{\infty} x_i \mathbf{e}^i. \tag{9–199}$$

This equation, together with Eq. (9–198), implies that

$$(\mathbf{x}, \mathbf{x}) = \sum_{i=1}^{\infty} |x_i|^2 \qquad [x_i = (\mathbf{e}^i, \mathbf{x})], \tag{9–200}$$

which is called the *completeness relation* for the set $\{\mathbf{e}^i\}$. If this relation is true for every vector $\mathbf{x}$ in the space, the set is complete.

Infinite-dimensional spaces are used frequently in quantum mechanics; indeed, we shall see that the vector spaces of quantum mechanics are necessarily infinite-dimensional. The results of the theory depend, therefore, upon infinite processes, and are contingent upon proofs of completeness. In practical cases, however, an appeal to the physical meaning of the theoretical conclusions will convince one of the plausibility of assuming that the mathematical limits exist, and a detailed mathematical investigation can frequently be omitted.

9–10 Hilbert space. The scalar product of two functions $f(x)$ and $g(x)$ defined in the interval (a, b) is (Section 6–5)

$$(f, g) = \int_a^b f^*(x) g(x) \, dx. \tag{9–201}$$

A function $f(x)$ is *square-integrable* in (a, b) if the norm

$$(f, f) = \int_a^b |f(x)|^2 \, dx \tag{9–202}$$

exists. The class of square-integrable functions defines a linear vector space, for the Schwarz inequality

$$|(f, g)|^2 \leq (f, f)(g, g) \tag{9–203}$$

shows that (f, g) exists if $f(x)$ and $g(x)$ are square integrable, and hence the linear combination

$$h(x) = \alpha f(x) + \beta g(x)$$

is square integrable. The vector space defined in this way is a *Hilbert space*.

The variable x, which here takes the place of the integer index i appearing in the preceding sections, varies continuously over the interval (a, b). Thus the vector **f** has infinitely many components $f(x)$, which are *not countable*. The fact that the components of a vector need not be countable is of fundamental importance. However, it raises certain new questions related to the concepts of orthogonality and normalization.

A linear transformation in Hilbert space, i.e., a correspondence **g** = **Af**, is defined by the equation

$$g(x) = \int_a^b A(x, x')f(x') \, dx', \tag{9–204}$$

where the continuous matrix $A(x, x')$ is a function of the two variables x, x', and is a representative of the operator **A**. Clearly, this transformation has all the formal properties of the corresponding transformation **y** = **Ax** in a finite-dimensional space.

It is important to define the meaning of the identity transformation **g** = **1f** = **f**. Formally, the matrix $1(x, x')$, which represents the unit operator, must satisfy

$$g(x) = \int_a^b 1(x, x')f(x') \, dx' = f(x) \tag{9–205}$$

for every function $f(x)$. By the definition of the Dirac δ-function (Appendix A–4), this implies that

$$1(x, x') = \delta(x - x'), \tag{9–206}$$

so that, for the Hilbert space, $\delta(x - x')$ takes the place of the Kronecker symbol δ_{ij} in the finite-dimensional space.

It is evident that every definition for the N-dimensional space has a formal analogue in Hilbert space. As a further generalization, the index x of a vector can be allowed to range over a partly discrete and partly continuous spectrum of eigenvalues. The integrals of the Hilbert space are then replaced by the sum of an integral over the continuous part of the spectrum and by a summation over the discrete part. Thus, if an operator **A** is represented by the matrix

$$A(ix; i'x'), \tag{9–207}$$

where i is the discrete index and x is the continuous index, the rule for matrix multiplication takes the form

$$(AB)(ix; i'x') = \sum_{i''} \int_{x''} A(ix; i''x'')B(i''x''; i'x')\, dx''. \qquad (9\text{-}208)$$

To illustrate the foregoing by a simple example, we shall investigate the properties of the operator $\mathbf{x}$. The equation

$$x\, \delta(x - x') = x'\, \delta(x - x') \qquad (9\text{-}209)$$

shows that the function $\delta(x - x')$ is an eigenfunction of $\mathbf{x}$ belonging to the eigenvalue x'. Thus, $\mathbf{x}$ has a continuous spectrum of real eigenvalues $-\infty < x' < \infty$. The identity

$$\int_{-\infty}^{\infty} \delta(x - x'')\, \delta(x'' - x')\, dx'' = \delta(x - x') \qquad (9\text{-}210)$$

indicates that these eigenfunctions are an orthonormal set, and the formula

$$f(x) = \int_{-\infty}^{\infty} \delta(x - x')f(x')\, dx' \qquad (9\text{-}211)$$

represents an arbitrary vector as a linear combination of the members of this set, which is therefore complete. The numbers $f(x')$, which appear in the integrand of Eq. (9–211), are the components of $\mathbf{f}$, expressed in the representation in which the operator $\mathbf{x}$ is diagonal. The Fourier transform

$$f(x) = \int_{-\infty}^{\infty} a(p)\, \frac{1}{\sqrt{2\pi\hbar}}\, e^{(i/\hbar)px}\, dp \qquad (9\text{-}212)$$

is, similarly, a representation of $\mathbf{f}$ as a linear combination of eigenfunctions of the operator $\mathbf{p}$ (cf. Section 2–8). The coefficients $a(p)$ in this integral are the components of $\mathbf{f}$ in a representation in which $\mathbf{p}$ is diagonal. The continuous matrix

$$u(x, p) = \frac{1}{\sqrt{2\pi\hbar}}\, e^{(i/\hbar)px}$$

is unitary, for

$$\int_{-\infty}^{\infty} u^*(x, p)u(x', p)\, dp = \frac{1}{2\pi\hbar} \int_{-\infty}^{\infty} e^{(i/\hbar)p(x-x')}\, dp = \delta(x - x'),$$

and the equation (9–212) defines a change of basis from the representation in which $\mathbf{p}$ is diagonal (the p-representation) to one in which $\mathbf{x}$ is diagonal (the x-representation).

The operator correspondence

$$\mathbf{p} \rightarrow \frac{\hbar}{i} \frac{d}{dx} = p_{\text{op}} \tag{9-213}$$

can now be considered to define the meaning of the operator $\mathbf{p}$ in the x-representation. It is equivalent to the formula

$$p_{\text{op}} f(x) = \frac{\hbar}{i} \frac{df}{dx}. \tag{9-214}$$

The element $p(x', x'')$ of the matrix which is the representative of $\mathbf{p}$ in the x-representation is

$$p(x', x'') = \int_{-\infty}^{\infty} \delta(x - x') \frac{\hbar}{i} \delta'(x - x'') \, dx. \tag{9-215}$$

Changing the basis by means of the matrix $u(x, p)$, one obtains, for the matrix element of $\mathbf{p}$ in the p-representation,

$$p(p', p'')$$

$$= \iint u(x', p') p(x', x'') u^*(x'', p'') \, dx' \, dx''$$

$$= \iiint u(x'p') \delta(x - x') \frac{\hbar}{i} \delta'(x - x'') u^*(x'', p'') \, dx \, dx' \, dx''$$

$$= -\iiint u(x', p') \delta(x - x') \frac{\hbar}{i} \delta(x - x'') \frac{d}{dx''} u^*(x'', p'') \, dx \, dx' \, dx''$$

$$= \int u(x, p') p'' u^*(x, p'') \, dx = p' \delta(p' - p''). \tag{9-216}$$

Thus $p(p', p'')$ is a diagonal matrix in this representation. The linear transformation

$$\int_{-\infty}^{\infty} p(p', p'') a(p'') \, dp''$$

$$= \int_{-\infty}^{\infty} p' \delta(p' - p'') a(p'') \, dp'' = p' a(p') \tag{9-217}$$

shows that, in the p-representation, the operation $\mathbf{p}$ is represented by multiplication by p (cf. Section 6–8).

The commutation rule

$$[\mathbf{x}, \mathbf{p}] = i\hbar \mathbf{1} \tag{9-218}$$

is an operator relation, and therefore valid in every representation. This has been verified for the p-representation in Problem 6–20. It is interesting that Eq. (9–218) is impossible for a finite-dimensional space, the trace of a commutator being always identically zero (Section 9–7). This shows that the preceding mathematical generalizations are essential for the physical theory; the vector spaces of quantum mechanics are Hilbert spaces.

9–11 Abbreviated notation for matrix elements. A complete set of commuting operators, A, B, C, . . . defines a set of vectors $\{\mathsf{u}^i\}$ which are eigenvectors of these operators, i.e.,

$$\mathsf{A}\mathsf{u}^i = \alpha\mathsf{u}^i, \qquad \mathsf{B}\mathsf{u}^i = \beta\mathsf{u}^i, \qquad \mathsf{C}\mathsf{u}^i = \gamma\mathsf{u}^i, \ldots, \qquad (9\text{–}219)$$

where $\alpha, \beta, \gamma, \ldots$ are the corresponding eigenvalues. The set $\{\mathsf{u}^i\}$ is a basis, and each member of the set is uniquely specified by the set of eigenvalues to which it belongs. Consequently, the eigenvalues themselves can be used as indices to number the vectors, and we can write

$$\mathsf{u}^i = \mathsf{u}(\alpha\beta\gamma \ldots). \qquad (9\text{–}220)$$

This notation dispenses with the arbitrary enumeration of the basis vectors by the index i and expresses more explicitly the meaning of u^i as a simultaneous eigenvector. A similar scheme has already been used in earlier chapters. The function $Y_l^m \, [= Y(lm)]$ is, for example, a spherical harmonic which is simultaneously an eigenfunction of $\mathbf{L}^2$ and L_z, and the indices or quantum numbers (lm) give the corresponding eigenvalues immediately. Similarly, the hydrogen function [Eq. (7–197)] $\psi_{nlm} \, [= \psi(nlm)]$ is a simultaneous eigenfunction of H, $\mathbf{L}^2$, and L_z, and the quantum numbers (nlm) are indices which denote the eigenvalues in a simple way. It should be noted that the quantum numbers are not necessarily the eigenvalues, but always determine the eigenvalues uniquely. The energy of the nth hydrogen state is given, for example, by Eq. (7–187).

In this notation, a matrix element of an operator G—in the representation in which A, B, C, . . . are diagonal—is [cf. Eq. (9–59)]

$$\big(\mathsf{u}(\alpha'\beta'\gamma' \ldots), \, \mathsf{G}\mathsf{u}(\alpha\beta\gamma \ldots)\big). \qquad (9\text{–}221)$$

Only the quantum numbers are required to specify this quantity; hence it is convenient to introduce an additional simplifying abbreviation and to write the matrix elements as

$$(\alpha'\beta'\gamma' \ldots |G|\alpha\beta\gamma \ldots). \qquad (9\text{–}222)$$

A unitary matrix representing a change of basis can be expressed in a similar form. Thus, let the vectors

$$\mathsf{u}'(\lambda\mu\nu \ldots) \qquad (9\text{–}223)$$

be a second basis for the vector space, defining a representation in which the commuting operators $\mathsf{L, M, N} \ldots$ are diagonal:

$$\mathsf{L}\mathsf{u}'(\lambda\mu\nu \ldots) = \lambda\mathsf{u}'(\lambda\mu\nu \ldots), \qquad \text{etc.}$$

The unitary transformation from the "$(A, B, C, \ldots)$-representation" to the "$(L, M, N, \ldots)$-representation" is given by a unitary matrix U, in accordance with Eq. (9–121). The matrix elements of U are the scalar products

$$\big(\mathsf{u}'(\lambda\mu\nu \ldots), \mathsf{u}(\alpha\beta\gamma \ldots)\big) \qquad (9\text{--}224)$$

or, in short,

$$(\lambda\mu\nu \ldots \mid \alpha\beta\gamma \ldots). \qquad (9\text{--}225)$$

This notation is part of a general scheme devised by Dirac.[1] It is introduced here as a convenient and useful abbreviated form for writing matrix elements, which will be employed occasionally in the remainder of this work.

9–12 Involutions and projection operators. An operator $\mathbf{I}$ which satisfies the equation

$$\mathbf{I}^2 = 1 \qquad (9\text{--}226)$$

is called an *involution*. The eigenvalues of an involution are ± 1, and the eigenvectors can be constructed by means of the associated operators

$$\epsilon_+ = \tfrac{1}{2}(1 + \mathbf{I}), \qquad \epsilon_- = \tfrac{1}{2}(1 - \mathbf{I}). \qquad (9\text{--}227)$$

(If we exclude the cases $\mathbf{I} = 1$ and $\mathbf{I} = -1$, then neither ϵ_+ nor ϵ_- is the zero operator.) Let x be any vector and write

$$\mathsf{x}_+ = \epsilon_+\mathsf{x}, \qquad \mathsf{x}_- = \epsilon_-\mathsf{x}; \qquad (9\text{--}228)$$

then, by Eq. (9–226), we obtain

$$\mathbf{I}\mathsf{x}_+ = \tfrac{1}{2}(\mathbf{I} + 1)\mathsf{x} = \mathsf{x}_+, \qquad \mathbf{I}\mathsf{x}_- = \tfrac{1}{2}(\mathbf{I} - 1)\mathsf{x} = -\mathsf{x}_-, \qquad (9\text{--}229)$$

whence x_+ and x_- are eigenvectors of $\mathbf{I}$ belonging to $+1$ and -1, respectively. Moreover, since

$$\epsilon_+ + \epsilon_- = 1, \qquad (9\text{--}230)$$

we have

$$\mathsf{x} = \mathsf{x}_+ + \mathsf{x}_-. \qquad (9\text{--}231)$$

---

[1] P. A. M. Dirac, *The Principles of Quantum Mechanics.* 2nd ed. Oxford: The Clarendon Press, 1935.

Consequently, every vector x can be written as a linear combination of eigenvectors of **I**. Furthermore, since **I** is Hermitian, the vectors x_+ and x_- are orthogonal. It can be shown that each of the sets of vectors $\{x_+\}$ and $\{x_-\}$ satisfies all the axioms defining a linear vector space (Problem 9–29). Hence, the original space $\{x\}$ is *reduced* by the Hermitian involution **I** into two distinct *subspaces* $\{x_+\}$ and $\{x_-\}$, such that each x is a unique linear combination of two vectors, one from each subspace [Eq. (9–231)]. The vectors x_+ and x_- are the *projections* of x onto the two subspaces, and ϵ_+ and ϵ_- are the corresponding *projection operators*. Each of the operators ϵ_+ and ϵ_- satisfies the equation

$$\epsilon^2 = \epsilon, \tag{9–232}$$

which may be regarded as the defining relation for a projection operator.

Now let the sets of vectors $\{u^i_+\}$ and $\{u^i_-\}$ be orthonormal bases in the two subspaces $\{x_+\}$ and $\{x_-\}$, respectively. Each u^i_+ is automatically orthogonal to each of the vectors u^i_-, and it follows from Eq. (9–231) that the combined set $\{u^i_+, u^i_-\}$ of basic vectors is an orthonormal basis for the whole space. Moreover, if **A** is any operator which commutes with **I**, we obtain

$$(x_+, Ax_-) = (Ix_+, Ax_-) = (x_+, IAx_-) = (x_+, AIx_-)$$
$$= -(x_+, Ax_-) = 0. \tag{9–233}$$

In particular, all matrix elements of the form (u^i_+, Au^j_-) are zero. Hence, in this representation, the matrix A is, schematically

$$A = \begin{pmatrix} (u_+, Au_+) & \vdots & 0 \\ \cdots\cdots\cdots & \vdots & \cdots\cdots\cdots \\ 0 & \vdots & (u_-, Au_-) \end{pmatrix}, \tag{9–234}$$

where (u_+, Au_+) denotes the matrix formed from the matrix elements of **A** which connect the u_+-vectors only, etc. We may say that this representation "partly diagonalizes" A.

These ideas can be extended considerably. If a basis $\{u^i\}$ is given, and we define a set of operators ϵ_i by specifying their matrix elements in this representation as

$$(u^j, \epsilon_i u^k) = (j|\epsilon_i|k) = \delta_{ij}\delta_{ik}, \tag{9–235}$$

then the operators ϵ_i are projection operators. Explicitly, the matrix elements of ϵ_i are all zero except for a single 1 on the diagonal at the intersection of the ith row and the ith column. Obviously, $\epsilon_i^2 = \epsilon_i$. Moreover, if the vector x is written in terms of its components, namely,

$$x = \sum_i (u^i, x)u^i, \tag{9–236}$$

then
$$\epsilon_i x = (u^i, x)u^i, \tag{9-237}$$

and this is the projection of x onto the "subspace" of vectors parallel to the basic vector u^i. Furthermore, it follows from Eqs. (9–236) and (9–237) that

$$x = \sum_i \epsilon_i x \tag{9-238}$$

for every vector x. Therefore,

$$\sum_i \epsilon_i = 1. \tag{9-239}$$

This is also clear from the matrix representations of the ϵ_i. Now a (non-degenerate) operator A determines a unique basis satisfying

$$Au^i = \alpha_i u^i, \tag{9-240}$$

and the projection operators ϵ_i, formed according to Eq. (9–235), are therefore uniquely associated with A. Moreover, if x is any vector, then

$$Ax = \sum_i (u^i, x)\alpha_i u^i = \sum_i \alpha_i \epsilon_i x, \tag{9-241}$$

whence A is represented in terms of the ϵ_i by

$$A = \sum_i \alpha_i \epsilon_i. \tag{9-242}$$

The essential value of these considerations derives from the fact that the ϵ_i are operators associated uniquely with the eigenvectors of A and, as operators, are independent of a representation. The entire theory of the eigenvalue problem can be developed in terms of projections, in a way which is completely independent of any representation; in this form, it is usually called the *spectral theory*. The right-hand member of Eq. (9–242) is a *spectral decomposition* of A in terms of its eigenvalues and projection operators, and Eq. (9–239) is the *resolution of the identity* determined by A.

The reader is referred to mathematical texts[1] for complete expositions of this theory. An outline has been presented here, because the basic concepts are useful in applications. In principle, it is always possible to find the spectral decomposition of an operator without reference to a co-ordinate system, and computations can often be shortened considerably by this means (cf. Section 10–8).

---

[1] P. R. Halmos, *Finite Dimensional Vector Spaces*. Princeton: Princeton University Press, 1942. J. von Neumann, *Mathematical Foundations of Quantum Mechanics*. Princeton: Princeton University Press, 1955.

9–13 Restatement of quantum-mechanical assumptions. The axiomatic basis of quantum mechanics, which has been described in previous chapters, will now be summarized in the language of vector theory. The initial assumption is that a unique correspondence exists between the properties of a physical system and the properties of a suitable vector space: *To every physical system there corresponds a linear vector space. Each state of the system is represented by a vector in this "state space" and, conversely, each vector is the representative of a possible state of the system.* This assumption implies the principle of superposition, namely, that a linear combination of state vectors represents a realizable state of the system.

Next, we assume that the observable properties of the system are represented by linear transformations: *To each physically measurable property (observable) of the system there corresponds a linear transformation of the state space, or a linear operator, such that the eigenvalues of the operator are the possible results of measurement of the corresponding observable.* The eigenvalues of such operators are all real, and it follows that an observable is represented by a Hermitian operator. Also, since only the eigenvalues are physically measurable, the "length" or norm of a state vector is unimportant. Two vectors which differ only in their norm represent the same state. For this reason, the states are more properly said to be represented by *rays* in the vector space, a ray being a directed line, considered independently of its length. In the mathematical theory, one is always allowed to choose any desired normalization of the state vectors, because their norms are not measurable quantities.

It is assumed that the eigenvectors of a Hermitian operator A span the state space. Consequently, any vector ψ can be represented as a linear combination of the eigenvectors of A, i.e.,

$$\psi = \sum_i c_i \mathsf{u}^i, \qquad \mathsf{A}\mathsf{u}^i = \alpha_i \mathsf{u}^i. \tag{9–243}$$

(The sum in this equation is, of course, infinite, and whenever the numbers α_i form a continuum, it is to be replaced by an integral.)

The third fundamental assumption assigns a physical meaning to the numbers c_i: *If an observation of a quantity corresponding to the operator A is made on the system in the state ψ, the relative probability of obtaining the result α_i is*

$$P(\alpha_i) = |c_i|^2 = |(\mathsf{u}^i, \psi)|^2. \tag{9–244}$$

This is a generalization of Born's postulate concerning the interpretation of the Schrödinger wave function. It expresses the statistical nature of the theory. The precise meaning of this postulate is: If the observation A is made on each of a large number of identical systems in identical states

ψ, the results will be statistically distributed with the relative probability $P(\alpha_i) = |c_i|^2$. If ψ is an eigenvector of A, $(\mathsf{A}\psi = \alpha\psi)$, then a measurement of A will certainly yield the result α. If ψ is not an eigenvector, the expectation, or weighted average, of A is

$$\langle \mathsf{A} \rangle = \frac{\sum_i \alpha_i P(\alpha_i)}{\sum_i P(\alpha_i)} = \frac{\sum_i \alpha_i |c_i|^2}{\sum_i |c_i|^2}. \qquad (9\text{-}245)$$

Since the vectors u^i are orthogonal, we have

$$(\psi, \mathsf{A}\psi) = \sum_i c_i(\psi, \mathsf{A}\mathsf{u}^i) = \sum_i c_i \alpha_i(\psi, \mathsf{u}^i) = \sum_i \alpha_i |c_i|^2, \quad (9\text{-}246)$$

and

$$(\psi, \psi) = \sum_i c_i(\psi, \mathsf{u}^i) = \sum_i |c_i|^2, \qquad (9\text{-}247)$$

whence

$$\langle \mathsf{A} \rangle = \frac{(\psi, \mathsf{A}\psi)}{(\psi, \psi)}. \qquad (9\text{-}248)$$

It should be noted that all of these statements have been made in a form which is independent of the representation, i.e., of the choice of the basic vectors in the state space.

A further consequence of the above assumption follows from the observation that the vector ψ can be a simultaneous eigenvector of two different operators A and B if and only if A and B commute. This means that simultaneous, precise measurements of the observables A and B are possible only if A and B are *commuting observables*. If they are not, the uncertainties ΔA and ΔB (to be expected in simultaneous measurements of A and B) are related by

$$\Delta A \, \Delta B \geqq \tfrac{1}{2}|\langle C \rangle|, \qquad (9\text{-}249)$$

where

$$[\mathsf{A}, \mathsf{B}] = i\mathsf{C}. \qquad (9\text{-}250)$$

We now have a mathematical structure with which to work. In order to use it, we need to establish a connection between the operators in the state space and the dynamical variables which describe the physical system. For this purpose, an appeal to the correspondence principle is necessary. The dynamical behavior of a classical particle is determined by the Hamiltonian function $H(p, x)$ of the momentum and coordinate of the particle. *We now assume that the operators corresponding to momentum and coordinate satisfy the commutation rule*

$$[\mathsf{x}, \mathsf{p}] = i\hbar\mathbf{1}, \qquad (9\text{-}251)$$

and that *the dynamical behavior of the system is given by the Schrödinger equation*

$$\mathsf{H}\psi = i\hbar\,\frac{\partial\psi}{\partial t}\,,\tag{9–252}$$

where H *is the Hamiltonian operator formed from the classical Hamiltonian by replacing* x *and* p *by the corresponding operators.* The state vector ψ of the system therefore depends upon the time according to Eq. (9–252), and the motion of the physical system is described by the change in time of the direction of the state vector. We have already seen that the operator relation (9–251) becomes, in the x-representation,

$$\mathsf{p} \to \frac{\hbar}{i}\,\frac{d}{dx}\,;\tag{9–253}$$

it follows from Eq. (9–252) that *the Schrödinger wave function of our earlier work is just the representative of the state vector of the system in the x-representation.* Equations (9–251) and (9–252) are, however, independent of the basis chosen for the description of the state vector. Generalization of the preceding argument to systems having more than one coordinate is immediate. It is left for the reader to make the necessary revisions.

An eigenvector ψ_E of the Hamiltonian operator represents a state for which the total energy of the system has the definite value E:

$$\mathsf{H}\psi_E = E\psi_E.\tag{9–254}$$

Thus, the determination of the energy levels of a given system is identical with the problem of finding a basis in the state space in which H is represented by a diagonal matrix. This is called the *energy representation*. This basically algebraic problem depends only upon the properties of the operator H, and is entirely independent of the representation chosen for the description of the state vectors. In the examples discussed thus far, we have always worked with the Schrödinger function, i.e., in the x-representation. We shall now show by an example how the problem of determining the energy levels of a system can be solved by algebraic methods, without the necessity of introducing a specific representation at the outset. In this treatment, the *intrinsic* properties of the system, as expressed by the Hamiltonian operator and the commutation rule (9–251), play the central role.

9–14 The one-dimensional harmonic oscillator in matrix mechanics. The Hamiltonian operator for the harmonic oscillator is

$$\mathsf{H} = \frac{1}{2m}\,\mathsf{p}^2 + \tfrac{1}{2}k\mathsf{x}^2.\tag{9–255}$$

We are required to find state vectors ψ_E, such that

$$\mathsf{H}\psi_E = E\psi_E, \tag{9-256}$$

the operators x and p being subject to the commutation rules

$$[\mathsf{x}, \mathsf{p}] = i\hbar 1. \tag{9-257}$$

For simplicity, we introduce units of mass, length and time such that $m = k = \hbar = 1$. Then we have

$$\mathsf{H} = \tfrac{1}{2}(\mathsf{p}^2 + \mathsf{x}^2), \tag{9-258}$$

and

$$[\mathsf{x}, \mathsf{p}] = i1. \tag{9-259}$$

It is useful to introduce, in addition to x and p, the operators[1]

$$\mathsf{a} = \frac{i}{\sqrt{2}}\,(\mathsf{p} - i\mathsf{x}), \qquad \mathsf{a}^\dagger = \frac{1}{\sqrt{2i}}\,(\mathsf{p} + i\mathsf{x}), \tag{9-260}$$

in terms of which

$$\mathsf{a}\mathsf{a}^\dagger = \tfrac{1}{2}[\mathsf{p}^2 - i(\mathsf{x}\mathsf{p} - \mathsf{p}\mathsf{x}) + \mathsf{x}^2] = \mathsf{H} + \tfrac{1}{2}1, \tag{9-261}$$

$$\mathsf{a}^\dagger\mathsf{a} = \tfrac{1}{2}[\mathsf{p}^2 + i(\mathsf{x}\mathsf{p} - \mathsf{p}\mathsf{x}) + \mathsf{x}^2] = \mathsf{H} - \tfrac{1}{2}1. \tag{9-262}$$

It follows that the commutator of a and $\mathsf{a}^\dagger$ is

$$[\mathsf{a}, \mathsf{a}^\dagger] = 1, \tag{9-263}$$

and

$$\mathsf{H} = \tfrac{1}{2}(\mathsf{a}\mathsf{a}^\dagger + \mathsf{a}^\dagger\mathsf{a}). \tag{9-264}$$

Also, it is easily verified that

$$[\mathsf{a}, \mathsf{H}] = \mathsf{a}, \qquad [\mathsf{a}^\dagger, \mathsf{H}] = -\mathsf{a}^\dagger. \tag{9-265}$$

From the definition of a scalar product, it follows that

$$(\psi_E, \mathsf{a}^\dagger\mathsf{a}\psi_E) = (\psi\mathsf{a}_E, \mathsf{a}\psi_E) \geq 0. \tag{9-266}$$

If we suppose that ψ_E is an eigenvector of H, then, by Eqs. (9–262) and (9–256),

$$(\psi_E, \mathsf{a}^\dagger\mathsf{a}\psi_E) = (\psi_E, (\mathsf{H} - \tfrac{1}{2})\psi_E) = (E - \tfrac{1}{2})(\psi_E, \psi_E), \tag{9-267}$$

from which it follows that

$$E \geq \tfrac{1}{2}. \tag{9-268}$$

---

[1] $\mathsf{a}^\dagger$ is the Hermitian conjugate of a, because p and x are Hermitian.

Applying the operator $\mathbf{a}$ to each member of Eq. (9–256) and using Eq. (9–265), we obtain

$$\mathbf{aH}\psi_E = (\mathbf{Ha} + \mathbf{a})\psi_E = E\mathbf{a}\psi_E, \qquad (9\text{–}269)$$

or

$$\mathbf{Ha}\psi_E = (E - 1)\mathbf{a}\psi_E. \qquad (9\text{–}270)$$

The vector $\mathbf{a}\psi_E$ is an eigenvector of $\mathbf{H}$ belonging to the eigenvalue $E - 1$, that is, the *lowering operator* $\mathbf{a}$ generates the vector $\psi_{E-1} = \mathbf{a}\psi_E$ from the vector ψ_E. Repeated application of $\mathbf{a}$ therefore lowers E indefinitely, so long as none of the ψ is zero:

$$\mathbf{a}\psi_E = \psi_{E-1}, \quad \mathbf{a}^2\psi_E = \psi_{E-2}, \quad \mathbf{a}^3\psi_E = \psi_{E-3}, \quad \text{etc.} \quad (9\text{–}271)$$

However, this result contradicts (9–268), unless the operation $\mathbf{a}$ produces the zero vector at some stage. Hence, there is an eigenvalue $E_0 = E - n$ for which $\psi_{E_0} \neq 0$, but

$$\mathbf{a}\psi_{E_0} = 0. \qquad (9\text{–}272)$$

For this eigenvalue we have, by Eq. (9–262),

$$\mathbf{a}^{\dagger}\mathbf{a}\psi_{E_0} = (\mathbf{H} - \tfrac{1}{2}\mathbf{1})\psi_{E_0} = (E_0 - \tfrac{1}{2})\psi_{E_0} = 0, \qquad (9\text{–}273)$$

whence, since ψ_{E_0} is not zero,

$$E_0 = \tfrac{1}{2}. \qquad (9\text{–}274)$$

The smallest eigenvalue of $\mathbf{H}$ is $\tfrac{1}{2}$; therefore,

$$E = n + \tfrac{1}{2}, \qquad n = 0, 1, 2, \ldots, \qquad (9\text{–}275)$$

which is exactly the result (5–114), obtained by working in the x-representation.

We now relabel the eigenfunctions ψ_E with the index n:

$$\psi_{E-n+1/2} = \text{constant} \times \psi_n, \qquad (9\text{–}276)$$

$$\mathbf{H}\psi_n = (n + \tfrac{1}{2})\psi_n. \qquad (9\text{–}277)$$

The algebraic properties of these eigenvectors follow easily from the properties of the matrices which represent $\mathbf{a}$ and $\mathbf{a}^{\dagger}$ in the energy representation. If we assume that the states ψ_n are normalized, so that

$$(\psi_{n'}, \psi_n) = \delta_{nn'}, \qquad (9\text{–}278)$$

then we have

$$\mathbf{a}\psi_n = \alpha_n\psi_{n-1}, \qquad \mathbf{a}^{\dagger}\psi_n = \beta_n\psi_{n+1}. \qquad (9\text{–}279)$$

The second of these equations is a consequence of Eq. (9–261). The matrix elements of $\mathbf{a}$ and $\mathbf{a}^\dagger$ are therefore

$$(n'|\mathbf{a}|n) = \alpha_n \, \delta_{n',n-1}, \tag{9–280}$$

$$(n'|\mathbf{a}^\dagger|n) = \beta_n \, \delta_{n',n+1}, \tag{9–281}$$

whence the constants α_n and β_n are related by

$$\alpha_{n+1} = \beta_n^*. \tag{9–282}$$

Moreover, from Eqs. (9–261) and (9–277), the matrix element of $\mathbf{a}\mathbf{a}^\dagger$ is

$$(n'|\mathbf{a}\mathbf{a}^\dagger|n) = (n+1) \, \delta_{nn'}. \tag{9–283}$$

Writing out the matrix product, we obtain

$$(n'|\mathbf{a}\mathbf{a}^\dagger|n) = \sum_{n''} (n'|\mathbf{a}|n'')(n''|\mathbf{a}^\dagger|n)$$

$$= \sum_{n''} \alpha_{n''} \, \delta_{n'n''-1}\beta_n \, \delta_{n'',n+1} = \alpha_{n+1}\beta_n \, \delta_{nn'}, \tag{9–284}$$

whence

$$\alpha_{n+1}\beta_n = n + 1, \tag{9–285}$$

and, by (9–282),

$$|\alpha_{n+1}|^2 = |\beta_n|^2 = n + 1. \tag{9–286}$$

Hence, the arbitrary phases of the vectors ψ_n can be chosen so that

$$\alpha_n = \sqrt{n}, \qquad \beta_n = \sqrt{n+1}, \tag{9–287}$$

and we have

$$\mathbf{a}\psi_n = \sqrt{n} \, \psi_{n-1}, \qquad \mathbf{a}^\dagger\psi_n = \sqrt{n+1} \, \psi_{n+1}. \tag{9–288}$$

The relations

$$\mathbf{a} + \mathbf{a}^\dagger = \sqrt{2}\mathbf{x}, \qquad \mathbf{a} - \mathbf{a}^\dagger = \sqrt{2}i\mathbf{p} \tag{9–289}$$

therefore lead to

$$\sqrt{n} \, \psi_{n-1} - \sqrt{2}\mathbf{x}\psi_n + \sqrt{n+1} \, \psi_{n+1} = 0, \tag{9–290}$$

$$\sqrt{n} \, \psi_{n-1} - \sqrt{2} \, i\mathbf{p}\psi_n - \sqrt{n+1} \, \psi_{n+1} = 0. \tag{9–291}$$

These recurrence relations, which were also derived in Section 5–10, lead immediately to the Schrödinger equation, and we know that in the x-representation, the vector ψ_n has the components

$$\psi_n(x) = \frac{1}{\sqrt{2^n n! \sqrt{\pi}}} \, e^{-(x^2/2)}H_n(x). \tag{9–292}$$

However, all the essential properties of the state vectors have been determined here in a manner which is independent of this representation.

The lowering and raising operators a and $a^\dagger$ play a fundamental role in the quantum theory of the electromagnetic field. It is well known that the Maxwell field can be represented as a linear combination of harmonic oscillators of different frequencies in various states of excitation.[1] Consequently, the electromagnetic field can be represented by a Hamiltonian function which describes an infinite set of harmonic oscillators, each of which represents a normal mode of the field oscillations (cf. Section 1–3). In quantum theory, each of these oscillators is "quantized" by imposing the condition (9–257) upon each of the pairs of conjugate coordinates in the Hamiltonian. A stationary state of the system can then be described by giving the quantum number n for each of the normal-mode oscillators. Thus, the total energy of the field is a sum of terms of the form $(n + \frac{1}{2})\hbar\omega$; there is one such term for each normal mode. In a transition in which a quantum of energy $\hbar\omega$ is absorbed by an atomic system, the corresponding quantum number n is reduced by unity. Similarly, the addition of a photon to the field, by emission of a quantum, corresponds to an increase of n by one. Now, according to Eqs. (9–279), the operators a and $a^\dagger$ have exactly this effect, and it is not surprising that, in the general theory, these operators appear as factors in the terms of the Hamiltonian of the system (atom + radiation field) which describe the interaction between the field and the charged particles. Indeed, it is this interaction which causes the absorption and emission of photons. When the operators a and $a^\dagger$ are used in this manner, they are called *destruction* and *creation* *operators* for photons. The essential properties of a and $a^\dagger$, which are required for the radiation theory, are expressed by the equations

$$(n'|a|n) = \sqrt{n}\, \delta_{n'n-1}, \qquad (n'|a^\dagger|n) = \sqrt{n+1}\, \delta_{n'n+1}. \quad (9\text{–}293)$$

The matrices which represent these operators are therefore:

$$a = \begin{pmatrix} 0 & \sqrt{1} & 0 & 0 & 0 & \cdots \\ 0 & 0 & \sqrt{2} & 0 & 0 \\ 0 & 0 & 0 & \sqrt{3} & 0 \\ 0 & 0 & 0 & 0 & \sqrt{4} \\ \vdots & & & & \end{pmatrix} \qquad a^\dagger = \begin{pmatrix} 0 & 0 & 0 & 0 & \cdots \\ \sqrt{1} & 0 & 0 & 0 \\ 0 & \sqrt{2} & 0 & 0 \\ 0 & 0 & \sqrt{3} & 0 \\ \vdots & & & & \end{pmatrix};$$

$$(9\text{–}294)$$

[1] L. Landau and E. Lifshitz, *The Classical Theory of Fields.* Reading, Mass.: Addison-Wesley Publishing Co., Inc., 1951, Chapter 6.

and, by Eqs. (9–289), x and p are the Hermitian matrices

$$x = \frac{1}{\sqrt{2}} \begin{pmatrix} 0 & \sqrt{1} & 0 & \cdots \\ \sqrt{1} & 0 & \sqrt{2} & \\ 0 & \sqrt{2} & 0 & \sqrt{3} \\ 0 & 0 & \sqrt{3} & 0 & \sqrt{4} \\ \vdots & & & & \end{pmatrix},$$

$$p = \frac{1}{\sqrt{2}} \begin{pmatrix} 0 & -i\sqrt{1} & 0 & \cdots \\ i\sqrt{1} & 0 & -i\sqrt{2} & \\ 0 & i\sqrt{2} & 0 & -i\sqrt{3} \\ 0 & 0 & i\sqrt{3} & 0 & -i\sqrt{4} \\ \vdots & & & & \end{pmatrix}.$$

$$(9\text{–}295)$$

The Hamiltonian **H** is represented by the diagonal matrix

$$H = \begin{pmatrix} \frac{1}{2} & 0 & 0 & \cdots \\ 0 & \frac{3}{2} & 0 \\ 0 & 0 & \frac{5}{2} \\ \vdots & & \end{pmatrix}. \qquad (9\text{–}296)$$

Note that the selection rules for the matrix elements of x and p, namely

$$\Delta n = \pm 1,$$

appear explicitly in Eq. (9–295), in that nonzero elements occur only in the two diagonals adjacent to the principal diagonal.

REFERENCES

BORN, M., and JORDAN, P., *Elementare Quantenmechanik*. Berlin: Julius Springer, 1930, Chapter 2.

COURANT, R., and HILBERT, D., *Methods of Mathematical Physics*. New York: Interscience Publishers, Inc., 1953. Chapter 1 describes the algebra of linear transformations in a form immediately adaptable to the solution of physical problems.

HALMOS, P. R., *Finite Dimensional Vector Spaces*. Princeton: Princeton University Press, 1942. A thorough and readable discussion of the mathematical concepts used in this chapter.

9–1. Show that the three-dimensional vectors

$$x^1 = [1, 0, -1], \qquad x^2 = [0, 1, 0], \qquad x^3 = [1, 0, 1]$$

are linearly independent (cf. Problem 7–8).

9–2. Write the vectors

$$[1, 0, 0], \qquad [0, 0, 1]$$

as linear combinations of the vectors of Problem 9–1.

9–3. Show that the vectors

$$[1, 0, 0, 0, -1], \qquad [0, 1, 0, 1, 0], \qquad [0, 1, 0, -1, 0],$$
$$[1, 0, 0, 0, 1], \qquad [0, 0, 1, 0, 0]$$

are mutually orthogonal. Normalize each.

9–4. The vectors of Problem 9–3 have an obvious relation to the second-degree polynomials of Problem 7–8. Show that the vector $[1, 0, \sqrt{6}, 0, 1]$ is related in the same way to the harmonic polynomial $z^2 - y^2$.

9–5. The vectors $\mathbf{r}$ in real three-dimensional space are subject to the transformation

$$\mathbf{r} = \mathbf{Ar} = \mathbf{a} \times \mathbf{r},$$

where $\mathbf{a}$ is a given, fixed vector. Show that $\mathbf{A}$ is a linear operator which satisfies

$$\mathbf{A}^3 + a^2\mathbf{A} = 0 \qquad (a = |\mathbf{a}|).$$

9–6. Show that the inequality

$$\sqrt{((\mathbf{x} - \mathbf{y}), (\mathbf{x} - \mathbf{y}))} \leq \sqrt{(\mathbf{x}, \mathbf{x})} + \sqrt{(\mathbf{y}, \mathbf{y})}$$

is satisfied by any pair of vectors $\mathbf{x}$ and $\mathbf{y}$. What is the geometric interpretation for real three-dimensional vectors? (Hint: Work out the square of the left-hand member and use the Schwarz inequality.) Show, conversely, that the Schwarz inequality is a consequence of the above inequality.

9–7. Construct the products AB and BA of the infinite matrices

$$A = \begin{pmatrix} 0 & 1 & 0 & 0 & \cdots \\ 0 & 0 & 1 & 0 & \cdots \\ 0 & 0 & 0 & 1 & \cdots \\ & & \text{etc.} & & \end{pmatrix} = (\delta_{i,j-1}),$$

$$B = \begin{pmatrix} 0 & 0 & 0 & 0 & \cdots \\ 1 & 0 & 0 & 0 & \cdots \\ 0 & 1 & 0 & 0 & \cdots \\ & & \text{etc.} & & \end{pmatrix} = (\delta_{i,j+1}).$$

Does either A or B have an inverse?

9–8. Calculate the vector $y = Ax$, where

$$A = \begin{pmatrix} 0 & 1 & 1 & 0 \\ 1 & 0 & 0 & 1 \\ 1 & 0 & 0 & 1 \\ 0 & 1 & 1 & 0 \end{pmatrix}, \qquad x = \begin{pmatrix} 0 \\ 1 \\ 1 \\ 0 \end{pmatrix}.$$

Show that $Ax = 0$ for every vector of the form

$$x = \begin{pmatrix} a \\ b \\ -b \\ -a \end{pmatrix},$$

and for no others. Find a vector such that $Ax = 2x$. Is the equation $Ax = x$ solvable?

9–9. Construct the matrix of a linear transformation in real two-dimensional space which doubles the length of every vector drawn from the origin and rotates it through a positive angle of 45°. Show that this matrix satisfies the equation $A^4 = -16$.

9–10. A linear transformation in an N-dimensional space is uniquely defined if the result of transforming each member of a set of N linearly independent vectors is known. Prove this statement, and show how a matrix representative of the transformation can be constructed.

9–11. Construct a 3×3 matrix A which transforms the vectors of Problem 9–1 into the vectors

$$y^1 = Ax^1 = [2, 0, 0], \qquad y^2 = Ax^2 = [1, 0, 1],$$
$$y^3 = Ax^3 = [1, 0, -1].$$

Is A singular?

9–12. Construct the products AB and BA, where A is the matrix of Problem 9–8 and

$$B = \begin{pmatrix} 0 & 1 & 1 & 0 \\ -1 & 0 & 0 & 1 \\ -1 & 0 & 0 & 1 \\ 0 & -1 & -1 & 0 \end{pmatrix}.$$

Show that $AB - BA = -4C$, where

$$C = \begin{pmatrix} 1 & 0 & 0 & 0 \\ 0 & 0 & 0 & 0 \\ 0 & 0 & 0 & 0 \\ 0 & 0 & 0 & -1 \end{pmatrix}.$$

9–13. Show that the six matrices

$$1 = \begin{pmatrix} 1 & 0 \\ 0 & 1 \end{pmatrix}, \qquad A = \begin{pmatrix} 1 & 0 \\ 0 & -1 \end{pmatrix}, \qquad B = \frac{1}{2}\begin{pmatrix} -1 & \sqrt{3} \\ \sqrt{3} & 1 \end{pmatrix},$$

$$C = \frac{1}{2}\begin{pmatrix} -1 & -\sqrt{3} \\ -\sqrt{3} & 1 \end{pmatrix}, \qquad D = \frac{1}{2}\begin{pmatrix} -1 & \sqrt{3} \\ -\sqrt{3} & -1 \end{pmatrix},$$

$$E = \frac{1}{2}\begin{pmatrix} -1 & -\sqrt{3} \\ \sqrt{3} & -1 \end{pmatrix}$$

satisfy the relations

$$A^2 = B^2 = C^2 = 1, \qquad AB = D, \qquad AC = BA = E.$$

9–14. Find the reciprocal of each of the matrices in Problem 9–13 and verify by direct multiplication that $D^{-1} = B^{-1}A^{-1}$.

9–15. Show that all the matrices of Problem 9–13 are unitary.

9–16. Prove the identity (9–131).

9–17. The Hermitian matrices

$$\sigma_1 = \begin{pmatrix} 0 & 1 \\ 1 & 0 \end{pmatrix}, \qquad \sigma_2 = \begin{pmatrix} 0 & -i \\ i & 0 \end{pmatrix}, \qquad \sigma_3 = \begin{pmatrix} 1 & 0 \\ 0 & -1 \end{pmatrix}$$

satisfy the equations

$$\sigma_1^2 = \sigma_2^2 = \sigma_3^2 = 1, \qquad \sigma_1\sigma_2 = i\sigma_3, \qquad \sigma_2\sigma_3 = i\sigma_1, \qquad \sigma_3\sigma_1 = i\sigma_2,$$

$$\sigma_1\sigma_2 - \sigma_2\sigma_1 = 2i\sigma_3, \qquad \sigma_1\sigma_2 + \sigma_2\sigma_1 = 0.$$

What are the eigenvalues of the matrices $\sigma_1, \sigma_2, \sigma_3$?

9–18. Show that every 2×2 matrix can be written in the form $a_0 \cdot 1 + a_1\sigma_1 + a_2\sigma_2 + a_3\sigma_3$, where the a_i are complex numbers, and the σ_i are the matrices defined in Problem 9–17.

9–19. Given the matrix

$$A = \begin{pmatrix} 2 & \sqrt{\frac{2}{3}} & 0 \\ \sqrt{\frac{2}{3}} & 2 & \sqrt{\frac{1}{3}} \\ 0 & \sqrt{\frac{1}{3}} & 2 \end{pmatrix},$$

show that the equation $(x, Ax) = 1$ represents an ellipsoid in real three-dimensional space, and find the lengths of the principal axes of the ellipsoid. What are the direction cosines of the principal axes? Construct the unitary matrix U which diagonalizes A and show by direct matrix multiplication that UAU^{-1} is a diagonal matrix.

9–20. Given

$$A = \begin{pmatrix} -\frac{1}{3} & \sqrt{\frac{2}{3}} & \frac{\sqrt{2}}{3} \\ \sqrt{\frac{2}{3}} & 0 & \frac{1}{\sqrt{3}} \\ \frac{\sqrt{2}}{3} & \frac{1}{\sqrt{3}} & -\frac{2}{3} \end{pmatrix}, \qquad B = \begin{pmatrix} \frac{5}{3} & \frac{1}{\sqrt{6}} & -\frac{1}{3\sqrt{2}} \\ \frac{1}{\sqrt{6}} & \frac{3}{2} & \frac{1}{2\sqrt{3}} \\ -\frac{1}{3\sqrt{2}} & \frac{1}{2\sqrt{3}} & \frac{11}{6} \end{pmatrix},$$

show that A and B commute. Find their eigenvalues and show that they determine a unitary transformation

$$U = \begin{pmatrix} \dfrac{1}{\sqrt{3}} & \dfrac{1}{\sqrt{2}} & \dfrac{1}{\sqrt{6}} \\ \dfrac{1}{\sqrt{3}} & 0 & -\dfrac{2}{\sqrt{6}} \\ \dfrac{1}{\sqrt{3}} & -\dfrac{1}{\sqrt{2}} & \dfrac{1}{\sqrt{6}} \end{pmatrix},$$

which diagonalizes both A and B. (Note that A and B are both degenerate.)

9–21. Verify Eqs. (9–180), (9–182), and (9–183) for the matrices A and B of Problem 9–19.

9–22. Verify that each of the matrices A and B of Problem 9–20 satisfies its characteristic equation.

9–23. Show that any matrix A can be written uniquely in the form $A = B + iC$, where B and C are Hermitian.

9–24. Prove the identities

$$(A^{\dagger})^{-1} = (A^{-1})^{\dagger}, \qquad (A^{*})^{-1} = (A^{-1})^{*}, \qquad (\widetilde{A})^{-1} = (\widetilde{A^{-1}}).$$

9–25. Show that if A is Hermitian, the matrix

$$U = \frac{A - i \cdot 1}{A + i \cdot 1} = (A - i \cdot 1)(A + i \cdot 1)^{-1}$$

is unitary. According to this formula, every Hermitian matrix gives rise to a unitary matrix. Prove this statement. (Hint: The essential point to be proved is that $A + i \cdot 1$ always has an inverse. Show that $(A + i \cdot 1)x = 0$ implies $x = 0$.)

9–26. Show that if U does not have the eigenvalue 1, the matrix

$$A = i \frac{1 + U}{1 - U}$$

exists and is Hermitian. This is the "Cayley transform" connecting a unitary and a Hermitian matrix.

9–27. Prove that a necessary and sufficient condition for U to be unitary is that, for every vector x, $(Ux, Ux) = (x, x)$; hence show that every eigenvalue of a unitary matrix has absolute magnitude unity.

9–28. Construct the matrix U representing a rotation through the angle ω about an axis in the direction of the unit vector $\hat{\mathbf{e}}$ in real three-dimensional space. Show that U is unitary and has the eigenvalues 1, $\exp(\pm i\omega)$. What are the corresponding eigenvectors?

9–29. Work out an entirely *algebraic* solution of the eigenvalue problem, which is independent of the analytic concept of extremum used in the text. We know already that the eigenvalues of A are the roots of the characteristic equation. Now the question is: What are the eigenvectors of A? The Laplace development

(9–86) is useful in the solution of this problem. Discuss the case of degeneracy from this point of view. (Hint: Look up the definitions of "rank" and "nullity" of a transformation and interpret them in terms of the degree of degeneracy of A.)

9–30. Show that each of the sets of vectors $\{\mathbf{x}_+\}$ and $\{\mathbf{x}_-\}$ (Section 9–12) satisfies all the axioms defining a linear vector space.

9–31. Construct the transformation $\mathbf{I}$ which represents rotation through $180°$ about the z-axis in real three-dimensional space. Form the projection operators $\epsilon_\pm = \frac{1}{2}(1 \pm \mathbf{I})$ and give a geometrical interpretation of the equation

$$\mathbf{x} = \epsilon_+\mathbf{x} + \epsilon_-\mathbf{x},$$

where $\mathbf{x}$ is any three-dimensional vector.

9–32. Verify equations (9–265).

9–33. Show that the projection operators $\frac{1}{2}(1 \pm \Pi)$, where Π is the operator of Section 6–2, correspond to the separation of a function $f(x)$ into its even and odd parts.

9–34. Verify, by direct computation, that the matrix A of Problem 9–19 satisfies the equation

$$(3 - A)(2 - A)(1 - A) = 0.$$

Use this fact to prove that each of the matrices

$$\epsilon_1 = \tfrac{1}{2}(2 - A)(1 - A), \qquad \epsilon_2 = -(1 - A)(3 - A),$$

$$\epsilon_3 = \tfrac{1}{2}(3 - A)(2 - A)$$

is a projection, i.e., $\epsilon_i^2 = \epsilon_i$. What is the form of the ϵ_i in a representation in which A is diagonal? How can one use these matrices in constructing the eigenvectors of A? Generalize these considerations for an arbitrary nondegenerate matrix.

9–35. Show that the projections ϵ_i of Problem 9–34 satisfy the equations

$$\epsilon_1 + \epsilon_2 + \epsilon_3 = 1,$$
$$3\epsilon_1 + 2\epsilon_2 + \epsilon_3 = A,$$
$$9\epsilon_1 + 4\epsilon_2 + \epsilon_3 = A^2,$$

and that, in general, any function $F(A)$ of the matrix A has the representation

$$F(A) = F(3)\epsilon_1 + F(2)\epsilon_2 + F(1)\epsilon_1.$$

Generalize.

9–36. Show that an $(N \times N)$ Hermitian matrix is degenerate if and only if it satisfies an algebraic equation whose degree is smaller than N. Construct the quadratic equations satisfied by the matrices A and B of Problem 9–20.

9–37. Prove that the matrices A and B of Problem 9–20 satisfy the equation

$$(1 + A)(2 - B) = 0.$$

What is the significance of this equation relative to the simultaneous eigenvectors of A and B?

9–38. Construct the matrices

$$\epsilon_1 = -\tfrac{1}{2}(1 + A)(1 - B),$$
$$\epsilon_2 = -\tfrac{1}{2}(1 - A)(1 - B),$$
$$\epsilon_3 = \tfrac{1}{2}(1 - A)(2 - B),$$

where A and B are the matrices of Problem 9–20. Prove that

$$\epsilon_1 + \epsilon_2 + \epsilon_3 = 1,$$
$$\epsilon_1 - \epsilon_2 - \epsilon_3 = A,$$
$$2\epsilon_1 + 2\epsilon_2 + \epsilon_3 = B,$$

and that each of the ϵ_i is a projection. How can these matrices be used to construct the simultaneous eigenvectors of A and B?

9–39. Prove that a matrix C commutes with both A and B of Problem 9–38 if and only if it has the form

$$C = \nu_1\epsilon_1 + \nu_2\epsilon_2 + \nu_3\epsilon_3,$$

where ν_1, ν_2, ν_3 are the eigenvalues of C.

9–40. The characteristic equation for a matrix A has the simple form (9–176) only if the basis of the representation is orthonormal. Write down the eigenvalue equation $\mathbf{A}\mathbf{x} = \lambda\mathbf{x}$, using an arbitrary basis $\{\mathbf{y}^i\}$, and show that the characteristic equation assumes the form

$$\det(a_{ij} - \lambda\,\Delta_{ij}) = 0,$$

where

$$a_{ij} = (\mathbf{y}^i, \mathbf{A}\mathbf{y}^j), \qquad \Delta_{ij} = (\mathbf{y}^i, \mathbf{y}^j).$$

9–41. Show that the set of all polynomials of degree smaller than N in a real variable μ can be regarded as a linear vector space of N dimensions. Let the scalar product in this space be defined as

$$(x, y) = \big(x(\mu), y(\mu)\big) = \int_{-1}^{1} x^*(\mu)y(\mu)\,d\mu.$$

Prove that the operator

$$A = -\frac{d}{d\mu}(1 - \mu^2)\frac{d}{d\mu}$$

is Hermitian.

9–42. Construct the characteristic equation for the operator A of Problem 9–41 in the special case $N = 3$, and show that it has the eigenvalues $0, 2, 6$. What are the corresponding eigenvectors?

CHAPTER 10

ANGULAR MOMENTUM AND SPIN

10–1 Rotations in three-dimensional space; angular momentum. It has been pointed out in Chapter 7 that the operators corresponding to angular momentum are infinitesimal rotation operators; the quantity

$$\frac{i}{\hbar} L_z \psi(\mathbf{r}) \cdot d\phi = \frac{\partial \psi}{\partial \phi} d\phi \tag{10-1}$$

is the change in the function $\psi(\mathbf{r})$ which is produced by rotation of the vector $\mathbf{r}$ through the angle $d\phi$ about the z-axis. This fact explains the importance of angular-momentum operators in describing the rotational symmetry properties of a physical system. It will now be shown that these operators are completely characterized by the fact that they are infinitesimal rotation operators. Thus, the quantum-mechanical theory of angular momentum is founded on the theory of rotations in three-dimensional space.

Let $\mathbf{r}$ be the position vector of the point P and consider the relation between $\mathbf{r}$ and the vector $\mathbf{r}'$, which is obtained from $\mathbf{r}$ by rotation through the small angle $d\omega = |d\boldsymbol{\omega}|$ about an axis in the direction of the vector $d\boldsymbol{\omega}$ (Fig. 10–1). To the first order of infinitesimals, the vector $\mathbf{r}'$ is given by

$$\mathbf{r}' = \mathbf{r} + d\boldsymbol{\omega} \times \mathbf{r}. \tag{10-2}$$

The result of two different infinitesimal rotations $d\boldsymbol{\omega}_1$ and $d\boldsymbol{\omega}_2$, performed in succession, is therefore

$$\mathbf{r}'' = \mathbf{r}' + d\boldsymbol{\omega}_2 \times \mathbf{r}' = \mathbf{r} + d\boldsymbol{\omega}_1 \times \mathbf{r} + d\boldsymbol{\omega}_2 \times \mathbf{r}, \tag{10-3}$$

so that the order of the rotations $d\boldsymbol{\omega}_1$ and $d\boldsymbol{\omega}_2$ is immaterial, to the first order of small quantities. However, this is certainly not true for *finite* rotations; it is therefore of interest to carry the analysis to terms of second

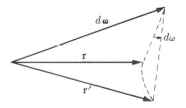

Fig. 10–1.   Rotation of the vector $\mathbf{r}$ through $d\omega$.

333

order in the angles. This can easily be done by more detailed consideration of the geometry of Fig. 10–1, but it is instructive to work instead with the differential equation

$$\mathbf{r}' - \mathbf{r} = d\mathbf{r} = d\boldsymbol{\omega} \times \mathbf{r}, \qquad (10\text{–}4)$$

and to obtain the desired result analytically, using the methods of vector theory.

Let $\hat{\boldsymbol{\kappa}}$ be a unit vector in the direction of the axis of rotation, so that

$$d\boldsymbol{\omega} = \hat{\boldsymbol{\kappa}}\, d\omega. \qquad (10\text{–}5)$$

Then we have

$$\frac{d\mathbf{r}}{d\omega} = \hat{\boldsymbol{\kappa}} \times \mathbf{r}. \qquad (10\text{–}6)$$

The operation denoted by $\hat{\boldsymbol{\kappa}}\times$ is a linear operation and thus can be represented by a matrix. We write $\mathbf{r}$ temporarily as a column vector,

$$\mathbf{r} = \begin{pmatrix} x \\ y \\ z \end{pmatrix},$$

and introduce the matrix

$$K = \begin{pmatrix} 0 & -\kappa_z & \kappa_y \\ \kappa_z & 0 & -\kappa_x \\ -\kappa_y & \kappa_x & 0 \end{pmatrix}, \qquad (10\text{–}7)$$

where κ_x, κ_y, κ_z are the components of the unit vector $\hat{\boldsymbol{\kappa}}$. The differential equation (10–6) now takes the form

$$\frac{d\mathbf{r}}{d\omega} = K\mathbf{r}, \qquad (10\text{–}8)$$

with the solution (Appendix A–7)

$$\mathbf{r}' = e^{K\omega}\mathbf{r} = e^{\boldsymbol{\omega}\times}\mathbf{r}. \qquad (10\text{–}9)$$

The operator $K\omega = \boldsymbol{\omega}\times$ represents a finite rotation through the angle ω, and $\mathbf{r}$ and $\mathbf{r}'$ are the initial and final values of $\mathbf{r}$, respectively. The meaning of the operator $e^{K\omega}$ is

$$e^{K\omega} = 1 + K\omega + \frac{1}{2!}(K\omega)^2 + \frac{1}{3!}(K\omega)^3 + \cdots, \qquad (10\text{–}10)$$

whence it follows that

$$\mathbf{r}' = \mathbf{r} + \boldsymbol{\omega} \times \mathbf{r} + \frac{1}{2!}\,\boldsymbol{\omega} \times (\boldsymbol{\omega} \times \mathbf{r}) + \frac{1}{3!}\,\boldsymbol{\omega} \times [\boldsymbol{\omega} \times (\boldsymbol{\omega} \times \mathbf{r})] + \cdots. \qquad (10\text{–}11)$$

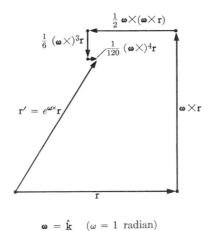

$$\boldsymbol{\omega} = \hat{\mathbf{k}} \quad (\omega = 1 \text{ radian})$$

Fig. 10–2.   Illustration of the series (10–11).

The geometrical meaning of this expansion is illustrated in Fig. 10–2. If the angle $d\boldsymbol{\omega}$ is small, Eq. (10–11) reduces in second order to

$$\mathbf{r}' = \mathbf{r} + d\boldsymbol{\omega} \times \mathbf{r} + \tfrac{1}{2} d\boldsymbol{\omega} \times (d\boldsymbol{\omega} \times \mathbf{r}). \qquad (10\text{--}12)$$

Now let the vector $\mathbf{r}$ be first subjected to the rotation $d\boldsymbol{\omega}_1$, so that it becomes

$$\mathbf{r}_1 = \mathbf{r} + d\boldsymbol{\omega}_1 \times \mathbf{r} + \tfrac{1}{2} d\boldsymbol{\omega}_1 \times (d\boldsymbol{\omega}_1 \times \mathbf{r}), \qquad (10\text{--}13)$$

and let this vector then be rotated by $d\boldsymbol{\omega}_2$, giving

$$\mathbf{r}_{21} = \mathbf{r}_1 + d\boldsymbol{\omega}_2 \times \mathbf{r}_1 + \tfrac{1}{2} d\boldsymbol{\omega}_2 \times (d\boldsymbol{\omega}_2 \times \mathbf{r}_1). \qquad (10\text{--}14)$$

Substituting Eq. (10–13) for $\mathbf{r}_1$ and dropping terms of order higher than two, one finds

$$\mathbf{r}_{21} = \mathbf{r} + d\boldsymbol{\omega}_1 \times \mathbf{r} + d\boldsymbol{\omega}_2 \times \mathbf{r} + \tfrac{1}{2} d\boldsymbol{\omega}_1 \times (d\boldsymbol{\omega}_1 \times \mathbf{r})$$
$$+ \tfrac{1}{2} d\boldsymbol{\omega}_2 \times (d\boldsymbol{\omega}_2 \times \mathbf{r}) + d\boldsymbol{\omega}_2 \times (d\boldsymbol{\omega}_1 \times \mathbf{r}). \qquad (10\text{--}15)$$

The result for the opposite order of the two rotations is obtained by interchanging the subscripts 1 and 2:

$$\mathbf{r}_{12} = \mathbf{r} + d\boldsymbol{\omega}_2 \times \mathbf{r} + d\boldsymbol{\omega}_1 \times \mathbf{r} + \tfrac{1}{2} d\boldsymbol{\omega}_2 \times (d\boldsymbol{\omega}_2 \times \mathbf{r})$$
$$+ \tfrac{1}{2} d\boldsymbol{\omega}_1 \times (d\boldsymbol{\omega}_1 \times \mathbf{r}) + d\boldsymbol{\omega}_1 \times (d\boldsymbol{\omega}_2 \times \mathbf{r}), \qquad (10\text{--}16)$$

and the difference between $\mathbf{r}_{12}$ and $\mathbf{r}_{21}$ is

$$\mathbf{r}_{12} - \mathbf{r}_{21} = d\boldsymbol{\omega}_1 \times (d\boldsymbol{\omega}_2 \times \mathbf{r}) - d\boldsymbol{\omega}_2 \times (d\boldsymbol{\omega}_1 \times \mathbf{r})$$
$$= (d\boldsymbol{\omega}_1 \times d\boldsymbol{\omega}_2) \times \mathbf{r}. \qquad (10\text{--}17)$$

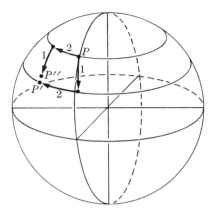

FIG. 10–3.   The rotations 1 and 2 do not commute: (1 2) leads from P to P', while (2 1) leads from P to P''.

This equation expresses the fact that the rotations do not commute, and shows that the rotations (2 1) and (1 2) differ (Fig. 10–3) by a rotation

$$d^2\boldsymbol{\omega} = d\boldsymbol{\omega}_1 \times d\boldsymbol{\omega}_2.$$

We now investigate the change in a function $\psi(\mathbf{r}) = \psi(x, y, z)$ of the coordinates of P, as the vector $\mathbf{r}$ is rotated about the origin. The function ψ is assumed to be continuous and differentiable, so that the value of ψ at the point $\mathbf{r}' = \mathbf{r} + d\mathbf{r}$ is given to first order in $d\mathbf{r}$ by

$$\psi(\mathbf{r}') = \psi(\mathbf{r}) + d\mathbf{r} \cdot \nabla\psi(\mathbf{r}). \quad (10\text{–}18)$$

If $\mathbf{r}$ and $\mathbf{r}'$ are related by an infinitesimal rotation $d\boldsymbol{\omega}$, then $d\mathbf{r} = d\boldsymbol{\omega} \times \mathbf{r}$, and

$$\begin{aligned}\psi(\mathbf{r}') &= \psi(\mathbf{r}) + d\boldsymbol{\omega} \times \mathbf{r} \cdot \nabla\psi(\mathbf{r}) \\ &= \psi(\mathbf{r}) + d\boldsymbol{\omega} \cdot (\mathbf{r} \times \nabla)\psi(\mathbf{r}),\end{aligned}$$

or

$$\psi(\mathbf{r}') = \psi(\mathbf{r}) + \frac{i}{\hbar} d\boldsymbol{\omega} \cdot \mathbf{L}\psi(\mathbf{r}), \quad (10\text{–}19)$$

where

$$\mathbf{L} = \frac{\hbar}{i} \mathbf{r} \times \nabla = \mathbf{r} \times \mathbf{p}$$

is the angular-momentum operator. We choose a system of units in which $\hbar = 1$ and write Eq. (10–19) as

$$\psi(\mathbf{r}') - \psi(\mathbf{r}) = d\psi = i\, d\boldsymbol{\omega} \cdot \mathbf{L}\psi. \quad (10\text{–}20)$$

By integration, we find for a finite rotation $\boldsymbol{\omega}$

$$\psi(\mathbf{r}') = \exp(i\boldsymbol{\omega} \cdot \mathbf{L})\psi(\mathbf{r}), \quad (10\text{–}21)$$

where $\mathbf{r}'$ and $\mathbf{r}$ are related by Eq. (10–9). As in the similar Eq. (10–10), the power series for the exponential operator $\exp(i\boldsymbol{\omega} \cdot \mathbf{L})$ gives Taylor's series for $\psi(\mathbf{r}')$ in terms of $\psi(\mathbf{r})$ and its derivatives at $\mathbf{r}$. The first three terms in this series are

$$\psi(\mathbf{r}') = [1 + i\, d\boldsymbol{\omega} \cdot \mathbf{L} + \tfrac{1}{2}(d\boldsymbol{\omega} \cdot \mathbf{L})^2]\psi(\mathbf{r}), \quad (10\text{–}22)$$

and, as before, we can find the effect of interchanging the infinitesimal rotations $d\boldsymbol{\omega}_1$ and $d\boldsymbol{\omega}_2$. A somewhat lengthy calculation yields, in the notation introduced in Eq. (10–14),

$$\psi(\mathbf{r}_{21}) = \psi + i\,d\boldsymbol{\omega}_1 \cdot \mathbf{L}\psi + i\,d\boldsymbol{\omega}_2 \cdot \mathbf{L}\psi + i\,(d\boldsymbol{\omega}_2 \times d\boldsymbol{\omega}_1) \cdot \mathbf{L}\psi$$
$$-d\boldsymbol{\omega}_2 \cdot \mathbf{L}\,d\boldsymbol{\omega}_1 \cdot \mathbf{L}\psi. \quad (10\text{--}23)$$

Upon interchanging 1 and 2 and subtracting, we obtain

$$\psi(\mathbf{r}_{12}) - \psi(\mathbf{r}_{21}) = 2i\,(d\boldsymbol{\omega}_1 \times d\boldsymbol{\omega}_2) \cdot \mathbf{L}\psi$$
$$- [d\boldsymbol{\omega}_1 \cdot \mathbf{L}\,d\boldsymbol{\omega}_2 \cdot \mathbf{L} - d\boldsymbol{\omega}_2 \cdot \mathbf{L}\,d\boldsymbol{\omega}_1 \cdot \mathbf{L}]\psi. \quad (10\text{--}24)$$

From the result (10–17) it is clear that this expression must also be equal to

$$i\,(d\boldsymbol{\omega}_1 \times d\boldsymbol{\omega}_2) \cdot \mathbf{L}\psi, \quad (10\text{--}25)$$

because, by Eq. (10–20), this is just the difference between the values of ψ at the points $\mathbf{r}_{12}$ and $\mathbf{r}_{21}$, correct to terms of second order. From Eqs. (10–24) and (10–25) it follows that

$$d\boldsymbol{\omega}_1 \cdot \mathbf{L}\,d\boldsymbol{\omega}_2 \cdot \mathbf{L} - d\boldsymbol{\omega}_2 \cdot \mathbf{L}\,d\boldsymbol{\omega}_1 \cdot \mathbf{L} = i\,(d\boldsymbol{\omega}_1 \times d\boldsymbol{\omega}_2) \cdot \mathbf{L}, \quad (10\text{--}26)$$

which is the *commutation rule* for the operators $d\boldsymbol{\omega}_1 \cdot \mathbf{L}$ and $d\boldsymbol{\omega}_2 \cdot \mathbf{L}$. Since $d\boldsymbol{\omega}_1$ and $d\boldsymbol{\omega}_2$ are arbitrary vectors, Eq. (10–26) implies that

$$\begin{aligned}
L_x L_y - L_y L_x &= iL_z, \\
L_y L_z - L_z L_y &= iL_x, \\
L_z L_x - L_x L_z &= iL_y.
\end{aligned} \quad (10\text{--}27)$$

These are exactly the commutation rules deduced in Chapter 7 on the basis of the definition $\mathbf{L} = \mathbf{r} \times \mathbf{p}$. The present derivation shows that the equations (10–27) are analytic expressions of the geometrical fact that rotations in three-dimensional space are not commutative.

10-2 Rotations of the coordinate frame. Equation (10–9) describes an *active* transformation, in which the point P is regarded as having moved from P to P'. The components of the vectors $\mathbf{r}$ and $\mathbf{r}'$ are the coordinates of P and P' in a frame of reference which is fixed in space. We now change our point of view and consider the effect of rotations of the coordinate frame. The *passive* transformation generated by rotating the coordinate frame through the angle $\boldsymbol{\omega}$ to a new position is

$$\mathbf{r}' = e^{-\boldsymbol{\omega}\times}\mathbf{r}. \quad (10\text{--}28)$$

The vectors $\mathbf{r}$ and $\mathbf{r}'$ are different representations of the same vector, namely the directed line from the origin to the fixed point P; their components are different because they are measured relative to different coordinate frames. Now let a function $\psi(P)$ be given at each point P of space. In the original coordinate frame, $\psi(P)$ is represented by a function

$\psi(\mathbf{r}) = \psi(x, y, z)$, and in the primed system, it is represented by a different function $\psi'(\mathbf{r}') = \psi'(x', y', z')$ of the coordinates of P relative to the rotated axes. If $\mathbf{r}$ and $\mathbf{r}'$ are related by Eq. (10–28), then clearly

$$\psi'(\mathbf{r}') = \psi(\mathbf{r}). \tag{10–29}$$

By Eq. (10–21), this relation can be written

$$\psi'(\mathbf{r}') = \psi'(\exp(-\boldsymbol{\omega}\times)\mathbf{r}) = \exp(-i\boldsymbol{\omega}\cdot\mathbf{L})\psi'(\mathbf{r}) = \psi(\mathbf{r}), \tag{10–30}$$

or

$$\psi'(\mathbf{r}) = \exp(i\boldsymbol{\omega}\cdot\mathbf{L})\psi(\mathbf{r}). \tag{10–31}$$

This equation describes explicitly, in terms of the operator $\mathbf{L}$, the connection between the two functions ψ' and ψ.

If $\psi(\mathbf{r})$ is the Schrödinger wave function for a one-particle system, then it is a component (in the coordinate representation) of the state vector ψ, and Eq. (10–31), which is of the form

$$\psi' = U\psi, \tag{10–32}$$

can be regarded as a unitary transformation in the state space, with matrix $U(\boldsymbol{\omega}) = \exp(i\boldsymbol{\omega}\cdot\mathbf{L})$. In other words, a rotation $\boldsymbol{\omega}$ of the coordinate frame in the real physical space of the system induces a change of basis in the state space.

Suppose that we are studying the energy levels of a given isolated system, i.e., we wish to find solutions of the matrix equation

$$H\psi = E\psi. \tag{10–33}$$

Since the system is isolated, the orientation of the coordinate frame cannot influence the results. Therefore, if $\psi(\mathbf{r})$ is an eigenvector, then $\psi'(\mathbf{r}) = U(\boldsymbol{\omega})\psi$ is also an eigenvector, because the latter is obtained from the former by a rotation. In other words, the matrix

$$H' = UHU^{\dagger}, \tag{10–34}$$

which represents the Hamiltonian operator in the primed representation, is identical with the matrix H:[1]

---

[1] The identity of the matrices H and H' is perhaps made more evident by the following physical example: Let H be the Hamiltonian function for a sodium atom. Then we can find the energy levels, i.e., the eigenvalues of H, by a term analysis of the sodium spectrum. This constitutes an experimental determination of the matrix elements of H in the energy representation. Obviously, the results of such a determination will not depend upon the orientation of the coordinate frame used to describe the positions of the particles in the sodium atom.

$$H = UHU^\dagger. \tag{10-35}$$

The matrix which represents the Hamiltonian of an isolated system is invariant to rotation of the coordinate system.

The equation (10–35) is equivalent to

$$[H, U] = HU - UH = 0, \tag{10-36}$$

i.e., H must commute with every rotation operator. In particular, H commutes with the infinitesimal operator

$$U (d\boldsymbol{\omega}) = 1 + i \, d\boldsymbol{\omega} \cdot \mathbf{L}, \tag{10-37}$$

whence, since $d\boldsymbol{\omega}$ is arbitrary,

$$[H, \mathbf{L}] = 0. \tag{10-38}$$

The angular momentum is a constant of the motion for a spherically symmetric system. This means that the expectation of each component of $\mathbf{L}$ is independent of the time for every state.

The Hamiltonian commutes with each component of $\mathbf{L}$. However, these components do not commute among themselves. Hence if we diagonalize H, we may expect to make one of the components of $\mathbf{L}$, say L_z, diagonal. Furthermore, it is seen that

$$[H, \mathbf{L}^2] = 0, \tag{10-39}$$

and since $\mathbf{L}^2$ also commutes with L_z, it is possible to find a representation in which H, L_z, and $\mathbf{L}^2$ are simultaneously diagonal. It was shown in Chapter 7 that the basic vectors in this representation are given by the functions[1]

$$\psi_l^m(r, \theta, \phi) = R_l(r) Y_l^m(\theta, \phi), \tag{10-40}$$

where $l = 0, 1, 2, \ldots$, and m is an integer in the range $[-l, l]$.

We shall now demonstrate that the eigenfunctions and eigenvalues of $\mathbf{L}^2$ and L_z are completely determined by the commutation relations (10–27), in a form which is independent of the representation. In order to have a convenient notation for later work, we define a generalized angular-momentum operator

$$\mathbf{J} = [J_x, J_y, J_z], \tag{10-41}$$

---

1 These functions are the components of the basic vectors expressed in the coordinate representation.

as any Hermitian operator whose components satisfy the commutation rules

$$[J_x, J_y] = iJ_z, \qquad [J_y, J_z] = iJ_x, \qquad [J_z, J_x] = iJ_y, \qquad (10\text{--}42)$$

or, equivalently,

$$\mathbf{J} \times \mathbf{J} = i\mathbf{J}. \qquad (10\text{--}43)$$

As we shall see, these rules are satisfied by quantities which are more general than the angular momentum of a single particle, and our results will therefore be applicable to systems of greater complexity.

10–3 Diagonalization of $\mathbf{J}^2$ and J_z. The representation in which $\mathbf{J}^2$ and J_z are simultaneously diagonal has basic vectors $\phi(\lambda m)$ which satisfy

$$\mathbf{J}^2 \phi(\lambda m) = \lambda \phi(\lambda m) \qquad \text{and} \qquad J_z \phi(\lambda m) = m \phi(\lambda m), \qquad (10\text{--}44)$$

where m and λ are real numbers ($\mathbf{J}^2$ and J_z are Hermitian).

In the algebraic work to follow, it is convenient to use the operators

$$J_+ = J_x + iJ_y, \qquad J_- = J_x - iJ_y, \qquad (10\text{--}45)$$

$$(J_+^\dagger = J_-), \qquad (10\text{--}46)$$

which satisfy the commutation rules

$$[J_+, J_z] = -J_+, \qquad (10\text{--}47)$$

$$[J_-, J_z] = J_-, \qquad (10\text{--}48)$$

$$[J_+, J_-] = 2J_z, \qquad (10\text{--}49)$$

and the identities

$$J_+ J_- = \mathbf{J}^2 - J_z^2 + J_z, \qquad (10\text{--}50)$$

$$J_- J_+ = \mathbf{J}^2 - J_z^2 - J_z. \qquad (10\text{--}51)$$

A relation between the eigenvalues m and λ is implied by the fact that the expectation value $(\phi, \mathbf{J}^2 \phi)$ cannot be smaller than $(\phi, J_z^2 \phi)$, that is,

$$(\phi, \mathbf{J}^2 \phi) = (\phi, J_x^2 \phi) + (\phi, J_y^2 \phi) + (\phi, J_z^2 \phi)$$
$$= (J_x \phi, J_x \phi) + (J_y \phi, J_y \phi) + (J_z \phi, J_z \phi) \geq (J_z \phi, J_z \phi). \qquad (10\text{--}52)$$

Substituting $\phi = \phi(\lambda m)$, we have

$$\lambda \geq m^2. \qquad (10\text{--}53)$$

Also, Eq. (10–47) yields

$$J_z(J_+\phi(\lambda m)) = J_+J_z\phi(\lambda m) + J_+\phi(\lambda m) = (m+1)(J_+\phi(\lambda m)), \quad (10\text{–}54)$$

and since $[\mathbf{J}^2, J_+] = 0$,

$$\mathbf{J}^2(J_+\phi(\lambda m)) = J_+\mathbf{J}^2\phi(\lambda m) = \lambda(J_+\phi(\lambda m)). \quad (10\text{–}55)$$

Consequently, the vector $J_+\phi(\lambda m)$ is an eigenvector of J_z belonging to the eigenvalue $m+1$, and of $\mathbf{J}^2$ belonging to the *same* eigenvalue λ of $\mathbf{J}^2$. J_+ is a "raising operator" for m. By iteration of J_+, we can construct eigenvectors of J_z belonging to larger and larger values of m. This process must stop, however, at some value $m = j$, where $\lambda \geq j^2$, because the inequality (10–53) would otherwise be violated. For this value of m, one has

$$J_+\phi(\lambda j) = 0, \qquad \phi(\lambda j) \neq 0. \quad (10\text{–}56)$$

By a calculation similar to that leading to Eqs. (10–54) and (10–55), one can also conclude that the vector $J_-\phi(\lambda m)$ is an eigenvector of J_z belonging to $m-1$, and of $\mathbf{J}^2$ belonging to λ. The operation J_- can therefore be applied iteratively to the function $\phi(\lambda j)$, yielding the sequence of eigenvalues of J_z,

$$m = j, j-1, j-2, \ldots.$$

Again, to prevent a violation of (10–53), there must be a value j' of m such that

$$J_-\phi(\lambda j') = 0, \qquad \phi(\lambda j') \neq 0, \quad (10\text{–}57)$$

and, by construction,

$$j = j' + \text{an integer.} \quad (10\text{–}58)$$

Multiplying the equations (10–56) and (10–57) by J_- and J_+, respectively, and using Eqs. (10–50) and (10–51), we obtain

$$J_-J_+\phi(\lambda j) = 0 = (\lambda - j^2 - j)\phi(\lambda j), \quad (10\text{–}59)$$

$$J_+J_-\phi(\lambda j') = 0 = (\lambda - j'^2 + j')\phi(\lambda j'). \quad (10\text{–}60)$$

It follows that

$$\lambda = j(j+1) = (-j')(-j'+1). \quad (10\text{–}61)$$

The only solution of the second of these equations, consistent with $j > j'$, is

$$j' = -j,$$

and we therefore conclude from Eq. (10–58) that

$$2j = \text{an integer.}$$

The eigenvalues of $\mathbf{J}^2$ are the numbers $j(j+1)$, where

$$j = 0, \tfrac{1}{2}, 1, \tfrac{3}{2}, \ldots,$$

and each of these is $(2j+1)$-fold degenerate, the eigenvalues of J_z for a given value of j being

$$m = -j, -j+1, \ldots, j-1, j.$$

The eigenvectors $\phi(\lambda m)$ for a given value of j, which we now denote by $\phi(jm)$, are a $(2j+1)$-dimensional basis for a representation of the angular-momentum operators. The matrix elements of J_z, which is diagonal in this representation, are determined by the equations

$$J_z\phi(jm) = m\phi(jm), \tag{10-62}$$

whence, if the $\phi(jm)$ are normalized,

$$\big(\phi(j'm'), J_z\phi(jm)\big) = (j'm'|J_z|jm) = m\,\delta_{mm'}\,\delta_{jj'}. \tag{10-63}$$

The matrix elements of J_+ and J_- are also easily found, for we have shown that

$$J_+\phi(jm) = a_m\phi(jm+1), \qquad J_-\phi(jm) = b_m\phi(jm-1), \tag{10-64}$$

whence

$$\big(\phi(jm+1), J_+\phi(jm)\big) = a_m = \big(J_-\phi(jm+1), \phi(jm)\big) = b^*_{m+1}. \tag{10-65}$$

The constant a_m can be found by means of Eq. (10-51):

$$J_-J_+\phi(jm) = a_m J_-\phi(jm+1) = a_m b_{m+1}\phi(jm) = |a_m|^2 \phi(jm)$$

$$= (\mathbf{J}^2 - J_z^2 - J_z)\phi(jm) = (j(j+1) - m(m+1))\phi(jm). \tag{10-66}$$

$$|a_m|^2 = j(j+1) - m(m+1) = (j-m)(j+m+1). \tag{10-67}$$

The arbitrary phase in the definition of the vectors ϕ_j^m is conventionally chosen in such a way[1] that

$$a_m = \sqrt{(j-m)(j+m+1)}. \tag{10-68}$$

---

[1] This normalization agrees, in the case $\mathbf{J} = \mathbf{L}$, with the normalization of the spherical harmonics adopted in Chapter 7, Eq. (7-44). Note that a factor of the form $e^{i\gamma(j)}$, in which $\gamma(j)$ depends upon j but *not upon* m, is still undetermined.

The equations (10–64) are therefore

$$J_+\phi(jm) = \sqrt{(j-m)(j+m+1)}\,\phi(jm+1), \qquad (10\text{–}69)$$

$$J_-\phi(jm) = \sqrt{(j+m)(j-m+1)}\,\phi(jm-1). \qquad (10\text{–}70)$$

The matrix elements of J_+ and J_- are

$$(j'm'|J_+|jm) = \sqrt{(j-m)(j+m+1)}\,\delta_{j'j}\,\delta_{m',m+1}, \qquad (10\text{–}71)$$

$$(j'm'|J_-|jm) = \sqrt{(j+m)(j-m+1)}\,\delta_{j'j}\,\delta_{m',m-1}. \qquad (10\text{–}72)$$

The equations (10–63), (10–71), and (10–72) provide a complete representation of the operator $\mathbf{J}$ with respect to the basis $\{\phi(jm)\}$, in which J^2 and J_z are diagonal. By means of this representation, a unitary matrix

$$U(\omega) = e^{i\omega J_z}, \qquad (10\text{–}73)$$

having the matrix elements

$$(j'm'|U(\omega)|jm) = e^{im\omega}\,\delta_{jj'}\,\delta_{mm'} \qquad (10\text{–}74)$$

can be constructed. This matrix corresponds to a finite rotation about the z-axis through the angle ω. Now the eigenvalue m can be either an integer or a half-integer.[1] If j is an integer, then m is one also, and the matrix $U(2\pi)$, corresponding to a complete rotation, is

$$(j'm'|U(2\pi)|jm) - e^{im\cdot 2\pi}\,\delta_{jj'}\,\delta_{mm'} \qquad (10\ 75)$$

i.e., $U(2\pi)$ is the unit matrix. A Schrödinger wave function $\psi(\mathbf{r})$ of the coordinate variables is single-valued[2] and therefore unchanged by a complete rotation. Consequently, the integral values of j are appropriate for the description of the transformation of such functions. In Chapter 5, we saw that, in the case $\mathbf{J} = \mathbf{L}$, the vectors $\phi(jm)$ in the coordinate representation are the spherical harmonics $Y_l^m(\theta, \phi)$, where $j = l =$ an integer.

On the other hand, the half-integral values of j lead to a matrix $U(\omega)$ for which

$$U(2\pi) = -1. \qquad (10\text{–}76)$$

The transformation induced in the state space is therefore such that a complete rotation of the coordinate frame produces a change of sign of the state vector. These transformations are therefore not applicable to the case $\mathbf{J} = \mathbf{L}$.

---

[1] "Half-integer" means "half of an odd integer."
[2] Section 7–1.

The representations of the rotation operators which arise from half-integral j are *double-valued*, because the identical rotations through the angles ω and $\omega + 2\pi$ correspond, according to Eq. (10–74), to different matrices

$$U(\omega) = -U(\omega + 2\pi) \qquad (j = \text{half-integer}). \qquad (10\text{–}77)$$

In contrast to this, we have for integral values of j

$$U(\omega) = U(\omega + 2\pi) \qquad (j = \text{integer}), \qquad (10\text{–}78)$$

so that only one matrix $U(\omega)$ corresponds to each rotation.

The double-valued representations are not excluded from consideration, however, because the physical properties of a system depend quadratically upon its state vector. As we shall see, these representations describe the intrinsic angular momentum or *spin* of the electron.

10–4 Explicit forms of the angular momentum matrices. The matrix elements of J_z, J_+ and J_- have been indexed by the eigenvalues of $\mathbf{J}^2$ and J_z to which they correspond. Thus, each of these matrices has the form

$$
\begin{array}{cc|ccc|c}
 & j & 0 & \tfrac{1}{2} & 1 & \\
 & m & 0 & \tfrac{1}{2} \;\; -\tfrac{1}{2} & 1 \;\; 0 \;\; -1 & \\
j' & m' & & & & \\
\hline
0 & 0 & (\quad) & (\quad 0 \quad) & (\quad 0 \quad) & \cdots \\
\tfrac{1}{2} & \begin{matrix}\tfrac{1}{2} \\ -\tfrac{1}{2}\end{matrix} & \begin{pmatrix}0\end{pmatrix} & (\qquad) & \begin{pmatrix}0\end{pmatrix} & \cdots \\
1 & \begin{matrix}1 \\ 0 \\ -1\end{matrix} & \begin{pmatrix}0\end{pmatrix} & \begin{pmatrix}0\end{pmatrix} & (\qquad) & \cdots \\
 & \vdots & & & &
\end{array}
\qquad (10\text{–}79)
$$

in which the only nonzero elements occur in the "submatrices" which lie along the principal diagonal. This expresses explicitly that every component of $\mathbf{J}$ is diagonal in j. Furthermore, every matrix function of matrices of the form (10–79) is also of this form, and elements belonging to different values of j are never mixed in any algebraic process that involves such matrices exclusively. In other words, each set of matrices J_x, J_y, J_z belonging to a given value of j is a representation of the operator $\mathbf{J}$. It is a finite-dimensional representation of dimension $2j + 1$. The first four such representations, for $j = 0, \frac{1}{2}, 1,$ and $\frac{3}{2}$, are given below:

$j = 0$:

$$J_+ = (0), \qquad J_- = (0), \qquad J_z = (0), \qquad J^2 = (0);$$

$j = \frac{1}{2}$:

$$J_+ = \begin{pmatrix} 0 & 1 \\ 0 & 0 \end{pmatrix}, \qquad J_- = \begin{pmatrix} 0 & 0 \\ 1 & 0 \end{pmatrix}, \qquad J_z = \begin{pmatrix} \frac{1}{2} & 0 \\ 0 & -\frac{1}{2} \end{pmatrix},$$

$$J^2 = \begin{pmatrix} \frac{3}{4} & 0 \\ 0 & \frac{3}{4} \end{pmatrix};$$

$j = 1$:

$$J_+ = \begin{pmatrix} 0 & \sqrt{2} & 0 \\ 0 & 0 & \sqrt{2} \\ 0 & 0 & 0 \end{pmatrix}, \qquad J_- = \begin{pmatrix} 0 & 0 & 0 \\ \sqrt{2} & 0 & 0 \\ 0 & \sqrt{2} & 0 \end{pmatrix},$$

$$J_z = \begin{pmatrix} 1 & 0 & 0 \\ 0 & 0 & 0 \\ 0 & 0 & -1 \end{pmatrix}, \qquad J^2 = \begin{pmatrix} 2 & 0 & 0 \\ 0 & 2 & 0 \\ 0 & 0 & 2 \end{pmatrix};$$

$j = \frac{3}{2}$:

$$J_+ = \begin{pmatrix} 0 & \sqrt{3} & 0 & 0 \\ 0 & 0 & 2 & 0 \\ 0 & 0 & 0 & \sqrt{3} \\ 0 & 0 & 0 & 0 \end{pmatrix}, \qquad J_- = \begin{pmatrix} 0 & 0 & 0 & 0 \\ \sqrt{3} & 0 & 0 & 0 \\ 0 & 2 & 0 & 0 \\ 0 & 0 & \sqrt{3} & 0 \end{pmatrix},$$

$$J_z = \begin{pmatrix} \frac{3}{2} & 0 & 0 & 0 \\ 0 & \frac{1}{2} & 0 & 0 \\ 0 & 0 & -\frac{1}{2} & 0 \\ 0 & 0 & 0 & -\frac{3}{2} \end{pmatrix}, \qquad J^2 = \begin{pmatrix} \frac{15}{4} & 0 & 0 & 0 \\ 0 & \frac{15}{4} & 0 & 0 \\ 0 & 0 & \frac{15}{4} & 0 \\ 0 & 0 & 0 & \frac{15}{4} \end{pmatrix}.$$

$$(10\text{–}80)$$

10–5 Effect of a magnetic field. The intrinsic degeneracy of the quantum states of an isolated system can only be removed by subjecting the system to an external force which is not spherically symmetric. For a system of charged particles, the forces arising from an externally applied magnetic field are of this nature and provide a means for the complete removal of the degeneracy. The study of magnetic effects is therefore important for understanding the physical nature and significance of angular-momentum states, and the present section is devoted to a development of their theory.

The Schrödinger equation (38–10) permits a complete description of the motion of a charged particle under an electrostatic force, because a potential-energy function always exists for such a force. If a magnetic field is also present, the total force on a particle of charge e is

$$\mathbf{F} = e\left(\boldsymbol{\varepsilon} + \frac{1}{c}\mathbf{v} \times \boldsymbol{\mathfrak{B}}\right),\qquad(10\text{–}81)$$

where $\boldsymbol{\varepsilon}$ is the intensity of the electric field, $\boldsymbol{\mathfrak{B}}$ the magnetic induction, and $\mathbf{v} = d\mathbf{r}/dt$ the velocity. This velocity-dependent force cannot be described by a potential-energy function, and it is therefore necessary to find a suitable generalization of the Schrödinger equation which will allow a quantum description of the effect of a magnetic field.

The required generalization is to be found, according to the program of Section 6–10, by an appeal to the correspondence principle. We first review the classical Hamiltonian formulation of the problem.[1] The equation of motion is

$$m\frac{d\mathbf{v}}{dt} = e\left(\boldsymbol{\varepsilon} + \frac{1}{c}\mathbf{v} \times \boldsymbol{\mathfrak{B}}\right),\qquad(10\text{–}82)$$

and the field intensities are given in terms of the scalar and vector potentials Φ and $\mathbf{A}$ by

$$\boldsymbol{\varepsilon} = -\nabla\Phi - \frac{1}{c}\frac{\partial \mathbf{A}}{\partial t},\qquad \boldsymbol{\mathfrak{B}} = \nabla \times \mathbf{A}.\qquad(10\text{–}83)$$

The equation (10–82) can be written in Lagrangian form, namely,

$$\frac{d}{dt}\frac{\partial \mathcal{L}}{\partial v_i} - \frac{\partial \mathcal{L}}{\partial x_i} = 0 \qquad \begin{array}{l}(\mathbf{r} = [x_1, x_2, x_3]),\\[4pt](\mathbf{v} = [v_1, v_2, v_3]),\end{array}\qquad(10\text{–}84)$$

by inserting Eq. (10–83) into Eq. (10–82) and making use of the vector identity[2]

$$\nabla(\mathbf{v} \cdot \mathbf{A}) = (\mathbf{v} \cdot \nabla)\mathbf{A} + \mathbf{v} \times (\nabla \times \mathbf{A}).\qquad(10\text{–}85)$$

One obtains

$$m\frac{d\mathbf{v}}{dt} = -e\nabla\Phi - \frac{e}{c}\left[\frac{\partial \mathbf{A}}{\partial t} + (\mathbf{v} \cdot \nabla)\mathbf{A}\right] + \frac{e}{c}\nabla(\mathbf{v} \cdot \mathbf{A}).\qquad(10\text{–}86)$$

The quantity

$$\frac{d\mathbf{A}}{dt} = \frac{\partial \mathbf{A}}{\partial t} + (\mathbf{v} \cdot \nabla)\mathbf{A}\qquad(10\text{–}87)$$

---

[1] H. Goldstein, *Classical Mechanics.* Reading, Mass.: Addison-Wesley Publishing Co., Inc., 1950, Section 7–3.
[2] The components of $\mathbf{r}$ and $\mathbf{v}$ are *independent* variables.

is the time rate of change of **A** as viewed from the moving particle, and therefore Eq. (10–86) can be written

$$\frac{d}{dt}\left(m\mathbf{v} + \frac{e}{c}\mathbf{A}\right) = -e\nabla\Phi + \frac{e}{c}\nabla(\mathbf{v}\cdot\mathbf{A}). \qquad (10\text{–}88)$$

Now the last term in this equation is the same as

$$\frac{1}{2m}\nabla\left[\left(m\mathbf{v} + \frac{e}{c}\mathbf{A}\right)^2 - \frac{e^2}{c^2}\mathbf{A}^2\right], \qquad (10\text{–}89)$$

and therefore Eq. (10–88) becomes

$$\frac{d}{dt}\left(m\mathbf{v} + \frac{e}{c}\mathbf{A}\right) - \nabla\left[\frac{1}{2m}\left(m\mathbf{v} + \frac{e}{c}\mathbf{A}\right)^2 - \frac{e^2}{2mc^2}\mathbf{A}^2 - e\Phi\right] = 0, \quad (10\text{–}90)$$

which is in the form of Eq. (10–84). The Lagrangian is

$$\mathcal{L} = \frac{1}{2m}\left(m\mathbf{v} + \frac{e}{c}\mathbf{A}\right)^2 - \frac{e^2}{2mc^2}\mathbf{A}^2 - e\Phi. \qquad (10\text{–}91)$$

The momentum conjugate to **r** is

$$\mathbf{p} = \left[\frac{\partial\mathcal{L}}{\partial\dot{x}_i}\right] = m\mathbf{v} + \frac{e}{c}\mathbf{A}. \qquad (10\text{–}92)$$

The Hamiltonian for the system is a function of **p** and **r** formed according to the rule

$$H = \mathbf{p}\cdot\mathbf{v} - \mathcal{L}. \qquad (10\text{–}93)$$

By substituting **v** from Eq. (10–92),

$$\mathbf{v} = \frac{1}{m}\left(\mathbf{p} - \frac{e}{c}\mathbf{A}\right),$$

and $\mathcal{L}$ from Eq. (10–91), we obtain

$$H(\mathbf{p},\mathbf{r}) = \frac{1}{2m}\left(\mathbf{p} - \frac{e}{c}\mathbf{A}\right)^2 + e\Phi, \qquad (10\text{–}94)$$

which is the Hamiltonian function for the (nonrelativistic) motion of a charged particle in an electromagnetic field. H represents the total energy of the particle, for its magnitude is $\frac{1}{2}mv^2 + e\Phi$.

The requirements of the correspondence principle will be satisfied if we now form the Hamiltonian operator from Eq. (10–94) by the substitutions

$$\mathbf{p} \to \mathbf{p}_{\text{op}} = \frac{\hbar}{i}\nabla; \qquad \mathbf{r} \to \mathbf{r}_{\text{op}}. \qquad (10\text{–}95)$$

The Schrödinger equation is thus

$$H_{\text{op}}\psi = i\hbar \frac{\partial \psi}{\partial t},$$ (10–96)

where

$$H_{\text{op}} = \frac{1}{2m}\left(\frac{\hbar}{i}\nabla - \frac{e}{c}\mathbf{A}\right)^2 + e\Phi.$$ (10–97)

The remainder of the work in this section will be limited to the case of a *static* electromagnetic field, in which H_{op} does not depend upon the time and it may be expected that there are stationary states of fixed energy E. The time-dependence of the wave function for such a state is given by Eq. (5–2), and the Schrödinger equation becomes

$$\frac{1}{2m}\left(\frac{\hbar}{i}\nabla - \frac{e}{c}\mathbf{A}\right)^2 \psi + e\Phi\psi = E\psi.$$ (10–98)

The vector identity

$$\nabla \cdot (\mathbf{A}\psi) = (\nabla \cdot \mathbf{A})\psi + \mathbf{A} \cdot \nabla\psi$$ (10–99)

may be used to rewrite Eq. (10–98) in the more explicit form

$$-\frac{\hbar^2}{2m}\nabla^2\psi + \frac{ie\hbar}{2mc}(\nabla \cdot \mathbf{A})\psi + \frac{ie\hbar}{mc}\mathbf{A} \cdot \nabla\psi + \frac{e^2}{2mc^2}\mathbf{A}^2\psi + e\Phi\psi = E\psi.$$ (10–100)

A uniform magnetic field of induction $\mathfrak{B}$ is described by the vector potential

$$\mathbf{A} = \tfrac{1}{2}\mathfrak{B} \times \mathbf{r},$$ (10–101)

for, since $\mathfrak{B}$ is a constant vector, we have

$$\nabla \times \mathbf{A} = \tfrac{1}{2}\nabla \times (\mathfrak{B} \times \mathbf{r}) = \tfrac{1}{2}[(\nabla \cdot \mathbf{r})\mathfrak{B} - (\mathfrak{B} \cdot \nabla)\mathbf{r}] = \tfrac{1}{2}(3\mathfrak{B} - \mathfrak{B}) = \mathfrak{B}.$$

Also,

$$\nabla \cdot \mathbf{A} = \tfrac{1}{2}(\mathbf{r} \cdot \nabla \times \mathfrak{B} - \mathfrak{B} \cdot \nabla \times \mathbf{r}) = 0,$$

and

$$\frac{ie\hbar}{mc}\mathbf{A} \cdot \nabla\psi = \frac{ie\hbar}{2mc}\mathfrak{B} \times \mathbf{r} \cdot \nabla\psi = -\frac{e}{2mc}\mathfrak{B} \cdot \left(\mathbf{r} \times \frac{\hbar}{i}\nabla\right)\psi$$

$$= -\frac{e}{2mc}\mathfrak{B} \cdot \mathbf{L}\psi,$$ (10–102)

where $\mathbf{L} = \mathbf{r} \times (\hbar/i)\nabla$ is the angular-momentum operator.   Equation

(10–100) is therefore reduced to

$$-\frac{\hbar^2}{2m}\nabla^2\psi - \frac{e}{2mc}\, \mathfrak{B}\cdot \mathbf{L}\psi + \frac{e^2}{8mc^2}(\mathfrak{B}\times\mathbf{r})^2\psi + e\Phi\psi = E\psi, \quad (10\text{–}103)$$

which is the required Schrödinger equation.

The physical meaning of the magnetic terms in this equation can be understood by considering the corresponding terms,

$$-\frac{e}{2mc}\,\mathfrak{B}\cdot\mathbf{L} + \frac{e^2}{8mc^2}(\mathfrak{B}\times\mathbf{r})^2, \quad (10\text{–}104)$$

in the classical Hamiltonian. The first term is the energy, $-\boldsymbol{\mu}\cdot\mathfrak{B}$, of a magnetic dipole of magnitude

$$\boldsymbol{\mu} = \frac{e\mathbf{L}}{2mc}. \quad (10\text{–}105)$$

For example, a particle of charge e and mass m moving in a circle of radius r with speed v has angular momentum $L = mrv$. At the same time, it constitutes a current $i = ev/2\pi r$ which flows around a loop of area $a = \pi r^2$. Consequently, the magnetic moment is

$$\mu = \frac{ia}{c} = \frac{ev}{2\pi r}\frac{\pi r^2}{c} = \frac{eL}{2mc}, \quad (10\text{–}106)$$

in agreement with the above.

The meaning of the first term in Eq. (10–104) is further clarified, from a different point of view, if we consider how the energy is acquired as the magnetic field increases from zero to $\mathfrak{B}$. The changing magnetic field induces an electromotive force in the orbit which is, by Faraday's law,

$$-\frac{1}{c}\frac{d}{dt}(\pi r^2\mathfrak{B}).$$

Now if the electrostatic force which acts to maintain the particle in its orbit is large compared to the magnetic force, r does not change appreciably as $\mathfrak{B}$ increases, and the emf is therefore $-(1/c)\pi r^2\dot{\mathfrak{B}}$, where $\dot{\mathfrak{B}}$ is the rate of increase of $\mathfrak{B}$, which for simplicity is assumed to be constant. Hence the work done per revolution of the particle is $-(e/c)\pi r^2\dot{\mathfrak{B}}$, and since the total number of revolutions in the time t is $vt/2\pi r$, the total work done by the induced electric field is

$$-\frac{e}{c}\pi r^2\dot{\mathfrak{B}}\frac{vt}{2\pi r} = -\frac{e}{2mc}\mathfrak{B}L. \quad (10\text{–}107)$$

Thus, the energy has its origin in the "betatron effect" produced by the changing magnetic field.

The effect of a weak magnetic field upon a particle moving in an electrostatic field can be very clearly described by means of *Larmor's theorem*. The force on such a particle is

$$\mathbf{F} = m\frac{d\mathbf{v}}{dt} = e\left(\boldsymbol{\varepsilon} + \frac{1}{c}\mathbf{v} \times \boldsymbol{\mathfrak{B}}\right). \tag{10--108}$$

In a frame of reference which rotates with constant angular velocity ω, let the apparent velocity and acceleration of the particle be denoted by $\dot{\mathbf{r}}$ and $\ddot{\mathbf{r}}$. Then we have

$$\mathbf{v} = \dot{\mathbf{r}} + \boldsymbol{\omega} \times \mathbf{r},$$

$$\frac{d\mathbf{v}}{dt} = \ddot{\mathbf{r}} + 2\boldsymbol{\omega} \times \dot{\mathbf{r}} + \boldsymbol{\omega} \times (\boldsymbol{\omega} \times \mathbf{r}),$$

and the equation of motion in the moving frame is

$$m[\ddot{\mathbf{r}} + 2\boldsymbol{\omega} \times \dot{\mathbf{r}} + \boldsymbol{\omega} \times (\boldsymbol{\omega} \times \mathbf{r})]$$

$$= e\left[\boldsymbol{\varepsilon} + \frac{1}{c}\dot{\mathbf{r}} \times \boldsymbol{\mathfrak{B}} + \frac{1}{c}(\boldsymbol{\omega} \times \mathbf{r}) \times \boldsymbol{\mathfrak{B}}\right]. \tag{10--109}$$

Let us now choose

$$\boldsymbol{\omega} = -\frac{e}{2mc}\boldsymbol{\mathfrak{B}}, \tag{10--110}$$

i.e., let the moving frame rotate about an axis in the direction of $\boldsymbol{\mathfrak{B}}$ with the angular speed $-e\mathfrak{B}/2mc$. If the magnetic force is sufficiently small compared to the electrostatic force, the last terms in the two members of Eq. (10–109) can be neglected, and we obtain

$$m\ddot{\mathbf{r}} = e\boldsymbol{\varepsilon}. \tag{10--111}$$

In this approximation, therefore, the motion in the rotating frame of reference takes place as it would if the frame were fixed and no magnetic field were present. If, for example, the particle rotates in a circle of radius r with angular velocity ω_0, its apparent kinetic energy is $E_0 = \frac{1}{2}mr^2\omega_0^2$, and the kinetic energy relative to the laboratory coordinates is $E = \frac{1}{2}mr^2(\omega_0 + \omega)^2$. In the approximation considered, namely $\omega \ll \omega_0$, this is

$$E \approx E_0 + mr^2\omega_0\omega. \tag{10--112}$$

The second term is the energy due to the magnetic field, and since $mr^2\omega_0 = L$ is the angular momentum, it amounts to

$$L\omega = -\frac{e}{2mc}\mathfrak{B}L. \tag{10--113}$$

10–6 The classical Zeeman effect. The preceding considerations form the basis for the classical theory of the Zeeman effect. In this theory, it is considered that atoms emit light of frequency ν because electrons within the atoms oscillate with a simple harmonic motion of that frequency. If a weak magnetic field is applied, the line of oscillation of the electrons will rotate about the direction of the field with an angular velocity $\omega = -(e/2mc)\mathfrak{B}$ (Larmor's theorem).

In analyzing the Zeeman effect from the classical point of view, we find it convenient to resolve the assumed simple harmonic motion of an atomic electron into components parallel and perpendicular to the magnetic field. The magnetic field does not affect the frequency of the parallel component of the motion, which continues to emit radiation of frequency ν. The component of the electron's motion perpendicular to the magnetic field can be further resolved into two circular motions of opposite directions, which in the absence of a field have the angular velocities $\pm 2\pi\nu$ (Fig. 10–4). In the presence of a magnetic field, the angular velocity of one of the circular motions is increased by $\omega = e\mathfrak{B}/2mc$, while the angular velocity of the opposite circular motion is decreased by the same amount. Consequently, the circular motions produce radiation of frequencies $\nu \pm \nu_L$, where $\nu_L = \omega/2\pi$ is the Larmor frequency. A weak magnetic field therefore causes a spectral line of frequency ν to split up into a *Lorentz triplet*, consisting of one (undeviated) line of frequency ν and two lines of frequencies $\nu \pm \nu_L$.

The classical theory, furthermore, correctly predicts the polarization of the three lines in a Lorentz triplet. When observed in a direction perpendicular to that of the magnetic field, the undeviated line of frequency

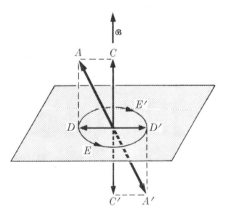

FIG. 10–4. Classical interpretation of the Zeeman effect. Resolution of the simple harmonic motion AA' into a component CC' in the direction of the magnetic field $\mathfrak{B}$ and into a component DD' at right angles to $\mathfrak{B}$. The latter component is resolved into two opposite circular motions E, E', about $\mathfrak{B}$.

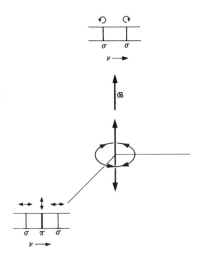

FIG. 10–5.   Polarization of light in the spectral lines produced by the Zeeman effect.

ν is linearly polarized, with its electric vector parallel to $\mathfrak{B}$ (π-component). This is indicated in Fig. 10–5. The displaced lines are linearly polarized in the direction normal to $\mathfrak{B}$ (σ-components). If, however, the light is viewed along the direction of $\mathfrak{B}$, the σ-components appear circularly polarized in opposite directions, while the π-component is not seen at all, since an oscillating charge does not radiate in the direction of its line of motion.

According to the classical theory, each spectral line of an atomic system should exhibit the *normal* Zeeman effect. However, many spectral lines are observed to have more than three Zeeman components, and the separation between components is frequently different from ν_L. In Section 11–12, it will be shown how a complete understanding of the *anomalous* Zeeman effect can be obtained from the quantum theory of the electron spin.

We shall now consider briefly the second term in Eq. (10–104) for the magnetic energy, which gives rise to the so-called *quadratic Zeeman effect*. This term also receives a simple interpretation if the motion is viewed in the rotating Lorentz frame. Thus, if terms of order $\mathfrak{B}^2$ are retained, the equation of motion in the moving frame becomes

$$m\ddot{\mathbf{r}} = e\boldsymbol{\mathcal{E}} + \frac{e}{c}\,(\boldsymbol{\omega} \times \mathbf{r}) \times \mathfrak{B} - m\boldsymbol{\omega} \times (\boldsymbol{\omega} \times \mathbf{r})$$

$$= e\boldsymbol{\mathcal{E}} - \frac{e^2}{4mc^2}\,(\mathfrak{B} \times \mathbf{r}) \times \mathfrak{B},$$

or, if ρ denotes the component of $\mathbf{r}$ perpendicular to $\mathfrak{B}$,

$$m\ddot{\mathbf{r}} = e\mathbf{\mathcal{E}} - \frac{e^2}{4mc^2}\mathfrak{B}^2\rho. \tag{10–114}$$

The last term in this expression represents a harmonic force $-k\rho$ in the plane normal to $\mathfrak{B}$, and the corresponding potential energy is

$$\tfrac{1}{2}k\rho^2 = \frac{e^2}{8mc^2}\mathfrak{B}^2\rho^2 = \frac{e^2}{8mc^2}(\mathfrak{B}\times\mathbf{r})^2, \tag{10–115}$$

which is exactly the second term in Eq. (10–104). Hence, this part of the magnetic energy can be represented by a potential-energy function. Indeed, the Larmor frame has special significance since, in this frame of reference, the apparent force is conservative, i.e., not velocity-dependent.

The energy given by Eq. (10–115) is negligibly small for most systems that can be realized experimentally. Compared with the linear term, the order of magnitude of this energy is

$$\frac{(e^2/8mc^2)\mathfrak{B}^2 r^2}{(e/2mc)\mathfrak{B}L} = \frac{(e^2\mathfrak{B}/4Lc)}{e/r^2}.$$

Since the angular momentum for an atomic electron is of order $\hbar$, while r is normally not much larger than the Bohr radius, this ratio is approximately

$$\frac{\mathfrak{B}}{548(e/a_0^2)} = \frac{\mathfrak{B}}{9\times10^9\text{ gauss}}. \tag{10–116}$$

Consequently, an extremely intense field is required if the quadratic effect is to be observed at all, and even then it is appreciable only for electronic orbits of unusually large extent. The quadratic Zeeman effect has actually been studied[1] in the absorption spectra of alkali metals, in which transitions to very highly excited states ($n = 20$ or 30) were observed in magnetic fields up to 27,000 gauss. However, except in very special circumstances, the quadratic effect can be disregarded.

10–7 Quantum theory of the normal Zeeman effect. The quantum states of energy E_0 for a particle in a spherically symmetric field are characterized by wave functions of the form (Section 7–1)

$$\psi = R(r)Y_l^m(\theta, \phi), \tag{10–117}$$

---

[1] F. A. Jenkins and E. Segrè, *Phys. Rev.* **55**, 52 (1939).

which are eigenfunctions of the Schrödinger equation

$$H_0\psi = E_0\psi, \qquad (10\text{--}118)$$

where

$$H_0 = -\frac{\hbar^2}{2m}\nabla^2 + e\Phi, \qquad (10\text{--}119)$$

and $e\Phi$ is the electrostatic potential energy of the particle. If this system is subjected to a magnetic field, the Schrödinger equation becomes

$$H\psi = H_0\psi - \frac{e}{2mc}\,\mathfrak{B}\cdot\mathbf{L}\psi = E\psi, \qquad (10\text{--}120)$$

where the quadratic term in Eq. (10–103) has been neglected. The direction of the z-axis is arbitrary and can be chosen parallel to $\mathfrak{B}$. It is customary to introduce, as a fundamental unit for magnetic moment, the *Bohr magneton*

$$\mu_0 = \frac{e\hbar}{2mc} = -0.9273 \times 10^{-20} \text{ erg/gauss.} \qquad (10\text{--}121)$$

The Schrödinger equation then becomes

$$H\psi = H_0\psi - \mu_0\mathfrak{B}L_z\psi = E\psi, \qquad (10\text{--}122)$$

where the z-component of angular momentum is measured in the unit $\hbar$. It is apparent that the wave function (10–117) is also a solution of this new Schrödinger equation. Since $L_z Y_l^m = mY_l^m$, we have

$$H\psi = (E_0 - m\mu_0\mathfrak{B})\psi = E\psi. \qquad (10\text{--}123)$$

The energy associated with this state is therefore

$$E = E_0 - m\mu_0\mathfrak{B}. \qquad (10\text{--}124)$$

This shows that the degenerate states associated with the orbital angular momentum l are separated by the magnetic field into $2l + 1$ distinct components, which are equally spaced on the energy scale.

The frequencies of the lines emitted in transitions between two states of the system are given by

$$\nu = \frac{E_1 - E_2}{h} = \frac{E_{01} - E_{02}}{h} - (m_1 - m_2)\frac{\mu_0\mathfrak{B}}{h}, \qquad (10\text{--}125)$$

or

$$\nu = \nu_0 - \frac{\mu_0\mathfrak{B}}{h}\,\Delta m,$$

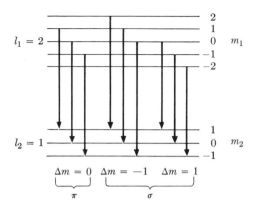

FIG. 10–6. Possible transitions between two levels with $l_1 = 2$ and $l_2 = 1$.

where $\Delta m = m_1 - m_2$ is the change of the *magnetic quantum number* m in the transition.

It will be shown in Section 11–11 that the transitions obey the selection rule

$$\Delta m = \pm 1 \quad \text{or} \quad 0, \tag{10–126}$$

and that the light emitted in transitions for which $\Delta m = 0$ is polarized parallel to $\mathfrak{B}$, while a change of m corresponds to polarization perpendicular to $\mathfrak{B}$. Thus we are led to the rules

$$\nu = \nu_0 \quad (\pi\text{-component}),$$

$$\nu = \nu_0 \pm \nu_L \quad (\sigma\text{-components}),$$

which are identical with those obtained classically by means of the Larmor theorem. A schematic diagram of the possible transitions between two levels with $l_1 = 2$ and $l_2 = 1$ is shown in Fig. 10–6.

10–8 Electron spin. The theory developed so far does not provide a correct description of the observed electronic states of an atom. It is not possible to specify the multiplicity of atomic states solely in terms of the orbital angular momentum of the electron. For instance, analysis of the spectra of the alkali metals in the absence of an external field shows that, while the s-terms are simple, the p-, d-, etc., terms are double. Thus, the D-line in the principal series of sodium is a doublet, split by about 6 cm^{-1}.

Furthermore, in a Stern-Gerlach experiment, a beam of neutral hydrogen atoms is split into two components, with equal and opposite deflections from the undeviated beam. Since the single electron in these atoms has zero angular momentum in the ground state, the magnetic quantum number must be $m = 0$, and no deflection at all is expected. The observed deviation, however, corresponds to a magnetic moment of exactly one Bohr magneton and hence cannot be due to the much smaller magnetic moment of the proton.

Finally, an even number of Zeeman levels is observed in many spectra. However, the quantum number of the orbital angular momentum is always an integer, so that the multiplicity $(2l + 1)$ is always odd.

These difficulties are resolved by the hypothesis[1] that an electron possesses an intrinsic angular momentum or *spin* $s = \frac{1}{2}\hbar$ and an intrinsic magnetic moment $\mu_s = e\hbar/2mc$, not associated with the orbital motion. The ratio of intrinsic magnetic moment to spin is

$$\frac{\mu_s}{s} = 2\,\frac{e}{2mc}, \tag{10-127}$$

which is twice the corresponding ratio for orbital motion, as expressed by Eq. (10–105).

In the early history of the electron-spin hypothesis, attempts were made to endow the electron with a dynamical structure that would lead to the postulated spin and intrinsic magnetic moment. All such attempts failed, however, mainly because of the abnormally large value of the magnetic moment. In simple terms, one might suppose that the electronic charge e circulates in a more or less spherical region of radius r. Such a moving charge would constitute a current of order of magnitude $ev/2\pi r$, where v is the velocity. This velocity, however, would have to be such that

$$\mu_s = \frac{1}{c}\,ia \approx \frac{1}{c}\,\frac{ev}{2\pi r}\,\pi r^2 = \frac{e\hbar}{2mc},$$

or

$$\frac{v}{c} \approx \frac{\hbar/mc}{r}. \tag{10-128}$$

Now, according to the classical theory of the electron, the radius is

$$r \approx \frac{e^2}{mc^2},$$

whence

$$\frac{v}{c} \approx \frac{\hbar c}{e^2} = 137;$$

this contradicts the requirements of special relativity. Consequently, no classical model can succeed in predicting the correct magnetic moment. Indeed, it has been shown by Dirac that the spin magnetic moment is a

[1] G. E. Uhlenbeck and S. Goudsmit, *Naturwissenschaften* **13**, 953 (1925) and *Nature* **117**, 264 (1926). F. R. Bichowsky and H. C. Urey, *Proc. Nat. Acad. Sci.*, **12**, 80 (1926).

purely relativistic property of the electron. Moreover, this magnetic moment arises from quantum effects which take place within a radius of order $\hbar/mc$ in the neighborhood of the position of the electron. Thus, the estimate (10–128) becomes

$$\frac{v}{c} \approx 1,$$

in conformity with the relativistic nature of the phenomenon. For the purposes of atomic theory, however, the question of the electronic structure can in general be ignored and the point-charge concept retained, without special inquiry into the origin of the spin.

The electron-spin hypothesis can be immediately incorporated into the theory of angular momentum developed in the early sections of this chapter. It is customary to denote the intrinsic spin angular momentum by **s** rather than by **J**. Since $j = s = \frac{1}{2}$ (in units $\hbar$), the components of **s** are represented by the matrices in the second line of Eqs. (10–80). The states of the electron cannot be represented simply in terms of functions $\psi(\mathbf{r})$ which depend only on the coordinates. Such a representation is not complete; the number of degrees of freedom of the electron has been increased (to four) by the introduction of the new observable **s**, and consequently a fourth coordinate is required for the complete specification of a state. This new coordinate is the eigenvalue of the observable s_z (or of any other component of **s**.) Hence, a component of the state vector, in the representation in which x, y, z, and s_z are diagonal, is a function of four variables, i.e.,

$$\psi = \psi(x, y, z, m_s). \qquad (10\text{–}129)$$

The coordinate m_s ranges over the values $\pm\frac{1}{2}$, the eigenvalues of s_z. There is, in reality, a fifth coordinate s, but since this has only the single value $\frac{1}{2}$, it can be suppressed.

The spin is an independent observable for the electron. Its components commute with position and momentum components. Consequently, every operator which depends upon the spatial variables alone has a two-fold degeneracy with respect to s_z. The states of hydrogen, for example, are doubly degenerate insofar as the energy is independent of s_z. The terms of the alkali spectra, on the other hand, are closely spaced doublets, showing that for these systems the energy (and therefore the Hamiltonian operator) is slightly dependent upon the direction of the electron spin.

We shall now consider in detail the representation of the spin states of an electron. The matrices representing the components of **s** have been shown to be

$$s_z = \begin{pmatrix} \frac{1}{2} & 0 \\ 0 & -\frac{1}{2} \end{pmatrix}, \qquad s_+ = \begin{pmatrix} 0 & 1 \\ 0 & 0 \end{pmatrix}, \qquad s_- = \begin{pmatrix} 0 & 0 \\ 1 & 0 \end{pmatrix}, \qquad (10\text{–}130)$$

whence

$$s_x = \tfrac{1}{2}(s_+ + s_-) = \begin{pmatrix} 0 & \frac{1}{2} \\ \frac{1}{2} & 0 \end{pmatrix},$$

$$s_y = \frac{1}{2i}(s_+ - s_-) = \begin{pmatrix} 0 & -\frac{i}{2} \\ \frac{i}{2} & 0 \end{pmatrix},$$

and

$$\mathbf{s}^2 = s_x^2 + s_y^2 + s_z^2 = \begin{pmatrix} \frac{3}{4} & 0 \\ 0 & \frac{3}{4} \end{pmatrix} = s(s+1) \cdot 1.$$

← unit matrix.

The eigenvectors $\phi(\tfrac{1}{2}, \tfrac{1}{2})$ and $\phi(-\tfrac{1}{2}, \tfrac{1}{2})$ of $\mathbf{s}^2$ and s_z will be denoted, respectively, by χ^+ and χ^-. They are evidently

$$\chi^+ = \begin{pmatrix} 1 \\ 0 \end{pmatrix}, \qquad \chi^- = \begin{pmatrix} 0 \\ 1 \end{pmatrix}, \tag{10–131}$$

and represent states in which the z-component of the spin is parallel $(m_s = \tfrac{1}{2})$ and anti-parallel $(m_s = -\tfrac{1}{2})$ to the z-axis.

It is also convenient to introduce the matrix $\boldsymbol{\sigma} = 2\mathbf{s}$, with components

$$\sigma_x = \begin{pmatrix} 0 & 1 \\ 1 & 0 \end{pmatrix}, \qquad \sigma_y = \begin{pmatrix} 0 & -i \\ i & 0 \end{pmatrix}, \qquad \sigma_z = \begin{pmatrix} 1 & 0 \\ 0 & -1 \end{pmatrix}, \tag{10–132}$$

which are called the *Pauli spin matrices* (cf. Problem 9–17). They satisfy the relations

$$\sigma_x^2 = \sigma_y^2 = \sigma_z^2 = 1, \qquad \sigma_x\sigma_y = i\sigma_z, \qquad \sigma_y\sigma_z = i\sigma_x, \qquad \sigma_z\sigma_x = i\sigma_y$$

$$\sigma_x\sigma_y + \sigma_y\sigma_x = \sigma_y\sigma_z + \sigma_z\sigma_y = \sigma_z\sigma_x + \sigma_x\sigma_z = 0, \tag{10–133}$$

or, in summary, $\boldsymbol{\sigma} \cdot \mathbf{a}\, \boldsymbol{\sigma} \cdot \mathbf{b} = \mathbf{a} \cdot \mathbf{b} + i\boldsymbol{\sigma} \cdot \mathbf{a} \times \mathbf{b}$, where $\mathbf{a}$ and $\mathbf{b}$ are arbitrary vectors.

As an exercise in the manipulation of the Pauli matrices, we shall work out the eigenvectors of the operator $\boldsymbol{\sigma} \cdot \hat{\mathbf{n}}$, where $\hat{\mathbf{n}}$ is a unit vector in an arbitrary direction in space. It is easily verified, by means of the identities (10–133), that

$$(\boldsymbol{\sigma} \cdot \hat{\mathbf{n}})^2 = (\sigma_x n_x + \sigma_y n_y + \sigma_z n_z)^2 = 1. \tag{10–134}$$

The eigenvalues of $\boldsymbol{\sigma} \cdot \hat{\mathbf{n}}$ are therefore ± 1. The corresponding eigenvectors, which we denote by χ_n^+ and χ_n^-, can now be found by solving the algebraic equations

$$\boldsymbol{\sigma} \cdot \hat{\mathbf{n}}\chi_n^{\pm} = \pm\chi_n^{\pm}.$$

However, relation (10–134) provides the basis for a simpler procedure (cf. Section 9–12). We construct the vectors

$$\tfrac{1}{2}(1 + \boldsymbol{\sigma} \cdot \hat{\mathbf{n}})\chi, \qquad \tfrac{1}{2}(1 - \boldsymbol{\sigma} \cdot \hat{\mathbf{n}})\chi,$$

where $\chi = a_+\chi^+ + a_-\chi^-$ is an arbitrary spin vector. If the operator $\boldsymbol{\sigma} \cdot \hat{\mathbf{n}}$ is applied to each of these vectors, Eq. (10–134) yields

$$\boldsymbol{\sigma} \cdot \hat{\mathbf{n}}\tfrac{1}{2}(1 + \boldsymbol{\sigma} \cdot \hat{\mathbf{n}})\chi = \tfrac{1}{2}(\boldsymbol{\sigma} \cdot \hat{\mathbf{n}} + 1)\chi = \tfrac{1}{2}(1 + \boldsymbol{\sigma} \cdot \hat{\mathbf{n}})\chi,$$

$$\boldsymbol{\sigma} \cdot \hat{\mathbf{n}}\tfrac{1}{2}(1 - \boldsymbol{\sigma} \cdot \hat{\mathbf{n}})\chi = \tfrac{1}{2}(\boldsymbol{\sigma} \cdot \hat{\mathbf{n}} - 1)\chi = -\tfrac{1}{2}(1 - \boldsymbol{\sigma} \cdot \hat{\mathbf{n}})\chi.$$

These quantities are therefore the desired solutions of Eq. (10–134). We may, for example, take $\chi = \chi^+$ and find, by means of the rules

$$\sigma_x\chi^\pm = \chi^\mp, \qquad \sigma_y\chi^\pm = \pm i\chi^\mp, \qquad \sigma_z\chi^\pm = \pm\chi^\pm, \quad (10\text{–}135)$$

the vectors

$$\tfrac{1}{2}(1 + \boldsymbol{\sigma} \cdot \hat{\mathbf{n}})\chi^+ = \tfrac{1}{2}(1 + \sigma_x n_x + \sigma_y n_y + \sigma_z n_z)\chi^+$$
$$= \tfrac{1}{2}\{(1 + n_z)\chi^+ + (n_x + in_y)\chi^-\},$$

$$\tfrac{1}{2}(1 + \boldsymbol{\sigma} \cdot \hat{\mathbf{n}})\chi^- = \tfrac{1}{2}(1 - \sigma_x n_x - \sigma_y n_y - \sigma_z n_z)\chi^+$$
$$= \tfrac{1}{2}\{(1 - n_z)\chi^+ - (n_x + in_y)\chi^-\}.$$

The squares of the norms of these vectors are

$$\tfrac{1}{4}(1 + 2n_z + n_z^2 + n_x^2 + n_y^2) = \tfrac{1}{2}(1 + n_z),$$

$$\tfrac{1}{4}(1 - 2n_z + n_z^2 + n_x^2 + n_y^2) = \tfrac{1}{2}(1 - n_z),$$

whence the required normalized eigenfunctions are

$$\chi_n^+ = \frac{1}{\sqrt{2(1 + n_z)}}\{(1 + n_z)\chi^+ + (n_x + in_y)\chi^-\},$$
$$\tag{10–136}$$
$$\chi_n^- = \frac{1}{\sqrt{2(1 - n_z)}}\{(1 - n_z)\chi^+ - (n_x + in_y)\chi^-\}.$$

In the state χ_n^+, for example, a measurement of the z-component of the spin will yield the values $\pm\tfrac{1}{2}$ with the relative probabilities

$$(1 + n_z)^2 : (n_x^2 + n_y^2) = (1 + \cos\theta)^2 : (1 - \cos^2\theta)$$

$$= \cos^2\frac{\theta}{2} : \sin^2\frac{\theta}{2}, \tag{10–137}$$

where θ is the angle between $\hat{\mathbf{n}}$ and z. This result, which gives an intuitively reasonable picture of the probability distribution, was first obtained by Pauli.

10–9 Electronic states in a central field. Consider an electron in a state of orbital angular momentum l. The dependence of the wave function on the angular coordinates θ, ϕ of the electron can be expressed by means of a linear combination of the spherical harmonics $Y_l^{m_l}$. The quantum number associated with the orbital motion is now written "m_l" to distinguish it from the spin quantum number "m_s." The spin, on the other hand, defines a complete set of spin functions $\chi(m_s)$, namely,

$$\chi(\tfrac{1}{2}) = \chi^+, \qquad \chi(-\tfrac{1}{2}) = \chi^-. \qquad (10\text{–}138)$$

From these, we may construct the functions

$$\psi(lm_lm_s) = Y_l^{m_l}\chi(m_s) \qquad (10\text{–}139)$$

which define a *product space* for the state vectors of the electron. The function $\psi(lm_lm_s)$ is simultaneously an eigenfunction of $\mathbf{L}^2$, L_z, $\mathbf{s}^2$, and s_z, according to

$$\mathbf{L}^2\psi = l(l+1)\psi, \qquad L_z\psi = m_l\psi, \qquad \mathbf{s}^2\psi = \tfrac{3}{4}\psi, \qquad s_z\psi = m_s\psi. \tag{10–140}$$

It is evident that there are $2(2l+1)$ linearly independent functions of this form.

Now under a rotation of the coordinate frame, the functions Y and χ are transformed according to

$$Y' = \exp{(i\boldsymbol{\omega}\cdot\mathbf{L})}Y, \qquad \chi' = \exp{(i\boldsymbol{\omega}\cdot\mathbf{s})}\chi; \qquad (10\text{–}141)$$

hence the transformation law for ψ is

$$\psi' = Y'\chi' = \exp{(i\boldsymbol{\omega}\cdot\mathbf{L})}Y\exp{(i\boldsymbol{\omega}\cdot\mathbf{s})}\chi = \exp{[i\boldsymbol{\omega}\cdot(\mathbf{L}+\mathbf{s})]}Y\chi, \quad (10\text{–}142)$$

or

$$\psi' = \exp{(i\boldsymbol{\omega}\cdot\mathbf{J})}\psi, \qquad (10\text{–}143)$$

where $\mathbf{J} = \mathbf{L} + \mathbf{s}$ is the total angular-momentum operator. Thus, $\mathbf{J}$ is the infinitesimal rotation operator in the state space of the electron.

The operators $\mathbf{L}^2$, L_z, $\mathbf{s}^2$, s_z are a set of commuting operators for the electron, and if the Hamiltonian commutes with all these, it may be adjoined to form a complete set. The functions $\psi(lm_lm_s)$ will then be eigenstates of the Hamiltonian and form a representation of the stationary states of the electron. However, if the Hamiltonian contains a part which does not commute with $\mathbf{L}$ and $\mathbf{s}$ separately, e.g., if there is "spin-orbit coupling," then the above operators cannot all be simultaneously diagonal. Nevertheless, if the system is isolated, the Hamiltonian must be invariant to rotations of the coordinate frame, which means that H and $\mathbf{J}$ must commute. Consequently, H, $\mathbf{J}^2$ and J_z may always be expected to be a commuting set, and it is of advantage to construct simultaneous eigenfunctions for these operators.

This is a typical example of the problem presented by the addition of two commuting angular momenta, which we shall now solve in general terms.

10–10 Addition of angular momenta. Consider two noninteracting systems (1) and (2) with angular momenta $\mathbf{J}_1$ and $\mathbf{J}_2$ and state vectors ψ_1 and ψ_2. A rotation of the coordinate frame induces the transformations

$$\psi_1' = \exp\,(i\boldsymbol{\omega}\cdot\mathbf{J}_1)\psi_1, \qquad \psi_2' = \exp\,(i\boldsymbol{\omega}\cdot\mathbf{J}_2)\psi_2,$$

in the state spaces of the two systems. If the systems (1) and (2) are now considered together as a single system, the state vector of the combination is the product

$$\psi = \psi_1\psi_2, \tag{10–144}$$

and the effect of a rotation is described by

$$\psi' = \exp\,(i\boldsymbol{\omega}\cdot\mathbf{J})\psi, \tag{10–145}$$

where

$$\mathbf{J} = \mathbf{J}_1 + \mathbf{J}_2 \tag{10–146}$$

is the total angular momentum. $\mathbf{J}$ is thus the infinitesimal rotation operator in the product space.

$\mathbf{J}$ is an angular-momentum operator in the generalized sense of Sections 10–2 and 10–3, for, from the commutation rules for $\mathbf{J}_1$ and $\mathbf{J}_2$, i.e.,

$$\mathbf{J}_1 \times \mathbf{J}_1 = i\mathbf{J}_1, \qquad \mathbf{J}_2 \times \mathbf{J}_2 = i\mathbf{J}_2,$$

we obtain

$$\mathbf{J} \times \mathbf{J} = \mathbf{J}_1 \times \mathbf{J}_1 + \mathbf{J}_1 \times \mathbf{J}_2 + \mathbf{J}_2 \times \mathbf{J}_1 + \mathbf{J}_2 \times \mathbf{J}_2$$
$$= \mathbf{J}_1 \times \mathbf{J}_1 + \mathbf{J}_2 \times \mathbf{J}_2 = i(\mathbf{J}_1 + \mathbf{J}_2) = i\mathbf{J},$$

in which the second equation follows from the fact that $\mathbf{J}_1$ commutes with $\mathbf{J}_2$.

Now suppose that we have a complete set of eigenvectors of $\mathbf{J}_1^2$ and J_{1z} and a similar complete set for system (2):

$$\mathbf{J}_1^2\psi_1(j_1\,m_1) = j_1(j_1 + 1)\psi_1(j_1\,m_1),$$

$$J_{1z}\psi_1(j_1\,m_1) = m_1\psi_1(j_1\,m_1),$$

$$\mathbf{J}_2^2\psi_2(j_2\,m_2) = j_2(j_2 + 1)\psi_2(j_2\,m_2), \tag{10–147}$$

$$J_{2z}\psi_2(j_2\,m_2) = m_2\psi_2(j_2\,m_2).$$

The products

$$\psi(j_1 j_2 m_1 m_2) = \psi_1(j_1 m_1)\psi_2(j_2 m_2) \qquad (10\text{–}148)$$

are then a complete set of $(2j_1 + 1)(2j_2 + 1)$ vectors in the space of the combined system. Also, these vectors are simultaneous eigenvectors of $\mathbf{J}_1^2$ and $\mathbf{J}_2^2$, but not of $\mathbf{J}^2$. However, since the set is complete, it is possible to form linear combinations which are eigenvectors of $\mathbf{J}^2$ and J_z. We will denote these by

$$\Phi(j_1 j_2 j m)$$

and show how to construct the unitary transformation from the $(j_1 j_2 m_1 m_2)$-representation to the $(j_1 j_2 j m)$-representation. (For definiteness, we assume $j_1 \geq j_2$.)

We begin by noting that $\psi(j_1 j_2 m_1 m_2)$ is an eigenvector of J_z belonging to the eigenvalue $m_1 + m_2$:

$$
\begin{aligned}
J_z\psi(j_1 j_2 m_1 m_2) &= (J_{1z} + J_{2z})\psi_1(j_1 m_1)\psi_2(j_2 m_2) \\
&= (m_1 + m_2)\psi_1(j_1 m_1)\psi_2(j_2 m_2) \\
&= (m_1 + m_2)\psi(j_1 j_2 m_1 m_2). \qquad (10\text{–}149)
\end{aligned}
$$

The numbers j_1 and j_2 are, respectively, the largest values of m_1 and m_2. Therefore, the largest possible value for m, the eigenvalue of J_z, is $j_1 + j_2$, and there is only one such function among the $\psi(j_1 j_2 m_1 m_2)$, namely,

$$\Phi(j_1 j_2 \quad j_1{+}j_2 \quad j_1{+}j_2) = \psi(j_1 j_2 j_1 j_2). \qquad (10\text{–}150)$$

Now there are $2(j_1 + j_2) + 1$ linearly independent eigenvectors of $\mathbf{J}^2$ belonging to $j = j_1 + j_2$, and clearly the eigenvector belonging to $m = j_1 + j_2 - 1$ is a linear combination of

$$\psi(j_1 j_2 \quad j_1{-}1 \quad j_2) \qquad \text{and} \qquad \psi(j_1 j_2 j_1 \quad j_2{-}1),$$

since these are the only ψ's for which $m_1 + m_2 = j_1 + j_2 - 1$. There is a second orthogonal combination of these two ψ's which also belongs to $m = j_1 + j_2 - 1$, and this can only be an eigenvector of $\mathbf{J}^2$ belonging to $j = j_1 + j_2 - 1$, i.e., it is the vector

$$\Phi(j_1 j_2 \quad j_1{+}j_2{-}1 \quad j_1{+}j_2{-}1). \qquad (10\text{–}151)$$

Proceeding to the next step, it is seen that by forming linear combinations of the three vectors

$$\psi(j_1 j_2 \quad j_1{-}2 \quad j_2), \qquad \psi(j_1 j_2 \quad j_1{-}1 \quad j_2{-}1), \qquad \psi(j_1 j_2 j_1 \quad j_2{-}2),$$

we obtain the vectors

$$\Phi(j_1 j_2 \quad j_1+j_2-2), \qquad \Phi(j_1 j_2 \quad j_1+j_2-1 \quad j_1+j_2-2) \left.\begin{array}{r}\\ \\ \\ \end{array}\right\}$$

and

$$\Phi(j_1 j_2 \quad j_1+j_2-2 \quad j_1+j_2-2), \qquad\qquad (10\text{–}152)$$

the last of which is the first in a set of $2(j_1 + j_2 - 2) + 1$ eigenvectors belonging to $j = j_1 + j_2 - 2$. Continuing in this way, we have at the k'th step the $k + 1$ eigenvectors

$$
\begin{aligned}
&\Phi(j_1 j_2 \quad j_1+j_2 \quad j_1+j_2-k), \\
&\Phi(j_1 j_2 \quad j_1+j_2-1 \quad j_1+j_2-k), \\
&\vdots \\
&\Phi(j_1 j_2 \quad j_1+j_2-k \quad j_1+j_2-k).
\end{aligned}
\qquad (10\text{–}153)
$$

Since the least value of m_2 is $-j_2$, the smallest value of j occurs for $j_2 - k = -j_2$, or $k = 2j_2$, the corresponding value of j being $j_1 + j_2 - 2j_2 = j_1 - j_2$. Hence, we obtain $2j + 1$ independent vectors for each of the values $j = j_1 + j_2, j_1 + j_2 - 1, j_1 + j_2 - 2, \ldots, j_1 - j_2$, in agreement with the "vector coupling rule" of spectroscopy. The total number of eigenstates of $\mathbf{J}^2$ is therefore

$$\sum_{j=j_1-j_2}^{j=j_1+j_2} (2j + 1) = (2j_1 + 1)(2j_2 + 1), \qquad (10\text{–}154)$$

which is, of course, the same as the number of independent ψ's.

The elements of the unitary matrix connecting the two representations are conventionally denoted by the symbol $(j_1 j_2 m_1 m_2 | j_1 j_2 jm)$ (Section 9–11), so that we have explicitly

$$\Phi(j_1 j_2 jm) = \sum_{m_1+m_2=m} (j_1 j_2 m_1 m_2 | j_1 j_2 jm)\psi(j_1 j_2 m_1 m_2). \quad (10\text{–}155)$$

The values of the coefficients in this expansion are completely determined by the above considerations, and their explicit form can be derived most conveniently by the methods of group theory.[1] However, the general formula is too cumbersome to be of value in most practical cases, and a method based upon the properties of the angular-momentum operator is more frequently employed.[2] This method will now be illustrated by examples.

---

[1] E. Wigner, *Group Theory and its Applications to the Theory of Atomic Spectra.* New York: Academic Press, 1959.

[2] N. M. Gray and L. A. Wills, *Phys. Rev.* **38**, 248 (1931). E. U. Condon and G. H. Shortley, *The Theory of Atomic Spectra.* Cambridge: Cambridge University Press, 1951, p. 226 ff.; see also Section 14^3.

10–11 The p-states of an electron. The orbital functions for the $l = 1$ states of a single electron are the three spherical harmonics Y_1^1, Y_1^0, and Y_1^{-1}, and the spin states are $\chi(\frac{1}{2}) = \chi^+$ and $\chi(-\frac{1}{2}) = \chi^-$. The total angular momentum of the system has the two possible values $l + s = \frac{3}{2}$ and $l - s = \frac{1}{2}$. If we drop the quantum numbers j_1, j_2, the state for $j = \frac{3}{2}$, $m = \frac{3}{2}$ can be written

$$\Phi(\tfrac{3}{2}, \tfrac{3}{2}) = Y_1^1\chi^+. \tag{10-156}$$

In order to construct the remainder of the $j = \frac{3}{2}$ states, we make use of the properties of the lowering operator

$$J_- = L_- + s_-, \tag{10-157}$$

for which [Eq. (10–70)]

$$J_-\Phi(\tfrac{3}{2}, m) = \sqrt{\tfrac{15}{4} - m(m - 1)}\ \Phi(\tfrac{3}{2}, m - 1). \tag{10-158}$$

Also,

$$L_-Y_1^{m_l} = \sqrt{2 - m_l(m_l - 1)}\ Y_1^{m_l-1}, \tag{10-159}$$

$$s_-\chi(m_s) = \sqrt{\tfrac{3}{4} - m_s(m_s - 1)}\ \chi(m_s - 1). \tag{10-160}$$

It follows that

$$J_-\Phi(\tfrac{3}{2}, \tfrac{3}{2}) = \sqrt{3}\ \Phi(\tfrac{3}{2}, \tfrac{1}{2}) = \sqrt{2}\ Y_1^0\chi^+ + Y_1^1\chi^-,$$

or

$$\Phi(\tfrac{3}{2}, \tfrac{1}{2}) = \frac{1}{\sqrt{3}}(\sqrt{2}\ Y_1^0\chi^+ + Y_1^1\chi^-). \tag{10-161}$$

Applying the lowering operator again, we have

$$\Phi(\tfrac{3}{2}, -\tfrac{1}{2}) = \frac{1}{\sqrt{3}}(Y_1^{-1}\chi^+ + \sqrt{2}\ Y_1^0\chi^-), \tag{10-162}$$

and

$$\Phi(\tfrac{3}{2}, -\tfrac{3}{2}) = Y_1^{-1}\chi^-. \tag{10-163}$$

The state $\Phi(\frac{1}{2}, \frac{1}{2})$ is a linear combination of $Y_1^0\chi^+$ and $Y_1^1\chi^-$, which is orthogonal to $\Phi(\frac{3}{2}, \frac{1}{2})$. Such a combination is easily formed by inspection of Eq. (10–161), that is,

$$\Phi(\tfrac{1}{2}, \tfrac{1}{2}) = \frac{1}{\sqrt{3}}(Y_1^0\chi^+ - \sqrt{2}\ Y_1^1\chi^-), \tag{10-164}$$

and the lowering operator can be applied once more to give

$$\Phi(\tfrac{1}{2}, -\tfrac{1}{2}) = \frac{1}{\sqrt{3}}(\sqrt{2}\ Y_1^{-1}\chi^+ - Y_1^0\chi^-), \tag{10-165}$$

which completes the analysis. All six of the states $\Phi(jm)$ are seen to be normalized and mutually orthogonal.

In the absence of a perturbation which destroys the spherical symmetry, the p-state corresponding to $j = \frac{3}{2}$ is four-fold degenerate with respect to the energy, while the degree of degeneracy of the state corresponding to $j = \frac{1}{2}$ is two. However, if the energy depends upon the relative orientation of $\mathbf{l}$ and $\mathbf{s}$, then the two sets of states belong to different values of E. An effect of this kind accounts for the separation of the D-lines of the sodium spectrum.

10–12 Spin states for two particles of spin one-half. As a second example of the general theory we shall discuss the system composed of two particles of spin $\frac{1}{2}$. The spin functions for the two particles will be denoted by $\chi_1(m_1)$ and $\chi_2(m_2)$. The vectors

$$\psi(m_1 m_2) = \chi_1(m_1)\chi_2(m_2) \tag{10–166}$$

describe the spin states of the combined system. They can be written as column vectors, i.e.,

$$\psi(\tfrac{1}{2}, \tfrac{1}{2}) = \begin{pmatrix} 1 \\ 0 \\ 0 \\ 0 \end{pmatrix}, \quad \psi(\tfrac{1}{2}, -\tfrac{1}{2}) = \begin{pmatrix} 0 \\ 1 \\ 0 \\ 0 \end{pmatrix}, \quad \psi(-\tfrac{1}{2}, \tfrac{1}{2}) = \begin{pmatrix} 0 \\ 0 \\ 1 \\ 0 \end{pmatrix}, \quad \psi(-\tfrac{1}{2}, -\tfrac{1}{2}) = \begin{pmatrix} 0 \\ 0 \\ 0 \\ 1 \end{pmatrix},$$

$$\tag{10–167}$$

in which the elements are numbered according to the following scheme:

$$\begin{array}{cc} m_1 & m_2 \\ \frac{1}{2} & \frac{1}{2} \\ \frac{1}{2} & -\frac{1}{2} \\ -\frac{1}{2} & \frac{1}{2} \\ -\frac{1}{2} & -\frac{1}{2} \end{array} \begin{pmatrix} a \\ b \\ c \\ d \end{pmatrix}. \tag{10–168}$$

The total spin is $\mathbf{S} = \mathbf{s}_1 + \mathbf{s}_2$, and according to the vector coupling rule, the operator $\mathbf{S}^2$ has the eigenvalues $S(S + 1)$, where $S = \frac{1}{2} + \frac{1}{2} = 1$ or $S = \frac{1}{2} - \frac{1}{2} = 0$. The first of these corresponds to a system of three distinct states, forming a *triplet*, while the second belongs to a single state, the *singlet*.

If A_1 and B_2 are operators in the spin spaces of particles 1 and 2, respectively, then the product $A_1 B_2$ has, in the $(m_1 m_2)$ representation, the matrix elements

$$(\psi(m_1' m_2'), A_1 B_2 \psi(m_1 m_2)) = (\chi_1^{m_1'}, A_1 \chi_1^{m_1})(\chi_2^{m_2'}, B_2 \chi_2^{m_2}), \tag{10–169}$$

i.e., the matrix A_1B_2 is the *direct product* of the corresponding matrices A_1 and B_2. For example, we have

$$S_z = s_{1z} + s_{2z} = s_{1z} \times 1_2 + 1_1 \times s_{2z}$$

$$= \tfrac{1}{2}\begin{pmatrix}1&0&0&0\\0&1&0&0\\0&0&-1&0\\0&0&0&-1\end{pmatrix} + \tfrac{1}{2}\begin{pmatrix}1&0&0&0\\0&-1&0&0\\0&0&1&0\\0&0&0&-1\end{pmatrix} = \begin{pmatrix}1&0&0&0\\0&0&0&0\\0&0&0&0\\0&0&0&-1\end{pmatrix}.$$

$$(10\text{--}170)$$

We may now work with the matrices exclusively. The square of the total spin is

$$\mathbf{S}^2 = \tfrac{1}{4}(\boldsymbol{\sigma}_1 + \boldsymbol{\sigma}_2)^2 = \tfrac{3}{2} + \tfrac{1}{2}\boldsymbol{\sigma}_1 \cdot \boldsymbol{\sigma}_2 \qquad (10\text{--}171)$$

and has the eigenvalue 2 in the triplet states and 0 in the singlet. It follows that the operator

$$P_\sigma = \mathbf{S}^2 - 1 = \tfrac{1}{2}(1 + \boldsymbol{\sigma}_1 \cdot \boldsymbol{\sigma}_2) \qquad (10\text{--}172)$$

has the eigenvalues $+1$ and -1, the former being triply degenerate. The matrix representation of P_σ is easily obtained by means of the rule (10–169). It is

$$P_\sigma = \begin{pmatrix}1&0&0&0\\0&0&1&0\\0&1&0&0\\0&0&0&1\end{pmatrix}. \qquad (10\text{--}173)$$

Evidently, we have

$$P_\sigma \begin{pmatrix}a\\b\\c\\d\end{pmatrix} = \begin{pmatrix}a\\c\\b\\d\end{pmatrix}, \qquad (10\text{--}174)$$

which shows that the operator P_σ has the effect of interchanging the $(\tfrac{1}{2}, -\tfrac{1}{2})$- and $(-\tfrac{1}{2}, \tfrac{1}{2})$-components of the state vector. This amounts to exchanging the roles of the two particles, and P_σ is therefore called the *spin-exchange operator*.

From Eqs. (10–174) and (10–170), it is seen that the normalized vectors

$$\Phi(1, 1) = \begin{pmatrix}1\\0\\0\\0\end{pmatrix} = {}^3\chi_1, \quad \Phi(1, 0) = \frac{1}{\sqrt{2}}\begin{pmatrix}0\\1\\1\\0\end{pmatrix} = {}^3\chi_0, \quad \Phi(1, -1) = \begin{pmatrix}0\\0\\0\\1\end{pmatrix} = {}^3\chi_{-1},$$

$$(10\text{--}175)$$

are eigenvectors of P_σ belonging to the eigenvalue 1, and of S_z belonging to 1, 0, -1, respectively. They are therefore the representatives of the three triplet states. The orthogonal vector

$$\Phi(0, 0) = \frac{1}{\sqrt{2}}\begin{pmatrix} 0 \\ 1 \\ -1 \\ 0 \end{pmatrix} = {}^1\chi_0 \tag{10–176}$$

satisfies the eigenvalue equations

$$P_\sigma{}^1\chi_0 = -{}^1\chi_0, \tag{10–177}$$

$$S_z{}^1\chi_0 = 0, \tag{10–178}$$

and represents the singlet state. In the notation of Section 10–10, these vectors are

$$\Phi(1, 1) = \psi(\tfrac{1}{2}, \tfrac{1}{2}),$$
$$\Phi(1, 0) = \frac{1}{\sqrt{2}}\left(\psi(\tfrac{1}{2}, -\tfrac{1}{2}) + \psi(-\tfrac{1}{2}, \tfrac{1}{2})\right),$$
$$\Phi(1, -1) = \psi(-\tfrac{1}{2}, -\tfrac{1}{2}), \tag{10–179}$$
$$\Phi(0, 0) = \frac{1}{\sqrt{2}}\left(\psi(\tfrac{1}{2}, -\tfrac{1}{2}) - \psi(-\tfrac{1}{2}, \tfrac{1}{2})\right);$$

they could have been obtained in this form by means of the lowering operator $S_- = s_{1-} + s_{2-}$.

The unitary transformation matrix U, connecting the (m_1, m_2)- and (S, m)-representations, can be formed according to the rule (9–161). It is

$$U = \frac{1}{\sqrt{2}}\begin{pmatrix} 0 & 1 & -1 & 0 \\ \sqrt{2} & 0 & 0 & 0 \\ 0 & 1 & 1 & 0 \\ 0 & 0 & 0 & \sqrt{2} \end{pmatrix}. \tag{10–180}$$

As a check, we note that the operators P_σ and S_z are diagonal in the (S, m)-representation:

$$S'_z = US_zU^\dagger = \begin{pmatrix} 0 & 0 & 0 & 0 \\ 0 & 1 & 0 & 0 \\ 0 & 0 & 0 & 0 \\ 0 & 0 & 0 & -1 \end{pmatrix}; \quad P'_\sigma = UP_\sigma U^\dagger = \begin{pmatrix} -1 & 0 & 0 & 0 \\ 0 & 1 & 0 & 0 \\ 0 & 0 & 1 & 0 \\ 0 & 0 & 0 & 1 \end{pmatrix}. \tag{10–181}$$

10–13 Other operators; selection rules. The various observables of a dynamical system can be classified according to their transformation properties under rotations. Such a classification is of great value in determining the matrix elements of the corresponding operators and, in particular, leads to selection rules which limit the number of nonzero matrix elements.

The rule for transformation of an operator O under rotations is

$$O' = UOU^\dagger, \tag{10–182}$$

where $U = \exp(i\boldsymbol{\omega} \cdot \mathbf{J})$ is the rotation operator. Operators which are invariant to this transformation (e.g., the Hamiltonian of an isolated system) are called *scalars*. A scalar operator S therefore satisfies

$$USU^\dagger = S, \tag{10–183}$$

or, what is the same,

$$US - SU = 0. \tag{10–184}$$

Thus, a scalar commutes with every rotation operator. In particular, it commutes with the operator $\exp(i\,d\boldsymbol{\omega} \cdot \mathbf{J}) \approx 1 + i\,d\boldsymbol{\omega} \cdot \mathbf{J}$, representing an infinitesimal rotation $d\boldsymbol{\omega}$. Since the direction of $d\boldsymbol{\omega}$ is arbitrary, this means that S commutes with each component of the total angular momentum $\mathbf{J}$:

$$[S, \mathbf{J}] = 0. \tag{10–185}$$

The equations

$$[S, J_z] = 0, \qquad [S, \mathbf{J}^2] = 0, \tag{10–186}$$

when written out in the (jm)-representation in terms of matrix elements, become

$$(j'm'|SJ_z - J_zS|jm) = (m - m')(j'm'|S|jm) = 0, \tag{10–187}$$

and

$$(j'm'|S\mathbf{J}^2 - \mathbf{J}^2S|jm) = \big(j(j+1) - j'(j'+1)\big)(j'm'|S|jm) = 0. \tag{10–188}$$

It follows that $(j'm'|S|jm)$ vanishes unless $m = m'$ and $j = j'$. Thus we have obtained the selection rules

$$\Delta m = 0, \qquad \Delta j = 0, \tag{10–189}$$

for the matrix elements of a scalar operator.

Furthermore, it may be noted that if S' denotes the diagonal element of S corresponding to the state $\Phi(jm)$, i.e.,

$$S\Phi(jm) = S'\Phi(jm), \tag{10–190}$$

then, since S commutes with J_+ and J_-, we have

$$SJ_+\Phi(jm) = J_+S\Phi(jm) = S'J_+\Phi(jm). \qquad (10\text{–}191)$$

Since the state $J_+\Phi(jm)$ belongs to the eigenvalue $m + 1$ of J_z, it follows that the eigenvalues of S are *independent of m* and depend upon j alone. These results are summarized in the equation

$$(j'm'|S|jm) = (j|S|j)\,\delta_{jj'}\,\delta_{mm'}. \qquad (10\text{–}192)$$

For example, there are $2j + 1$ linearly independent energy states of an isolated system of total angular momentum j.

Equation (10–192) only describes the properties of S which are associated with its scalar character. The states of a quantum-mechanical system will, in general, depend upon other quantum numbers in addition to j and m. If these are denoted collectively by α, then Eq. (10–192) can be written more explicitly as

$$(\alpha'j'm'|S|\alpha jm) = (\alpha'j|S|\alpha j)\,\delta_{jj'}\,\delta_{mm'}. \qquad (10\text{–}193)$$

It may happen, of course, that S is not diagonal in α.

The next class of operators which we shall consider is the class of *vectors*. A vector operator $\mathbf{V}$ has three components, $[V_x, V_y, V_z]$, which transform under rotations like the coordinates of a point [cf. Eq. (10–9)]. The transformation formula is therefore[1]

$$\exp(i\boldsymbol{\omega} \cdot \mathbf{J})\mathbf{V}\exp(-i\boldsymbol{\omega} \cdot \mathbf{J}) = \exp(\boldsymbol{\omega}\times)\mathbf{V} \qquad (10\text{–}194)$$

or, for an infinitesimal transformation,

$$(1 + i\,d\boldsymbol{\omega} \cdot \mathbf{J})\mathbf{V}(1 - i\,d\boldsymbol{\omega} \cdot \mathbf{J}) = \mathbf{V} + d\boldsymbol{\omega} \times \mathbf{V}. \qquad (10\text{–}195)$$

Performing the indicated multiplications and neglecting terms in $d\omega^2$, this expression is reduced to

$$i\,d\boldsymbol{\omega} \cdot \mathbf{J}\,\mathbf{V} - i\mathbf{V}\,d\boldsymbol{\omega} \cdot \mathbf{J} = d\boldsymbol{\omega} \times \mathbf{V}. \qquad (10\text{–}196)$$

Since $d\boldsymbol{\omega}$ is an arbitrary vector, this equation gives the commutator of $\mathbf{V}$ with any component of $\mathbf{J}$. It is equivalent to the set of nine *commutation rules*:

$$[J_x, V_x] = 0, \qquad [J_y, V_x] = -iV_z, \qquad [J_z, V_x] = iV_y,$$

$$[J_x, V_y] = iV_z, \qquad [J_y, V_y] = 0, \qquad [J_z, V_y] = -iV_x, \qquad (10\text{–}197)$$

$$[J_x, V_z] = -iV_y, \qquad [J_y, V_z] = iV_x, \qquad [J_z, V_z] = 0.$$

---

[1] It is important to distinguish $\mathbf{r}$ from $\mathbf{r}_{op}$.

As in the case of the operator $\mathbf{J}$, certain properties of $\mathbf{V}$ are expressed more conveniently by means of the combinations

$$V_+ = V_x + iV_y, \qquad V_- = V_x - iV_y. \qquad (10\text{--}198)$$

One readily finds

$$[J_+, V_+] = 0, \qquad [J_z, V_+] = V_+, \qquad [J_-, V_+] = -2V_z,$$

$$[J_+, V_z] = -V_+, \qquad [J_z, V_z] = 0, \qquad [J_-, V_z] = V_-, \qquad (10\text{--}199)$$

$$[J_+, V_-] = 2V_z, \qquad [J_z, V_-] = -V_-, \qquad [J_-, V_-] = 0.$$

Selection rules and matrix elements of $\mathbf{V}$ in the (jm)-representation will now be worked out. Selection rules for m are obtained immediately from the commutators containing J_z:

$$(j'm'|J_zV_+ - V_+J_z|jm) = (m' - m)(j'm'|V_+|jm) = (j'm'|V_+|jm).$$
$$(10\text{--}200)$$

The matrix element of V_+ is therefore zero unless

$$m' - m = \Delta m = 1. \qquad (10\text{--}201)$$

In a similar way, one finds for V_z

$$\Delta m = 0, \qquad (10\text{--}202)$$

and for V_-

$$\Delta m = -1. \qquad (10\text{--}203)$$

Certain vector equations, which follow from Eqs. (10–197) and the commutation rules for the components of $\mathbf{J}$, are useful in summarizing the properties of $\mathbf{V}$. They are

$$\mathbf{J} \times \mathbf{V} + \mathbf{V} \times \mathbf{J} = 2i\mathbf{V} \qquad (10\text{--}204)$$

and

$$[\mathbf{J}^2, \mathbf{V}] = i(\mathbf{V} \times \mathbf{J} - \mathbf{J} \times \mathbf{V}) = 2i\mathbf{K}, \qquad (10\text{--}205)$$

where $\mathbf{K}$ is the Hermitian vector operator

$$\mathbf{K} = \mathbf{V} \times \mathbf{J} - i\mathbf{V} = i\mathbf{V} - \mathbf{J} \times \mathbf{V}. \qquad (10\text{--}206)$$

The operator $\mathbf{K}$ satisfies the equations

$$\mathbf{J} \cdot \mathbf{K} = 0$$

and

$$\mathbf{J} \times \mathbf{K} = \mathbf{J}^2\mathbf{V} - \mathbf{J} \cdot \mathbf{VJ}, \qquad \mathbf{K} \times \mathbf{J} = \mathbf{J} \cdot \mathbf{VJ} - \mathbf{VJ}^2. \quad (10\text{--}207)$$

Since $\mathbf{J}$ commutes with $\mathbf{J}^2$, it follows from Eq. (10–205) that

$$[\mathbf{J}^2, \mathbf{J} \times \mathbf{V}] = 2i\mathbf{J} \times \mathbf{K}. \qquad (10\text{--}208)$$

Now the commutator in this equation has no matrix elements which are diagonal in j. Consequently,

$$(jm'|\mathbf{J} \times \mathbf{K}|jm) = (jm'|\mathbf{J}^2\mathbf{V} - \mathbf{J} \cdot \mathbf{VJ}|jm) = 0.$$

The operator $\mathbf{J} \cdot \mathbf{V}$ is a *scalar* and therefore diagonal in both m and j [cf. Eq. (10–192)], whence

$$(jm'|\mathbf{V}|jm) = \frac{(j|\mathbf{J} \cdot \mathbf{V}|j)}{j(j+1)} (jm'|\mathbf{J}|jm). \qquad (10\text{--}209)$$

The diagonal elements of $\mathbf{V}$ are proportional to those of $\mathbf{J}$, and the constant of proportionality is independent of m. For brevity, we write

$$\rho(j) = \frac{(j|\mathbf{J} \cdot \mathbf{V}|j)}{j(j+1)}. \qquad (10\text{--}210)$$

Equation (10–209) is the quantum-mechanical form of the *vector rule*, which was deduced empirically in the early study of atomic spectra. If it is imagined that the vector $\mathbf{V}$ (Fig. 10–7) moves around the total angular-

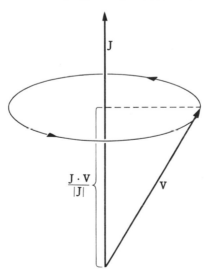

Fig. 10–7.   Precession of the vector $\mathbf{V}$ about the angular-momentum vector $\mathbf{J}$.

momentum vector $\mathbf{J}$ with a steady precessional motion, then the time-average value of the component normal to $\mathbf{J}$ is zero. The time average of $\mathbf{V}$ is therefore parallel to $\mathbf{J}$ and has the magnitude $\mathbf{J} \cdot \mathbf{V}/|\mathbf{J}|$. Consequently, on this model, the average is

$$\langle \mathbf{V} \rangle = \frac{\mathbf{J} \cdot \mathbf{V}}{\mathbf{J}^2} \mathbf{J}. \tag{10-211}$$

This result corresponds to Eq. (10–209) if $j(j+1)$ is substituted for $\mathbf{J}^2$ and the classical vectors are replaced by their quantum expectations.

The above matrix elements correspond to the selection rule

$$\Delta j = 0. \tag{10-212}$$

Also, it is clear that the matrix element of $\mathbf{V}$ in the state $j = 0$ vanishes. This is frequently symbolized by writing

$$\text{no } 0 \rightarrow 0. \tag{10-213}$$

In general, a vector operator also has matrix elements connecting states with different values of j. Before evaluating these explicitly, we shall complete the list of selection rules by means of a device due to Dirac.[1]

The first of the equations (10–205) holds for any vector operator and can therefore be written for $\mathbf{K}$ as

$$[\mathbf{J}^2, \mathbf{K}] = i[\mathbf{K} \times \mathbf{J} - \mathbf{J} \times \mathbf{K}]. \tag{10-214}$$

By substituting from Eqs. (10–205) and (10–207), this expression is reduced to

$$[\mathbf{J}^2, [\mathbf{J}^2, \mathbf{V}]] = 2(\mathbf{J}^2 \mathbf{V} - 2\mathbf{J} \cdot \mathbf{V}\mathbf{J} + \mathbf{V}\mathbf{J}^2). \tag{10-215}$$

The (jj) matrix element of the left-hand member of this equation is zero, and we are led once more to Eq. (10–209). However, if we assume $j' \neq j$, then since $\mathbf{J}$ is diagonal in j, the second term in the right-hand member drops out and we obtain

$$(j'm'|[\mathbf{J}^2, [\mathbf{J}^2, \mathbf{V}]]|jm) = 2(j'm'|\mathbf{J}^2\mathbf{V} + \mathbf{V}\mathbf{J}^2|jm) \qquad (j' \neq j), \tag{10-216}$$

which is easily found to be

$$\{[j'(j'+1)]^2 - 2[j'(j'+1)j(j+1)] + [j(j+1)]^2\}(j'm'|\mathbf{V}|jm)$$
$$= 2\{j'(j'+1) + j(j+1)\}(j'm'|\mathbf{V}|jm), \tag{10-217}$$

---

[1] P. A. M. Dirac, *The Principles of Quantum Mechanics.* 3rd ed. Oxford: The Clarendon Press, 1947, Section 40.

or, after some algebraic manipulation,

$$\{(j' - j)^2 - 1\}\{(j' + j + 1)^2 - 1\}(j'm'|\mathbf{V}|jm) = 0. \quad (10\text{-}218)$$

Since j' and j are unequal and non-negative, the second factor cannot vanish; hence $(j'm'|\mathbf{V}|jm)$ is zero unless

$$j' - j = \Delta j = \pm 1. \quad (10\text{-}219)$$

This completes the list of selection rules for $\mathbf{V}$.

Explicit expressions for the matrix elements can now be found from the commutation rules (10–199) and the matrix elements of $\mathbf{J}$. Let us consider the case $j' = j + 1$. Of course, we need only be concerned with values of Δm which satisfy the selection rules (10–201), (10–202), and (10–203). The equation

$$[J_+, V_+] = 0$$

becomes, in terms of matrix elements,

$$(j{+}1 \ m{+}1 \ |J_+V_+ - V_+J_+| \ j \ m{-}1)$$
$$= \sqrt{(j + m + 2)(j - m + 1)} \ (j{+}1 \ m \ |V_+| \ j \ m{-}1)$$
$$- \sqrt{(j - m + 1)(j + m)} \ (j{+}1 \ m{+}1 \ |V_+| \ j m) = 0.$$

This equation, which may be regarded as a recurrence relation for the matrix elements, is equivalent to

$$\frac{(j{+}1 \ m \ |V_+| \ j \ m{-}1)}{\sqrt{(j + m + 1)(j + m)}} = \frac{(j{+}1 \ m{+}1 \ |V_+| \ j \ m)}{\sqrt{(j + m + 2)(j + m + 1)}} = -\sigma(j\,m),$$
$$(10\text{-}220)$$

in which the function $\sigma(j\,m)$ is defined as the negative of either of the equal ratios on the left. Inspection of this equation shows that $\sigma(j\,m)$ is independent of m, i.e.,

$$\sigma(j\,m) = \sigma(j \ m{-}1) = \sigma(j \ m{-}2) = \cdots = \sigma(j). \quad (10\text{-}221)$$

Consequently,

$$(j{+}1 \ m{+}1 \ |V_+| \ j\,m) = -\sigma(j)\sqrt{(j + m + 2)(j + m + 1)}. \quad (10\text{-}222)$$

The matrix element of V_z can now be obtained from the commutation relation

$$-2V_z = [J_-, V_+].$$

Thus,

$$-2(j{+}1 \ m \ |V_z| \ j \ m) = (j{+}1 \ m \ |J_-V_+ - V_+J_-| \ j \ m)$$

$$= \sqrt{(j-m+1)(j+m+2)} \ (j{+}1 \ m{+}1 \ |V_+| \ j \ m)$$

$$- \sqrt{(j{+}m)(j-m+1)} \ (j{+}1 \ m \ |V_+| \ j \ m{-}1)$$

$$= -\sigma(j)\{(j+m+2)\sqrt{(j+m+1)(j-m+1)}$$

$$- (j{+}m)\sqrt{(j+m+1)(j-m+1)}\}$$

$$= -2\sigma(j)\sqrt{(j+m+1)(j-m+1)},$$

or

$$(j{+}1 \ m \ |V_z| \ j \ m) = \sigma(j)\sqrt{(j+m+1)(j-m+1)}. \quad (10\text{--}223)$$

Similarly, the matrix elements for V_- can be found from the equation $[J_-, V_z] = V_-$. These results, together with those for $\Delta j = 0$ and $\Delta j = -1$, are summarized in Table 10–1. The factors $\rho(j)$, $\sigma(j)$ and $\tau(j)$ depend upon the definition of $\mathbf{V}$ and are not determined by its rotational character. They are also functions of the other quantum numbers (α) of the system and can be written more explicitly as

$$\rho(j) = \rho(j\alpha'\alpha) = \frac{(\alpha'j|\mathbf{J}\cdot\mathbf{V}|\alpha j)}{j(j+1)}, \quad (10\text{--}224)$$

TABLE 10–1

THE NONZERO MATRIX ELEMENTS OF $\mathbf{V}$

$\Delta j = 1$:

$(j{+}1 \ m{+}1 \ |V_+| \ j \ m) = -\sigma(j)\sqrt{(j+m+2)(j+m+1)}$

$(j{+}1 \ m \ |V_z| \ j \ m) = \sigma(j)\sqrt{(j+m+1)(j-m+1)}$

$(j{+}1 \ m{-}1 \ |V_-| \ j \ m) = \sigma(j)\sqrt{(j-m+1)(j-m+2)}$

$\Delta j = 0$:

$(j \ m{+}1 \ |V_+| \ j \ m) = \rho(j)\sqrt{(j-m)(j+m+1)}$

$(j \ m|V_z|j \ m) = \rho(j)m$

$(j \ m{-}1 \ |V_-| \ j \ m) = \rho(j)\sqrt{(j+m)(j-m+1)}$

$\Delta j = -1$:

$(j{-}1 \ m{+}1 \ |V_+| \ j \ m) = \tau(j)\sqrt{(j-m-1)(j-m)}$

$(j{-}1 \ m \ |V_z| \ j \ m) = \tau(j)\sqrt{(j-m)(j+m)}$

$(j{-}1 \ m{-}1 \ |V_-| \ j \ m) = -\tau(j)\sqrt{(j+m)(j+m-1)}$

and so forth. If $\mathbf{V}$ is Hermitian, then $V_+^\dagger = V_-$, and the functions σ and τ are related by

$$\tau(j\alpha'\alpha) = \sigma^*(j{-}1 \quad \alpha\,\alpha'). \tag{10–225}$$

The classification of operators by their transformation properties under rotations can be extended to tensors of any rank, and in each case the form of the matrix elements is determined, in the (αjm)-representation, except for factors which depend only upon α and j. More advanced works may be consulted for complete accounts of this theory.[1]

References

BRINKMAN, H. C., *Applications of Spinor Invariants in Atomic Physics.* Amsterdam: North-Holland Publishing Co., 1956. A summary of an important technique, due to H. A. Kramers, which facilitates calculations in the theory of angular momentum.

CONDON, E. U., and G. H. SHORTLEY, *The Theory of Atomic Spectra.* Cambridge: Cambridge University Press, 1951. The standard treatise on the subject. A thorough discussion of angular momenta is contained in Chapter III.

EDMONDS, A. R., *Angular Momentum in Quantum Mechanics.* Princeton: Princeton University Press, 1957. A complete summary, on a somewhat advanced level, including a very complete bibliography.

FEENBERG, E., and G. E. PAKE, *Notes on the Quantum Theory of Angular Momentum.* Reading, Mass.: Addison-Wesley Publishing Co., Inc., 1953. A handy reference work on the elementary properties of the angular momentum matrices.

ROSE, M. E., *Elementary Theory of Angular Momentum.* New York: John Wiley and Sons, Inc., 1957. A thorough discussion of angular momentum on the basis of transformation properties under rotations, with applications to fields of current interest.

WIGNER, E., *Group Theory and its Applications to the Theory of Atomic Spectra.* New York: Academic Press, 1959. The theory of angular momentum finds its proper mathematical setting in the theory of group representations. This text is by far the best presentation of group theory for physical applications.

---

[1] See list of references.

Problems

10–1. Show that the vector $\mathbf{r}'$ which results from rotation of $\mathbf{r}$ through a finite angle ω about an axis in the direction of the unit vector $\hat{\mathbf{\kappa}}$ is

$$\mathbf{r}' = \cos \omega\, \mathbf{r} + (1 - \cos \omega)\hat{\mathbf{\kappa}}\hat{\mathbf{\kappa}} \cdot \mathbf{r} + \sin \omega\, \hat{\mathbf{\kappa}} \times \mathbf{r}.$$

Express this transformation as $\mathbf{r}' = R\mathbf{r}$, where R is a 3×3 matrix, and show that $R = \exp(\omega K)$, where K is the matrix 10–7.

10–2. Given that the real 3×3 matrix $R = (R_{ij})$ is a rotation matrix, i.e., is orthogonal, show how to find the axis of rotation and the angle.

10–3. Let $x(\alpha)$ be a column vector whose components are analytic functions of a parameter α:

$$x(\alpha) = \begin{pmatrix} x_1(\alpha) \\ x_2(\alpha) \\ \vdots \\ x_n(\alpha) \end{pmatrix},$$

and let $dx/d\alpha$ denote the derivative of x, i.e.,

$$\frac{dx}{d\alpha} = \lim_{\delta\alpha \to 0} \frac{x(\alpha + \delta\alpha) - x(\alpha)}{\delta\alpha}.$$

Prove that if

$$\frac{dx}{d\alpha} = Ax,$$

where A is a matrix independent of α, then

$$x(\alpha) = e^{(\alpha - \alpha_0)A}\, x(\alpha_0),$$

where α_0 is some fixed value of α. [Hint: Remember that $\exp[(\alpha - \alpha_0)A]$ is defined by its Taylor series.]

10–4. Show that every finite rotation is equivalent to the product of two involutions, i.e., to two successive rotations through 180°. Find two involutions which are together equivalent to a rotation through 90° about the z-axis. Give a geometrical interpretation.

10–5. Show that the matrix $R(\omega)$, representing a rotation through the angle ω about any given axis, can be written in the form

$$R(\omega) = U R_z(\omega) U^{-1},$$

where $R_z(\omega)$ is a rotation through the same angle about the z-axis, and U is a suitable rotation. How is U related to the axis of rotation for R?

10–6. Show that every real orthogonal 3×3 matrix representing a rotation through the angle ω satisfies the matrix equation

$$(1 - R)(1 - 2\cos \omega\, R + R^2) = 0,$$

and that therefore R can be written

$$R = e^{i\omega J},$$

where J is a matrix which satisfies

$$(J - 1)(J)(J + 1) = 0.$$

10-7. Prove that every matrix representative of a component of a vector $\mathbf{J}$ which satisfies

$$\mathbf{J} \times \mathbf{J} = i\mathbf{J}$$

has zero trace.

10-8. Verify Eqs. (10-47), (10-48), and (10-49).

10-9. Verify Eqs. (10-50) and (10-51).

10-10. Show that under the transformation

$$\mathbf{r}' = e^{(-\boldsymbol{\omega} \times)}\mathbf{r},$$

corresponding to the rotation $\boldsymbol{\omega}$ of the coordinate frame, the complex-valued functions

$$-\frac{1}{\sqrt{2}}(x + iy), \qquad z, \qquad \frac{1}{\sqrt{2}}(x - iy)$$

of the coordinates of a point undergo the same transformation as the eigenfunctions $\phi(1, m)$ of the operators $\mathbf{J}^2$ and J_z. In other words, prove that the above transformation is identical with $\phi' = \exp(i\boldsymbol{\omega} \cdot \mathbf{J})\phi$ if we make the correspondence

$$\phi(1, 1) \to -\frac{1}{\sqrt{2}}(x + iy), \qquad \phi(1, 0) \to z, \qquad \phi(1, -1) \to \frac{1}{\sqrt{2}}(x - iy).$$

10-11. Show that the three functions (cf. Problem 10-10)

$$\phi_x = -\frac{1}{\sqrt{2}}\big(\phi(1, 1) - \phi(1, -1)\big),$$

$$\phi_y = \frac{i}{\sqrt{2}}\big(\phi(1, 1) + \phi(1, -1)\big),$$

$$\phi_z = \phi(1, 0),$$

transform under rotations in the same way as the components $[x, y, z]$ of the vector $\mathbf{r}$. Hence prove that the wave functions for the $^3P_1$ states of a system of two particles with spin $\frac{1}{2}$ (e.g., the $n\text{-}p$ system) are proportional to the components of the vector $\boldsymbol{\phi} \times \boldsymbol{\chi}$, where $\boldsymbol{\phi}$ and $\boldsymbol{\chi}$ are the vectors formed in the above manner from the orbital P-wave functions and the triplet spin functions, respectively.

10–12. Construct the ^3P_j functions for the n-p system by the method of lowering operators. Compare the functions for $j = 1$ to the functions obtained in Problem 10–11.

10–13. Devise a method, analogous to that of Problem 10–11, for the construction of the $^3P_2$ and $^3P_0$ states obtained in Problem 10–12.

10–14. Prove the identities

$$[J_x^2, J_y^2] = [J_y^2, J_z^2] = [J_z^2, J_x^2],$$

and show that these commutators are all zero in states for which $j = 0$, $\frac{1}{2}$, or 1.

10–15. Construct a representation of the three states of unit angular momentum in which J_x^2, J_y^2, and J_z^2 are simultaneously diagonal.

10–16. Derive the relations

$$J_x\phi_x = 0, \qquad J_y\phi_x = -i\phi_z, \qquad J_z\phi_x = i\phi_y,$$

$$J_x\phi_y = i\phi_z, \qquad J_y\phi_y = 0, \qquad J_z\phi_y = -i\phi_x,$$

$$J_x\phi_z = -i\phi_y, \qquad J_y\phi_z = i\phi_x, \qquad J_z\phi_z = 0,$$

where ϕ_x, ϕ_y, ϕ_z are the functions of Problem 10–11.

10–17. Any wave function representing a triplet state of a system of two particles with spin $\frac{1}{2}$ can be expressed, in the center of mass system, as

$$\psi = \psi_x\chi_x + \psi_y\chi_y + \psi_z\chi_z,$$

where χ_x, χ_y and χ_z are formed from the functions (10–175) according to the procedure outlined in Problem 10–11. The quantities ψ_x, ψ_y, and ψ_z depend only upon the coordinates $[x, y, z]$ of the vector $\mathbf{r}$ joining the two particles. These functions can be used to define formally a vector

$$\psi = \psi_x\hat{\mathbf{i}} + \psi_y\hat{\mathbf{j}} + \psi_z\hat{\mathbf{k}}$$

in three-dimensional space, so that a one-to-one correspondence between wave functions and spatial vectors is established. On the basis of Problem 10–16, show that the functions $S_x\psi$, $S_y\psi$, $S_z\psi$ correspond to the vectors $i\hat{\mathbf{i}} \times \psi$, $i\hat{\mathbf{j}} \times \psi$, and $i\hat{\mathbf{k}} \times \psi$, respectively.

10–18. The total angular momentum of the system described in Problem 10–17 is $\mathbf{J} = \mathbf{L} + \mathbf{S}$. Show that the equations

$$J_z\psi = L_z\psi + S_z\psi,$$

and

$$\mathbf{J}^2\psi = \mathbf{L}^2\psi + \mathbf{S}^2\psi + 2\mathbf{L}\cdot\mathbf{S}\psi$$

become, in vector notation,

$$J_z\psi = L_z\psi + i\hat{\mathbf{k}} \times \psi,$$

and

$$\mathbf{J}^2\psi = \mathbf{L}^2\psi + 2\psi + 2i\mathbf{L} \times \psi,$$

and that the eigenvalue equation defining the simultaneous eigenstates of $\mathbf{J}^2$, $\mathbf{L}^2$ and $\mathbf{S}^2 = 2$ is [1]

$$\{J(J+1) - L(L+1) - 2\}\boldsymbol{\psi} = 2i\mathbf{L} \times \boldsymbol{\psi}.$$

10–19. Show that the eigenvalue equation of Problem 10–18 can be satisfied only by $L = J$, $J - 1$ and $J + 1$. For $L = J$, prove that

$$\mathbf{L}^2(\mathbf{L} \cdot \boldsymbol{\psi}) = J(J+1)(\mathbf{L} \cdot \boldsymbol{\psi}),$$

and therefore that

$$\psi(J, m, J) = \frac{1}{\sqrt{J(J+1)}} \mathbf{L} Y_J^m(\mathbf{r})$$

is a simultaneous (vector) eigenfunction of $\mathbf{J}^2$, J_z, and $\mathbf{L}^2$, normalized in the sense that

$$\int_\omega \boldsymbol{\psi}^* \boldsymbol{\psi} \, d\omega = 1.$$

Show also that for $L = J \pm 1$, $\mathbf{L} \cdot \boldsymbol{\psi} = 0$ and $\mathbf{L}^2(\mathbf{r} \cdot \boldsymbol{\psi}) = J(J+1)(\mathbf{r} \cdot \boldsymbol{\psi})$, whence the functions

$$\psi(J, m, J+1) = \frac{1}{\sqrt{(J+1)(2J+1)}} \left\{(J+1)\frac{\mathbf{r}}{r} + i\frac{\mathbf{r}}{r} \times \mathbf{L}\right\} Y_J^m,$$

$$\psi(J, m, J-1) = \frac{1}{\sqrt{J(2J+1)}} \left\{-J\frac{\mathbf{r}}{r} + i\frac{\mathbf{r}}{r} \times \mathbf{L}\right\} Y_J^m,$$

are the normalized eigenfunctions for these values of L.

10–20. Show that the operator $\boldsymbol{l} \cdot \mathbf{s}$ has the eigenvalues $\frac{1}{2}$ and -1 in the p-states of an electron. Also, derive the formula

$$\boldsymbol{l} \cdot \mathbf{s} = \frac{1}{2}(l_+ s_- + l_- s_+ + 2l_z s_z),$$

and verify directly that

$$\boldsymbol{l} \cdot \mathbf{s} \, \Phi(\tfrac{3}{2}, m) = \tfrac{1}{2}\Phi(\tfrac{3}{2}, m),$$

$$\boldsymbol{l} \cdot \mathbf{s} \, \Phi(\tfrac{1}{2}, m) = -\Phi(\tfrac{1}{2}, m),$$

using the representation (10–156), etc. of the states $\Phi(j, m)$.

10–21. The operator $s_r = \mathbf{s} \cdot (\mathbf{r}/r)$ is the component of the electron spin in the direction of the vector $\mathbf{r}$. Since s_r is a scalar operator, it commutes with each component of the total angular momentum $\mathbf{J} = \boldsymbol{l} + \mathbf{s}$; hence,

$$[\mathbf{J}, s_r] = 0.$$

---

[1] H. C. Corben and J. Schwinger [*Phys. Rev.* **58**, 953 (1940)] devised this representation of triplet states in a discussion of the motion of a spin-1 particle in a central field. See especially their Appendix.

Verify this by calculating $[l, s_r]$ and $[s, s_r]$ by means of the commutation rules for l and r; i.e., show that

$$[s, s_r] = -[l, s_r] = i\frac{r}{r} \times s.$$

10–22. Construct the simultaneous eigenvectors of J^2 and $J \cdot \hat{n}$ for the states $j = 1$. Show that if a measurement of J_z is made on a state in which $J \cdot \hat{n}$ is certainly unity, the values $1, 0, -1$ are obtained with relative probabilities $\cos^4 (\theta/2)$, $2 \sin^2 (\theta/2) \cos^2 (\theta/2)$, $\sin^4 (\theta/2)$, respectively (θ is the angle between $\hat{n}$ and the z-axis).

10–23. Prove that every unitary 2×2 matrix with determinant one has the form

$$\cos \theta + i\boldsymbol{\sigma} \cdot \hat{n} \sin \theta = \exp (i\theta\boldsymbol{\sigma} \cdot \hat{n}),$$

where θ and the components of the unit vector $\hat{n}$ are suitably chosen parameters. What are the eigenvalues and eigenvectors of this matrix?

10–24. Show that the operator $P_\sigma = \frac{1}{2}(1 + \boldsymbol{\sigma}_1 \cdot \boldsymbol{\sigma}_2)$ [Eq. (10–172)] satisfies

$$\boldsymbol{\sigma}_1 P_\sigma = P_\sigma \boldsymbol{\sigma}_2,$$

and deduce from this equation that P_σ is the "spin exchange" operator, i.e., that

$$P_\sigma \psi(1, 2) = \psi(2, 1),$$

where $\psi(1, 2)$ is any function depending upon the spin coordinates of particles 1 and 2.

CHAPTER 11

PERTURBATION THEORY

11-1 Introduction. Almost all the applications of quantum mechanics which we have studied so far have led to mathematical problems that can be solved exactly. However, in practice, exactly solvable problems are rare, and one must frequently resort to approximations. In the present chapter, we shall discuss a general method which is applicable whenever the problem at hand is sufficiently similar to one that has an exact solution. In such cases, the Hamiltonian can be broken up into two parts, one of which is large and characterizes a system for which the Schrödinger equation can be solved exactly, while the other part is small and can be treated as a *perturbation*. There are many problems of this kind. If the potential energy of a system is changed by the influence of additional forces, the energy levels are shifted, and for a weak perturbation, the amount of the shift can be estimated if the original unperturbed states are known. An example of this type has already been discussed in connection with Eq. (5–90). The effects of perturbations on degenerate systems are also of the greatest importance. We have seen, for example, in Section 10–7 that a weak magnetic field removes the degeneracy associated with the angular momentum of a system of charged particles.

Interactions can also very often be treated as perturbations, so that the time behavior of systems which exert weak forces on one another can be described in terms of the properties of the unperturbed states of the noninteracting systems. This leads to the theory of quantum transitions. Perturbation theory plays a very fundamental role here, since it provides the connection between the observable properties of a system and its stationary quantum states. On this basis, we shall obtain (Section 11–12) the selection and intensity rules for the optical lines in atomic spectra and complete the theory of the anomalous Zeeman effect.

11-2 Perturbation of nondegenerate stationary states. We shall begin by deriving, in operator form, the result of Section 5–8, giving the effect of a small perturbation on the energy levels of a system with discrete stationary states. The Hamiltonian operator for a given unperturbed system will be designated by $H^{(0)}$, and it is assumed that the energy spectrum is discrete and nondegenerate. The state vectors $\psi_n^{(0)}$ for the system are specified by the eigenvalue equations

$$H^{(0)}\psi_n^{(0)} = E_n^{(0)}\psi_n^{(0)}, \tag{11-1}$$

in which $E_n^{(0)}$ denotes the energy of the nth level of the system.

381

The perturbation consists in a small addition to the Hamiltonian, for which we write $\epsilon H^{(1)}$. The small parameter ϵ is introduced to facilitate comparison of quantities as to order of magnitude. The perturbed Hamiltonian is then

$$H = H^{(0)} + \epsilon H^{(1)}. \tag{11-2}$$

The eigenvectors ψ_n for the perturbed system satisfy

$$H\psi_n = E_n\psi_n, \tag{11-3}$$

and are a complete set of state vectors. The numbers E_n are the energy eigenvalues of the modified Hamiltonian.

If ϵ is sufficiently small, it is reasonable to expect that the vectors ψ_n and the eigenvalues E_n do not differ very much from the corresponding quantities $\psi_n^{(0)}$ and $E_n^{(0)}$ for the unperturbed system. That is, in first approximation they can be written

$$\psi_n \approx \psi_n^{(0)} + \epsilon\psi_n^{(1)}, \tag{11-4}$$

$$E_n \approx E_n^{(0)} + \epsilon E_n^{(1)}, \tag{11-5}$$

in which the second terms on the right-hand sides are small corrections of order ϵ.

Substituting these expressions into the eigenvalue equations (11–3), we obtain

$$(H^{(0)} + \epsilon H^{(1)})(\psi_n^{(0)} + \epsilon\psi_n^{(1)}) = (E_n^{(0)} + \epsilon E_n^{(1)})(\psi_n^{(0)} + \epsilon\psi_n^{(1)}). \tag{11-6}$$

Dropping terms in ϵ^2, we have

$$H^{(0)}\psi_n^{(0)} + \epsilon(H^{(1)}\psi_n^{(0)} + H^{(0)}\psi_n^{(1)}) = E_n^{(0)}\psi_n^{(0)} + \epsilon(E_n^{(1)}\psi_n^{(0)} + E_n^{(0)}\psi_n^{(1)}),$$
$$\tag{11-7}$$

and because of Eq. (11–1), this expression reduces to

$$H^{(0)}\psi_n^{(1)} + H^{(1)}\psi_n^{(0)} = E_n^{(1)}\psi_n^{(0)} + E_n^{(0)}\psi_n^{(1)}. \tag{11-8}$$

If we now form the scalar product of the two sides of this equation with $\psi_n^{(0)}$ and cancel the term

$$(\psi_n^{(0)}, H^{(0)}\psi_n^{(1)}) = (H^{(0)}\psi_n^{(0)}, \psi_n^{(1)})$$

on the left against the term

$$E_n^{(0)}(\psi_n^{(0)}, \psi_n^{(1)})$$

on the right, we obtain

$$(\psi_n^{(0)}, H^{(1)}\psi_n^{(0)}) = E_n^{(1)}(\psi_n^{(0)}, \psi_n^{(0)}), \tag{11–9}$$

or

$$E_n^{(1)} = \frac{(\psi_n^{(0)}, H^{(1)}\psi_n^{(0)})}{(\psi_n^{(0)}, \psi_n^{(0)})}. \tag{11–10}$$

This is the principal result of this part of the theory: *The change in energy of the nth state of the system is approximately the expectation of the perturbation operator in the nth unperturbed state.*

In certain special cases, the above treatment is inadequate to give a complete description of the effect of the perturbation. Obviously, this is so if the matrix element $(n|H^{(1)}|n)$ is zero, which occurs most frequently for reasons of symmetry. For example, the energies of the states of a one-dimensional harmonic oscillator are unaffected, in this approximation, by the addition of a perturbation which is an odd function of x, because every unperturbed state has a definite parity.[1] It also happens occasionally that a closer approximation to the energy is required.

One could attempt to obtain a more complete theory by extending Eqs. (11–4) and (11–5) to terms of higher order in ϵ, i.e., by assuming that ψ_n and E_n can be expressed as power series in ϵ. However, if one proceeds directly from such expansions, the calculation soon becomes very involved, principally because the functions ψ_n contain normalization factors which must be computed anew for each order in ϵ. It is therefore better to adopt a method in which the normalization of ψ is automatically specified at each stage. This can be done most easily by using the ideas outlined in Chapter 9 and treating the connection between the functions $\psi_n^{(0)}$ and ψ_n as a unitary transformation in the state space.[2]

For generality, let us assume that the Hamiltonian is given by a power series in the small parameter ϵ:

$$H = H^{(0)} + \epsilon H^{(1)} + \epsilon^2 H^{(2)} + \cdots. \tag{11–11}$$

As before, the unperturbed wave functions $\psi_n^{(0)}$ satisfy

$$H^{(0)}\psi_n^{(0)} = E_n^{(0)}\psi_n^{(0)} \tag{11–12}$$

and determine a basis in the state space which we shall assume to be orthonormal:

$$(\psi_{n'}^{(0)}, \psi_n^{(0)}) = \delta_{n'n}. \tag{11–13}$$

---

[1] This corresponds to the well-known classical result that the frequency of an oscillator is unaffected, in first approximation, by the introduction of an anharmonic force which acts always in one direction.

[2] Cf. also S. Epstein, *Am. J. Phys.* **22**, 613 (1955).

We shall work in the representation defined by these vectors, so that the representative of $H^{(0)}$ is a diagonal matrix, i.e.,

$$(\psi_{n'}^{(0)}, H^{(0)}\psi_n^{(0)}) = H_{n'n}^{(0)} = E_n^{(0)} \, \delta_{n'n}. \qquad (11\text{--}14)$$

Now the vectors ψ_n, which satisfy

$$H\psi_n = E_n\psi_n,$$

are also, by hypothesis, a basis for the state space, and the unitary matrix U, defined by

$$U_{n'n} = (\psi_{n'}, \psi_n^{(0)}), \qquad (11\text{--}15)$$

transforms the matrix H to diagonal form:

$$H' = UHU^{-1} = \text{diagonal matrix}. \qquad (11\text{--}16)$$

The problem is solved when we determine the elements of U; for the energies of the perturbed levels are simply the elements of the matrix (11–16), and the ψ_n are given in terms of the wave functions of the unperturbed system by

$$\psi_n = \sum_{n'} U_{nn'}^* \psi_{n'}^{(0)}. \qquad (11\text{--}17)$$

For formal calculation, it is convenient to write U in the exponential form,

$$U = e^{iS}, \qquad (11\text{--}18)$$

where S is a Hermitian matrix (cf. Appendix, Section A–7). Since in the limit $\epsilon \to 0$, U becomes the unit matrix, we may expect that S can be expanded in the form

$$S = \epsilon S^{(1)} + \epsilon^2 S^{(2)} + \cdots. \qquad (11\text{--}19)$$

Of course, we have no assurance that such an expansion is convergent,[1] but we can nevertheless find the formal series for S and the corresponding series for H', namely,

$$H' = E^{(0)} + \epsilon E^{(1)} + \epsilon^2 E^{(2)} + \cdots. \qquad (11\text{--}20)$$

(The matrices $E^{(0)}$, $E^{(1)}$, $E^{(2)}$ are diagonal matrices with elements $E_n^{(0)}$, $E_n^{(1)}$, etc.) The series are easily obtained by means of the formal expansion

---

[1] The potential function $V(x) = x^2 + \epsilon x^3$ describes a system which has *no* stationary states. (Why?) In this case, the series (11–19) cannot be convergent for any value of ϵ.

(Appendix, Section A–7)

$$e^{iS} H e^{-iS} = H + i[S, H] + \frac{i^2}{2!} [S, [S, H]] + \frac{i^3}{3!} [S, [S, [S, H]]] + \cdots$$

$$= \sum_{n=0}^{\infty} \frac{i^n}{n!} [S, [S, \ldots, (n) \cdot [S, H]] \ldots]. \tag{11–21}$$

Inserting the expansions (11–11), (11–19), and (11–20) into Eq. (11–21) and collecting the coefficients of the various powers of ϵ, one finds

$$E^{(0)} = H^{(0)}, \tag{11–22}$$

$$E^{(1)} = H^{(1)} + i[S^{(1)}, H^{(0)}], \tag{11–23}$$

$$E^{(2)} = H^{(2)} + i[S^{(2)}, H^{(0)}] + i[S^{(1)}, H^{(1)}] + \frac{i^2}{2!} [S^{(1)}, [S^{(1)}, H^{(0)}]], \tag{11–24}$$

$$E^{(3)} = H^{(3)} + i[S^{(3)}, H^{(0)}] + i[S^{(2)}, H^{(1)}] + i[S^{(1)}, H^{(2)}]$$

$$+ \frac{i^2}{2!} [S^{(2)}, [S^{(1)}, H^{(0)}]] + \frac{i^2}{2!} [S^{(1)}, [S^{(2)}, H^{(0)}]]$$

$$+ \frac{i^2}{2!} [S^{(1)}, [S^{(1)}, H^{(1)}]] + \frac{i^3}{3!} [S^{(1)}, [S^{(1)}, [S^{(1)}, H^{(0)}]]], \tag{11–25}$$

and so on.

These equations can be written for any order by inspection and then solved successively for the elements of the matrices $E^{(k)}$ and $S^{(k)}$. Equation (11–22) is an identity, since we are using the representation in which $H^{(0)}$ is diagonal.

In terms of matrix elements, Eq. (11–23) is

$$E_n^{(1)} \delta_{n'n} = H_{n'n}^{(1)} + i(S^{(1)} H^{(0)} - H^{(0)} S^{(1)})_{n'n},$$

or, since

$$H_{n'n}^{(0)} = E_n^{(0)} \delta_{n'n},$$

$$E_n^{(1)} \delta_{n'n} = H_{n'n}^{(1)} + i(E_n^{(0)} - E_{n'}^{(0)}) S_{n'n}^{(1)}. \tag{11–26}$$

The diagonal element of $E^{(1)}$ is therefore

$$E_n^{(1)} = H_{nn}^{(1)},$$

which is the same as Eq. (11–10); (the zero-order functions have now been assumed to be normalized).

The nondiagonal elements in Eq. (11–26) satisfy

$$(E_n^{(0)} - E_{n'}^{(0)}) S_{n'n}^{(1)} = i H_{n'n}^{(1)}, \tag{11–27}$$

and since, by assumption, the unperturbed states are not degenerate, the first-order elements of $S^{(1)}$ are

$$S_{n'n}^{(1)} = \frac{1}{i} \frac{H_{n'n}^{(1)}}{E_{n'}^{(0)} - E_n^{(0)}}. \qquad (11\text{--}28)$$

It is clear from equation (11–27) that the case of degeneracy requires special consideration. Leaving this aside for the moment, we note that the diagonal elements of $S^{(1)}$ are not determined. They may be chosen arbitrarily, which corresponds, in this approximation, to the arbitrary choice of the phase of the perturbed wave functions. To see this explicitly, we write, to order ϵ,

$$\psi_n = \sum_{n'} U_{nn'}^* \psi_{n'}^{(0)} \approx \sum_{n'} (\delta_{nn'} - i\epsilon S_{nn'}^{(1)}) \psi_{n'}^{(0)}$$

$$= \psi_n^{(0)} (1 - i\epsilon S_{nn}^{(1)}) - i\epsilon \sum_{(n' \neq n)} S_{nn'}^{(1)} \psi_{n'}^{(0)}$$

$$\approx \left\{ \psi_n^{(0)} - i\epsilon \sum_{n' \neq n} S_{nn'}^{(1)} \psi_{n'}^{(0)} \right\} (1 - i\epsilon S_{nn}^{(1)})$$

$$\approx \left\{ \psi_n^{(0)} - i\epsilon \sum_{n' \neq n} S_{nn'}^{(1)} \psi_{n'}^{(0)} \right\} e^{-i\epsilon S_{nn}^{(1)}}.$$

Thus, we may choose to set $S_{nn}^{(1)} = 0$, and the perturbed wave function becomes

$$\psi_n = \psi_n^{(0)} + \epsilon \sum_{n' \neq n} \frac{H_{n'n}^{(1)}}{E_n^{(0)} - E_{n'}^{(0)}} \psi_{n'}^{(0)}. \qquad (11\text{--}29)$$

The second-order equation (11–24) can be simplified by means of Eq. (11–23) and written in the form

$$E^{(2)} = H^{(2)} + i[S^{(2)}, H^{(0)}] + \frac{i}{2} [S^{(1)}, H^{(1)}] + \frac{i}{2} [S^{(1)}, E^{(1)}],$$

and since $H^{(0)}$ and $E^{(1)}$ are diagonal matrices, the diagonal element of this equation is

$$E_n^{(2)} = H_{nn}^{(2)} + \frac{i}{2} (S^{(1)} H^{(1)} - H^{(1)} S^{(1)})_{nn}. \qquad (11\text{--}30)$$

Substituting for the matrix elements of $S^{(1)}$ from Eq. (11–28) and writing out the matrix products, Eq. (11–30) reduces to

$$E_n^{(2)} = H_{nn}^{(2)} + \sum_{n' \neq n} \frac{H_{nn'}^{(1)} H_{n'n}^{(1)}}{E_n^{(0)} - E_{n'}^{(0)}}. \tag{11–31}$$

Thus, the energy of the nth perturbed level is, to second order in ϵ,

$$E_n = E_n^{(0)} + \epsilon H_{nn}^{(1)} + \epsilon^2 \left\{ H_{nn}^{(2)} + \sum_{n' \neq n} \frac{H_{nn'}^{(1)} H_{n'n}^{(1)}}{E_n^{(0)} - E_{n'}^{(0)}} \right\}. \tag{11–32}$$

The second term is important when $H^{(1)}$ has no diagonal elements.

Because of their complexity, higher orders of the perturbation theory are seldom useful in practice. For completeness, we give here the expression for $E_n^{(3)}$:

$$E_n^{(3)} = H_{nn}^{(3)} + \sum_{n' \neq n} \left\{ \frac{H_{nn'}^{(1)} H_{n'n}^{(2)} + H_{nn'}^{(2)} H_{n'n}^{(1)}}{E_n^{(0)} - E_{n'}^{(0)}} + \frac{H_{nn'}^{(1)} H_{n'n}^{(1)} (H_{n'n'}^{(1)} - H_{nn}^{(1)})}{(E_n^{(0)} - E_{n'}^{(0)})^2} \right\}$$

$$+ \frac{1}{3} \sum_{n' \neq n} \sum_{n'' \neq n, n'} \left\{ \frac{1}{E_{n'}^{(0)} - E_{n''}^{(0)}} + \frac{1}{E_n^{(0)} - E_{n''}^{(0)}} \right\}$$

$$\times \frac{H_{nn''}^{(1)} H_{n''n'}^{(1)} H_{n'n}^{(1)} + H_{nn'}^{(1)} H_{n'n''}^{(1)} H_{n''n}^{(1)}}{E_n^{(0)} - E_{n'}^{(0)}}. \tag{11–33}$$

It may be noted that the kth approximation to E_n can be found as soon as the $(k-1)$th approximation to S is known. This is typical of perturbation calculations and is equivalent to the statement that the perturbation energy is determined by the wave function of the unperturbed state. The formula (11–32) could have been obtained also by calculating the expectation of the second-order Hamiltonian, $H^{(0)} + \epsilon H^{(1)} + \epsilon^2 H^{(2)}$, in the state defined by the first-order wave function (11–29).

11–3 Example: anharmonic oscillator. We shall illustrate the preceding method by calculating the shift of the energy levels of the harmonic oscillator, produced by a perturbation of the form ϵx^4. In the system of units introduced in Section 9–14, the Hamiltonian is

$$H = \tfrac{1}{2}(p^2 + x^2) + \epsilon x^4, \tag{11–34}$$

and the energy of the nth unperturbed state is

$$E_n^{(0)} = n + \tfrac{1}{2} \qquad (n = 0, 1, 2, \ldots).$$

The matrix elements $(n'|x^4|n)$ are zero unless $(n' - n) = 0, \pm 2, \pm 4;$

in these cases, Eq. (9–295) yields the following values:

$$(n - 4|x^4|n) = \tfrac{1}{4}\sqrt{n(n - 1)(n - 2)(n - 3)},$$

$$(n - 2|x^4|n) = \tfrac{1}{2}(2n - 1)\sqrt{n(n - 1)},$$

$$(n|x^4|n) = \tfrac{3}{2}(n^2 + n + \tfrac{1}{2}),$$

$$(n + 2|x^4|n) = \tfrac{1}{2}(2n + 3)\sqrt{(n + 1)(n + 2)},$$

$$(n + 4|x^4|n) = \tfrac{1}{4}\sqrt{(n + 1)(n + 2)(n + 3)(n + 4)}.$$

(11–35)

The energy, to second order in ϵ, can now be obtained by substitution in Eq. (11–32). The sum $E_n^{(2)}$ [Eq. (11–31)] contains four nonzero terms and reduces to a polynomial in n. The result is

$$E_n = n + \tfrac{1}{2} + \tfrac{3}{2}\epsilon(n^2 + n + \tfrac{1}{2}) - \frac{\epsilon^2}{8}(34n^3 + 51n^2 + 59n + 21).$$

(11–36)

The angular frequency of the radiation emitted in a quantum jump from the state $n + 1$ to the state n is, by the Bohr frequency rule,[1]

$$\omega_{n+1,n} = E_{n+1} - E_n = 1 + 3\epsilon(n + 1) - \frac{\epsilon^2}{4}(51n^2 + 102n + 72).$$

(11–37)

This equation is valid provided that the correction terms are small compared to unity. The frequency $\omega_{n+1,n}$ is now dependent upon n, so that the spectrum no longer consists of a single line of frequency ω_0. This is illustrated in Fig. 11–1, in which the function $\omega_{n+1,n}/\omega_0$ is plotted against ϵ for the first four transitions.

According to the correspondence principle, the angular frequency $\omega_{n+1,n}$ for the highly excited states $(n \gg 1)$ should approach that of a classical oscillator having the Hamiltonian (11–34). If n is large, the energy [Eq. (11–36)] becomes

$$E_n \rightarrow E = n + \tfrac{3}{2}\epsilon n^2 - \tfrac{17}{4}\epsilon^2 n^3,$$

(11–38)

and the frequency of the emitted radiation approaches

$$\omega_{n+1,n} \rightarrow \omega = 1 + 3\epsilon n - \tfrac{51}{4}\epsilon^2 n^2.$$

(11–39)

By eliminating n between these equations, one obtains, to order ϵ^2,

$$\omega = 1 + 3\epsilon E - \tfrac{69}{4}\epsilon^2 E^2.$$

(11–40)

---

[1] The angular frequency is measured in the unit $\omega_0 = \sqrt{k/m}$.

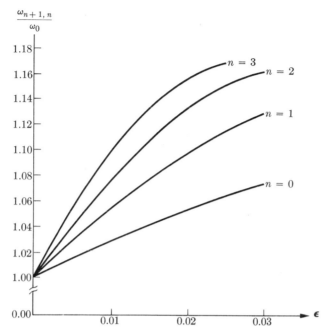

FIG. 11–1.   The frequency emitted by the anharmonic oscillator in a quantum jump from the state $n + 1$ to the state n (in units of the frequency $\omega_0 = \sqrt{k/m}$ of the harmonic oscillator) as a function of the strength of the perturbation ϵx^4.

We shall verify that this is the correct relation between frequency and energy for the classical oscillator. The kinetic energy is

$$\tfrac{1}{2}mv^2 = E - V, \qquad (11\text{–}41)$$

whence

$$\tfrac{1}{2}v^2 = E - \tfrac{1}{2}x^2 - \epsilon x^4. \qquad (11\text{–}42)$$

The amplitude a of the oscillation is the value of x at $v = 0$, whence a^2 is the positive root of the equation

$$E - \tfrac{1}{2}x^2 - \epsilon x^4 = 0, \qquad (11\text{–}43)$$

namely,

$$a^2 = \frac{1}{4\epsilon} \left(\sqrt{1 + 16\epsilon E} - 1\right). \qquad (11\text{–}44)$$

The negative root is

$$-b^2 = -\frac{1}{4\epsilon} \left(\sqrt{1 + 16\epsilon E} + 1\right), \qquad (11\text{–}45)$$

and v^2 can be written

$$v^2 = 2\epsilon(a^2 - x^2)(b^2 + x^2). \qquad (11\text{–}46)$$

The period of the oscillation can be expressed as

$$P = \frac{2\pi}{\omega} = 4 \int_0^a \frac{dx}{v} = \frac{4}{\sqrt{2\epsilon}} \int_0^a \frac{dx}{\sqrt{(a^2 - x^2)(b^2 + x^2)}}. \qquad (11\text{-}47)$$

By the substitution $x = a \cos \theta$, Eq. (11–47) is reduced to

$$\frac{2\pi}{\omega} = \frac{4}{\sqrt{2\epsilon}\,\sqrt{a^2 + b^2}} \int_0^{\pi/2} \frac{d\theta}{\sqrt{1 - \kappa^2 \sin^2 \theta}} = \frac{4K}{(1 + 16\epsilon E)^{1/4}}, \qquad (11\text{-}48)$$

where

$$\kappa^2 = \frac{a^2}{a^2 + b^2} = \frac{1}{2}\left(1 - \frac{1}{\sqrt{1 + 16\epsilon E}}\right),$$

and K is the complete elliptic integral of modulus κ. Solving for ω, we have

$$\omega = \frac{(1 + 16\epsilon E)^{1/4}}{(2/\pi)K}, \qquad (11\text{-}49)$$

which remains to be expanded in powers of ϵ. Working to order ϵ^2, we find[1]

$$\kappa^2 = 4\epsilon E(1 - 12\epsilon E),$$

$$(1 + 16\epsilon E)^{1/4} = 1 + 4\epsilon E - 24\epsilon^2 E^2,$$

$$\frac{2}{\pi} K = 1 + \tfrac{1}{4}\kappa^2 + \tfrac{9}{64}\kappa^4$$

$$= 1 + \epsilon E - \tfrac{39}{4}\epsilon^2 E^2,$$

and finally,

$$\omega = 1 + 3\epsilon E - \tfrac{69}{4}\epsilon^2 E^2.$$

This expression is seen to be identical with the correspondence limit (11–40).

11–4 Perturbation of degenerate stationary states. The argument of Section 11–2 fails when the unperturbed Hamiltonian has degenerate eigenstates. The difficulty appears first in Eq. (11–27), which, if $E_n^{(0)} = E_{n'}^{(0)}$, becomes

$$H_{n'n}^{(1)} = 0 \qquad (n' \neq n).$$

Now if the eigenfunctions $\psi_n^{(0)}$ and $\psi_{n'}^{(0)}$ have been arbitrarily chosen,

---

[1] E. Jahnke and F. Emde, *Tables of Functions.* 4th ed. New York: Dover Publications, 1945, p. 73.

this equation will not, in general, be satisfied by the nondiagonal matrix elements of $H^{(1)}$. However, it can always be satisfied by properly selected unperturbed eigenfunctions. The reason is that, in the subspace defined by the degenerate states, $H^{(0)}$ is a multiple of the unit matrix and thus commutes with that part of the matrix $H^{(1)}$ which belongs to the degenerate states. By constructing unperturbed eigenfunctions which diagonalize $H^{(1)}$, we *adapt* the representation to the perturbation, and the difficulty is resolved. Furthermore, the matrix elements of S are thereby determined.

To examine the details of this process, we suppose that

$$E_1^{(0)} = E_2^{(0)} = \cdots = E_g^{(0)} = E^{(0)}, \qquad (11\text{--}50)$$

so that there is a g-fold degeneracy of the zero-order states. We now form linear combinations of the functions $\psi_1^{(0)}, \psi_2^{(0)}, \ldots, \psi_g^{(0)}$ such that $H^{(1)}$ is diagonal in these states, i.e., we construct g functions $\phi_i^{(0)}$ according to

$$\phi_i^{(0)} = \sum_{j=1}^{g} U_{ji} \psi_j^{(0)} \qquad (i = 1, 2, \ldots, g) \qquad (11\text{--}51)$$

and require that

$$H^{(1)} \phi_i^{(0)} = E_i^{(1)} \phi_i^{(0)} + \Phi, \qquad (11\text{--}52)$$

where the function Φ is orthogonal to each of the degenerate states. Substituting Eq. (11–51) and forming the scalar product with $\psi_k^{(0)}$, we find

$$\sum_{j=1}^{g} U_{ji}(\psi_k^{(0)}, H^{(1)}\psi_j^{(0)}) = E_i^{(1)} \sum_{j=1}^{g} U_{ji}(\psi_k^{(0)}, \psi_j^{(0)}) + (\psi_k^{(0)}, \Phi). \quad (11\text{--}53)$$

By hypothesis, the last term in this equation is zero, and since the $\psi_k^{(0)}$ are orthonormal, we have

$$\sum_{j=1}^{g} (H_{kj}^{(1)} - E_i^{(1)} \delta_{kj}) U_{ji} = 0 \qquad (k = 1, 2, \ldots, g). \qquad (11\text{--}54)$$

For a given value of i, Eq. (11–54) is a set of g homogeneous linear equations which determine the g coefficients U_{ji}. The numbers $E_i^{(1)}$ are the roots of the determinantal equation

$$\det (H_{kj}^{(1)} - E^{(1)} \delta_{kj}) = 0. \qquad (11\text{--}55)$$

This is an equation of gth degree in $E^{(1)}$, and for each of its roots a function $\phi_i^{(0)}$ of the form (11–51) can be determined by solving the equations (11–54).

We now *adapt the representation to the perturbation* by replacing the functions $\psi_i^{(0)}$ with the $\phi_i^{(0)}$ in the representation of the unperturbed states. The perturbation $H^{(1)}$ is now represented by a new matrix, for which the troublesome equation (11–27) is identically satisfied whenever $E_n^{(0)} = E_{n'}^{(0)}$, and the rest of the theory can be developed very much as before. The wave functions are given, to first order in ϵ, by

$$\phi_n = \phi_n^{(0)} + \epsilon {\sum}' \frac{H_{n'n}^{(1)}}{E_n^{(0)} - E_{n'}^{(0)}} \psi_{n'}^{(0)}, \qquad (11\text{–}56)$$

where the sum, denoted by ${\sum}'$, contains only terms for which $E_{n'}^{(0)} \neq E_n^{(0)}$. Similarly, the energy of the perturbed state ϕ_n is

$$E_n = (\phi_n, H\phi_n) = (\phi_n, (H^{(0)} + \epsilon H^{(1)})\phi_n) = E_n^{(0)} + \epsilon E_n^{(1)}. \qquad (11\text{–}57)$$

If $H^{(1)}$ is not itself degenerate, the degeneracy in $H^{(0)}$ is completely removed by the perturbation; the g-fold degenerate level is separated into g distinct levels, the separation being proportional to the strength of the perturbation. This is shown graphically in Fig. 11–2.

The representation which is adapted to a given perturbation can frequently be determined from considerations of symmetry. For example, a degenerate system with spherical symmetry, such as the hydrogen atom, is best described in terms of eigenfunctions of $\mathbf{L}^2$ and L_z if one is interested in the effect of spherically symmetric departures from the Coulomb field. A perturbation of the x-component of force in the three-dimensional oscillator can be calculated by means of the (n_x, n_y, n_z)-representation.

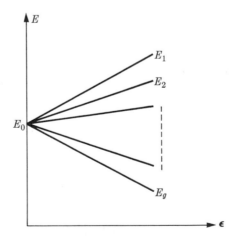

FIG. 11–2.   First-order splitting of degenerate level.

An example of some importance in the shell theory of nuclear structure[1] is the perturbation of the levels of the harmonic oscillator by a field of force concentrated near the origin. We shall work out the change in the nth level of the three-dimensional oscillator produced by a perturbation of the form

$$H^{(1)} = e^{-\beta r^2}. \tag{11–58}$$

Since $H^{(1)}$ is spherically symmetric, the wave functions (7–149) are adapted to the perturbation. The diagonal element of $H^{(1)}$ in the state $\psi(nlm)$ is

$$(nlm|H^{(1)}|nlm) = H^{(1)}(nl) = \int_0^\infty \frac{2}{r} e^{-\beta r^2} \{\Lambda_k^{l+1/2}(r^2)\}^2 r^2 \, dr, \tag{11–59}$$

where $k = \frac{1}{2}(n - l)$, and the function $\Lambda_k^{l+1/2}$ is defined in Eq. (7–145). By means of the substitutions

$$t = (1 + \beta)r^2, \qquad \lambda = \frac{1}{1 + \beta}, \tag{11–60}$$

the integral (11–59) can be written in the form

$$H^{(1)}(nl) = \frac{\lambda^{n+3/2}}{N^2} \int_0^\infty e^{-t} t^\alpha \left\{ \frac{1}{\lambda^k} L_k^\alpha(\lambda t) \right\}^2 dt,$$

where N is the normalization constant of Eq. (7–148), and $\alpha = l + \frac{1}{2}$. The identity[2]

$$\frac{1}{\lambda^k} L_k^\alpha(\lambda t) = \sum_{m=0}^k \binom{k + \alpha}{m} \lambda^{-m}(1 - \lambda)^m L_{k-m}^\alpha(t)$$

can now be used, together with the orthogonality theorem (7–146), to evaluate the integral explicitly. One obtains

$$H^{(1)}(nl) = \frac{1}{(1 + \beta)^{n+3/2}} \sum_{m=0}^k \binom{k}{m}\binom{k + \alpha}{m} \beta^{2m}. \tag{11–61}$$

In particular,

$$H^{(1)}(0, 0) = \frac{1}{(1 + \beta)^{3/2}}, \qquad H^{(1)}(1, 1) = \frac{1}{(1 + \beta)^{5/2}},$$

$$H^{(1)}(2, 0) = \frac{1}{(1 + \beta)^{7/2}}(1 + \tfrac{3}{2}\beta^2), \quad H^{(1)}(2, 2) = \frac{1}{(1 + \beta)^{7/2}},$$

---

[1] M. Goeppert Mayer and J. H. D. Jensen, *Elementary Theory of Nuclear Shell Structure*. New York: John Wiley and Sons, Inc., 1955, Section IV.3, p. 52.
[2] A. Erdélyi, *Higher Transcendental Functions*, Bateman Manuscript Project. New York: McGraw-Hill Book Co., Inc., 1953, Vol. 2, p. 192, Eq. (40).

$$H^{(1)}(3, 1) = \frac{1}{(1 + \beta)^{9/2}} (1 + \tfrac{5}{2}\beta^2), \quad H^{(1)}(3, 3) = \frac{1}{(1 + \beta)^{9/2}}, \quad (11\text{–}62)$$

$$H^{(1)}(4, 0)$$

$$= \frac{1}{(1 + \beta)^{11/2}} (1 + 5\beta^2 + \tfrac{15}{8}\beta^4), \quad H^{(1)}(4, 2) = \frac{1}{(1 + \beta)^{11/2}} (1 + \tfrac{7}{2}\beta^2),$$

$$H^{(1)}(4, 4) = \frac{1}{(1 + \beta)^{11/2}}.$$

The perturbed levels are conventionally labelled by the symbol (Nl), where $N = k + 1$ is the number of radial nodes in the wave function. The order of the levels corresponding to Eq. (11–61) is as follows:[1]

$$\underbrace{(1s)}_{n\,=\,0} \quad \underbrace{(1p)}_{n\,=\,1} \quad \underbrace{(1d)(2s)}_{n\,=\,2} \quad \underbrace{(1f)(2p)}_{n\,=\,3} \quad \underbrace{(1g)(2d)(3s)}_{n\,=\,4} \quad \underbrace{(1h)(2f)(3p)}_{n\,=\,5}, \text{ etc.}$$

Each of these levels is, of course, $(2l + 1)$-fold degenerate, because the perturbed system is spherically symmetric, but the accidental degeneracy is completely removed.

11–5 Atomic Zeeman levels; Russell-Saunders coupling. In Section 10–7 it was shown that the energy of an atom placed in a magnetic field $\mathfrak{B}$ is increased by the amount $-\mu_0 \mathfrak{B} \cdot \mathbf{L}$. This change of energy arises through the interaction of the magnetic field with the moving electrons in the atom. The quantity μ_0 is the Bohr magneton, and $\mathbf{L}$ is the total orbital angular momentum of the electrons:

$$\mathbf{L} = \sum_{i=1}^{Z} l_i. \quad (11\text{–}63)$$

In addition, each electron has an intrinsic magnetic moment associated with its spin. The magnetic energy arising from interaction with the spin is $-2\mu_0 \mathfrak{B} \cdot \mathbf{S}$, where

$$\mathbf{S} = \sum_{i=1}^{Z} s_i \quad (11\text{–}64)$$

is the total spin angular momentum. The factor 2 enters because of the anomalous gyromagnetic ratio for the intrinsic magnetic moment (cf.

---

[1] M. Goeppert Mayer and J. H. D. Jensen, *Elementary Theory of Nuclear Shell Structure*. New York: John Wiley and Sons, Inc., 1955, Fig. IV.1, p. 53.

Section 10–8). The perturbation produced by a weak magnetic field is, therefore,

$$H^{(1)} = -\mu_0 \mathcal{B} \cdot (\mathbf{L} + 2\mathbf{S}) = -\mu_0 \mathcal{B} \cdot (\mathbf{J} + \mathbf{S}). \qquad (11\text{--}65)$$

In many cases, it is a good approximation to assume that both $\mathbf{L}$ and $\mathbf{S}$ are constants of the motion, i.e., that they are good quantum numbers for the classification of atomic energy levels. This amounts essentially to the approximation in which only the electrostatic interaction of the electrons is considered, the effects of the spin-orbit interaction being neglected. This approximation holds quite accurately for atoms whose electronic configurations are nearly *closed-shell* configurations (Russell-Saunders coupling).

The diagonal element of $H^{(1)}$ for a given LSJ-state is the weak-field perturbation of the corresponding Russell-Saunders term. Since $\mathbf{S}$ is a vector operator, the diagonal element of $\mathbf{S}$ in such a state is

$$\langle \mathbf{S} \rangle = \left\langle \frac{\mathbf{J} \cdot \mathbf{S}}{J(J+1)} \, \mathbf{J} \right\rangle,$$

and from

$$\mathbf{J} - \mathbf{S} = \mathbf{L},$$

we have

$$\mathbf{J} \cdot \mathbf{S} = \tfrac{1}{2}(\mathbf{J}^2 + \mathbf{S}^2 - \mathbf{L}^2),$$

whence the perturbation energy is

$$\Delta E = (LSJM|H^{(1)}|LSJM)$$

$$= \mu_0(M|\mathcal{B} \cdot \mathbf{J}|M) \left\{ 1 + \frac{J(J+1) + S(S+1) - L(L+1)}{2J(J+1)} \right\},$$

or

$$\Delta E = -\mu_0 \mathcal{B} M g, \qquad (11\text{--}66)$$

provided that the axis of quantization is taken in the direction of $\mathcal{B}$. The factor

$$g = 1 + \frac{J(J+1) + S(S+1) - L(L+1)}{2J(J+1)} \qquad (11\text{--}67)$$

is the *Landé g-factor*, which was constructed from empirical analysis of Zeeman spectra before the advent of quantum mechanics.

The quantum number M in Eq. (11–66) has the $2J + 1$ values J, $J - 1, \ldots, -J$, so that the perturbed levels are evenly spaced about the unperturbed eigenvalue. Only in the case $g = 1$ does the formula (11–66) agree with the expression (10–124), which leads to the normal Lorentz triplet. This always occurs, for example, for singlet states

$(S = 0)$, so that the normal effect is always observed in transitions between singlet levels.   In Section 11–13, we shall see how the formula (11–67), together with the selection rules for dipole transitions, gives a complete account of the (weak-field) anomalous Zeeman effect.

11–6 The variational method.   The perturbation method, described in the preceding sections, can only be applied to problems that are similar to others for which exact solutions exist.   For systems differing substantially from those subject to exact treatment, it is sometimes possible to write a trial wave function, on the basis of an intelligent guess.   Then a powerful method can be applied that leads to an approximation to the lowest energy eigenvalue and ground-state wave function for the system. This method is based upon the following *variational principle:*

The expectation of the Hamiltonian in any state ψ, namely

$$\langle H \rangle = \frac{(\psi, H\psi)}{(\psi, \psi)},$$

is never smaller than the energy E_0 of the ground state of the system. If ψ differs from the eigenfunction ψ_0 of the ground state by a small quantity of order ϵ, then $(\psi, H\psi)/(\psi, \psi)$ differs from E_0 by a small quantity of order ϵ^2.

This principle follows immediately from the stationary property of the eigenvectors of H, which formed the basis of the diagonalization theorem discussed in Section 9–7; that is, E_0 is the least value of $\langle H \rangle$ for all normalized functions ψ.   From a slightly different point of view, it can be understood in the following way: Let us define

$$\grave{E}' = \frac{(\psi, H\psi)}{(\psi, \psi)} \tag{11–68}$$

to be the approximate value of E_0, obtained as the value of $\langle H \rangle$ for a wave function

$$\psi = \psi_0 + \epsilon\phi \tag{11–69}$$

which differs slightly from ψ_0.   Inserting this expression for ψ into Eq. (11–68) and expanding, we find

$$\{(\psi_0, \psi_0) + \epsilon(\psi_0, \phi) + \epsilon(\phi, \psi_0) + \epsilon^2(\phi, \phi)\}\grave{E}'$$
$$= (\psi_0, H\psi_0) + \epsilon(\psi_0, H\phi) + \epsilon(\phi, H\psi_0) + \epsilon^2(\phi, H\phi). \tag{11–70}$$

If we now assume that $\grave{E}'$ can be expanded in powers of ϵ,

$$\grave{E}' = E_0 + \epsilon E^{(1)} + \epsilon^2 E^{(2)}, \tag{11–71}$$

we obtain, on equating like powers of ϵ in Eq. (11–70),

$$(\psi_0, \psi_0)E_0 = (\psi_0, H\psi_0), \tag{11–72}$$

$$E_0(\psi_0, \phi) + E_0(\phi, \psi_0) + (\psi_0, \psi_0)E^{(1)} = (\psi_0, H\phi) + (\phi, H\psi_0). \tag{11–73}$$

The first of these expressions is an identity because of the definition of ψ_0:

$$H\psi_0 = E_0\psi_0. \tag{11–74}$$

Also, since H is Hermitian, it follows from Eq. (11–73) that $E^{(1)}$ vanishes, and hence,

$$\backslash E' = E_0 + \epsilon^2 E^{(2)}. \tag{11–75}$$

The quantity $\backslash E'$ contains no term proportional to ϵ. The coefficients of ϵ^2 in Eq. (11–70) give the further relation

$$E_0(\phi, \phi) + E^{(2)}(\psi_0, \psi_0) = (\phi, H\phi),$$

or

$$E^{(2)}(\psi_0, \psi_0) = (\phi, (H - E_0)\phi). \tag{11–76}$$

If we now expand the "error function" ϕ in terms of the normalized eigenfunctions ψ_n of H, i.e.,

$$\phi = \sum_n a_n\psi_n, \qquad H\psi_n = E_n\psi_n,$$

Eq. (11–76) becomes

$$E^{(2)} = \sum_{n>0} |a_n|^2(E_n - E_0). \tag{11–77}$$

Since E_0 is the smallest of the E_n, the sum is non-negative, that is,

$$E^{(2)} \geq 0. \tag{11–78}$$

The quantity $\backslash E'$ is therefore an approximation in excess.

The variational principle is useful when an approximate ground-state wave function ψ can be obtained. In practice, one frequently has information about the general form of the wave function, which can be expressed in terms of a suitably chosen function containing one or more parameters. Equation (11–68) can then be used to compute $\backslash E'$ as a function of these parameters, and the parameters can be adjusted to minimize $\backslash E'$. This minimum value of $\backslash E'$ is the best value of E_0 obtainable with a wave function of the chosen form. Because of the second-order nature of the approximation, the value of $\backslash E'$ is, in general, a much better approximation to E_0 than the corresponding ψ to the ground-state wave function ψ_0.

We shall illustrate the method by applying it to a problem in the theory of the deuteron.[1] The force between a neutron and a proton can be described, in good approximation, by a potential-energy function of the form

$$V(r) = -V_0 \frac{\exp\left[-(r/r_0)\right]}{(r/r_0)}.\tag{11-79}$$

This form arises in the meson theory of nuclear forces and is called the *Yukawa potential*. The range of the force, r_0, is related to the mass μ of the associated mesons by the formula

$$r_0 = \frac{\hbar}{\mu c},\tag{11-80}$$

and the strength V_0, or depth of the potential well, is connected with the strength of the coupling between the meson and nucleon fields. In the center-of-mass coordinates, the Hamiltonian for the *s*-state of the neutron-proton system is

$$H = \frac{p^2}{2m} + V(r),\tag{11-81}$$

in which the reduced mass, m, is one-half the mass of a nucleon,[2] and r is the neutron-proton separation. The binding energy, E_0, of the deuteron is the minimum of the quantity (11-68). We shall estimate this by means of the variational principle, using the simple trial function

$$\psi = e^{-\alpha r/r_0},\tag{11-82}$$

in which we treat α as a variable parameter. For convenience, we introduce the variable $x = r/r_0$, and obtain

$$`E' = \frac{(\psi,\ (p^2/2m)\psi) + (\psi,\ V(r)\psi)}{(\psi,\ \psi)},\tag{11-83}$$

in which the scalar products are of the form

$$(\psi,\ \phi) = 4\pi \int_0^\infty \psi^* \phi r^2\, dr = 4\pi r_0^3 \int_0^\infty \psi^* \phi x^2\, dx.\tag{11-84}$$

---

[1] R. G. Sachs, *Nuclear Theory*. Reading, Mass.: Addison-Wesley Publishing Co., Inc., 1953, Chapter 3. J. M. Blatt and V. F. Weisskopf, *Theoretical Nuclear Physics*. New York: John Wiley and Sons, 1952, Chapter II.
[2] The difference of the masses of the neutron and proton is negligible for these considerations.

Equations (11–79) and (11–82) now yield, in succession,

$$(\psi, \psi) = 4\pi r_0^3 \int_0^\infty e^{-2\alpha x} x^2 \, dx = \frac{\pi r_0^3}{\alpha^3} \; ; \qquad (11\text{–}85)$$

$$\left(\psi, \frac{p^2}{2m} \psi\right) = \frac{\hbar^2}{2m} \cdot 4\pi r_0 \int_0^\infty \left(\frac{d\psi}{dx}\right)^2 x^2 \, dx = \frac{\hbar^2 r_0}{2m} \frac{\pi}{\alpha} \; ; \qquad (11\text{–}86)$$

$$(\psi, V(r)\psi) = -V_0 \cdot 4\pi r_0^3 \int_0^\infty e^{-(2\alpha+1)x} x \, dx = -V_0 \frac{4\pi r_0^3}{(2\alpha + 1)^2} \cdot \qquad (11\text{–}87)$$

Substituting into Eq. (11–83) and simplifying, we find

$$\frac{2m r_0^2}{\hbar^2} (-\mathstrut^\backprime E') = \alpha^2 \left\{ \frac{4\gamma^2 \alpha}{(2\alpha + 1)^2} - 1 \right\}, \qquad (11\text{–}88)$$

where

$$\gamma^2 = \frac{2m V_0 r_0^2}{\hbar^2}.$$

The best approximation is obtained by minimizing the expression (11–88) with respect to α. By differentiation, one easily finds that α must satisfy

$$\gamma^2 = \frac{1}{2\alpha} \frac{(2\alpha + 1)^3}{(2\alpha + 3)}, \qquad (11\text{–}89)$$

and that, for this value of α,

$$\frac{2m r_0^2}{\hbar^2} (-\mathstrut^\backprime E') = \alpha^2 \frac{2\alpha - 1}{2\alpha + 3}. \qquad (11\text{–}90)$$

If the quantities V_0 and r_0 are given, the corresponding value of $\mathstrut^\backprime E'$ can be found by solving Eq. (11–89) for α and substituting in Eq. (11–90). The experimentally determined quantity, however, is [1]

$$-E_0 = 2.226 \text{ Mev}; \qquad (11\text{–}91)$$

this can be substituted for $-\mathstrut^\backprime E'$ in Eq. (11–90), which then allows an approximate calculation of the relation between V_0 and r_0 (*range-depth relation*) that must hold if the potential function (11–79) is to give the value (11–91) for the binding energy.

The result of such a calculation is shown in Fig. 11–3, in which the function

$$V_0 r_0^2 = 41.37 \gamma^2 \text{ Mev } (10^{-26} \text{ cm}^2) \qquad (11\text{–}92)$$

---

[1] R. C. Mobley and R. A. Laubenstein, *Phys. Rev.* **80**, 309 (1950).

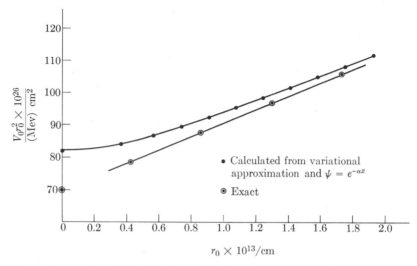

Fɪɢ. 11–3.　Range-depth relation for Yukawa potential (deuteron).

is plotted as a function of r_0. The circled points on the graph have been obtained from an exact calculation and are given for comparison with the approximate result.　In the neighborhood of the range $r_0 = 1.4 \times 10^{-13}$ cm, which corresponds to a meson of mass (270 × electron mass),[1] the approximate result is in error by about one-fourth of one percent.[2]

　　The variational principle (11–68) can also be used to obtain approximate values for the energy of an excited state of the system provided that the wave functions of states of lower energy are accurately known. If ψ is orthogonal to all states below a given one, say the kth, Eq. (11–69) may be replaced by

$$\psi = \psi_k + \epsilon\phi, \qquad (11\text{–}93)$$

and Eqs. (11–75) and (11–77) by

$$\grave{}E\grave{} = E_k + \epsilon^2 \sum_{n>k} |a_n|^2 (E_n - E_k), \qquad (11\text{–}94)$$

whence the variational expression $\grave{}E\grave{}$ is an approximation in excess to E_k.

---

　　[1] The mass of the pi-meson, which is believed to be the meson associated with the nuclear force, is 273 m_e. See H. A. Bethe and F. de Hoffmann, *Mesons and Fields.* Evanston: Row Peterson and Co., 1955, Vol. II, p. 4.)
　　[2] It is possible to construct a variational expression for V_0, in which E_0 is treated as a given fixed constant. See L. H. Thomas, *Phys. Rev.* **51**, 202 (1937); R. G. Sachs, *Nuclear Theory.* Reading, Mass.: Addison-Wesley Publishing Co., Inc., 1953, p. 35.

This observation, however, is rarely of great practical utility, because it is usually quite difficult to obtain accurate wave functions for the lower states. A different situation arises, however, when the eigenfunctions can be classified according to the eigenvalues α of a commuting operator A. One can then work entirely within the subspace of functions belonging to a given value of α, and thus obtain an approximation to the lowest energy level within this subset of levels.

There is a simple relationship between the variational principle (11–68) and the formula (11–10) of perturbation theory. If the Hamiltonian

$$H = H^{(0)} + \epsilon H^{(1)}$$

differs only slightly from $H^{(0)}$, then it is natural to use the unperturbed wave function ψ_0 as a trial function. The variational approximation obtained in this way is

$$`E' = \frac{(\psi_0, (H^{(0)} + \epsilon H^{(1)})\psi_0)}{(\psi_0, \psi_0)} = E_0 + \langle \epsilon H^{(1)} \rangle_0, \qquad (11\text{–}95)$$

that is, exactly the first-order perturbation formula. It is frequently possible, however, to obtain an improvement by slight modification of the trial function.

11–7 Time-dependent perturbations; transition probability. So far we have been concerned with isolated systems for which the Hamiltonian is independent of the time. The energy eigenstates of such a system are stationary; the time enters only in the phases, according to

$$\psi_n(t) = \psi_n e^{(-i/\hbar)E_n t}, \qquad (11\text{–}96)$$

in which ψ_n is the time-independent energy eigenfunction for the nth state. If the system is in an arbitrary state

$$\psi(0) = \sum_n a_n \psi_n \qquad (11\text{–}97)$$

at time $t = 0$, its wave function at any other time is given by

$$\psi(t) = \sum_n a_n \psi_n e^{(-i/\hbar)E_n t}, \qquad (11\text{–}98)$$

and the amplitude of the state ψ_n in this expansion is

$$a_n(t) = (\psi_n, \psi(t)) = a_n e^{(-i/\hbar)E_n t}, \qquad (11\text{–}99)$$

from which follows

$$|a_n(t)|^2 = \text{a constant.} \qquad (11\text{–}100)$$

Thus, only the relative phases of the component states change with time, and the probability of finding the value E_n in a measurement of the energy is constant. The same is true for the values of any other observables which commute with H, i.e., which are constants of the motion.

If external forces act upon the system, this situation is changed. The Hamiltonian then contains a time-dependent part, representing the external influence, and the quantities $a_n(t)$ depend upon the time in amplitude as well as phase. In particular, it is possible that a state not initially present in the expansion of ψ will *grow*. Thus, a system which is in an eigenstate ψ_n at time $t = 0$ can change its character under the external force, so that at a later time the predominant component of ψ is a different state $\psi_{n'}$. This change is described by saying that the external influence produces a *transition* from ψ_n to $\psi_{n'}$. It is implied that the expectation of the energy of the system also changes from E_n to $E_{n'}$. The difference in energy is accounted for by the work done by the external force. It will be seen below that the law of conservation of energy is an automatic result of the quantum-mechanical treatment.

In the strictest sense, the description of external forces by means of a time-dependent term in the Hamiltonian is always an approximation, because an external force always arises through interaction with another system whose time behavior is regarded as given in advance and not influenced by the interaction. This assumption can only be true if the coupling between the two systems is weak. If the energy exchange between the systems during a time characteristic of their unperturbed motion is an appreciable fraction of the total energy, then the motions of both systems are strongly affected, and the approximation fails. In the case of an atom in interaction with electromagnetic radiation, for example, the forces on the charged particles in the atom can be approximately described by an additive term in the atomic Hamiltonian, of the form $V_0 \sin \omega t$, in which ω is the frequency of the radiation. When this is done, however, it is implicitly assumed that the influence of the atom itself, as a source of electromagnetic waves, can be neglected. An important consequence of this approximation is that the theory describes only absorption and induced emission,[1] but does not account for spontaneous emission. A thorough treatment of the latter phenomenon can be obtained only if the entire system (atom + radiation field) is treated as a single entity. The Hamiltonian is then independent of the time, and a transition between atomic levels is described, in a sense, as a redistribution of the energy between the two parts of the coupled system. A quantum theory

---

[1] Cf. Section 1–3.

of the electromagnetic field is required for a development of this theory, which will not be given in this book.[1]

A system subject to a weak time-dependent perturbation can be described approximately by a Hamiltonian of the form

$$H = H^0 + V(t), \tag{11-101}$$

where $V(t)$ is a small term arising from interaction with a second system, and H^0 is the Hamiltonian of the unperturbed system.

We assume that the eigenfunctions ψ_n of H^0, which satisfy

$$H^0\psi_n = E_n\psi_n, \tag{11-102}$$

are known, and that the spectrum of energy eigenvalues E_n is discrete. (The superscripts $^{(0)}$, which were employed in the preceding sections, are unnecessary here and have been dropped.) A state of the unperturbed system which corresponds to ψ_n at $t = 0$ is given, at a later time t, by Eq. (11–96).

Since the perturbation is time-dependent, the perturbed system does not, in general, have stationary states. The time behavior of its wave function is determined by the Schrödinger equation

$$H\psi = i\hbar \frac{\partial \psi}{\partial t}. \tag{11-103}$$

The functions $\psi_n(t)$ are, by hypothesis, a complete set for the system, and ψ can therefore be expanded in the form

$$\psi(t) - \sum_n c_n(t)\psi_n(t) - \sum_n c_n(t)\psi_n e^{(-i/\hbar)E_n t}, \tag{11-104}$$

in which the expansion coefficients are

$$c_n = c_n(t) = (\psi_n(t), \psi). \tag{11-105}$$

Substitution in the Schrödinger equation (11–103) yields

$$i\hbar \sum_n \left(\frac{dc_n}{dt} - \frac{i}{\hbar} E_n c_n \right) \psi_n e^{(-i/\hbar)E_n t} = \sum_n c_n H\psi_n e^{(-i/\hbar)E_n t},$$

and, by taking the scalar product with ψ_m in each member, we find

$$\left(i\hbar \frac{dc_m}{dt} + E_m c_m \right) e^{(-i/\hbar)E_m t} = \sum_n c_n [(\psi_m, H^0\psi_n) + (\psi_m, V(t)\psi_n)] e^{(-i/\hbar)E_n t}. \tag{11-106}$$

---

[1] W. Heitler, *The Quantum Theory of Radiation*. London: Oxford University Press, 1936. J. M. Jauch and F. Rohrlich, *The Theory of Photons and Electrons*. Reading, Mass.: Addison-Wesley Publishing Co., Inc., 1955.

Since the ψ_n are orthonormal, Eq. (11–102) gives

$$(\psi_m, H^0\psi_n) = E_n(\psi_m, \psi_n) = E_n\,\delta_{mn},$$

whence Eq. (11–106) reduces to

$$i\hbar\,\frac{dc_m}{dt} = \sum_n c_n V_{mn}(t)e^{(i/\hbar)(E_m-E_n)t}, \qquad (11\text{–}107)$$

where

$$V_{mn}(t) = (\psi_m, V(t)\psi_n). \qquad (11\text{–}108)$$

The equations (11–107) are equivalent to the Schrödinger equation, and their solution, subject to the initial conditions

$$c_m(0) = (\psi_m, \psi(0)), \qquad (11\text{–}109)$$

determines the behavior of the system for all time. It is not possible, in general, to find an exact solution of this infinite set of equations. However, if the perturbation is small, the equations show that the c_i change slowly with the time, so that an approximate solution can be obtained by neglecting the time variation of c_n in the right-hand members. For definiteness, we assume that the initial state is a particular one, ψ_I, of the states ψ_n. The initial conditions (11–109) are then

$$c_n(0) = \delta_{nI}, \qquad (11\text{–}110)$$

and the equations (11–107) become, approximately,

$$i\hbar\,\frac{dc_n}{dt} = V_{nI}(t)e^{(i/\hbar)(E_n-E_I)t}. \qquad (11\text{–}111)$$

The solution of these equations, namely

$$c_I(t) = 1 + \frac{1}{i\hbar}\int_0^t V_{II}(t)\,dt, \qquad (11\text{–}112)$$

$$c_n(t) = \frac{1}{i\hbar}\int_0^t V_{nI}(t)e^{(i/\hbar)(E_n-E_I)t}\,dt, \qquad (n \neq I) \qquad (11\text{–}113)$$

is the basis for the first-order time-dependent perturbation theory.

11–8 Constant perturbation. We consider a perturbation $V(t)$ that is zero before the initial instant $t = 0$ and constant thereafter. The system, which has been prepared in the initial unperturbed state ψ_I, is suddenly subjected, at $t = 0$, to a weak perturbation, which persists at a constant value. The functions $c_n(t)$ are then

$$c_I(t) = 1 + \frac{1}{i\hbar}\, V_{II} t, \qquad (11\text{--}114)$$

$$c_n(t) = \frac{V_{nI}}{E_n - E_I}\, (1 - e^{(i/\hbar)(E_n - E_I)t}) \quad (n \neq I). \qquad (11\text{--}115)$$

The first of these equations shows that the term in ψ_I in the expansion of the wave function is

$$c_I(t)\psi_I e^{(-i/\hbar)E_I t} = \psi_I e^{(-i/\hbar)E_I t}\left(1 - \frac{i}{\hbar}\, V_{II} t\right). \qquad (11\text{--}116)$$

Since V is small, we can write approximately

$$c_I(t)\psi_I e^{(-i/\hbar)E_I t} \approx \psi_I e^{(-i/\hbar)(E_I + V_{II})t}, \qquad (11\text{--}117)$$

provided that t is not too large. It follows that, so far as the initial state is concerned, the principal influence of the perturbation is to change the phase. This change is in accord with the fact that the energy of the perturbed state is

$$E = E_I + V_{II},$$

where V_{II} is the expectation of V in the state ψ_I. This is in satisfactory agreement with the result (11–10) of the approximate theory of stationary states.

The quantity $|c_n(t)|^2$ measures the probability that the state ψ_n is occupied by the system at the time t. If the perturbation is turned off at the time t_1, and a measurement of the energy is subsequently made, the probability of obtaining the value E_n is

$$|c_n(t_1)|^2 = \frac{|V_{nI}|^2}{(E_n - E_I)^2}\left\{2 \sin \frac{E_n - E_I}{2\hbar}\, t_1\right\}^2. \qquad (11\text{--}118)$$

This result is not consistent with the approximation made in its derivation unless each of the coefficients $c_n(t)$ is small. However, it follows from Eq. (11–118) that

$$\sum_{n \neq I} |c_n(t)|^2 \leq \sum_{n \neq I} \frac{4|V_{nI}|^2}{(E_n - E_I)^2};$$

hence the condition that the $c_n(t)$ be small can be fulfilled if V is sufficiently small, provided none of the states E_n is degenerate with the initial state.

The quantity $|c_n(t)|^2$ is a periodic function of t, with angular frequency $(E_n - E_I)/2\hbar$ and amplitude

$$|c_n|^2_{\max} = \frac{4|V_{nI}|^2}{(E_n - E_I)^2} . \qquad (11\text{--}119)$$

Consequently, the probability of finding the system in the state n is small unless the energy of this state is close to the energy of the initial state. In other words, the probability of a transition from state I to state n is small unless the energy is approximately conserved in the process. Furthermore, if $E_n - E_I$ is large compared to $|V_{nI}|$, the probability oscillates rapidly between zero and the small maximum given by Eq. (11–119). The period of this oscillation, that is, the time interval within which $|c_n(t)|^2$ has an appreciable magnitude, is of the order $\Delta t = \hbar/(E_n - E_I)$, or

$$\Delta E \, \Delta t \sim \hbar. \qquad (11\text{--}120)$$

One can say that the energy change ΔE can occur only for a time of order Δt, given by the uncertainty relation (11–120); the only states which are excited for an appreciable time are those that conserve the energy of the system.

11–9 Transitions to the continuum. The above considerations are of special importance when the states of energy E_n, in the neighborhood of the initial energy, are very closely spaced on the energy scale, thus constituting a spectrum which is practically continuous. Such a case arises, for example, in the scattering problem. To satisfy the formal requirement imposed above, that the spectrum of H^0 be discrete, we may imagine the system to be enclosed in a very large cubical box, at the walls of which periodic boundary conditions are imposed on the wave functions (cf. Section 1–3). The stationary states of the system will then be discrete, but separated in energy by an interval which is inversely proportional to the volume of the box. If we pass to the limit of a box of infinite size, the levels within a given energy interval increase in number and merge into a continuum. This limiting process is, of course, only formal; however, it permits direct application of the above results in calculating the probability for transitions that involve states in the continuum.

We are now interested in the total probability for transitions into a continuum of final states n with energy in the neighborhood of E_I. Let the density of these states on the energy scale be denoted by $\rho(E_n)$, so that the quantity $\rho(E_n) \, dE_n$ measures the number of final states in an interval dE_n containing the energy E_n. The total probability for transitions into these states is obtained by multiplying the expression (11–118) by $\rho(E_n) \, dE_n$ and integrating with respect to E_n.

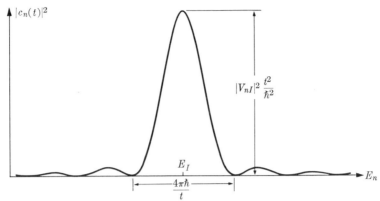

FIG. 11-4.   The transition probability $|c_n(t)|^2$.

The function $|c_n(t)|^2$, considered as a function of E_n, is illustrated in Fig. 11-4. The amplitude of the maximum at $E_n = E_I$ is proportional to t^2, and the function is small except in an interval of order $\hbar/t$ in the neighborhood of E_I. In the limit $t \to \infty$, the graph has the character of a Dirac delta function. The area under the graph, in this limit, is

$$\int |c_n(t)|^2 \, dE_n = 4|V_{nI}|^2 \int_{-\infty}^{\infty} \sin^2 \frac{(E_n - E_I)t}{2\hbar} \frac{dE_n}{(E_n - E_I)^2}$$

$$= 4|V_{nI}|^2 \frac{t}{2\hbar} \int_{-\infty}^{\infty} \frac{\sin^2 x}{x^2} \, dx = \frac{2\pi}{\hbar} |V_{nI}|^2 \cdot t.$$

Hence we may write

$$|c_n(t)|^2 = \frac{2\pi}{\hbar} |V_{nI}|^2 t \, \delta(E_n - E_I). \tag{11–121}$$

The total transition probability to a final state F of energy $E_F = E_I$ is therefore

$$\int |c_n(t)|^2 \rho(E_n) \, dE_n = \frac{2\pi t}{\hbar} \int |V_{nI}|^2 \, \delta(E_n - E_F)\rho(E_n) \, dE_n$$

$$= \frac{2\pi t}{\hbar} |V_{FI}|^2 \rho(E_F).$$

This quantity increases linearly with t; hence we can say that, for a transition from state I to state F, the transition probability per unit time, w_{FI}, is

$$w_{FI} = \frac{2\pi}{\hbar} |V_{FI}|^2 \rho(E_F). \tag{11–122}$$

This formula has wide application in quantum physics. We shall illustrate it by an example from the theory of scattering.

11–10 The Born approximation. The perturbation theory can be applied in a simple way to the theory of scattering if the scattering potential $V(\mathbf{r})$ is regarded as a perturbation of the states of the free-particle Hamiltonian

$$H^0 = \frac{p^2}{2m}. \tag{11–123}$$

In order to have a discrete spectrum, we imagine that the particle is confined within a large cubical box of volume L^3, as discussed in the preceding section. The wave functions,

$$\psi_I = \frac{1}{\sqrt{L^3}} \exp\,(i\mathbf{k}\cdot\mathbf{r}), \qquad \psi_F = \frac{1}{\sqrt{L^3}} \exp\,(i\mathbf{k}'\cdot\mathbf{r}), \tag{11–124}$$

represent initial and final states in which the particle moves in the directions of $\mathbf{k}$ and $\mathbf{k}'$, respectively. We shall assume that $V(\mathbf{r})$ is constant in time, so that the scattering is elastic, and therefore $|\mathbf{k}| = |\mathbf{k}'|$. The functions (11–124) are normalized, in the sense that

$$\int |\psi|^2\,d\mathbf{r} = 1,$$

where the integration extends over the volume of the box. If we consider the final states to be those for which the direction of $\mathbf{k}'$ is within the solid angle $d\Omega$, while the magnitude $|\mathbf{k}'| = k'$ is within the interval $(k',\, k' + dk')$, then the number of such states is the number of lattice points with integral coordinates (n_x, n_y, n_z) which are contained within the volume element $n^2\, dn\, d\Omega$, where the number $n = (n_x^2 + n_y^2 + n_z^2)^{1/2}$ is related to k' through

$$k'L = 2\pi n. \tag{11–125}$$

(These considerations are identical with those described in detail in Section 1–3, in connection with the normal modes of oscillation of the electromagnetic field confined within a cubical box of sides L.) Thus, we have

$$\rho(E')\,dE' = n^2\,dn\,d\Omega = \frac{k'^2 L^3\,dk'}{(2\pi)^3}\,d\Omega. \tag{11–126}$$

The energy of the final states is

$$E' = \frac{p'^2}{2m} = \frac{\hbar^2 k'^2}{2m},$$

whence the increments dE' and dk' are related by

$$dE' = \frac{\hbar^2 k'}{m}\,dk'. \tag{11–127}$$

By combining Eqs. (11–126) and (11–127) and cancelling common factors, we find

$$\rho(E') = \frac{mL^3}{\hbar^2} \, k' \, \frac{d\Omega}{(2\pi)^3} \cdot \qquad (11\text{–}128)$$

The matrix element V_{FI} is

$$V_{FI} = (\psi_F, V\psi_I) = \frac{1}{L^3} \int \exp\left(-i\mathbf{k}' \cdot \mathbf{r}\right) V(\mathbf{r}) \exp\left(i\mathbf{k} \cdot \mathbf{r}\right) d\mathbf{r}.$$

Substitution into Eq. (11–122) yields, for the transition probability per unit time,

$$w_{FI} = \frac{2\pi m}{\hbar^3 L^3} \, k' \left| \int \exp\left(-i\mathbf{k}' \cdot \mathbf{r}\right) V(\mathbf{r}) \exp\left(i\mathbf{k} \cdot \mathbf{r}\right) d\mathbf{r} \right|^2 \frac{d\Omega}{(2\pi)^3}. \qquad (11\text{–}129)$$

The initial state consists of a single particle within the box, moving with the speed $v = \hbar k/m$. The incident flux is therefore $S_i = v/L^3$ particles per square centimeter per second. Hence, if $\sigma(\mathbf{k}, \mathbf{k}')$ denotes the cross section per unit solid angle for the scattering from $\mathbf{k}$ to $\mathbf{k}'$, the transition rate, or number of events per second, is

$$w_{FI} = S_i \sigma(\mathbf{k}, \mathbf{k}') \, d\Omega = \frac{\hbar k}{mL^3} \, \sigma(\mathbf{k}, \mathbf{k}') \, d\Omega. \qquad (11\text{–}130)$$

Combining this expression with Eq. (11–129), we find

$$\sigma(\mathbf{k}, \mathbf{k}') = \left| \left\{ \frac{1}{4\pi} \frac{2m}{\hbar^2} \int V(\mathbf{r}) \exp\left[i(\mathbf{k} - \mathbf{k}') \cdot \mathbf{r}\right] d\mathbf{r} \right\} \right|^2 \cdot \qquad (11\text{–}131)$$

The quantity within the braces is recognized as the Born approximation to the scattering amplitude [Eq. (8–167)].

This result is consistent with the perturbation approximation, in which only effects of first order in $V(\mathbf{r})$ are taken into account. We have seen in Section 8–14 that it is equivalent, in a sense, to the assumption that only "single" scattering is important.

We shall now develop an important generalization of Eq. (11–131), which can be applied to cases of inelastic scattering by complex systems. In the scattering of an electron by an atom or molecule, for example, the Coulomb interaction between the incident electron and the atomic electrons may produce atomic excitation, with a corresponding loss of energy of the scattered particle. If the electron energy is large compared to the ionization energy of the scattering system, its initial and final states are approximately those of a free particle, and the Born approximation is valid. Another application can be made to the scattering of slow neutrons by systems of nuclei. In this case, the interaction energy is too large to allow direct application of the Born approximation, but since only s-scattering is important, the effect of the interaction can be expressed entirely in terms of the scattering length (Section 8–7). Consequently, a fictitious weak interaction, of relatively long range, can be substituted for the actual one, provided that it results in the same scattering length. This device, which is due to Fermi,[1] permits the use of the Born approximation.

The typical problem which we wish to solve is to find the cross section for a process in which a particle of initial momentum $\mathbf{k}$ is scattered by a system composed of N particles, the resultant momentum being directed into a small solid angle $d\Omega$ in a given direction $\mathbf{k}'/k'$, and the resultant energy being in the interval $(E', E' + dE')$. We denote this cross section by $\sigma(\mathbf{k}, \mathbf{k}', E')\, dE'\, d\Omega$. Then the total cross section for scattering into this energy range is

$$\sigma(\mathbf{k}, E')\, dE' = \int_\Omega \sigma(\mathbf{k}, \mathbf{k}', E')\, d\Omega\, dE',$$

and the total cross section for the process is

$$\sigma(\mathbf{k}) = \int_0^\infty \sigma(\mathbf{k}, E')\, dE'.$$

The Hamiltonian can be written in the form

$$H = \frac{p^2}{2m} + H^0(\xi_1, \xi_2, \ldots, \xi_N) + V(\mathbf{r}, \mathbf{r}_1, \mathbf{r}_2, \ldots, \mathbf{r}_N), \qquad (11\text{–}132)$$

where the first term is the kinetic energy of the scattered particle, H^0 is the Hamiltonian of the scattering system consisting of N interacting particles, and V is the part of the energy associated with the interaction between this system and the scattered particle. The Hamiltonian H^0 is an operator in the coordinates (denoted collectively by ξ_i) of the particles composing the scatterer. The eigenfunctions ϕ_n of H^0 are the stationary states of this system and satisfy

$$H^0 \phi_n = E_n \phi_n. \qquad (11\text{–}133)$$

The interaction, which will be treated as a perturbation, is represented by V in Eq. (11–132), and is assumed to have the form

$$V = V(\mathbf{r} - \mathbf{r}_1) + V(\mathbf{r} - \mathbf{r}_2) + \cdots + V(\mathbf{r} - \mathbf{r}_N), \qquad (11\text{–}134)$$

---

[1] E. Fermi, *Ricerca sci.* **7**, 13 (1936). Application to the chemical binding effect in $(n\text{-}p)$-scattering is discussed in detail by R. G. Sachs, *Nuclear Theory.* Reading, Mass.: Addison-Wesley Publishing Co., 1954, Chapter 5.

where $V(\mathbf{r} - \mathbf{r}_i)$ is a static potential that depends only upon the vector $\mathbf{r} - \mathbf{r}_i$ joining the ith particle of the scatterer to the incident particle. For simplicity, we assume that the particles composing the scatterer all interact in the same way with the scattered particle. The modifications required in the calculation to remove this restriction will be obvious. It may be remarked that $V(\mathbf{r} - \mathbf{r}_i)$ need not have spherical symmetry with respect to the position of the ith particle.

Since the scatterer and scattered particle do not interact in the unperturbed states, the wave functions for the initial and final states can be written in the form (cf. Section 7–6)

$$\psi_I = \frac{1}{\sqrt{L^3}} e^{i\mathbf{k}\cdot\mathbf{r}} \phi_n,$$

(11–135)

$$\psi_F = \frac{1}{\sqrt{L^3}} e^{i\mathbf{k}'\cdot\mathbf{r}} \phi_{n'},$$

in which ϕ_n and $\phi_{n'}$ represent the unperturbed states of the scatterer before and after the collision, respectively. The energies E_n and $E_{n'}$ of these states are related by

$$E_n + \frac{\hbar^2 k^2}{2m} = E_{n'} + \frac{\hbar^2 k'^2}{2m}.$$

(11–136)

The states ϕ_n are assumed to be normalized, i.e.,

$$(\psi_n, \psi_{n'}) = \delta_{nn'}.$$

(11–137)

The scalar-product notation implies integration over the spatial coordinates and summation over the spin coordinates of the scatterer. The wave functions (11–135) are therefore normalized.

The probability per unit time [Eq. (11–122)] for a transition in which the final energy of the scattered particle is $E' = \hbar^2 k'^2/2m$ can now be written as

$$w = \frac{2\pi}{\hbar} \sum_{n'} |V_{FI}|^2 \rho(E') \, \delta(E' - E + E_{n'} - E_n).$$

(11–138)

The sum over final states n' of the scatterer is introduced to account for the fact that a range dE' of final energies may include more than one of the unperturbed states of the scatterer; the integration implied by the delta function is to be taken with respect to the energy E' of the scattered particle, and as before, the density of final plane-wave states is

$$\rho(E') = \frac{mL^3 k'}{(2\pi)^3 \hbar^2} \, d\Omega.$$

(11–139)

By means of Eq. (11–130), Eq. (11–138) is transformed to an expression for the cross section, and one obtains

$$\sigma(\mathbf{k}, \mathbf{k}', E') = \frac{m^2 L^6}{4\pi^2 \hbar^4} \frac{k'}{k} \sum_{n'} |V_{FI}|^2 \, \delta(E' - E + E_{n'} - E_n). \quad (11\text{–}140)$$

The matrix element V_{FI} is

$$V_{FI} = \frac{1}{L^3} \int \exp[i(\mathbf{k} - \mathbf{k}') \cdot \mathbf{r}] \left\{ \sum_{i=1}^{N} (\phi_{n'}, V(\mathbf{r} - \mathbf{r}_i)\phi_n) \right\} d\mathbf{r}$$

$$= \frac{1}{L^3} \sum_{i=1}^{N} \left(\phi_{n'}, \left\{ \int V(\mathbf{r} - \mathbf{r}_i) \exp[i(\mathbf{k} - \mathbf{k}') \cdot \mathbf{r}] \, d\mathbf{r} \right\} \phi_n \right). \quad (11\text{–}141)$$

The integral contained here can be conveniently rewritten by changing the integration variable from $\mathbf{r}$ to $\mathbf{r} - \mathbf{r}_i$; writing

$$\mathbf{K} = \mathbf{k} - \mathbf{k}', \quad (11\text{–}142)$$

we have

$$\int V(\mathbf{r} - \mathbf{r}_i) \exp[i(\mathbf{k} - \mathbf{k}') \cdot \mathbf{r}] \, d\mathbf{r}$$

$$= \int V(\mathbf{r} - \mathbf{r}_i) \exp[i\mathbf{K} \cdot (\mathbf{r} - \mathbf{r}_i)] \exp(i\mathbf{K} \cdot \mathbf{r}_i) \, d(\mathbf{r} - \mathbf{r}_i)$$

$$= F(\mathbf{K}) \exp(i\mathbf{K} \cdot \mathbf{r}_i),$$

where

$$F(\mathbf{K}) = \int V(\mathbf{r}) \exp(i\mathbf{K} \cdot \mathbf{r}) \, d\mathbf{r}. \quad (11\text{–}143)$$

The expression (11–141) for the matrix element now assumes the form

$$V_{FI} = \frac{1}{L^3} F(\mathbf{K}) \left(\phi_{n'}, \left[\sum_{i=1}^{N} \exp(i\mathbf{K} \cdot \mathbf{r}_i) \right] \phi_n \right). \quad (11\text{–}144)$$

In this expression, the effects which depend upon the details of the interaction are contained in the *form factor* $F(\mathbf{K})$, while the second factor depends only upon the operator

$$\sum_{i=1}^{N} \exp(i\mathbf{K} \cdot \mathbf{r}_i) \quad (11\text{–}145)$$

and upon the unperturbed states ϕ_n, $\phi_{n'}$ of the scatterer. An analogous situation is encountered in physical optics, in which the amplitude of the wave scattered by an obstacle in the path of a light beam is the product of two factors, one giving the intensity of the light scattered from a small area, the other accounting for the difference in phase of the light scattered from the different parts of the obstacle.[1] The analogy is incomplete, however, since the wavelength of the light is not changed by the scattering.

A final modification of Eq. (11–140) is required in the case that the scattering system may initially occupy one of several different states ϕ_n. A gaseous system at temperature T, for example, occupies a state of energy E_n with a probability proportional to the Boltzmann factor $e^{-E_n/kT}$. In calculating the total scattering cross section, therefore, this state must be included with the *statistical weight*

$$ g_n = \frac{e^{-E_n/kT}}{\sum_{n'} e^{-E_{n'}/kT}}, $$

and a sum must be taken over all possible initial states. Introducing such statistical factors, together with the expression (11–144), into Eq. (11–140), we have finally

$$ \sigma(\mathbf{k}, \mathbf{k}', E') = \frac{m^2}{4\pi^2\hbar^5} \frac{k'}{k} |F(\mathbf{K})|^2 \sum_n g_n \sum_{n'} \left| \sum_{i=1}^N (\phi_{n'}, \exp(i\mathbf{K} \cdot \mathbf{r}_i)\phi_n) \right|^2 $$

$$ \times \delta\left(\omega - \frac{E_{n'} - E_n}{\hbar}\right), \tag{11–146} $$

where $\omega = (E - E')/\hbar$ is the angular frequency corresponding to the energy transfer. This formula is the basis for a rather extensive theory of scattering by complex systems.[2]

The conservation of energy in any scattering process is, as we have seen, an automatic consequence of the perturbation theory. It is of some interest to examine how the conservation of momentum is also implied by the above results. If $\mathbf{P}$ and $\mathbf{P}'$ represent the momenta of the center of gravity of the scattering system before and after the collision, respectively, then we expect that no transitions will occur unless the momenta satisfy

$$ \mathbf{P} + \hbar\mathbf{k} = \mathbf{P}' + \hbar\mathbf{k}'. \tag{11–147} $$

To verify this, we introduce the position vector

$$ \mathbf{R} = \frac{\sum_{i=1}^N m_i\mathbf{r}_i}{\sum_{i=1}^N m_i} \tag{11–148} $$

---

[1] B. Rossi, *Optics*. Reading, Mass.: Addison-Wesley Publishing Co., Inc., 1957, Chapter 4.

[2] N. F. Mott and H. S. W. Massey, *The Theory of Atomic Collisions*. Oxford: The Clarendon Press, 1949, Chapter IX. G. Placzek, *Phys. Rev.* **86**, 377 (1952). L. Van Hove, *Phys. Rev.* **95**, 249 (1954).

of the center of mass of the scatterer, and the coordinates

$$\boldsymbol{\rho}_i = \mathbf{r}_i - \mathbf{R} \qquad (11\text{–}149)$$

of the particles of the scatterer relative to the center of mass. If the scattering system in an unperturbed state ϕ_n is isolated, its wave function can be written as a product of the form [cf. Eq. (7–173)]

$$\phi_n = \frac{1}{(2\pi\hbar)^{3/2}} \exp\left(\frac{i}{\hbar}\mathbf{P}\cdot\mathbf{R}\right) \bar{\phi}_n(\boldsymbol{\rho}_i), \qquad (11\text{–}150)$$

in which the function $\bar{\phi}_n$ depends only upon the internal coordinates $\boldsymbol{\rho}_i$. Also, the operator (11–145) can be written

$$\sum_{i=1}^{N} \exp(i\mathbf{K}\cdot\mathbf{r}_i) = \exp(i\mathbf{K}\cdot\mathbf{R}) \sum_{i=1}^{N} \exp(i\mathbf{K}\cdot\boldsymbol{\rho}_i),$$

whence the matrix element appearing in Eq. (11–144) becomes

$$\left(\phi_{n'}, \sum_{i=1}^{N} \exp(i\mathbf{K}\cdot\mathbf{r}_i)\phi_n\right)$$

$$= \left\{\frac{1}{(2\pi\hbar)^3}\int \exp\left[\frac{i}{\hbar}(\mathbf{P}+\hbar\mathbf{K}-\mathbf{P}')\cdot\mathbf{R}\right]d\mathbf{R}\right\}\left(\bar{\phi}_{n'}, \sum_{i=1}^{N}\exp(i\mathbf{K}\cdot\boldsymbol{\rho}_i)\bar{\phi}_n\right)$$

$$= \delta(\mathbf{P}+\hbar\mathbf{k}-\mathbf{P}'-\hbar\mathbf{k}')\left(\bar{\phi}_{n'}, \sum_{i=1}^{N}\exp(i\mathbf{K}\cdot\boldsymbol{\rho}_i)\bar{\phi}_n\right).$$

This quantity is zero unless the condition (11–147) is satisfied. Since the matrix element V_{FI} vanishes automatically if the momentum of the system (particle + scatterer) is not conserved, it is common practice to omit all such vanishing matrix elements at the outset, and to include in the sum in Eq. (11–146) only those transitions for which (11–147) holds.

11–11 Perturbation harmonic in time. Radiative transitions. The effect of electromagnetic radiation of frequency $\omega/2\pi$ upon a system of charged particles can be represented by a perturbation that changes sinusoidally in time. The perturbation is Hermitian and can therefore be written as

$$V(t) = \mathcal{V}e^{i\omega t} + \mathcal{V}^\dagger e^{-i\omega t}, \qquad (11\text{–}151)$$

in which the time-independent operator $\mathcal{V}$ determines the amplitude and phase of $V(t)$ at $t = 0$.

The integral (11–113) for the expansion coefficient $c_n(t)$ gives

$$c_n(t) = \mho_{nI} \frac{1 - \exp\left[(i/\hbar)(\hbar\omega + E_n - E_I)t\right]}{(\hbar\omega + E_n - E_I)}$$
$$- \mho_{nI}^{\dagger} \frac{1 - \exp\left[-(i/\hbar)(\hbar\omega - E_n + E_I)t\right]}{(\hbar\omega - E_n + E_I)}. \qquad (11\text{–}152)$$

Each term in this equation is, in general, a periodic function of small amplitude. If, however, the energy E_n is close to the value $E_I - \hbar\omega$, the first term becomes large and represents a steadily growing probability of transition to a state of energy smaller than that of the initial state. The energy difference is $\hbar\omega_{nI}$, where

$$\omega_{nI} = \frac{E_I - E_n}{\hbar}. \qquad (11\text{–}153)$$

This is the Bohr frequency rule. Thus, under the action of the perturbation (11–151), the system can do work on the electromagnetic field and lose energy by *emission*. As in the similar discussion in Section 11–8, the probability that the state ψ_n is occupied at the time t is

$$|c_n(t)|^2 = \frac{|\mho_{nI}|^2}{(\hbar\omega + E_n - E_I)^2} \left\{ 2 \sin \frac{(\hbar\omega + E_n - E_I)t}{2\hbar} \right\}^2 .$$

The transition rate w_{FI} therefore is

$$w_{FI} = \frac{2\pi}{\hbar} |\mho_{FI}|^2 \, \delta(\hbar\omega + E_F - E_I) \qquad \text{(Emission).} \qquad (11\text{–}154)$$

Similarly, the second term in Eq. (11–152) becomes significant if the frequency is close to the value

$$\omega_{In} = \frac{E_n - E_I}{\hbar}, \qquad (11\text{–}155)$$

in which case the energy E_n must be larger than that of the initial state, and energy is *absorbed* by the system from its surroundings. The transition rate for absorption is

$$w_{FI} = \frac{2\pi}{\hbar} |\mho_{FI}^{\dagger}|^2 \, \delta(\hbar\omega - E_F + E_I) \qquad \text{(Absorption).} \qquad (11\text{–}156)$$

The foregoing relations provide theoretical justification for the assumptions made by Einstein and Bohr which were basic in the early development of quantum mechanics (see Chapter 1). In particular, we can use Eqs. (11–154) and (11–156) to calculate theoretically the Einstein coefficients B_{12}, B_{21}, introduced in Section 1–3. This leads to an interpretation of the transition rates in terms of the internal structure of the radiating atom, and at the same time, yields formulae for the intensities of spectral lines and provides a means for deducing selection rules.

The Hamiltonian for a single particle in interaction with the electromagnetic field has been given in Section 10–5, Eq. (10–97). For a system containing N particles, we have

$$H = \sum_{i=1}^{N} \left\{ \frac{1}{2m_i} \left(\mathbf{p}_i - \frac{e_i}{c} \mathbf{A}_i \right)^2 + e_i \Phi_i \right\} + U, \qquad (11\text{–}157)$$

in which $\mathbf{A}_i$ and Φ_i are the vector and scalar potentials of the radiation field, evaluated at the position of the ith particle. Our aim is to deduce the form of the perturbation which arises from the interaction between the light field and the charges e_i; we have therefore neglected smaller terms, e.g., those arising from magnetic dipole moments of the individual particles. The term U contains the interactions between the N particles of the system, including Coulomb and nonelectrical (e.g., nuclear) forces.

It can be shown[1] that a plane electromagnetic wave of propagation vector $\mathbf{k}$ and frequency $\omega = c|\mathbf{k}|$ is described by the potentials

$$\mathbf{A}(\mathbf{r}) = \mathbf{A}_0 \exp[i(\mathbf{k} \cdot \mathbf{r} - \omega t)] + \mathbf{A}_0^* \exp[-i(\mathbf{k} \cdot \mathbf{r} - \omega t)]; \qquad \Phi = 0, \qquad (11\text{–}158)$$

in which $\mathbf{A}_0$ is a constant complex vector, normal to the direction of propagation $\mathbf{k}$. The Hamiltonian of the unperturbed system is obtained by setting $\mathbf{A}_i = 0$ and $\Phi_i = 0$ in Eq. (11–157):

$$H^0 = \sum_{i=1}^{N} \frac{p_i^2}{2m} + U. \qquad (11\text{–}159)$$

Expanding Eq. (11–157) and retaining only terms linear in $\mathbf{A}$, we find

$$H = H^0 + V(t),$$

where

$$V(t) = - \sum_i \frac{e_i}{2m_i c} [\mathbf{A}(\mathbf{r}_i) \cdot \mathbf{p}_i + \mathbf{p}_i \cdot \mathbf{A}(\mathbf{r}_i)] \qquad (11\text{–}160)$$

[cf. Eq. (10–100)]. The vector potential for the transverse wave (11–158) has been chosen to satisfy the equation

$$\nabla \cdot \mathbf{A} = 0,$$

1 L. Landau and E. Lifshitz, *The Classical Theory of Fields*. Reading, Mass.: Addison-Wesley Publishing Co., Inc., 1951, Section 6–3.

whence the operators $\mathbf{p}$ and $\mathbf{A}$ commute under scalar (dot) multiplication. Equation (11–160) therefore reduces to

$$V(t) = - \sum_{i=1}^{N} \frac{e_i}{m_i c} \mathbf{A}(\mathbf{r}_i) \cdot \mathbf{p} = \left\{ - \sum_{i=1}^{N} \frac{e_i}{m_i c} \mathbf{A}_0 \cdot \mathbf{p}_i \exp\left(i\mathbf{k} \cdot \mathbf{r}_i\right) \right\} e^{-i\omega t}$$

$$+ \left\{ - \sum_{i=1}^{N} \frac{e_i}{m_i c} \mathbf{A}_0^* \cdot \mathbf{p}_i \exp\left(-i\mathbf{k} \cdot \mathbf{r}_i\right) \right\} e^{i\omega t},$$

which has the form (11–151), with

$$\mathcal{V} = - \sum_{i=1}^{N} \frac{e_i}{m_i c} \mathbf{A}_0 \cdot \mathbf{p}_i \exp\left(i\mathbf{k} \cdot \mathbf{r}_i\right). \tag{11–161}$$

The coordinates $\mathbf{r}_i$ in the Hamiltonian (11–157) are defined with respect to the laboratory frame of reference; hence, the equations of motion derived from H include the motion of the center of mass of the system, as well as the relative motions of the particles. The center-of-mass motion is separable from H^0 in the way described in Section 7–7, but it is necessary to examine the effect of the perturbation (11–161) in this respect. The momentum operators $\mathbf{P}$ and $\boldsymbol{\pi}_i$ associated with the coordinates $\mathbf{R}$ and $\boldsymbol{\rho}_i$ [Eqs. (11–148) and (11–149)], respectively, are

$$\mathbf{P} = \left(\sum_{i=1}^{N} m_i \right) \dot{\mathbf{R}} = M\dot{\mathbf{R}} \quad \text{and} \quad \boldsymbol{\pi}_i = \mathbf{p}_i - m_i \dot{\mathbf{R}}, \tag{11–162}$$

where $\dot{\mathbf{R}} = (i/\hbar)\lfloor H, \mathbf{R} \rfloor$ is the velocity of the center of mass, and $M = \sum_{i=1}^{N} m_i$ is the total mass of the system. Making these substitutions in Eq. (11–161), we obtain

$$\mathcal{V} = -\exp\left(i\mathbf{k} \cdot \mathbf{R}\right)$$

$$\times \left\{ \sum_{i=1}^{N} \frac{e_i}{m_i c} \mathbf{A}_0 \cdot \boldsymbol{\pi}_i \exp\left(i\mathbf{k} \cdot \boldsymbol{\rho}_i\right) + \mathbf{A}_0 \cdot \mathbf{P} \sum_{i=1}^{N} \frac{e_i}{Mc} \exp\left(i\mathbf{k} \cdot \boldsymbol{\rho}_i\right) \right\}. \tag{11–163}$$

The factor $\exp\left(i\mathbf{k} \cdot \mathbf{R}\right)$ in this equation does not affect the transition probabilities and is of no consequence. The first term on the right, which contains only the relative coordinates, represents the effect of the internal motion of the system and gives rise to transitions among the internal quantum states. It becomes identical in form with Eq. (11–161) when the center-of-mass motion is neglected. The second term accounts for the radiative effects produced by the motion of the charged system as a whole. In general, we shall be concerned with radiation whose wavelength is long compared to the dimensions of the radiating system, so that the exponential factors are nearly unity. The sum is then very closely proportional to the total charge, $\sum_i e_i$. The effects of the second term, in any event, are small compared to those of the first term, when the total mass of the system is large compared to the mass of one of its component particles. In the work that follows, we shall disregard the motion of the center of mass and assume it to be at rest at the origin of coordinates. The vectors $\mathbf{r}_i$ and $\boldsymbol{\rho}_i$ are then identical, and the operator (11–161) can be regarded as dependent upon the internal coordinates only.

Since the strength of the perturbation (11–161) is proportional to the amplitude of the electromagnetic field, the transition probabilities are proportional to the intensity, which is given by the Poynting vector[1]

$$\mathbf{S} = \frac{c}{4\pi}\, \boldsymbol{\varepsilon} \times \boldsymbol{\mathcal{B}}, \qquad (11\text{–}164)$$

in which the electromagnetic fields $\boldsymbol{\varepsilon}$ and $\boldsymbol{\mathcal{B}}$ are [2] given by

$$\boldsymbol{\varepsilon} = -\frac{1}{c}\frac{\partial \mathbf{A}}{\partial t} = \frac{i\omega}{c}\,\{\mathbf{A}_0 \exp[i(\mathbf{k}\cdot\mathbf{r} - \omega t)] - \mathbf{A}_0^* \exp[-i(\mathbf{k}\cdot\mathbf{r} - \omega t)]\},$$

$$\boldsymbol{\mathcal{B}} = \nabla \times \mathbf{A} = i\mathbf{k} \times \{\mathbf{A}_0 \exp[i(\mathbf{k}\cdot\mathbf{r} - \omega t)] - \mathbf{A}_0^* \exp[-i(\mathbf{k}\cdot\mathbf{r} - \omega t)]\}$$

$$= \frac{\mathbf{k} \times \boldsymbol{\varepsilon}}{k}, \qquad (11\text{–}165)$$

whence

$$\mathbf{S} = \frac{c}{4\pi}\frac{\mathbf{k}}{k}\,\boldsymbol{\varepsilon}^2$$

$$= \frac{\omega^2}{4\pi c}\,\{2|\mathbf{A}_0|^2 + \mathbf{A}_0^2 \exp[2i(\mathbf{k}\cdot\mathbf{r} - \omega t)] + \mathbf{A}_0^{*2} \exp[-2i(\mathbf{k}\cdot\mathbf{r} - \omega t)]\}\frac{\mathbf{k}}{k}.$$

$$(11\text{–}166)$$

The intensity, or time-average magnitude of the Poynting vector, is therefore

$$I = \frac{\omega^2}{2\pi c}\,|\mathbf{A}_0|^2. \qquad (11\text{–}167)$$

If we now substitute the expression (11–161) into Eq. (11–154) and introduce the intensity I given by Eq. (11–167), we find the transition rate for induced emission due to a single plane wave of angular frequency ω, i.e.,

$$w_{FI} = \frac{4\pi^2 c}{\hbar\omega^2}\, I|v_{FI}|^2\, \delta(\hbar\omega + E_F - E_I), \qquad (11\text{–}168)$$

---

[1] L. Landau and E. Lifshitz, *The Classical Theory of Fields*. Reading, Mass.: Addison-Wesley Publishing Co., Inc., 1951, Section 4–6.
[2] Section 10–5, Eq. (10–83).

in which
$$v = -\sum_i \frac{e_i}{m_i c}\, \mathbf{a}\cdot\mathbf{p}_i \exp\,(i\mathbf{k}\cdot\mathbf{r}_i),$$ (11–169)

and $\mathbf{a} = \mathbf{A}_0/|\mathbf{A}_0|$ is a unit complex vector describing the state of polarization of the incident field.

In an experimental arrangement for observing spectra, the incident light is not a single, monochromatic plane wave, but an incoherent mixture of waves in different states of polarization, propagated in various directions. In Chapter 1, a radiation field of this kind was characterized by the energy density per unit frequency, u_ν. If the radiation field is isotropic, an equivalent description can be given in terms of the intensity per unit solid angle and per unit frequency, $I(\omega)$, which is such that the quantity $I(\omega)\,d\omega\,d\Omega$ is the energy transported in unit time across an element of area dA, in directions lying within a small solid angle $d\Omega$ about the normal to dA, by waves whose frequency lies in the small interval $d\omega$.

The connection between $u_\nu = 2\pi u_\omega$ and $I(\omega)$ is obtained by considering Fig. 11–5: The energy transported normally across the area element dA into the small cone of directions $d\Omega$ in the time dt is the fraction $d\Omega/4\pi$ of the total energy, $u_\omega\,d\omega\,dA\cdot c\,dt$, contained in the volume of base dA and height $c\,dt$. Hence, by definition of $I(\omega)$,

$$u_\omega\,d\omega\,dA\cdot c\,dt\cdot\frac{d\Omega}{4\pi} = 2I(\omega)\,d\omega\,dA\,dt.$$

The factor 2 appears in the right-hand member to account for the two linearly independent states of polarization for each frequency and direction. Cancelling common factors, we have

$$I(\omega) = \frac{cu_\omega}{8\pi}.$$ (11–170)

By substitution of $I(\omega)\,d\omega\,d\Omega$ for I and integration over ω, we find the total rate for transitions induced by radiation which is propagated in

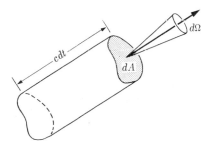

FIG. 11–5. Construction for deriving the relation between the intensity $I(\omega)$ and the energy density u_ω.

directions within the solid angle $d\Omega$ and polarized in the direction $\mathbf{a}$. We denote this quantity by dw_{FI}:

$$dw_{FI} = \frac{4\pi^2 c}{\hbar^2 \omega_{FI}^2} \, I(\omega_{FI})|v_{FI}|^2 \, d\Omega \qquad \text{(Emission)}. \qquad (11\text{--}171)$$

The corresponding expression for the rate of absorption is

$$dw_{IF} = \frac{4\pi^2 c}{\hbar^2 \omega_{IF}^2} \, I(\omega_{IF})|v_{IF}^\dagger|^2 \, d\Omega \qquad \text{(Absorption)}. \qquad (11\text{--}172)$$

Since the operator v [Eq. (11–169)] is Hermitian,

$$v_{IF}^\dagger = v_{FI}^*. \qquad (11\text{--}173)$$

It follows that, except for the interchange of the initial and final states, the two expressions (11–171) and (11–172) are identical. Thus, the *principle of detailed balance*, invoked in the discussion of Einstein's derivation of the blackbody-radiation law, is a fundamental consequence of the quantum theory. This principle, which plays so important a role in statistical theory, is inherent in the mathematical structure of quantum mechanics, and can be shown to follow, as in classical mechanics, from the symmetry with respect to a reversal of the sense of time.

11–12 Atomic radiation. For atomic spectra, only those terms in the interaction that refer to the Z atomic electrons are of importance. During an optical transition, the nucleus remains essentially at rest because of its large mass, and due to the very large nuclear binding energy, its internal structure is entirely unaffected. The operator v can therefore be written

$$v = -\frac{e}{mc} \sum_{i=1}^{Z} \mathbf{a} \cdot \mathbf{p}_i \exp (i\mathbf{k} \cdot \mathbf{r}_i), \qquad (11\text{--}174)$$

in which $\mathbf{r}_i$ is the coordinate of the ith electron with respect to the nucleus.

A well-known rule of spectroscopy, stating that in an optical transition "only one electron jumps at a time," follows immediately from the structure of Eq. (11–174). The rule refers to the fact that, in most optical transitions, the initial and final states differ only in the motion of one electron, and that transitions in which several electrons change their states of motion are rare.[1] In order to derive this rule, suppose that the wave func-

[1] G. Herzberg, *Atomic Spectra and Atomic Structure*. New York: Dover Publications, 1944, p. 153. E. U. Condon and G. H. Shortley, *The Theory of Atomic Spectra*. Cambridge: Cambridge University Press, 1953, Section 6⁶.

tions ψ_I and ψ_F have the forms

$$\psi_I = \phi_1(1)\phi_2(2)\phi(3, 4, \ldots, Z),$$
$$\psi_F = \bar{\phi}_1(1)\bar{\phi}_2(2)\phi(3, 4, \ldots, Z),$$
$$(11\text{–}175)$$

so that they differ only in their dependence on the coordinates of the electrons numbered 1 and 2. The factorized form of these expressions is implied in the assumption that it is meaningful to speak of a transition in which only one electron jumps. If the motions of 1 and 2 were coupled, the wave function would not be separable, and hence, it would not be possible to change the state of one electron without affecting that of the other. Now the operator v [Eq. (11–169)] can be written

$$v = v_1 + v_2 + \sum_{i=3}^{Z} v_i,$$

that is, as a sum of individual terms for each electron. The matrix element v_{FI} therefore has the form

$$v_{FI} = (\bar{\phi}_1, v_1\phi_1)(\bar{\phi}_2, \phi_2)(\phi, \phi) + (\bar{\phi}_1, \phi_1)(\bar{\phi}_2, v_2\phi_2)(\phi, \phi)$$
$$+ (\bar{\phi}_1, \phi_1)(\bar{\phi}_2, \phi_2)\left(\phi, \sum_{i=3}^{Z} v_i\phi\right). \quad (11\text{–}176)$$

The single-particle states ϕ_1 and $\bar{\phi}_1$ have, by hypothesis, different quantum numbers, and are therefore orthogonal; the same is true of ϕ_2 and $\bar{\phi}_2$. Hence, each term on the right in Eq. (11–176) contains a zero factor, and $v_{FI} = 0$. Transitions between the states (11–175) can therefore appear only in the higher approximations of the perturbation theory and are correspondingly improbable. However, if $\phi_2 = \bar{\phi}_2$, so that only the state of particle 1 is changed in the transition, the first term in v_{FI} is not zero, and a transition is possible in first order.[1]

Since the forces between the optical electrons in an atom are relatively weak compared to the forces arising from the nuclear charge and from the electrons in closed shells, the single-particle picture implied by the form (11–175) of the wave functions always has an approximate validity and is very useful in the interpretation of spectra.

Atomic spectra are almost entirely due to *dipole radiation*. This will now be briefly explored in the light of the preceding discussion. The frequency of the light emitted in an optical transition is of the order $e^2/\hbar r$, where r is a length of the order of the atomic dimensions. Consequently,

---

[1] In Chapter 12, it will be indicated how this proof is to be modified to account properly for the Pauli principle.

the magnitude of the exponent $i\mathbf{k}\cdot\mathbf{r}$ in Eq. (11–174) is of the order $r/\lambda \sim e^2/\hbar c$. It is therefore a good approximation to expand the exponential,

$$\exp(i\mathbf{k}\cdot\mathbf{r}) = 1 + i\mathbf{k}\cdot\mathbf{r} + \tfrac{1}{2}(i\mathbf{k}\cdot\mathbf{r})^2 + \cdots, \qquad (11\text{–}177)$$

and to consider the successive terms separately. The result of replacing the exponential by unity is

$$v_{FI} = -\frac{e}{mc}\sum_{i=1}^{Z} \mathbf{a}\cdot(\mathbf{p}_i)_{FI}.$$

Since the states ψ_I and ψ_F are eigenstates of H^0, the matrix element of the momentum can be written [Eq. (6–87)]

$$(\mathbf{p}_i)_{FI} = m(\dot{\mathbf{r}}_i)_{FI} = m\frac{i}{\hbar}(E_F - E_I)(\mathbf{r}_i)_{FI} = -m\omega_{FI}(\mathbf{r}_i)_{FI}, \quad (11\text{–}178)$$

whence

$$v_{FI} = \frac{\omega_{FI}}{c}\mathbf{a}\cdot\mathbf{D}_{FI}, \qquad (11\text{–}179)$$

where

$$\mathbf{D} = \sum_{i=1}^{Z} e\,\mathbf{r}_i \qquad (11\text{–}180)$$

is the operator which represents the *electric dipole moment* of the atom. The transitions to which the quantity (11–179) gives rise result in dipole radiation and are responsible for almost all lines observed in atomic and molecular spectra, as indicated before. The properties of the matrix elements of the dipole moment are therefore decisive for the character of such spectra. They account for all major regularities which are observed, and lead immediately to selection and intensity rules which are exactly verified by experiment.

The transition rate for induced emission [Eq. (11–171)] can now be written, in the dipole approximation, as

$$dw_{FI} = \frac{4\pi^2}{\hbar^2 c} I(\omega_{FI})|\mathbf{a}\cdot\mathbf{D}_{FI}|^2\,d\Omega. \qquad (11\text{–}181)$$

In order to find the total rate of emission induced by an isotropic field of spectral density u_ν, it is necessary to sum this expression over the two possible states of polarization for each direction of propagation and to integrate over the solid angle. As independent polarization vectors, we choose two perpendicular real vectors $\mathbf{e}_1$ and $\mathbf{e}_2$, representing plane-polarized waves. Every possible polarization can be formed from these vectors by superposition, with appropriate phase and amplitude. The vectors $\mathbf{e}_1$, $\mathbf{e}_2$, and $\mathbf{k}/k$ are a triplet of orthogonal unit vectors, and we have, for any fixed vector $\mathbf{D}$,

$$\sum_{\text{polarizations}} |\mathbf{a} \cdot \mathbf{D}|^2 = |\mathbf{e}_1 \cdot \mathbf{D}|^2 + |\mathbf{e}_2 \cdot \mathbf{D}|^2 = |\mathbf{D}|^2 - \left|\frac{\mathbf{k}}{k} \cdot \mathbf{D}\right|^2.$$

Integration over the solid angle yields

$$\int \sum_{\text{polarizations}} |\mathbf{a} \cdot \mathbf{D}|^2 \, d\Omega = |\mathbf{D}|^2 \left\{ 4\pi - \int_\Omega \cos^2 \theta \, d\Omega \right\} = \frac{8\pi}{3} |\mathbf{D}|^2. \quad (11\text{–}182)$$

Combining Eqs. (11–181), (11–182), and (11–170), we have

$$B_{FI} u_\nu = \frac{2\pi}{3\hbar^2} |\mathbf{D}_{FI}|^2 u_\nu, \quad (11\text{–}183)$$

where B_{FI} is the Einstein coefficient for induced transitions.

In the dipole approximation, the rate of occurrence of spontaneous transitions can now be obtained by combining the result (11–183) with the expression

$$A_{FI} = \frac{8\pi h \nu_{FI}^3}{c^3} B_{FI} = \frac{2\hbar \omega_{FI}^3}{\pi c^3} B_{FI},$$

obtained in the Einstein derivation of the Planck radiation law [Eqs. (1–16) and (1–10)]. We find

$$A_{FI} = \frac{4(\omega_{FI})^3}{3\hbar c^3} |\mathbf{D}_{FI}|^2. \quad (11\text{–}184)$$

An example will illustrate how this fundamental relation is applied: Let us consider a one-dimensional harmonic oscillator, for which the matrix elements of the dipole moment [Section 5–11, Eqs. (5–148), (5–149), and (5–150)] are

$$ex_{nn'} = \begin{cases} e\sqrt{\dfrac{n+1}{2}}\sqrt{\dfrac{\hbar}{m\omega}} & (n' = n + 1), \\[2ex] e\sqrt{\dfrac{n}{2}}\sqrt{\dfrac{\hbar}{m\omega}} & (n' = n - 1), \qquad (11\text{–}185) \\[2ex] 0 & \text{otherwise,} \end{cases}$$

where ω is the classical frequency. The only transitions which occur are those for which the frequency of the light emitted is $\omega_{FI} = \omega$, and according to Eq. (11–184), the rate of spontaneous emission is

$$A = \frac{4\omega^3}{3\hbar c^3} e^2 \left(\frac{n}{2}\frac{\hbar}{m\omega}\right) = \frac{2\omega^2 e^2}{3mc^3} n. \quad (11\text{–}186)$$

In the limit of large n, the correspondence between this expression and the classical law of radiation by a harmonic oscillator is easily traced.

The amplitude of the classical oscillator (cf. Section 5–11) is related to n by the formula

$$a^2 \sim 2n \frac{\hbar}{m\omega}.$$

The average energy radiated per unit time is obtained by multiplying the expression (11–186) by $\hbar\omega$, the energy of the emitted quantum:

$$A \cdot \hbar\omega = \frac{\omega^4 e^2}{3c^3} 2n \frac{\hbar}{m\omega} \sim \frac{\omega^4 e^2}{3c^3} a^2. \tag{11–187}$$

It is shown in electrodynamics that the instantaneous rate at which an accelerated charge radiates power is given by Larmor's formula

$$P = \frac{2}{3} \frac{e^2}{c^3} \ddot{x}^2,$$

where $\ddot{x}$ is the acceleration. For an oscillator of frequency ω, we have $\ddot{x} = -\omega^2 x$, whence

$$P = \frac{2}{3} \frac{e^2}{c^3} \omega^4 x^2.$$

If the oscillator amplitude is a, the time average value of x^2 is $a^2/2$. The mean rate of radiation is therefore

$$\bar{P} = \frac{\omega^4 e^2}{3c^3} a^2,$$

in exact agreement with Eq. (11–187).

We shall conclude this section with some comments on selection rules. The selection rule $\Delta n = \pm 1$, which governs radiative transitions of the harmonic oscillator, is typical of many such rules which follow from the properties of the matrix elements of $\mathbf{D}$ for special systems. Most important are the selection rules governing the change of angular-momentum quantum numbers.

The states of any isolated system can be classified according to their symmetry properties under rotation. The total angular momentum J and its projection M are always good quantum numbers, and every transition produces certain definite changes in these numbers. By its definition, the dipole moment operator $\mathbf{D}$ is a vector operator, and its nonzero matrix elements are therefore given by the expressions derived in Section 10–13. From these it follows that radiative transitions can only occur between states for which the difference in the values of J does not exceed unity. The same is true for the quantum number M. In addition, the vector rule [Eq. (10–209)] shows that the matrix element $(J, M|\mathbf{D}|J, M')$ vanishes for $J = 0$ and for $M = M' = 0$. In summary, we have

$$\Delta J = 0, \pm 1, \qquad \Delta M = 0, \pm 1, \qquad \text{no } J = 0 \to J = 0,$$
$$\text{no } M = 0 \to M = 0 \quad \text{for} \quad \Delta J = 0. \tag{11–188}$$

These selection rules are *strict* in the sense that they originate purely from the symmetry properties of the quantum states; a transition which is forbidden by these rules cannot be due to dipole radiation unless the radiating system is subject to external forces, e.g., to an electric or magnetic field. The only other rule of this nature is the *Laporte rule*, which is based on the fact that the parity of any state of an isolated system is definite. Since the operator $\mathbf{D}$ is a linear combination of the position vectors of the particles of the system, it changes sign under the parity operation Π (Section 6–9). In operator language, this means that

$$\Pi \mathbf{D} \Pi^{-1} = -\mathbf{D}.$$

Therefore, if ψ_I and ψ_F are states of the same parity,

$$\mathbf{D}_{FI} = (\psi_F, \mathbf{D}\psi_I) = -(\psi_F, \Pi \mathbf{D} \Pi^{-1} \psi_I)$$
$$= -(\Pi \psi_F, \mathbf{D}\Pi\psi_I) = -(\psi_F, \mathbf{D}\psi_I) = 0.$$

In a dipole transition, the parity must change. This is the Laporte rule.

11–13 Zeeman effect—polarization. The quantum states of an atom in a uniform magnetic field depend upon the orientation of the atom's angular momentum with respect to the direction of the field (Section 11–5). In the absence of a field, the initial and final states for a given transition are eigenstates of the total angular momentum, conveniently denoted by (α, J, M) and (α', J', M'), respectively (α and α' denote the quantum numbers associated with quantities other than angular momentum). The $(2J + 1)$-fold degeneracy of each state is removed by a weak field, which causes the single line $(\alpha, J) \to (\alpha', J')$ to split into a *Zeeman pattern*. Allowed dipole transitions occur for $J' = J$ and $J' = J \pm 1$.

To discuss the intensity and polarization of the light emitted in various directions with respect to the applied field, it is necessary to refer to Eq. (11–181), which gives the differential induced transition probability per unit solid angle for a definite polarization $\mathbf{a}$. The equivalent formula for spontaneous emission[1] is

$$dA = \frac{(\omega_{FI})^3}{2\pi \hbar c^3} |\mathbf{a} \cdot \mathbf{D}_{FI}|^2 \, d\Omega. \tag{11–189}$$

---

[1] Problem 11–14. W. Heitler, *The Quantum Theory of Radiation*. 2nd ed. London: Oxford University Press, 1944, Chapter III, Section 11, Eq. (10).

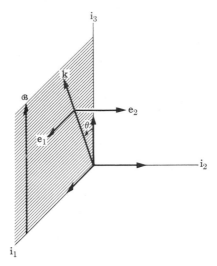

F$_{\text{IG}}$. 11–6.   The unit vectors $\mathbf{e}_1$ and $\mathbf{e}_2$ and the coordinate system $(\mathbf{i}_1, \mathbf{i}_2, \mathbf{i}_3)$.

We denote by $\mathbf{e}_1$ the unit vector normal to the propagation vector $\mathbf{k}$, lying in the plane of $\mathbf{\mathfrak{G}}$ and $\mathbf{k}$. If $\mathbf{a} = \mathbf{e}_1$, then at $\mathbf{r} = 0$, the electric vector of the emitted wave [Eq. (11–165)] is

$$\boldsymbol{\varepsilon} = \frac{2\omega}{c}\, |\mathbf{A}_0| \mathbf{e}_1 \sin \omega t,$$

so that the light is plane-polarized in the direction $\mathbf{e}_1$. Similarly, a wave plane-polarized in a direction normal to the plane of $\mathbf{k}$ and $\mathbf{\mathfrak{G}}$ is represented by $\mathbf{a} = \mathbf{e}_2$ (Fig. 11–6).   In general, a given state of (elliptical) polarization[1] is represented by a polarization vector of the form

$$\mathbf{b} = \alpha_1 \mathbf{e}_1 + \alpha_2 \mathbf{e}_2, \tag{11–190}$$

in which the complex numbers α_1 and α_2 satisfy the relation

$$\mathbf{b}^* \cdot \mathbf{b} = |\alpha_1|^2 + |\alpha_2|^2 = 1. \tag{11–191}$$

The vector

$$\mathbf{a} = \alpha_2^* \mathbf{e}_1 - \alpha_1^* \mathbf{e}_2 = \mathbf{b}^* \times \frac{\mathbf{k}}{k}, \tag{11–192}$$

which obviously satisfies $\mathbf{a}^* \cdot \mathbf{b} = 0$, represents a polarization state *orthogonal* to that represented by $\mathbf{b}$. Every possible polarization vector can be expressed as a linear combination of $\mathbf{a}$ and $\mathbf{b}$, as well as of $\mathbf{e}_1$ and $\mathbf{e}_2$.

---

[1] B. Rossi, *Optics*.   Reading, Mass.: Addison-Wesley Publishing Co., Inc., 1957, Section 6–2.

The components of the matrix element $\mathbf{D}_{FI}$ in Eq. (11–189) depend upon the angular-momentum quantum numbers of the initial and final states, as given in Table 10–1. We take the axis of quantization for the unperturbed states in the direction $\mathbf{i}_3$ of the magnetic field $\mathfrak{B}$, and choose the axis $\mathbf{i}_1$ of the right-handed coordinate system $(\mathbf{i}_1, \mathbf{i}_2, \mathbf{i}_3)$ in the plane of $\mathfrak{B}$ and $\mathbf{k}$. Writing the polarization vector as

$$\mathbf{b} = b_1\mathbf{i}_1 + b_2\mathbf{i}_2 + b_3\mathbf{i}_3, \tag{11–193}$$

we find

$$\begin{aligned} \mathbf{b} \cdot \mathbf{D} &= b_1\, D_1 + b_2\, D_2 + b_3\, D_3 \\ &= \tfrac{1}{2}(b_1 - ib_2)\, D_+ + \tfrac{1}{2}(b_1 + ib_2)\, D_- + b_3\, D_3, \end{aligned} \tag{11–194}$$

where $D\pm = D_1 \pm iD_2$. It follows that the matrix elements for the transitions $\Delta M = \pm 1, 0$ for the polarization $\mathbf{b}$ are

$$\begin{aligned} &(\alpha', J', M \pm 1|\mathbf{b} \cdot \mathbf{D}|\alpha, J, M) \\ &\qquad = \tfrac{1}{2}(b_1 \mp ib_2)(\alpha', J', M \pm 1|D_\pm|\alpha, J, M), \end{aligned} \tag{11–195}$$

$$(\alpha', J', M|\mathbf{b} \cdot \mathbf{D}|\alpha, J, M) = b_3(\alpha', J', M|D_3|\alpha, J, M). \tag{11–196}$$

Furthermore, comparison of the forms (11–190) and (11–193) yields

$$b_1 = \alpha_1 \cos\theta, \qquad b_2 = \alpha_2, \qquad b_3 = -\alpha_1 \sin\theta. \tag{11–197}$$

Now it is clear from Eq. (11–195) that the transition $\Delta M = 1$ does not occur at all if the components of $\mathbf{b}$ satisfy

$$b_1 - ib_2 = \alpha_1 \cos\theta - i\alpha_2 = 0. \tag{11–198}$$

This equation, together with Eq. (11–191), gives (apart from an unimportant phase factor)

$$\alpha_1 = \frac{i}{\sqrt{1 + \cos^2\theta}}, \qquad \alpha_2 = \frac{\cos\theta}{\sqrt{1 + \cos^2\theta}}. \tag{11–199}$$

The state of polarization of the light that is actually emitted in the transition $\Delta M = 1$ is therefore orthogonal to the state given by Eq. (11–199); the corresponding orthogonal polarization vector is, by Eq. (11–192),

$$\mathbf{a}_+ = \frac{1}{\sqrt{1 + \cos^2\theta}}(\cos\theta\, \mathbf{e}_1 + i\, \mathbf{e}_2), \tag{11–200}$$

and, by replacing $\mathbf{b}$ with $\mathbf{a}_+$ in Eq. (11–195), we find

$$(\alpha', J', M + 1|\mathbf{a}_+ \cdot \mathbf{D}|\alpha, J, M)$$
$$= \tfrac{1}{2}\sqrt{1 + \cos^2\theta}\,(\alpha', J', M + 1|D_+|\alpha, J, M). \quad (11\text{–}201)$$

Equation (11–200) gives the polarization of the light emitted in the direction θ in a transition for which $\Delta M = +1$. At $\theta = \pi/2$ (transverse observation), the light is evidently plane-polarized in the direction $\mathbf{e}_2$, i.e., normal to the field and to the direction of observation. For longitudinal observation, on the other hand, we have

$$\mathbf{a}_+ = \frac{1}{\sqrt{2}}\,(\mathbf{e}_1 + i\mathbf{e}_2) \qquad (\theta = 0), \qquad\qquad (11\text{–}202)$$

so that the electric vector in the emitted light at $\mathbf{r} = 0$ is

$$\boldsymbol{\varepsilon} = \frac{i\omega}{c}\,|\mathbf{A}_0|(\mathbf{a}_+ e^{-i\omega t} - \mathbf{a}_+^* e^{i\omega t})$$

$$= \frac{2\omega}{c}\,|\mathbf{A}_0|\,\frac{1}{\sqrt{2}}\,(\mathbf{e}_1 \sin\omega t - \mathbf{e}_2 \cos\omega t).$$

Thus $\boldsymbol{\varepsilon}$ is a vector of constant length, which rotates in a counterclockwise direction with respect to an observer looking toward the source; the light is left-circularly polarized. This radiation, referred to as a σ-*component* in Section 10–6, will now be denoted specifically by σ_+. The polarizations for other directions of observation relative to $\mathcal{B}$ are summarized in the following table:

Direction	Sense of rotation	Polarization
$\theta = 0$	left	circular
$0 < \theta < \pi/2$	left	elliptical
$\theta = \pi/2$	–	plane (perpendicular to $\mathcal{B}$)
$\pi/2 < \theta < \pi$	right	elliptical
$\theta = \pi$	right	circular

A similar analysis of the σ_--components ($\Delta M = -1$) leads to

$$\mathbf{a}_- = \frac{1}{\sqrt{1 + \cos^2\theta}}\,(\cos\theta\,\mathbf{e}_1 - i\,\mathbf{e}_2); \qquad (11\text{–}203)$$

$$(J', M - 1|\mathbf{a}_- \cdot \mathbf{D}|J, M) = \tfrac{1}{2}\sqrt{1 + \cos^2 \theta}\,(J', M - 1|D_-|J, M).$$

$$(11\text{--}204)$$

The π-component of the Zeeman pattern corresponds to $\Delta M = 0$, and is plane-polarized:

$$\mathbf{a}_\pi = \mathbf{e}_1; \qquad (11\text{--}205)$$

$$(J', M|\mathbf{a}_\pi \cdot \mathbf{D}|J, M) = -\sin \theta (J', M|D_3|J, M). \qquad (11\text{--}206)$$

The factor $\sin \theta$ shows that the π-component is not observed in longitudinal observation (cf. Fig. 10–5).

As shown in Section 11–5, the effect of the magnetic field is to split an initially degenerate level into $2J + 1$ states characterized by different values of M. In the absence of a magnetic field, the total radiation resulting from the unperturbed level is the incoherent superposition of all radiations given by the equations (11–201), (11–204), and (11–206). In the following discussion, we shall show that this radiation is isotropic and unpolarized. This result is a consequence of the fact that, in the absence of a field, there is no uniquely defined direction, and all observable effects must therefore be spherically symmetric.

By Eq. (11–189), the transition probabilities corresponding to the various transitions discussed above are

$$\sigma_+: \quad dA_+ = \frac{\omega^3}{2\pi\hbar c^3}\,\tfrac{1}{4}(1 + \cos^2 \theta)|(\alpha', J', M + 1|D_+|\alpha, J, M)|^2\,d\Omega,$$

$$\sigma_-: \quad dA_- = \frac{\omega^3}{2\pi\hbar c^3}\,\tfrac{1}{4}(1 + \cos^2 \theta)|(\alpha', J', M - 1|D_-|\alpha, J, M)|^2\,d\Omega,$$

$$\pi: \quad dA_\pi = \frac{\omega^3}{2\pi\hbar c^3}\,\sin^2 \theta|(\alpha', J', M|D_3|\alpha, J, M)|^2\,d\Omega. \qquad (11\text{--}207)$$

In a statistical assemblage of radiating atoms, each of the $(2J + 1)$ states of the initial level has the same population, and the mean intensity radiated is the sum of the quantities

$$I_+ = \frac{\hbar\omega}{2J + 1}\sum_M dA_+, \qquad I_- = \frac{\hbar\omega}{2J + 1}\sum_M dA_-,$$

$$I_\pi = \frac{\hbar\omega}{2J + 1}\sum_M dA_\pi, \qquad (11\text{--}208)$$

obtained by multiplying the average transition probabilities by the quantum energy $\hbar\omega$.

For each of the transitions $\Delta J = \pm 1, 0$, the sums of the matrix elements which occur in these expressions can be calculated from Table 10–1

(Problem 11–18).[1] However, the essential property of the sums needed in the present discussion can be expressed by

$$\sum_M |(\alpha', J', M|D_3|\alpha, J, M)|^2 = \tfrac{1}{2} \sum_M |(\alpha', J', M + 1|D_+|\alpha, J, M)|^2$$

$$= \tfrac{1}{2} \sum_M |(\alpha', J', M - 1|D_-|\alpha, J, M)|^2,$$

$$(11\text{–}209)$$

and it is instructive to demonstrate this relation by matrix methods. Because of the selection rule $\Delta M = +1$, the second sum in the above expression can be written

$$\sum_M |(\alpha', J', M + 1|D_+|\alpha, J, M)|^2 = \sum_{MM'} |(\alpha', J', M'|D_+|\alpha, J, M)|^2,$$

$$(11\text{–}210)$$

which is the sum of the squared matrix elements for all transitions connecting the two levels (α, J) and (α', J').

The proof of Eq. (11–209) which we wish to construct is based upon a matrix identity[2] that can be stated in the following general form: Let a set of orthonormal functions ϕ_k and a unitary operator U be given, such that the functions ϕ_k "transform among themselves" under the operation U,

$$U\phi_k = \sum_l U_{lk}\phi_l,$$

$$(11\text{–}211)$$

and let ψ be an arbitrary function for which the scalar products (ψ, ϕ_k) exist; then

$$\sum_k |(\psi, \phi_k)|^2 = \sum_k |(\psi, U\phi_k)|^2.$$

$$(11\text{–}212)$$

The proof is immediate and is left as an exercise (Problem 11–19).

Now the functions $\psi(\alpha, J, M)$, which belong to the level (α, J), transform among themselves under the operation $U = \exp(i\boldsymbol{\omega} \cdot \mathbf{J})$, which represents a rotation of the coordinate frame (Section 10–10), and the same is true of the functions $\psi(\alpha', J', M')$ belonging to (α', J'). It follows from two applications of Eq. (11–212) that the equation

$$\sum_{MM'} |(\alpha', J', M'|D_+|\alpha, J, M)|^2 = \sum_{MM'} |(\alpha', J', M'|U\, D_+ U^{-1}|\alpha, J, M)|^2$$

$$(11\text{–}213)$$

---

[1] E. U. Condon and G. H. Shortley, *The Theory of Atomic Spectra*. Cambridge: Cambridge University Press, 1953, Section 13³.

[2] Section 2² of Condon and Shortley.

is an identity. We can therefore expand the right-hand member in powers of the rotation angle ω and equate coefficients. Writing $\omega = [\omega_1, \omega_2, \omega_3]$ and using Eq. (10–194), we easily find, to second order in $|\omega|$,

$$
\begin{aligned}
U\,D_+U^{-1} &= \exp\,(i\omega \cdot \mathbf{J})\,D_+\exp\,(-i\omega \cdot \mathbf{J}) \\
&= (1 + i\omega_3 - \tfrac{1}{2}\omega^2)\,D_+ - i(\omega_1 + i\omega_2)\,D_3 \\
&\qquad + \tfrac{1}{2}(\omega_1 + i\omega_2)\omega \cdot \mathbf{D}.
\end{aligned}
$$

If this result is substituted into Eq. (11–213) and note is taken of the relations

$$
\sum_{M\,M'} (\alpha', J', M'|D_+|\alpha, J, M)(\alpha, J, M|D_3|\alpha', J', M') = 0,
$$

$$
\sum_{M\,M'} (\alpha', J', M'|D_\pm|\alpha, J, M)(\alpha, J, M|D_\pm|\alpha', J', M') = 0,
$$

which are immediate consequences of the selection rules, one obtains, after some reduction,

$$
(\omega_1^2 + \omega_2^2) \left\{ \sum_{M\,M'} |(\alpha', J', M'|D_3|\alpha, J, M)|^2 \right.
$$

$$
\left. - \tfrac{1}{2} \sum_{M\,M'} |(\alpha', J', M'|D_+|\alpha, J, M)|^2 \right\} = 0,
$$

from which the first of Eqs. (11–209) follows. The proof of the second equation is similar.

Equations (11–207) and (11–208) now reduce to the form

$$
I_+ = I_- = \frac{\omega^4}{2\pi c^3} \frac{S}{2J + 1} \tfrac{1}{2}(1 + \cos^2 \theta)\,d\Omega,
$$

$$
I_\pi = \frac{\omega^4}{2\pi c^3} \frac{S}{2J + 1} \sin^2 \theta\,d\Omega,
$$
(11–214)

where S is the common value of the sums (11–209).

The total mean intensity, namely,

$$
I = I_+ + I_- + I_\pi = \frac{\omega^4}{\pi c^3} \frac{S}{2J + 1}\,d\Omega,
$$
(11–215)

is independent of θ: the radiation is isotropic. It now remains to be shown that the radiation in the absence of a field is unpolarized.

The polarization of the incoherently superimposed components is most conveniently described by means of the complex amplitude of the electric field, which we write in the form

$$
\mathbf{E} = E\mathbf{a}e^{i\phi}.
$$
(11–216)

Since, by its definition, the complex vector **a** satisfies

$$\mathbf{a^*} \cdot \mathbf{a} = 1, \tag{11-217}$$

the real positive number E and the phase ϕ are given by

$$\mathbf{a^*} \cdot \mathbf{E} = E e^{i\phi}. \tag{11-218}$$

The real electric field vector $\boldsymbol{\varepsilon}$ is taken to be the real part of $\mathbf{E}$, and it therefore follows from Eqs. (11-165) and (11-167) that the intensity is

$$I = \frac{c}{8\pi} \mathbf{E^*} \cdot \mathbf{E} = \frac{c}{8\pi} E^2, \tag{11-219}$$

whence

$$E = \sqrt{\frac{8\pi I}{c}}$$

is the magnitude of $\mathbf{E}$.

If two waves $\mathbf{E}_1$ and $\mathbf{E}_2$ are superimposed, the complex amplitude of the resultant wave is

$$\mathbf{E} = \mathbf{E}_1 + \mathbf{E}_2 = E_1 \mathbf{a}_1 e^{i\phi_1} + E_2 \mathbf{a}_2 e^{i\phi_2}, \tag{11-220}$$

so that the intensity is

$$\frac{c}{8\pi} \mathbf{E^*} \cdot \mathbf{E} = \frac{c}{8\pi} \{ E_1^2 + E_2^2 + E_1 E_2 (\mathbf{a}_1^* \cdot \mathbf{a}_2 e^{i(\phi_2 - \phi_1)} + \mathbf{a}_2^* \cdot \mathbf{a}_1 e^{-i(\phi_2 - \phi_1)}) \}.$$

The first two terms in this expression are the intensities of the individual components, and the third term accounts for the modification of the total intensity due to *interference*. It is seen immediately that the interference term is zero, *independently of the phases of the two components*, if and only if $\mathbf{a}_1$ and $\mathbf{a}_2$ are orthogonal, i.e., if

$$\mathbf{a}_1^* \cdot \mathbf{a}_2 = 0. \tag{11-221}$$

Hence, if $\mathbf{a}_1$ and $\mathbf{a}_2$ are orthogonal, it is meaningful to divide the intensity into the parts $(c/8\pi)E_1^2$ and $(c/8\pi)E_2^2$, associated with the two components. Also, if $\mathbf{a}_1$ and $\mathbf{a}_2$ satisfy Eq. (11-221), an arbitrary vector $\mathbf{E}$ can be expressed in the form (11-220), where the amplitudes and phases are given by

$$E_1 e^{i\phi_1} = \mathbf{a}_1^* \cdot \mathbf{E}, \qquad E_2 e^{i\phi_2} = \mathbf{a}_2^* \cdot \mathbf{E}. \tag{11-222}$$

Finally, we consider the intensity arising from the superposition of a large number N of waves of various amplitudes and phases. The complex amplitude is

$$\mathbf{E} = \mathbf{E}^{(1)} + \mathbf{E}^{(2)} + \cdots + \mathbf{E}^{(N)}, \tag{11-223}$$

and each amplitude can be written in the form (11–220), so that $\mathbf{E}$ becomes

$$\mathbf{E} = (E_1^{(1)}e^{i\phi_1^{(1)}} + E_1^{(2)}e^{i\phi_1^{(2)}} + \cdots)\mathbf{a}_1 + (E_2^{(1)}e^{i\phi_2^{(1)}} + E_2^{(2)}e^{i\phi_2^{(2)}} + \cdots)\mathbf{a}_2,$$

where, by Eqs. (11–222),

$$E_1^{(i)}e^{i\phi_1^{(i)}} = \mathbf{a}_1^* \cdot \mathbf{E}^{(i)} \quad \text{and} \quad E_2^{(i)}e^{i\phi_2^{(i)}} = \mathbf{a}_2^* \cdot \mathbf{E}^{(i)}.$$

The intensities in the two components are

$$I_1 = \frac{c}{8\pi} |E_1^{(1)}e^{i\phi_1^{(1)}} + E_1^{(2)}e^{i\phi_1^{(2)}} + \cdots|^2,$$

$$I_2 = \frac{c}{8\pi} |E_2^{(1)}e^{i\phi_2^{(1)}} + E_2^{(2)}e^{i\phi_2^{(2)}} + \cdots|^2. \tag{11–224}$$

If the waves $\mathbf{E}^{(1)}$ originate with randomly distributed phases, the intensity recorded by an instrument designed to measure I_1, for example, is obtained by averaging the above expression with respect to the phases $\phi_1^{(1)}$. This average is[1]

$$\overline{I_1} = \frac{c}{8\pi N} (E_1^{(1)^2} + E_1^{(2)^2} + \cdots). \tag{11–225}$$

Similarly, the average intensity in the component $\mathbf{a}_2$ is[2]

$$\overline{I_2} = \frac{c}{8\pi N} (E_2^{(1)^2} + E_2^{(2)^2} + \cdots). \tag{11–226}$$

If these two intensities are equal, the light is *unpolarized*, but if, for example, $\overline{I_2} = 0$, the superposition (11–223) is composed entirely of waves polarized in the state $\mathbf{a}_1$, that is, the light is *completely polarized*. If $\overline{I_1}$ and $\overline{I_2}$ are unequal, and neither is zero, the light is *partially polarized*.

The analysis of a statistical mixture of waves is facilitated by the introduction of *statistical factors* ϵ_i, which by definition satisfy the relations[3]

$$\overline{\epsilon_i^* \epsilon_j} = \delta_{ij}/N. \tag{11–227}$$

The bar denotes the "average over phases" discussed above. In this

---

[1] J. W. Strutt, Baron Rayleigh, *The Theory of Sound*. New York: Dover Publications, 1945, Section 42a.

[2] Note that the phases $\phi_1^{(i)}$ and $\phi_2^{(i)}$ are not independent, but are in a fixed relation to each other, determined by the Eqs. (11–222). However, since the polarizations $\mathbf{a}_1$ and $\mathbf{a}_2$ are orthogonal, this does not matter for the computation of the mean intensities.

[3] G. Breit, *Phys. Rev.* **71**, 402 (1947).

notation, the superposition (11–223) is represented by

$$\mathbf{E} = \mathbf{E}^{(1)}\epsilon_1 + \mathbf{E}^{(2)}\epsilon_2 + \cdots + \mathbf{E}^{(N)}\epsilon_N, \qquad (11\text{–}228)$$

and the Eqs. (11–224) become

$$I_1 = \frac{c}{8\pi}|E_1^{(1)}\epsilon_1 + E_1^{(2)}\epsilon_2 + \cdots + E_1^{(N)}\epsilon_N|^2,$$

$$I_2 = \frac{c}{8\pi}|E_2^{(1)}\epsilon_1 + E_2^{(2)}\epsilon_2 + \cdots + E_2^{(N)}\epsilon_N|^2. \qquad (11\text{–}229)$$

The factors $e^{i\phi_1^{(i)}}$, etc., in Eqs. (11–224) have been suppressed: If the numbers ϵ_i satisfy Eqs. (11–227), then the numbers $\epsilon_i e^{i\phi_i}$, in which the phases ϕ_i are arbitrary, also satisfy this relation. Because of (11–227), the averages of the expressions (11–229) are identical with (11–225) and (11–226). This, of course, justifies the notation.

We now return to the discussion of the dipole radiation by a statistical assemblage of atoms. We have found that the σ-components have the average intensities [Eqs. (11–214), (11–215)]

$$I_{\pm} = I \cdot \tfrac{1}{4}(1 + \cos^2\theta), \qquad (11\text{–}230)$$

and the polarization vectors [Eqs. (11–200), (11–203)]

$$\mathbf{a}_{\pm} = \frac{1}{\sqrt{1 + \cos^2\theta}}(\cos\theta\,\mathbf{e}_1 \pm i\,\mathbf{e}_2). \qquad (11\text{–}231)$$

The random superposition of these two radiations is represented by the complex vector

$$\mathbf{E}_+ + \mathbf{E}_- = \sqrt{\frac{8\pi I_+}{c}}\,\mathbf{a}_+\epsilon_+ + \sqrt{\frac{8\pi I_-}{c}}\,\mathbf{a}_-\epsilon_-$$

$$= \sqrt{\frac{8\pi I}{c}}\{\tfrac{1}{2}(\cos\theta\,\mathbf{e}_1 + i\,\mathbf{e}_2)\epsilon_+ + \tfrac{1}{2}(\cos\theta\,\mathbf{e}_1 - i\,\mathbf{e}_2)\epsilon_-\},$$

in which ϵ_+ and ϵ_- are statistical factors. The intensity of the light polarized in the direction $\mathbf{e}_1$ is therefore

$$\overline{I_\sigma(\mathbf{e}_1)} = I|\tfrac{1}{2}\cos\theta\,\epsilon_+ + \tfrac{1}{2}\cos\theta\,\epsilon_-|^2$$

$$= I(\tfrac{1}{4}\cos^2\theta + \tfrac{1}{4}\cos^2\theta) = \tfrac{1}{2}I\cos^2\theta, \qquad (11\text{–}232)$$

and in the direction $\mathbf{e}_2$,

$$\overline{I_\sigma(\mathbf{e}_2)} = I|\tfrac{1}{2}i\epsilon_+ - \tfrac{1}{2}i\epsilon_-|^2 = \tfrac{1}{2}I. \qquad (11\text{–}233)$$

Thus the superposition of these two components is partially polarized in the direction $\mathbf{e}_2$.

The π-component is completely polarized in the direction $\mathbf{e}_1$; by Eqs. (11–205) and (11–214), we have

$$I_\pi = I \cdot \tfrac{1}{2} \sin^2 \theta, \qquad \mathbf{a}_\pi = \mathbf{e}_1, \qquad (11\text{–}234)$$

whence

$$\overline{I_\pi(\mathbf{e}_1)} = \tfrac{1}{2}I \sin^2 \theta, \qquad (11\text{–}235)$$

$$\overline{I_\pi(\mathbf{e}_2)} = 0. \qquad (11\text{–}236)$$

Thus, when this component is added to the σ-components, we find equal intensities in the two directions:

$$\overline{I_\sigma(\mathbf{e}_1)} + \overline{I_\pi(\mathbf{e}_1)} = \overline{I_\sigma(\mathbf{e}_2)} + \overline{I_\pi(\mathbf{e}_2)} = \tfrac{1}{2}I.$$

The total light emitted from all transitions in a statistical assemblage is therefore unpolarized.

We shall conclude this section with an example, calculating the relative intensities and the weak-field Zeeman patterns for the $^2P \rightarrow {}^2S$ transitions that occur in the spectra of the alkalis.[1] The ^2P term has two sublevels, corresponding to $J = 3/2$ and $J = 1/2$. Because of the spin-orbit interaction, these levels are slightly separated in energy. By Eq. (11–67), the corresponding g-factors are

$$g(^2P_{3/2}) = \tfrac{4}{3}, \qquad g(^2P_{1/2}) = \tfrac{2}{3}, \qquad g(^2S_{1/2}) = 2.$$

The possible transitions, obeying the selection rules $\Delta M = \pm 1$, 0, are shown in Fig. 11–7. The difference between the frequencies of the lines in the Zeeman pattern and the frequency ω_0, which appears in the absence of a field, is given by Eq. (11–66):

$$\Delta\omega = \omega - \omega_0 = -\frac{e}{2mc} \, \mathfrak{B}(gM - g'M').$$

The primed and unprimed quantities refer to the final and initial states, respectively.[2]

The intensity ratios for the various lines are given by the squares of the appropriate factors in Table 10–1. Thus, for the $J = 3/2 \rightarrow J = 1/2$ transitions, we have $\Delta J = -1$, and by Eq. (11–207), the intensities for transverse observation are found to be related in the ratios 1:3:4:4:3:1. These intensity ratios are indicated by the length of the bars on the right bottom of Fig. 11–7. Similar calculations show that the Zeeman pattern for $\Delta J = 0$ transitions contains four equally intense lines: Two are

---

[1] E. U. Condon and G. H. Shortley, *The Theory of Atomic Spectra*. Cambridge: Cambridge University Press, 1953, Chapters V and XVI.

[2] Note that the electronic charge is a negative quantity.

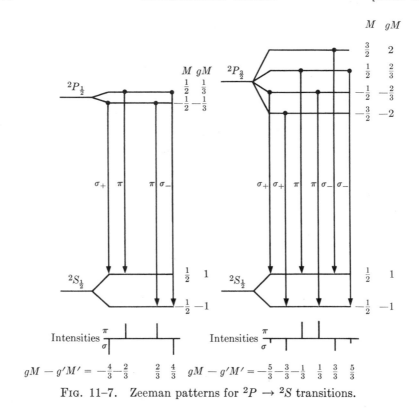

FIG. 11–7. Zeeman patterns for $^2P \rightarrow \,^2S$ transitions.

π-components and two are σ-components (cf. left bottom of Fig. 11–7). All these predictions of the theory are accurately confirmed by experiment, and give a complete account of the "anomalous" Zeeman effect which is now seen not to be anomalous at all, but subject to detailed explanation on the basis of the electron-spin hypothesis.

The behavior of the Zeeman pattern for fields which are sufficiently strong to invalidate the assumption that J is a "good quantum number," i.e., that the system is approximately isolated, is also explained by a suitable extension of the theory. The reader is referred to works on atomic spectra for discussions of this case, known as the *Paschen-Back effect*.[1]

11–14 Forbidden transitions. Higher multipoles. Transitions which are forbidden by the dipole selection rules (11–188) may nevertheless occur, with a finite probability, through radiation of a higher *multipole*

---

[1] E. U. Condon and G. H. Shortley, *The Theory of Atomic Spectra*. Cambridge: Cambridge University Press, 1953.

order. Such transitions arise from the higher-order terms in the expansion (11–177), which we now examine briefly.

The contribution of the second term, $i\mathbf{k} \cdot \mathbf{r}$, to the operator v [Eq. (11–174)] is

$$v' = -\frac{e}{mc} \sum_{i=1}^{Z} i(\mathbf{a} \cdot \mathbf{p}_i)(\mathbf{k} \cdot \mathbf{r}_i).$$ (11–237)

The quantity $(\mathbf{a} \cdot \mathbf{p}_i)(\mathbf{k} \cdot \mathbf{r}_i)$, which appears in the typical term of this sum, is a component of a second-rank tensor.[1] It can be expressed as a sum of symmetric and antisymmetric parts,

$$\mathbf{p}_i \mathbf{r}_i = \tfrac{1}{2}(\mathbf{p}_i \mathbf{r}_i - \mathbf{r}_i \mathbf{p}_i) + \tfrac{1}{2}(\mathbf{p}_i \mathbf{r}_i + \mathbf{r}_i \mathbf{p}_i),$$ (11–238)

so that Eq. (11–237) becomes

$$v' = v_M + v_Q,$$

where

$$v_M = -\frac{ie}{2mc} \sum_{i=1}^{Z} [(\mathbf{a} \cdot \mathbf{p}_i)(\mathbf{k} \cdot \mathbf{r}_i) - (\mathbf{a} \cdot \mathbf{r}_i)(\mathbf{k} \cdot \mathbf{p}_i)]$$

$$= -\frac{ie}{2mc} \sum_{i=1}^{Z} (\mathbf{k} \times \mathbf{a}) \cdot (\mathbf{r}_i \times \mathbf{p}_i) = -\frac{ie}{2mc} \sum_{i=1}^{Z} (\mathbf{k} \times \mathbf{a}) \cdot \mathbf{l}_i,$$ (11–239)

$$v_Q = -\frac{ie}{2mc} \sum_{i=1}^{Z} [(\mathbf{a} \cdot \mathbf{p}_i)(\mathbf{k} \cdot \mathbf{r}_i) + (\mathbf{a} \cdot \mathbf{r}_i)(\mathbf{k} \cdot \mathbf{p}_i)],$$ (11–240)

and $\mathbf{l}_i = \mathbf{r}_i \times \mathbf{p}_i$ is the orbital angular momentum of the ith particle. Now we have seen in Section 10–5 that the quantity

$$\boldsymbol{\mu}_i = \frac{e}{2mc} \mathbf{l}_i$$ (11–241)

is the magnetic dipole moment resulting from the orbital motion of a particle of charge e. Consequently, the operator v_M can be written

$$v_M = \frac{\omega}{c} \mathbf{a}' \cdot \mathbf{M},$$ (11–242)

where $\mathbf{M}$ is the total magnetic dipole moment of the system,

$$\mathbf{M} = \sum_{i=1}^{Z} \boldsymbol{\mu}_i,$$

and $\mathbf{a}' = -i(\mathbf{k}/k) \times \mathbf{a}$ is a unit polarization vector perpendicular to $\mathbf{a}$

---

[1] W. K. H. Panofsky and M. Phillips, *Classical Electricity and Magnetism.* Reading, Mass.: Addison-Wesley Publishing Co., Inc., 1955, Section 1–7.

and $\mathbf{k}$. Transitions corresponding to the matrix elements of the operator v_M give rise to *magnetic dipole radiation*, for which the angular distribution is the same as that of the electric dipole radiation discussed in Section 11–12 [cf. Eq. (11–179)]. The two electromagnetic fields differ qualitatively only in their states of polarization.

The selection rules for the change of angular-momentum quantum numbers in magnetic dipole transitions are the same as those given for electric dipole transitions in Eq. (11–188). However, the rule requiring a change in parity between the initial and final states for electric dipole transitions does not apply for magnetic dipole transitions. The magnetic moment vector does not change sign under the parity operation, i.e.,

$$\Pi \mathbf{M} \Pi^{-1} = \mathbf{M},$$

whence a transition is allowed for magnetic dipole radiation only if the initial and final states have the same parity. For this reason, electric and magnetic dipole transitions cannot both occur between the same pairs of states.

The infrequent occurrence of observable magnetic dipole radiation in atomic spectra is due simply to the fact that, by comparison with $\mathbf{D}$, $\mathbf{M}$ is of relatively small magnitude. Since l_i is of the order mrv, where v is the velocity of an atomic electron, the ratio $\mathbf{M}/\mathbf{D}$ has the order of magnitude

$$\frac{e}{mc} \, mrv/er = v/c.$$

Magnetic dipole transition probabilities are therefore smaller than electric dipole transition probabilities by the factor v^2/c^2.[1]

The operator v_Q can be simplified by a calculation similar to that which led to Eq. (11–179). Writing $\mathbf{p}_i = m\dot{\mathbf{r}}_i$, we have

$$\mathbf{p}_i \mathbf{r}_i + \mathbf{r}_i \mathbf{p}_i = \frac{im}{\hbar} \{[H, \mathbf{r}_i]\mathbf{r}_i + \mathbf{r}_i[H, \mathbf{r}_i]\} = \frac{im}{\hbar}[H, \mathbf{r}_i\mathbf{r}_i],$$

whence

$$(v_Q)_{FI} = \frac{1}{2\hbar c}(E_F - E_I)\mathbf{a} \cdot Q_{FI} \cdot \mathbf{k} = \pm \frac{\omega^2}{2c^2}\mathbf{a} \cdot Q_{FI} \cdot \frac{\mathbf{k}}{k}, \quad (11\text{–}243)$$

where

$$Q = \sum_{i=1}^{Z} e_i \mathbf{r}_i \mathbf{r}_i \qquad\qquad (11\text{–}244)$$

_____

[1] The above argument is not valid, in general, for multipole radiations of relatively high order observed in nuclear transitions. See J. M. Blatt and V. F. Weisskopf, *Theoretical Nuclear Physics*. New York: John Wiley and Sons, 1952, Chapter XII, Section 5A.

is the *electric quadrupole moment*. (The discussion of electric quadrupole radiation is left to the problems.) The principal selection rules are $\Delta J = \pm 2, \pm 1, 0$; $\Delta M = \pm 2, \pm 1, 0$; no parity change. The transition probability for an electric quadrupole transition is smaller than that for the electric dipole transition by a factor of order $(r/\lambda)^2$, where λ is the wavelength of the emitted radiation.

The discussion of the higher multipole orders is best carried out by means of a description of the radiation field which exploits its symmetry properties in a more systematic way; it will not be given here. For a complete development of this interesting and important topic, the reader is referred to appropriate sources.[1]

References

Bohm, D., *Quantum Theory.* New York: Prentice-Hall, Inc., 1951, Chapters 18, 19, and 20.

Condon, E. U., and Shortley, G. H., *The Theory of Atomic Spectra.* Cambridge: Cambridge University Press, 1951. The standard treatise on the application of the theory of radiative transitions to atomic spectra. See especially Chapter IV.

Dirac, P. A. M., *The Principles of Quantum Mechanics.* 3rd ed. Oxford: The Clarendon Press, 1947, Chapter VII. A short, concise summary of the essential elements of perturbation theory.

Kemble, E. C., *The Fundamental Principles of Quantum Mechanics.* New York: McGraw-Hill Book Co., 1937, Chapters XI and XII. Clear discussions of perturbation theory with remarks on its relationship to quantum-statistical mechanics.

Schiff, L. I., *Quantum Mechanics.* New York: McGraw-Hill Book Co., Inc., 1955, 2nd ed., Chapters VII and VIII. Interesting examples of the application of perturbation theory are given in this text.

---

[1] J. M. Blatt and V. F. Weisskopf, *Theoretical Nuclear Physics.* New York: John Wiley and Sons, 1952, Chapter XII and Appendix B. M. E. Rose, *Multipole Fields.* New York: John Wiley and Sons, Inc., 1955.

PROBLEMS

11–1. Work out the eigenvalues and eigenvectors of the matrix

$$H = \begin{pmatrix} 1 & 2\epsilon & 0 \\ 2\epsilon & 2 + \epsilon & 3\epsilon \\ 0 & 3\epsilon & 3 + 2\epsilon \end{pmatrix},$$

to second order in the small parameter ϵ. Expand the roots of the characteristic equation and show that the characteristic values can be obtained to third order by substitution of the second-order eigenvectors into Eq. (11–10).

11–2. Find the first-order eigenfunctions of the matrix

$$H = \begin{pmatrix} 1 & \epsilon & 0 \\ \epsilon & 1 & \epsilon \\ 0 & \epsilon & 2 \end{pmatrix},$$

and use them to calculate the eigenvalues of H to second order. Draw a graph showing the eigenvalues as a function of ϵ for $-0.1 \leq \epsilon \leq 0.1$. Calculate a few values exactly, and compare with the perturbation-theory approximations.

11–3. Derive Eqs. (11–32) and (11–33) by using the first- and second-order wave functions in Eq. (11–10).

11–4. Calculate the shift of the energy levels of a harmonic oscillator, produced by a perturbation of the form ϵx^3. Examine the approach to classical behavior for large n and comment on this calculation in the light of the remark in the footnote on p. 384.

11–5. A one-dimensional harmonic oscillator is subjected to a constant force F. Classically, its motion is unaffected, except for a displacement of the equilibrium position. Solve the corresponding quantum-mechanical problem by the perturbation method, and compare with the exact solution.

11–6. Calculate the expectation of the induced dipole moment for an atom in a weak electric field. Assume that the energy states of the atom are non-degenerate. The perturbation energy may be taken to be $-\sum_i e_i \mathcal{E} z_i$, where $\mathcal{E}$ is the electric field. The z-component of the dipole moment is $\sum_i e_i z_i$. Derive an expression for the polarizability of the atom, i.e., for the induced dipole moment per unit field strength.

11–7. Draw an energy-level diagram for the 3-dimensional harmonic oscillator and show the effect of the perturbation (11–58) in splitting the degenerate levels. Using the perturbation theory and the variational principle, try to devise a method which will permit one to follow the changes in the energy levels as the potential-energy function is changed continuously from that of the harmonic oscillator to the infinitely deep square well

$$V = \begin{cases} 0, & 0 \leq r < R, \\ \infty, & R < r. \end{cases}$$

How do the results depend upon the parameter $\hbar/m\omega R^2$?

11–8. Apply the variational method to the deuteron problem discussed in Section 11–6, taking as the trial function the expression

$$\psi = e^{-\alpha r/r_0} - e^{-\beta r/r_0},$$

in which β is an additional variational parameter [cf. Eq. (11–82)]. In comparison with the exact result given in Fig. 11–3, what is the error in the function $V_0 r_0^2$ at $r_0 = 1.4 \times 10^{-13}$ cm?

11–9. Use the trial function (11–82) to find the lowest energy eigenvalue for the potential-energy function (8–105). Adjust the parameters to give the binding energy of the deuteron and derive a range-depth relation for this potential. Solve the problem exactly and compare with the approximate result.[1]

11–10. The nuclear charge e of a hydrogenic atom in its ground state $\psi_0(e)$ is changed abruptly to $2e$ by β-decay at the instant $t = 0$. Show that the probability of finding the electron in the state $\psi_n(2e)$ after the change is

$$|c_n|^2 = 2^9 n^5 \frac{(n-2)^{2n-4}}{(n+2)^{2n+4}},$$

where n is the principal quantum number for the s-states of the electron in the Coulomb field of the charge $2e$. Show that $\sum_{n=0}^{\infty} |c_n|^2 \approx 0.974$, whence the probability for ionization of the atom is approximately 2.6%. Neglect the perturbation caused by the emitted β-particle.

11–11. The states of an electron in a central field can be written in the form

$$\psi(nl, m_l m_s) = R_{nl}(r) Y_l^{m_l}(\theta, \phi) \chi(m_s).$$

The energy of such a state depends upon the orbital angular momentum l and the principal quantum number n, but is independent of m_l and m_s (Section 10–9). Hence each energy level has a $2(2l+1)$-fold degeneracy. Show that the possible values for the total angular momentum in these states are $j = l \pm 1/2$, and that the corresponding eigenfunctions of $\mathbf{J}^2$ and J_z are $\phi(nl, j, m)$, where

$$\phi(nl, l+\tfrac{1}{2}, m) = \sqrt{\frac{l+m+\tfrac{1}{2}}{2l+1}} \, \psi(nl, m-\tfrac{1}{2}, \tfrac{1}{2})$$

$$+ \sqrt{\frac{l-m+\tfrac{1}{2}}{2l+1}} \, \psi(nl, m+\tfrac{1}{2}, -\tfrac{1}{2}),$$

$$\phi(nl, l-\tfrac{1}{2}, m) = \sqrt{\frac{l-m+\tfrac{1}{2}}{2l+1}} \, \psi(nl, m-\tfrac{1}{2}, \tfrac{1}{2})$$

$$- \sqrt{\frac{l+m+\tfrac{1}{2}}{2l+1}} \, \psi(nl, m+\tfrac{1}{2}, -\tfrac{1}{2}).$$

---

[1] R. G. Sachs, *Nuclear Theory*. Reading, Mass.: Addison-Wesley Publishing Company, Inc., 1953, Section 3–2.

[*Hint:* Show from the equations $\mathbf{J} = \mathbf{L} + \mathbf{s} = \mathbf{L} + \frac{1}{2}\boldsymbol{\sigma}$ that a simultaneous eigenfunction of $\mathbf{J}^2$ and $\mathbf{L}^2$ is also an eigenfunction of the operator $\boldsymbol{\sigma} \cdot \mathbf{L}$; then, by means of the identity $\mathbf{L}^2 = \boldsymbol{\sigma} \cdot \mathbf{L}(\boldsymbol{\sigma} \cdot \mathbf{L} + 1)$ (proof?), demonstrate that the eigenvalues of $\boldsymbol{\sigma} \cdot \mathbf{L}$ are l, $-(l + 1)$, corresponding to $j = l + 1/2$ and $j = l - 1/2$, respectively. Finally, evaluate the coefficients in the above equations by writing $\boldsymbol{\sigma} \cdot \mathbf{L} = L_+ s_- + L_- s_+ + 2L_z s_z$ and using the relations (10–69) and (10–70).]

11–12. The *spin-orbit* interaction, mentioned in Section 10–9, arises from the interaction between the electron's magnetic moment and the magnetic field which appears in the electron's frame of reference, due to its motion through the static electric field of potential $V(r)$. It introduces a term in the Hamiltonian of the form[1]

$$H^{(1)} = \xi(r)\mathbf{L} \cdot \mathbf{s},$$

where $\xi(r) = (1/2m^2c^2)[(1/r)(dV/dr)]$. Show that the eigenfunctions of Problem 11–11 are adapted to this perturbation, and that the energies of the perturbed states are

$$E(nlj) = E_0(nl) + \begin{cases} \frac{1}{2}l\zeta_{nl}, & j = l + \frac{1}{2}, \\ -\frac{1}{2}(l + 1)\zeta_{nl}, & j = l - \frac{1}{2}, \end{cases}$$

where

$$\zeta_{nl} = \hbar^2 \int_0^\infty R_{nl}^2(r)\xi(r)r^2 \, dr,$$

and $E_0(nl)$ is the energy of the unperturbed level.

11–13. Calculate the quantity ζ_{nl}, defined in the preceding problem, for the potential function $V(r) = -Ze^2/r$.

Answer: $\zeta_{nl} = \dfrac{e^2\hbar^2}{2m^2c^2a_0^3} \dfrac{Z^4}{n^3l(l + \frac{1}{2})(l + 1)}.$ (Cf. Problem 7–40.)

Estimate the magnitude of the spin-orbit splitting for the terms of hydrogen. The above expression is invalid for *s*-states, which must be treated relativistically (why?).

11–14. Justify formula (11–189).

11–15. Describe the polarization represented by Eq. (11–190). Give formulae for the ellipticity and inclination of the major axis in terms of the parameters α_1 and α_2. Also show how to find the sense of rotation of the electric vector.

11–16. Derive the equations (11–203), (11–204), (11–205), and (11–206).

11–17. Obtain Eq. (11–215) directly from Eq. (11–189), by summing over the polarizations $\mathbf{e}_1$ and $\mathbf{e}_2$.

11–18. Using Table 10–1, calculate the sum S and verify Eq. (11–209).

11–19. Prove the identity (11–212). Give a physical interpretation of this identity on the basis of the fundamental assumptions set forth in Section 9–13.

---

[1] E. U. Condon and G. H. Shortley, *The Theory of Atomic Spectra*. Cambridge: Cambridge University Press, 1953, Sec. 4[5].

11–20. Supply the details of the proof of Eq. (11–209), given in the text.

11–21. If the vectors $\mathbf{E}^{(1)}$ and $\mathbf{E}^{(2)}$ in Eq. (11–228) represent waves in the same state of polarization, i.e.

$$\mathbf{E}^{(1)} = E^{(1)}\mathbf{a}, \qquad \mathbf{E}^{(2)} = E^{(2)}\mathbf{a},$$

then the terms $\mathbf{E}^{(1)}\epsilon_1 + \mathbf{E}^{(2)}\epsilon_2$ can be replaced by $\sqrt{(E^{(1)})^2 + (E^{(2)})^2}\, \mathbf{a}\epsilon$, where ϵ is a suitably defined statistical factor. Prove this statement. What is its physical significance?

11–22. Let a given state of polarization be described by the vector

$$\mathbf{b} = \alpha_1\mathbf{a}_1 + \alpha_2\mathbf{a}_2 = \alpha'_1\mathbf{a}'_1 + \alpha'_2\mathbf{a}'_2,$$

in which $(\mathbf{a}_1, \mathbf{a}_2)$ and $(\mathbf{a}'_1, \mathbf{a}'_2)$ are two different pairs of orthogonal polarization vectors. Show that the coefficients (α_1, α_2) and (α_1, α'_2) are related by a unitary transformation.

11–23. Work out the frequencies and intensities in the Zeeman patterns, and draw figures similar to Fig. 11–7 for $^3S_1 \to {}^3P_1$ and $^2P_{3/2} \to {}^2D_{5/2}$ transitions. Try to find experimental data which confirm your results.

11–24. Discuss the polarization of the radiations produced in a magnetic dipole transition.

11–25. Show that the prohibition of $J = 0 \to J = 0$ transitions cannot be violated in radiative transitions of any multipole order. (Radiative transitions of this type are strictly forbidden.)

11–26. Show that the six components of the quadrupole moment tensor [Eq. (11–244)] are linear combinations of the six quantities

$$Q_m = \sum_i e_i r_i^2 Y_2^m(\theta_i, \phi_i), \qquad m = 0, \pm 1, \pm 2,$$

$$T = \sum_i e_i r_i^2,$$

in which (r_i, θ_i, ϕ_i) are the polar coordinates of the ith electron. Prove that the component T makes no contribution to the matrix element (11–243). Give a physical reason for this fact.

11–27. Prove that the five quantities Q_m have the following commutation relations with the angular-momentum operator $\mathbf{J}$:

$$[J_+, Q_m] = \sqrt{(2 - m)(3 + m)}\, Q_{m+1},$$

$$[J_-, Q_m] = \sqrt{(2 + m)(3 - m)}\, Q_{m-1},$$

$$[J_z, Q_m] = mQ_m.$$

[Hint: Express the spherical harmonics in Problem 11–26 in terms of (x, y, z) and use the commutation relations (10–199).]

11–28. Show that the operator Q_m gives rise to transitions for which $\Delta M = m$.

11–29. Use the results of Problems 11–26 and 11–27, in conjunction with Table 10–1, to obtain the matrix elements and selection rules for the operators Q_m. In particular, show that the transitions $J = 1/2 \to J = 1/2$ and $J = 0 \to J = \pm 1$ are forbidden for quadrupole radiation.[1]

11–30. Discuss the polarization of the light emitted in an electric quadrupole transition.

11–31. The electrostatic energy of a system of charges e_i in a static electric field of potential Φ is

$$V = \sum_i e_i \Phi(\mathbf{r}_i).$$

If the electric field varies slowly within the volume containing the charges, the potential can be expanded in the form

$$\Phi(\mathbf{r}_i) = \Phi(0) + \mathbf{r}_i \cdot \nabla \Phi(\mathbf{r})|_{\mathbf{r}=0} + \tfrac{1}{2}(\mathbf{r}_i \cdot \nabla)^2 \Phi(\mathbf{r})|_{\mathbf{r}=0} + \cdots.$$

Show that this expansion leads to the expression

$$V = \sum_i e_i \Phi(0) - \mathbf{D} \cdot \boldsymbol{\varepsilon} + \tfrac{1}{2} \sum_{\alpha,\beta} Q_{\alpha\beta} \frac{\partial^2 \Phi}{\partial x_\alpha \, \partial x_\beta}\bigg|_{\mathbf{r}=0} + \cdots,$$

where $\mathbf{D}$ is the electric dipole moment, and $Q_{\alpha\beta} = \sum_i e_i x_{i\alpha} x_{i\beta}$ is the quadrupole moment tensor (11–244). Express the components of Q in terms of the components defined in Problem 11–26 and show that T makes no contribution to the potential energy.

11–32. Use the expression derived in Problem 11–31 to calculate the first-order shift of the energy levels of a system of charged particles in a weak electric field. Give a physical interpretation of your answers. In particular, show that the contribution of the second term is zero for any unperturbed nondegenerate state which has a definite parity. Comment upon the statement: "Quantum-mechanical systems in stationary states do not have permanent electric dipole moments."[2]

11–33. Show that the diagonal matrix element of the quadrupole interaction energy for a state (α, J, M) is[3]

$$\frac{1}{4}\left(\frac{\partial^2 \Phi}{\partial x_3^2}\right)_{\mathbf{r}=0} \frac{Q}{J(2J-1)} \{3M^2 - J(J+1)\},$$

[1] S. A. Moszkowski, *Theory of Multipole Radiation*, in K. Siegbahn, Ed., *Beta- and Gamma-Ray Spectroscopy*. Amsterdam: North-Holland Publishing Co., 1955, Chapter XIII.

[2] J. M. Blatt and V. F. Weisskopf, *Theoretical Nuclear Physics*. New York: John Wiley and Sons, 1952, p. 25. See also L. I. Schiff, *Quantum Mechanics*. 2nd. ed. New York: McGraw-Hill Book Co., Inc., 1955, p. 160.

[3] E. Feenberg and G. E. Pake, *Notes on the Quantum Theory of Angular Momentum*. Reading, Mass.: Addison-Wesley Publishing Co., Inc., 1953, p. 46.

where

$$Q = (\alpha JJ| \sum_i e_i(3x_{i3}^2 - r_i^2)|\alpha JJ).$$

Observe that the dependence of the quadrupole energy on the quantum numbers α of the state is contained entirely in the single quantity Q, the *quadrupole moment* of the system. Explain physically why the above expression depends upon the electric field only through the single derivative $(\partial^2\Phi/\partial x_3^2)_{r=0}$.

11–34. The deuteron is known to have a quadrupole moment

$$Q = (0.002738 \pm 0.000014) \times 10^{-24}e \text{ cm}^2.[1]$$

Show that, in the coordinate system introduced in Section 11–6, the quadrupole moment is

$$Q = \frac{e}{4}(\alpha JJ|3z^2 - r^2|\alpha JJ).$$

The total angular momentum of the deuteron is $J = 1$, and in the ground state the spins of the neutron and proton are parallel. In the discussion of the binding energy (Section 11–6), the orbital state was assumed to be "purely S," i.e., $L = 0$. However, since the system has a measurable quadrupole moment, this cannot be strictly true, and it is usual to assume that there is, in addition, a small admixture of the D-state, for which $L = 2$. In the notation of Problems 10–17, 10–18, and 10–19, the ground-state wave function can be represented as the vector

$$\psi = a\frac{f(r)}{r}\psi(1, M, 0) + b\frac{g(r)}{r}\psi(1, M, 2),$$

in which a and b are real numbers which satisfy

$$a^2 + b^2 = 1,$$

and the real radial functions $f(r)$ and $g(r)$ are normalized, i.e.,

$$\int_0^\infty f^2(r)\,dr = \int_0^\infty g^2(r)\,dr = 1.$$

Show that the quadrupole moment for this state is

$$Q = \frac{e}{20}\int_0^\infty \left\{2\sqrt{2}\,abf(r)g(r) - b^2g^2(r)\right\}r^2\,dr.$$

---

[1] C. H. Townes, *Determination of Nuclear Quadrupole Moments*, in S. Flügge, Ed., *Handbuch der Physik*. Berlin: Springer-Verlag, 1958, Vol. 38, Part 1, p. 443.

CHAPTER 12

IDENTICAL PARTICLES

12–1 Principle of indistinguishability of identical particles. Modern physical theory rests upon the basic fact that matter is composed of relatively few types of elementary particles (electrons, positrons, protons, neutrons, etc.).[1] Each type is characterized by a few properties, such as mass, charge, and intrinsic angular momentum or spin, which enter into the equations of the theory as invariable parameters. Within the limits of our knowledge, these parameters are exactly the same for each particle of a given type. All electrons, for example, are intrinsically identical in every respect. It follows that the substitution of one electron for another in the theoretical description of a system cannot affect any predictions as to its motions.

In the Hamiltonian formulation of dynamics, this *principle of indistinguishability* implies that the Hamiltonian for any system is not affected by an interchange of the symbols that assign dynamical quantities to one or the other member of any pair of identical particles. For example, the Hamiltonian for the electronic system of the helium atom is

$$H = \frac{p_1^2}{2m} + \frac{p_2^2}{2m} - \frac{Ze^2}{r_1} - \frac{Ze^2}{r_2} + \frac{e^2}{|\mathbf{r}_1 - \mathbf{r}_2|}, \qquad (12\text{--}1)$$

in which $\mathbf{p}_1$ and $\mathbf{p}_2$ are the momenta of the two electrons, and $\mathbf{r}_1$ and $\mathbf{r}_2$ are their position vectors with respect to the nucleus. The parameters e and m need not carry subscripts, since they are the same for each electron. If we denote the above Hamiltonian, for brevity, by $H(1, 2)$, then clearly

$$H(1, 2) = H(2, 1). \qquad (12\text{--}2)$$

This equation expresses the symmetry of the Hamiltonian with respect to a permutation of the labels attached to identical particles.

In classical mechanics, no particularly unusual consequences follow from the symmetry property (12–2). The motions of a system described by the Hamiltonian (12–1) are merely special cases of those derived from the more general Hamiltonian

$$H = \frac{p_1^2}{2m_1} + \frac{p_2^2}{2m_2} - \frac{Zee_1}{r_1} - \frac{Zee_2}{r_2} + \frac{e_1e_2}{|\mathbf{r}_1 - \mathbf{r}_2|}, \qquad (12\text{--}3)$$

---

[1] A. M. Shapiro, *Revs. Modern Phys.* **28**, 164 (1956).

in which the two "electrons," of charges e_1, e_2 and masses m_1, m_2, are treated as distinguishable particles. A complete solution of Eq. (12–3) is a set of equations that give the position and velocity of each of the particles as functions of the time, for all possible initial conditions. If such equations are known, a complete solution of (12–1) can be obtained by setting $e_1 = e_2$ and $m_1 = m_2$. The only special consequence of the symmetry of Eq. (12–1) is that if, in a particular solution, the orbits of particles 1 and 2 are denoted by A and B, respectively, then there is another solution, in which particle 1 follows the orbit B and particle 2 follows the orbit A, in such a way that corresponding points of the orbits are occupied at the same time in the two motions. If, however, the particle which is in orbit A at the initial instant is called No. 1, then it is possible to say at a later time that the particle in orbit A is No. 1, and not No. 2, since the motion of No. 1 could, in principle, be followed from instant to instant. In this sense, therefore, the two solutions are distinguishable, even though one is obtained from the other by an interchange of identical particles.

In quantum mechanics, the classical equations that describe the orbits of the particles are replaced by the wave function $\psi(1, 2)$, which determines a probability distribution for the positions of the particles at each instant of time. The wave function is a solution of the Schrödinger equation

$$H\psi = i\hbar \frac{\partial \psi}{\partial t}. \tag{12–4}$$

It is, of course, still true that every solution of the Schrödinger equation with the Hamiltonian (12–3) leads to a solution of the equation with the symmetric Hamiltonian (12–1) when the charges and masses of the two particles are identified. The wave functions $\psi(1, 2)$ and $\psi(2, 1)$, which correspond to the two classical motions discussed in the last paragraph, are both solutions of Eq. (12–4). Explicitly, if in the equation

$$H(1, 2)\psi(1, 2) = i\hbar \frac{\partial}{\partial t} \psi(1, 2) \tag{12–5}$$

the labels 1 and 2 are interchanged, it becomes

$$H(2, 1)\psi(2, 1) = i\hbar \frac{\partial}{\partial t} \psi(2, 1), \tag{12–6}$$

whence, by Eq. (12–2),

$$H(1, 2)\psi(2, 1) = i\hbar \frac{\partial}{\partial t} \psi(2, 1). \tag{12–7}$$

The quantum-mechanical principle of superposition, however, implies a much deeper significance for the symmetry expressed by Eq. (12–2).

Every linear combination of the functions $\psi(1, 2)$ and $\psi(2, 1)$ is also a solution of Eq. (12–4), and this makes it possible to separate the states of the system into two distinct classes. The identity

$$\psi(1, 2) = \tfrac{1}{2}[\psi(1, 2) + \psi(2, 1)] + \tfrac{1}{2}[\psi(1, 2) - \psi(2, 1)] \qquad (12\text{–}8)$$

shows that every state is a superposition of a *symmetric state*

$$\psi_S = \tfrac{1}{2}[\psi(1, 2) + \psi(2, 1)] \qquad (12\text{–}9)$$

and an *antisymmetric state*

$$\psi_A = \tfrac{1}{2}[\psi(1, 2) - \psi(2, 1)]. \qquad (12\text{–}10)$$

Moreover, this separation is permanent in time; a state which is initially symmetric can never change, either wholly or in part, into an antisymmetric one. This follows immediately from Eq. (12–4), which shows that $\partial\psi/\partial t$ has the same symmetry as ψ itself, whence the increment in ψ in time Δt, namely $(\partial\psi/\partial t)\Delta t$, is symmetric if ψ is symmetric, and antisymmetric if ψ is antisymmetric. Two states of different symmetry are obviously orthogonal; consequently, the relative amplitudes of the symmetric and antisymmetric components of a general wave function (12–8) are determined by their values at the initial instant and do not change. This classification of the states has no classical analogue.

The coordinates symbolized by the numerals 1 and 2 must be all the coordinates of the electrons, including the spin coordinates. If, for example, the Hamiltonian (12–1) is augmented by terms describing the spin-orbit coupling (Problems 11–12 and 11–13), then these terms must also be symmetric in the coordinates and spin variables of the two electrons.

To express the foregoing argument in the operator formalism, we introduce the *exchange operator P*, which is defined by

$$P\psi(1, 2) = \psi(2, 1). \qquad (12\text{–}11)$$

The sequence of equations

$$PH(1, 2)\psi(1, 2) = H(2, 1)\psi(2, 1) = H(1, 2)\psi(2, 1) = H(1, 2)P\psi(1, 2),$$
$$(12\text{–}12)$$

which follows from Eq. (12–2) and the definition (12–11), expresses the fact that P commutes with the Hamiltonian:

$$PH = HP. \qquad (12\text{–}13)$$

The operator P is an involution (Section 9–12):

$$P^2 = 1, \tag{12–14}$$

whence its eigenvalues are $+1$ and -1. The corresponding eigenfunctions are, respectively, the symmetric and antisymmetric functions (12–9) and (12–10):

$$P\psi_S = \psi_S, \qquad P\psi_A = -\psi_A. \tag{12–15}$$

The orthogonality of ψ_S and ψ_A is in accord with the general rule which states that eigenfunctions of a Hermitian operator, belonging to different eigenvalues, are orthogonal (Section 6–7).

The eigenfunctions $\psi_E(1, 2)$ of the Hamiltonian, which satisfy

$$H\psi_E(1, 2) = E\psi_E(1, 2),$$

are the stationary states of the system. Equation (12–13) shows that the symmetric and antisymmetric parts of $\psi_E(1, 2)$, as given by Eqs. (12–9) and (12–10), are linearly independent functions belonging to the same value of E. This fundamental *exchange degeneracy*, which is resolved by specifying the eigenvalues of P, is a consequence of the symmetry of H with respect to interchange of 1 and 2.

It is not difficult to generalize the preceding considerations to systems containing more than two identical particles. The Hamiltonian for such systems is invariant to the interchange of any pair of particles. If the coordinates of the particles are denoted by the numerals $1, 2, 3, 4, \ldots, N$, the Hamiltonian can be written symbolically as

$$H = H(1, 2, 3, 4, \ldots, N). \tag{12–16}$$

The indistinguishability of the particles is expressed by equations of the form

$$H(1, 2, 3, 4, \ldots, N) = H(2, 1, 3, 4, \ldots, N) = H(4, 3, N, 1, \ldots, 2), \text{ etc.} \tag{12–17}$$

In general, the Hamiltonian is unchanged by any permutation of the particle labels. It follows, by reasoning similar to that for the two-particle system, that if the function

$$\psi(1, 2, 3, 4, \ldots, N) \tag{12–18}$$

is a solution of the Schrödinger equation, then the functions

$$\psi(2, 1, 3, 4, \ldots, N), \qquad \psi(2, 1, N, 4, \ldots, 3), \qquad \text{etc.,} \tag{12–19}$$

are also solutions, and additional solutions can be built up by superposition of these.

The number of simultaneous eigenfunctions of H which can be formed from a single one is $N!$, the number of permutations of the N labels. Since every permutation can be achieved by a succession of transpositions, the symmetry of the wave function can be adequately described by means of the *transposition operators* $P_{ij} = P_{ji}$. By definition, the operator P_{ij} interchanges the ith and jth labels:

$$P_{ij}\psi(a, b, \ldots, f, \ldots, h, \ldots, p) = \psi(a, b, \ldots, h, \ldots, f, \ldots, p).$$

For example,

$$(12\text{--}20)$$

$$P_{23}\psi(a, b, c, d) = \psi(a, c, b, d).$$

The transposition operators, which are $\frac{1}{2}N(N-1)$ in number, are all involutions:

$$P_{ij}^2 = 1. \tag{12--21}$$

These operators therefore have the eigenvalues $+1$ and -1. With respect to the interchange of the ith and jth labels, the eigenfunctions belonging to the eigenvalue $+1$ are symmetric, and those belonging to -1 are antisymmetric. The symmetry of the Hamiltonian is expressed by

$$P_{ij}H = HP_{ij}; \tag{12--22}$$

that is, the Hamiltonian commutes with every transposition operator, and is therefore invariant to every permutation.

The operators P_{ij} do not commute with one another. We have, for example,

$$P_{13}P_{12}\psi(1, 2, 3) = P_{13}\psi(2, 1, 3) = \psi(3, 1, 2)$$
$$= P_{23}\psi(3, 2, 1) = P_{23}P_{13}\psi(1, 2, 3),$$

whence

$$P_{13}P_{12} = P_{23}P_{13} \neq P_{12}P_{13}. \tag{12--23}$$

Consequently, except for $N = 2$, it is not possible to find a complete set of the N degenerate eigenfunctions of H which are all simultaneously eigenfunctions of the transposition operators. There are, however, always exactly two eigenfunctions which do have this property: One is the *totally symmetric* function ψ_S, which belongs to the eigenvalue $+1$ for every P_{ij},

$$P_{ij}\psi_S = \psi_S; \tag{12--24}$$

the other is the *totally antisymmetric* function ψ_A, for which

$$P_{ij}\psi_A = -\psi_A. \tag{12--25}$$

This fact is easily proved by means of the identity (12–23). Assuming that $\psi(1, 2, 3)$ is a simultaneous eigenfunction of the three operators P_{12}, P_{13}, and P_{23}, we have

$$P_{12}\psi(1, 2, 3) = \lambda_1\psi(1, 2, 3); \qquad P_{13}\psi(1, 2, 3) = \lambda_2\psi(1, 2, 3);$$

$$P_{23}\psi(1, 2, 3) = \lambda_3\psi(1, 2, 3), \qquad (12\text{--}26)$$

where each of the numbers λ_i is either $+1$ or -1. Equation (12–23) now yields the relation

$$\lambda_2\lambda_1 = \lambda_3\lambda_2,$$

whence $\lambda_1 = \lambda_3$. Similarly, the relation $P_{13}P_{12} = P_{12}P_{23}$ shows that $\lambda_2 = \lambda_3$. Hence, a function which satisfies each of the equations (12–26) either remains unchanged by every transposition, i.e., it is totally symmetric, or it changes sign under every transposition, i.e., it is totally antisymmetric. This proof can be generalized by means of the identities

$$P_{ij}P_{ik} = P_{jk}P_{ij}, \qquad (12\text{--}27)$$

and is seen to hold for the wave functions of an arbitrary number N of identical particles.

Any wave function $\psi(1, 2, \ldots, N)$ is to be interpreted as a probability amplitude for the system: The quantity $|\psi(1, 2, \ldots, N)|^2$ is a probability function for the coordinates of the N particles. In quantum mechanics, the principle of indistinguishability of identical particles is expressed by the requirement that the states represented by ψ and $P\psi$ are in fact the same state, where P is the operator for any permutation of the labels of the particles. This is equivalent to requiring that for every permutation P, ψ and $P\psi$ differ only by a phase factor,

$$P\psi = e^{i\delta(P)}\psi. \qquad (12\text{--}28)$$

Thus the quantities $|\psi(1, 2, \ldots, N)|^2$ and $|P\psi(1, 2, \ldots, N)|^2$ are the same. This means that the probability

$$|\psi(1, 2, \ldots, N)|^2\, d\mathbf{r}_1\, d\mathbf{r}_2 \ldots d\mathbf{r}_N$$

is proportional to the probability that one of the particles is in the element $d\mathbf{r}_1\, d\mathbf{r}_2 \ldots d\mathbf{r}_N$ of configuration space, but nothing can be said as to which of the particles this may be. It should be noted that the interpretation of $|\psi|^2$ as a probability density requires that it be normalized to unity,

$$\int |\psi|^2\, d\mathbf{r}_1\, d\mathbf{r}_2 \ldots d\mathbf{r}_N = 1;$$

whereas if an interpretation in terms of particle density is desired (Section 2–9), then the proper normalization is

$$\int |\psi|^2 \, d\mathbf{r}_1 \, d\mathbf{r}_2 \ldots d\mathbf{r}_N = N.$$

Every ψ which corresponds to a physical state of the system must be a simultaneous eigenfunction of all the transposition operators. The result of the last paragraph therefore leads to the conclusion that *every wave function which describes a physical state of a system containing any number of identical particles must be either totally symmetric or totally antisymmetric with respect to permutations of the labels of the particles.* Of the $N!$ wave functions that can be formed from a given solution of Schrödinger's equation, only two linear combinations can describe physical states. Thus, the number of states accessible to the system is much smaller than the number accessible to an equivalent system in which the particles are not identical. This result is in striking contrast to the classical situation, in which every solution of the equations of motion for a system of distinguishable particles leads, upon making the particles identical, to a possible solution for a system of identical particles.

The symmetry character of a solution of the Schrödinger equation is permanent for an N-particle system as well as for the two-particle system discussed earlier, and we now see that every physical eigenstate is a *pure* one, in that it is either symmetric or antisymmetric. Linear combinations of the two symmetry types do not occur, and a system which is at one instant in a symmetric (or antisymmetric) state remains so for all time. Every perturbation of such a system is a symmetric function; transitions between symmetric and antisymmetric states are impossible. The special significance of these remarks stems from the empirical fact that *only one of the two symmetry types ever occurs for a given type of particle.* The states of any system of electrons are always of the antisymmetric type. The same is true for systems of protons or of neutrons. Pi-mesons, on the other hand, are known to be described by symmetric wave functions.

12–2 Statistics of identical particles. It has been shown in the preceding section that the quantum-mechanical states accessible to a system of identical particles are fewer in number than those for a similar system of distinguishable particles. This can be illustrated simply in the imaginary case of a system of two noninteracting particles, for each of which there are just two quantum states, say ϕ and $\bar{\phi}$. Then three different situations can arise:

(1) If the two particles are not identical, then the combined system has four linearly independent states, represented by the four products

$$\phi(1)\phi(2), \qquad \phi(1)\bar{\phi}(2), \qquad \phi(2)\bar{\phi}(1), \qquad \bar{\phi}(1)\bar{\phi}(2). \qquad (12\text{–}29)$$

(2) If the particles are identical and of a type for which every wave function is symmetric, only three states can be formed by superposition of the product wave functions, namely,

$$\phi(1)\phi(2), \qquad \phi(1)\bar{\phi}(2) + \phi(2)\bar{\phi}(1), \qquad \bar{\phi}(1)\bar{\phi}(2). \qquad (12\text{–}30)$$

(3) If the two particles are identical and of the type for which every wave function is antisymmetric, then they can only exist in the single state described by the antisymmetric wave function

$$\phi(1)\bar{\phi}(2) - \phi(2)\bar{\phi}(1). \qquad (12\text{–}31)$$

If the single-particle states ϕ and $\bar{\phi}$ happen to be degenerate, then each of the above states has the same energy, and the statistical weights associated with this energy are 4, 3, and 1, in the three cases.

The statistical properties of a system composed of a large number of particles are determined, to a considerable extent, by the number of degrees of freedom, or number of possible states of motion, of the system for each value of the total energy. The three types of systems obey quite different statistical laws. For a system of distinguishable particles, every permutation of the particles corresponds to a different state, just as in classical theory, and the statistical behavior of such a system is therefore *classical*. Systems of identical particles with symmetric wave functions obey different laws, called *Bose-Einstein* statistics. For this reason, such particles have been named *bosons* by Dirac.[1] Identical particles with antisymmetric wave functions compose still another type of system, obeying *Fermi-Dirac* statistics. Particles of this type are called *fermions*.[2]

There is a close connection between the value of the intrinsic spin of an elementary particle and the symmetry character of the wave functions for systems of such particles: *Particles whose spin is an integral multiple of $\hbar$ are bosons, and particles whose spin is a half-integral multiple of $\hbar$ are fermions.* This is an empirical law, to which no exception is known.[3] The law applies to composite particles (e.g., atoms or nuclei), as well as to elementary ones. Thus, a system of composite particles whose inter-

---

[1] P. A. M. Dirac, *The Principles of Quantum Mechanics*. 3rd ed. Oxford: Clarendon Press, 1947, Chapter IX.

[2] J. C. Slater, *Introduction to Chemical Physics*. New York: McGraw-Hill Book Co., Inc., 1939, Chapter V. T. L. Hill, *Statistical Mechanics*. New York: McGraw-Hill Book Co., Inc., 1956, Sec. 16. D. ter Haar, *Elements of Statistical Mechanics*. New York: Rinehart and Company, Inc., 1954, Chapter IV.

[3] In the quantum field theory of elementary particles, this law is shown to follow from the basic postulates of quantum mechanics and relativity. W. Pauli, *Phys. Rev.* **58,** 716 (1940).

actions are sufficiently weak so that the internal structure of the individual particles is not disturbed, can be treated as if it were a system of elementary particles, for which the symmetry of the wave function is determined by the above rule.[1] A composite particle which contains an even number of fermions obeys Bose-Einstein statistics, while an odd number of fermions per composite particle leads to Fermi-Dirac statistics. A composite system which contains more than one type of elementary particles must have the proper symmetry character with respect to exchange of any pair of identical particles. The nucleus Li^7, for example, is composed of four neutrons and three protons. Its wave function must therefore change sign if the members of any pair of protons or of any pair of neutrons are interchanged, and since Li^7 contains seven spin-$\frac{1}{2}$ particles, it is a fermion. The helium atom, on the other hand, is composed of two electrons (fermions) and one alpha-particle (boson); hence, a collection of helium atoms obeys Bose-Einstein statistics if the interactions among the atoms do not affect their internal structure, as at very low temperatures. This accounts for the unique properties of liquid helium.[2]

12–3 The helium atom. As a first example dealing with identical particles, we shall discuss briefly the properties of the electronic states of helium and its optical spectrum. The approximate Hamiltonian for this system has been given in Eq. (12–1):

$$H = \frac{p_1^2}{2m} + \frac{p_2^2}{2m} - \frac{Ze^2}{r_1} - \frac{Ze^2}{r_2} + \frac{e^2}{|\mathbf{r}_1 - \mathbf{r}_2|}.$$

The first four terms on the right-hand side of this equation represent the kinetic energies of the electrons and their potential energies in the Coulomb field of the nucleus. The last term is the energy associated with the electrostatic repulsion between the electrons. As a first approximation, this term can be treated as a perturbation. The zero-order orbital wave functions for the energy states are therefore products of hydrogen wave functions [Eq. (7–196)], modified to take account of the doubled nuclear charge. We denote these wave functions by $\phi_\alpha(\mathbf{r})$, in which the symbol α represents the quantum numbers n, l, m_l:

$$\phi_\alpha(\mathbf{r}) = \sqrt{\frac{2Z^2}{n^3 a_0^2 r}} \, \Lambda_{n-l-1}^{2l+1} \left(\frac{2Zr}{na_0}\right) Y_l^{m_l}(\theta, \phi). \tag{12–32}$$

Two functions of this form, say ϕ_α and ϕ_β ($\beta = n'$, l', m_l'), can be used to form two linearly independent zero-order eigenfunctions, namely,

---

[1] P. Ehrenfest and J. R. Oppenheimer, *Phys. Rev.* **37**, 333 (1931).
[2] F. London, *Superfluids*. New York: John Wiley and Sons, 1950.

$\phi_\alpha(1)\phi_\beta(2)$ and $\phi_\alpha(2)\phi_\beta(1)$. Each of these eigenfunctions belongs to the same energy of the unperturbed system,

$$E_0 = E_\alpha + E_\beta, \tag{12-33}$$

where E_α and E_β are given by Eq. (1–30).[1] For the present discussion, it is simpler to use instead the equivalent normalized functions

$$\phi_S = \frac{1}{\sqrt{2}}[\phi_\alpha(1)\phi_\beta(2) + \phi_\alpha(2)\phi_\beta(1)], \tag{12-34}$$

$$\phi_A = \frac{1}{\sqrt{2}}[\phi_\alpha(1)\phi_\beta(2) - \phi_\alpha(2)\phi_\beta(1)]. \tag{12-35}$$

The wave function ϕ_S is symmetric, and ϕ_A is antisymmetric, with respect to interchange of the spatial coordinates of the electrons.

The spin states for the system have been given in Section 10–12. They consist of a set of three triplet functions, $^3\chi_1$, $^3\chi_0$, $^3\chi_{-1}$, which are symmetric with respect to interchange of the spin coordinates of the electrons, and of one singlet function, $^1\chi_0$, which is antisymmetric.

Of the eight possible products which can be formed by multiplying the wave functions (12–34) and (12–35) by each of the four spin functions, four are totally symmetric under interchange of all the coordinates of the electrons, namely,

$$\phi_S{}^3\chi_1, \qquad \phi_S{}^3\chi_0, \qquad \phi_S{}^3\chi_{-1}, \qquad \phi_A{}^1\chi_0. \tag{12-36}$$

These states are not allowed for a system of fermions, and must be discarded. The remaining four functions,

$$\phi_S{}^1\chi_0, \qquad \phi_A{}^3\chi_1, \qquad \phi_A{}^3\chi_0, \qquad \phi_A{}^3\chi_{-1}, \tag{12-37}$$

are totally antisymmetric, and are the correct zero-order wave functions for the perturbation calculation.

It can be noted immediately that all four of the wave functions (12–37) vanish if the sets of quantum numbers (n, l, m_l, m_s) and (n', l', m_l', m_s') are alike. This is an example of Pauli's *exclusion principle* which will be discussed in the next section.

The effect of the electrostatic interaction between the two electrons can now be obtained by calculating the diagonal element of the operator

$$H^{(1)} = \frac{e^2}{|\mathbf{r}_1 - \mathbf{r}_2|} \tag{12-38}$$

---

[1] Here, the symbol m in Eq. (1–30) stands for the reduced mass of an electron in the helium atom.

for each of the wave functions (12–37). Since this operator does not involve the spins, it is clear that, under this perturbation, the three triplet states remain degenerate. The singlet and triplet states, however, are separated. We obtain, for the energy change due to the perturbation,

$$E^{(1)} = J_{\alpha\beta} \pm K_{\alpha\beta}, \qquad (12\text{–}39)$$

where

$$J_{\alpha\beta} = (\phi_\alpha(1)\phi_\beta(2),\, H^{(1)}\phi_\alpha(1)\phi_\beta(2)) = \iint |\phi_\alpha(1)|^2\, \frac{e^2}{|\mathbf{r}_1 - \mathbf{r}_2|}\, |\phi_\beta(2)|^2\, d\mathbf{r}_1\, d\mathbf{r}_2,$$

$$K_{\alpha\beta} = \begin{cases} 0, & \alpha = \beta \\ (\phi_\alpha(1)\phi_\beta(2),\, H^{(1)}\phi_\alpha(2)\phi_\beta(1)) \end{cases}$$

$$= \iint \phi_\alpha^*(1)\phi_\beta(1)\, \frac{e^2}{|\mathbf{r}_1 - \mathbf{r}_2|}\, \phi_\alpha(2)\phi_\beta^*(2)\, d\mathbf{r}_1\, d\mathbf{r}_2, \qquad \alpha \neq \beta.$$

The upper sign in Eq. (12–39) applies to the singlet state, and the lower sign to the triplets. The first term in this expression would have been obtained if we had used the unsymmetrized function $\phi_\alpha(1)\phi_\beta(2)$ as the zero-order eigenfunction. The second term, which produces the separation of the states of different total spin, is a result of the symmetry imposed upon the wave functions. It should be noticed especially that this splitting is present in spite of the fact that the perturbation (12–38) has no relation to the spins of the particles. It can be shown that $K_{\alpha\beta}$ is always positive, so that, for the same configuration, the singlet level is always higher in energy than the triplet level.

In the conventional spectroscopic notation,[1] the electronic configuration in the ground state of helium is $1s^2$. The quantum numbers $n = 1$, $l = 0$, $m_l = 0$ are the same for the two electrons, whence the only possible state is the singlet, and since the wave function is spherically symmetric, the state is $^1S_0$. In this case, we have

$$\phi_s = \frac{Z^3}{\pi a_0}\, e^{(-Z/a_0)(r_1 + r_2)}, \qquad (12\text{–}40)$$

and the perturbation energy (Problem 12–2) is

$$E^{(1)} = J_{1s^2} = \frac{5}{4}\frac{Ze^2}{2a_0}, \qquad (12\text{–}41)$$

whence, taking $Ze^2/2a_0 = 27.19$ ev, we find for the ground-state energy

---

[1] G. Herzberg, *Atomic Spectra and Atomic Structure.* New York: Dover Publications, 1944, Chapter III. D. E. Gray, Ed., *American Institute of Physics Handbook.* New York: McGraw-Hill Book Co., Inc., 1957, Sec. 7d.

$$E_{1s^2} = (-2Z + \tfrac{5}{4})\frac{Ze^2}{2a_0} = -74.77 \text{ ev.} \qquad (12\text{--}42)$$

This checks satisfactorily with the experimentally determined value of −78.983 ev. We note, incidentally, that the energy obtained by means of the perturbation calculation is larger (by about 5%) than the correct value (see Section 11–6). More refined calculations, based upon the variational principle, can be made to reproduce the experimental binding energy with negligible error.[1]

Methods for evaluating the integrals J and K for configurations in which one electron is excited are discussed in the problems; the results are:

$$J_{1s2s} = \frac{34}{81}\frac{Ze^2}{2a_0} = 11.42 \text{ ev}, \qquad K_{1s2s} = \frac{2^5}{3^6}\frac{Ze^2}{2a_0} = 1.20 \text{ ev},$$

$$J_{1s2p} = \frac{118}{243}\frac{Ze^2}{2a_0} = 13.22 \text{ ev}, \qquad K_{1s2p} = \frac{7\cdot 2^5}{3^8}\frac{Ze^2}{2a_0} = 0.94 \text{ ev.}$$

$$(12\text{--}43)$$

The energies of the first five states of the helium atom, derived from Eqs. (12–33) and (12–39), are shown in Table 12–1, together with the corresponding experimental values.[2] The energy level diagram is shown in Fig. 12–1.

The singlet and triplet states of helium do not combine in optical transitions, because their wave functions are orthogonal in the spin variables,

TABLE 12–1

ENERGIES OF THE FIVE LOWEST STATES
OF THE HELIUM ATOM

(in electron volts)

State	Energy	
	Calculated	Experimental
$1s^2\ {}^1S_0$	−74.80	−78.98
$1s2s\ {}^1S_0$	−55.38	−58.37
$1s2p\ {}^1P_1$	−53.86	−57.77
$1s2s\ {}^3S_1$	−57.77	−59.17
$1s2p\ {}^3P_{0,1,2}$	−55.71	−58.02

[1] E. Hylleraas, Z. Physik 54, 347 (1929); 65, 209 (1930). J. Traub and H. M. Foley, Phys. Rev. 111, 1098 (1958). C. L. Pekeris, Phys. Rev. 112, 1649 (1958).
[2] C. E. Moore, Atomic Energy Levels. Washington: National Bureau of Standards Circular 467, 1949.

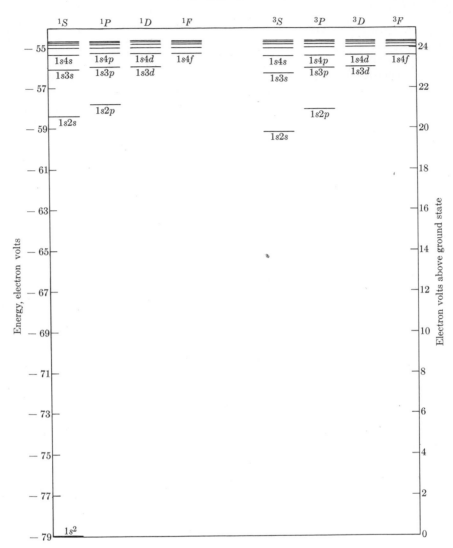

FIG. 12–1. Energy level diagram for the helium atom.

whereas the operators for optical transitions are independent of the spin. Spectroscopically, the singlet and triplet systems are independent of one another, and have been given the names *parahelium* (singlets) and *ortho-helium* (triplets). The ground state of the orthohelium system, namely the state $1s2s\ {}^3S_1$, is therefore stable with respect to optical transitions of the usual kind, and has a correspondingly long lifetime. For this reason, it is frequently referred to as a *metastable* state.

12–4 The Pauli exclusion principle. The spin quantum number for a spin-$\frac{1}{2}$ particle can have only the two values $\pm\frac{1}{2}$. Hence it is not possible for more than two electrons to have identical orbital wave functions. This is the *exclusion principle* of Pauli,[1] which is fundamental for the understanding of atomic structure and of the properties of the elements, as represented in the periodic table.[2]

The effect of the exclusion principle in determining atomic energy states is best illustrated in the approximate theory in which the wave function of the Z electrons can be written in the form assumed for helium in the last section, that is, as a product of "single-particle" states. The basic wave functions in this approximate theory have the form

$$\psi_1(1)\psi_2(2)\ldots\psi_N(N). \tag{12–44}$$

If each of the functions ψ_i is different from all the others, then just one antisymmetric combination can be formed, namely,

$$\psi_A(1, 2, \ldots, N) = \frac{1}{\sqrt{N!}} \sum_P (-)^P P \psi_1(1)\psi_2(2)\ldots\psi_N(N), \tag{12–45}$$

in which the symbol $\sum_P$ denotes the sum of the $N!$ terms formed by permutation of the particle labels, and $(-)^P$ is $+1$ or -1, depending on whether the permutation P can be represented as an even or an odd number of transpositions. An alternative form of Eq. (12–45) is

$$\psi_A(1, 2, \ldots, N) = \frac{1}{\sqrt{N!}} \begin{vmatrix} \psi_1(1) & \psi_1(2) & \cdots & \psi_1(N) \\ \psi_2(1) & \psi_2(2) & \cdots & \psi_2(N) \\ \vdots & & & \\ \psi_N(1) & \psi_N(2) & \cdots & \psi_N(N) \end{vmatrix}. \tag{12–46}$$

In this form, it is immediately evident that ψ_A vanishes whenever two of the single-particle functions are alike, because the determinant then has two identical rows and is zero.

For a system of integral spin, the equation corresponding to (12–45) is

$$\psi_S(1, 2, \ldots, N) = \frac{1}{\sqrt{N!}} \sum_P P \psi_1(1)\psi_2(2)\ldots\psi_N(N). \tag{12–47}$$

This function, however, is never zero; there is no exclusion principle for bosons.

---

[1] W. Pauli, *Z. Physik* **31**, 765 (1925).

[2] M. Born, *Atomic Physics*. 6th ed. New York: Hafner Publishing Co., 1957, Chapter VI. F. K. Richtmyer and E. H. Kennard, *Introduction to Modern Physics*. 4th ed. New York: McGraw-Hill Book Co., Inc., 1947, Chapter VIII.

We shall now consider briefly the effect of the exclusion principle on the classification of atomic states according to their total orbital and spin angular momentum. We first note that each of the operators

$$\mathbf{L} = \mathbf{l}_1 + \mathbf{l}_2 + \cdots + \mathbf{l}_N,$$

$$\mathbf{S} = \mathbf{s}_1 + \mathbf{s}_2 + \cdots + \mathbf{s}_N, \qquad (12\text{–}48)$$

$$\mathbf{J} = \mathbf{j}_1 + \mathbf{j}_2 + \cdots + \mathbf{j}_N$$

commutes with every transposition P_{ij} (as every observable must), so that an antisymmetric wave function can be represented as a linear combination of the antisymmetric eigenfunctions of these operators. Indeed, every nondegenerate eigenfunction of $\mathbf{L}^2$ and L_z, or of $\mathbf{S}^2$ and S_z, is either symmetric or antisymmetric in the corresponding orbital or spin quantum numbers of the one-electron states. The angular-momentum states of a two-particle system are never degenerate (cf. Section 12–1), and this remark is therefore sufficient to obtain a complete characterization of the states. For illustration, we consider the case of two electrons in p-states (Section 10–11). The wave functions in a central field can be written in the form

$$\psi_1(nlm_lm_s) = R_1(nl)\, Y_l^{m_l}\, \chi(m_s),$$

$$\psi_2(n'l'm_l'm_s') = R_2(n'l')\, Y_{l'}^{m_l'}\, \chi(m_s'), \qquad (12\text{–}49)$$

in which $l = l' = 1$, and m_l and m_l' each take the values $-1, 0, 1$. For a given configuration, namely, for given values of n and n', there are altogether $(3 \times 2) \times (3 \times 2) = 36$ functions of the form (12–44). By means of the rules for combination of angular momenta, these can be combined to form $3 \times 5 = 15\ ^3D$-states, $1 \times 5 = 5\ ^1D$-states, $3 \times 3 = 9\ ^3P$-states, $1 \times 3 = 3\ ^1P$-states, $3 \times 1 = 3\ ^3S$-states, and $1 \times 1 = 1\ ^1S$-state. The D- and S-states are symmetric with respect to interchange of the spatial coordinates of electrons, and the P-states are antisymmetric. Similarly, the triplet spin functions are symmetric, and the singlet spin functions antisymmetric. Finally, from the two functions $R_1(nl)$ and $R_2(n'l)$, we can form the symmetric and antisymmetric combinations

$$R_S = \frac{1}{\sqrt{2}}\,[R_1(nl)\,R_2(n'l) + R_1(n'l)\,R_2(nl)],$$

$$\qquad (12\text{–}50$$

$$R_A = \frac{1}{\sqrt{2}}\,[R_1(nl)\,R_2(n'l) - R_1(n'l)\,R_2(nl)].$$

Totally antisymmetric eigenfunctions of $\mathbf{L}^2, \mathbf{S}^2, \mathbf{J}^2$ and J_z can now be formed as follows:

$$R_A \, {}^3D_{3,2,1} \quad \text{(15 states)}, \qquad R_S \, {}^1D_2 \qquad \text{(5 states)},$$
$$R_A \, {}^1P_1 \qquad \text{(3 states)}, \qquad R_S \, {}^3P_{2,1,0} \quad \text{(9 states)}, \qquad (12\text{–}51)$$
$$R_A \, {}^3S_1 \qquad \text{(3 states)}, \qquad R_S \, {}^1S_0 \qquad \text{(1 state)}.$$

All these states are accessible provided that the electrons are *not equivalent* $(n \neq n')$. In the np^2 configuration, however, the antisymmetric radial functions R_1 and R_2 are alike, and R_A consequently vanishes. In this case, only the states listed in the right-hand column above exist, and their number is reduced from 36 to 15. The terms for two equivalent p-electrons are seen to be ^1D, ^3P and ^1S.

An alternative form of the eigenfunctions of $\mathbf{L}^2$ can be used to arrive at the above conclusions regarding equivalent electrons, in a way which illustrates more clearly the essential role of symmetry principles. In Chapter 10, Problem 10–11, it was shown that the p-states for a single particle transform under rotations like the coordinates (x_1, x_2, x_3) of the position vector of the particle. Aside from the radial factors, therefore, the p-states of electrons 1 and 2 can be symbolized by the vectors $\mathbf{r}_1$ and $\mathbf{r}_2$, and the product functions by $x_{1i}x_{2j}$. There are, in all, $3 \times 3 = 9$ products of this form. Now the **only** scalar function which can be formed from these vectors is $\mathbf{r}_1 \cdot \mathbf{r}_2$. Hence, this is the S-function, and since it is obviously symmetric, the S-state must be a singlet. Similarly, the P-functions must transform as the components of a vector; they are therefore the three components of the vector $\mathbf{r}_1 \times \mathbf{r}_2$, and are antisymmetric; the P-states are therefore triplets. Finally, the remaining five functions are the components of the symmetric tensor

$$x_{1i}x_{2j} + x_{1j}x_{2i} - \tfrac{2}{3}(\mathbf{r}_1 \cdot \mathbf{r}_2)\, \delta_{ij}, \qquad (12\text{–}52)$$

representing the five D-functions. This expression is symmetric and thus belongs to the singlet states. Considerations of this kind can frequently be used to avoid the complications associated with the combination of angular-momentum states.

To illustrate a particularly easy application of this principle, we can show that the term corresponding to a *closed shell* is always $^1S_0$. A closed shell is a configuration in which each of the states corresponding to a given nl is occupied by an electron. The number of electrons in a closed-shell configuration is therefore $2(2l + 1)$. Now we have seen that there is only one totally antisymmetrized product of N functions of N variables. Hence, if such a function is an eigenstate of the total angular momentum, it must be invariant to rotation, that is, $J = 0$. Since the total spin is obviously 0 (the spins cancel in pairs), we must also have $L = 0$; the state is $^1S_0$. An extension of this argument can be made to show that the term system for a configuration in which x electrons are "missing" from a closed shell must be the same as that for the configuration nl^x. The configuration np^4 leads, for example, to the same system of terms, $({}^1S, {}^3P, {}^1D)$, as the configuration np^2. Note that the total number of states for the configuration nl^x is the binomial coefficient $\binom{2(2l+1)}{x}$, which is unchanged if x is replaced by $2(2l + 1) - x$.

12–5 Scattering of identical particles. In the scattering of identical particles, there are two indistinguishable situations for every scattering angle, as shown in Fig. 12–2, in which the incident particle (1) and the scatterer (2) have equal and opposite velocities in the center-of-mass frame of reference. In Fig. 12–2(a), the particle observed in the detector at D is the incident particle (1), and in (b), it is the "recoiling" scatterer (2). Since the particles observed in the two cases are identical and have equal energy, they cannot be distinguished experimentally. If we suppose that the cross section for the process (a) is given classically by the function $\sigma(\mathbf{k}, \mathbf{k}')$, then the cross section for the process (b) is $\sigma(\mathbf{k}, -\mathbf{k}')$, since a recoil particle is observed in the direction $-\mathbf{k}'$ for each scattering in which the scattered particle emerges in the direction $\mathbf{k}'$. The observed classical cross section is therefore

$$\sigma_c(\theta, \phi) = \sigma(\mathbf{k}, \mathbf{k}') + \sigma(\mathbf{k}, -\mathbf{k}'). \tag{12–53}$$

In the quantum-mechanical theory, the cross section for the process (a) alone is $|f(\mathbf{k}, \mathbf{k}')|^2$, where $f(\mathbf{k}, \mathbf{k}')$ is the amplitude of the outgoing scattered wave (Section 8–4). The asymptotic form of the wave function is

$$\psi(\mathbf{k}, \mathbf{k}', \mathbf{r}) \sim e^{i\mathbf{k} \cdot \mathbf{r}} + f(\mathbf{k}, \mathbf{k}') \frac{e^{ikr}}{r}. \tag{12–54}$$

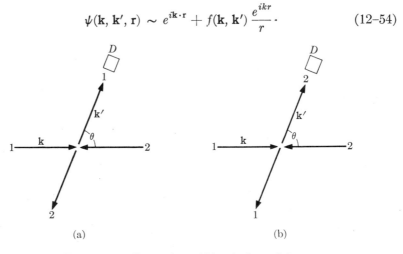

(a)   (b)

FIG. 12–2.   Scattering of identical particles.

The function $\psi(\mathbf{k}, \mathbf{k}')$ can be obtained by solving the Schrödinger equation for the scattering of distinguishable particles, i.e., by ignoring the exchange degeneracy. However, the correct wave function for the degenerate system is obtained from this expression by symmetrization with respect to interchange of 1 and 2. Since $\mathbf{r} = \mathbf{r}_2 - \mathbf{r}_1$, and since $\mathbf{k}'$ has, by definition, the same direction as $\mathbf{r}$, the position vector of the point of observation, we have for a given direction of incidence $\mathbf{k}$,

$$P_{12}\psi(\mathbf{k}, \mathbf{k}', \mathbf{r}) = \psi(\mathbf{k}, -\mathbf{k}', \mathbf{r}). \qquad (12\text{-}55)$$

It follows that the functions

$$\psi_S = \psi(\mathbf{k}, \mathbf{k}', \mathbf{r}) + \psi(\mathbf{k}, -\mathbf{k}', \mathbf{r}), \qquad (12\text{-}56)$$

$$\psi_A = \psi(\mathbf{k}, \mathbf{k}', \mathbf{r}) - \psi(\mathbf{k}, -\mathbf{k}', \mathbf{r}) \qquad (12\text{-}57)$$

are, respectively, the symmetric and antisymmetric spatial wave functions. By Eq. (12-54), their asymptotic forms are

$$\psi \sim [e^{i\mathbf{k}\cdot\mathbf{r}} \pm e^{i\mathbf{k}\cdot\mathbf{r}}] + [f(\mathbf{k}, \mathbf{k}') \pm f(\mathbf{k}, -\mathbf{k}')]\frac{e^{ikr}}{r}. \qquad (12\text{-}58)$$

[Note that if the angular coordinates of $\mathbf{k}'$ are (θ, ϕ), then the coordinates of $-\mathbf{k}'$ are $(\pi - \theta, \pi + \phi)$.] The mean value of $|\psi|^2$ in the incident wave is $\langle |4\cos^2 \mathbf{k}\cdot\mathbf{r}|\rangle = 2$, in agreement with the fact that ψ is the wave function for a system of two particles.[1]

The scattering of a pair of identical particles of spin zero is described entirely by the wave function ψ_S. The first terms in the asymptotic form (12-58) represent unit intensities of particles traveling in the directions $\mathbf{k}$ and $-\mathbf{k}$. The probability for observing an outgoing particle in the volume element $d\mathbf{r}_1 d\mathbf{r}_2$ of the configuration space is derived from the second term; it is

$$|f(\mathbf{k}, \mathbf{k}') + f(\mathbf{k}, -\mathbf{k}')|^2 \frac{d\mathbf{r}_1\, d\mathbf{r}_2}{r^2} = |f(\mathbf{k}, \mathbf{k}') + f(\mathbf{k}, -\mathbf{k}')|^2 \frac{d\mathbf{r}\, d\mathbf{R}}{r^2}, \qquad (12\text{-}59)$$

in which $d\mathbf{r}\, d\mathbf{R}$ is the volume element expressed in terms of the relative coordinates $\mathbf{r}$ and the coordinates $\mathbf{R}$ of the center of mass. It follows that the cross section, or intensity of particles traveling outward per unit solid angle and per unit incident flux, is[2]

$$\sigma_0(\mathbf{k}, \mathbf{k}') = |f(\mathbf{k}, \mathbf{k}') + f(\mathbf{k}, -\mathbf{k}')|^2$$

$$= |f(\mathbf{k}, \mathbf{k}')|^2 + |f(\mathbf{k}, -\mathbf{k}')|^2 + 2\,\mathrm{Re}\,[f^*(\mathbf{k}, \mathbf{k}')f(\mathbf{k}, -\mathbf{k}')]. \quad (12\text{-}60)$$

---

[1] See N. F. Mott and H. S. W. Massey, *The Theory of Atomic Collisions.* 2nd. ed. Oxford: Clarendon Press, 1949, Chapter V, Section 4.

[2] The intensity considerations leading to the expression (12-60) can be considerably clarified by a discussion of the scattering process in terms of localized wave packets. N. F. Mott and H. S. W. Massey, *loc. cit.*

Thus we see that, in addition to the classical cross section, expressed by the first two terms of Eq. (12–60), the symmetrization of the wave function introduces a term arising from interference between the scattering amplitudes for the cases (a) and (b) of Fig. 12–2. If $f(\mathbf{k}, \mathbf{k}')$ is independent of the azimuthal angle ϕ, Eq. (12–60) can be written

$$\sigma_0(\theta) = |f(\theta) + f(\pi - \theta)|^2, \qquad (12–61)$$

and it follows that the scattering at $\theta = \pi/2$ is

$$\sigma_0\left(\frac{\pi}{2}\right) = \left|2f\left(\frac{\pi}{2}\right)\right|^2 = 4\left|f\left(\frac{\pi}{2}\right)\right|^2, \qquad (12–62)$$

which is twice the classical cross section at the same angle.

The case of particles of spin $\frac{1}{2}$ can be treated similarly. Since the singlet spin-state is antisymmetric, it must be combined with the symmetric function ψ_S, giving the cross section

$$^1\sigma_{1/2}(\mathbf{k}, \mathbf{k}') = |f(\mathbf{k}, \mathbf{k}') + f(\mathbf{k}, -\mathbf{k}')|^2. \qquad (12–63)$$

The triplet state, on the other hand, is symmetric and requires the antisymmetric orbital function, leading to

$$^3\sigma_{1/2}(\mathbf{k}, \mathbf{k}') = |f(\mathbf{k}, \mathbf{k}') - f(\mathbf{k}, -\mathbf{k}')|^2. \qquad (12–64)$$

In the absence of forces which cause a preferential orientation of the spins of the particles, the four spin states are occupied equally, and the total cross section is therefore

$$\sigma_{1/2}(\mathbf{k}, \mathbf{k}') = \tfrac{1}{4}\,^1\sigma_{1/2}(\mathbf{k}, \mathbf{k}') + \tfrac{3}{4}\,^3\sigma_{1/2}(\mathbf{k}, \mathbf{k}')$$
$$= |f(\mathbf{k}, \mathbf{k}')|^2 + |f(\mathbf{k}, -\mathbf{k}')|^2 - \operatorname{Re} f^*(\mathbf{k}, \mathbf{k}')f(\mathbf{k}, -\mathbf{k}'). \qquad (12–65)$$

The factors $\tfrac{1}{4}$ and $\tfrac{3}{4}$ are the statistical weights of the singlet and triplet states, respectively. The scattering at $\theta = \pi/2$ is easily found to be

$$\sigma_{1/2}\left(\frac{\pi}{2}\right) = \left|f\left(\frac{\pi}{2}\right)\right|^2, \qquad (12–66)$$

which is one-half as large as the classical cross section. Other things being equal, the three cross sections at $\theta = \pi/2$ are in the ratios

$$\sigma_0\left(\frac{\pi}{2}\right) : \sigma_c\left(\frac{\pi}{2}\right) : \sigma_{1/2}\left(\frac{\pi}{2}\right) = 4 : 2 : 1. \qquad (12–67)$$

It is therefore possible, in principle, to measure the spin by means of a scattering experiment, provided the interaction between the particles is known, so that σ_c can be determined. This is the case for the Coulomb scattering of low-energy protons, for which the classical cross section is given by the Rutherford formula. The interpretation of such scattering experiments provides direct evidence that the proton spin is $\frac{1}{2}$. Similar experiments show that the alpha particle has zero spin.

It is instructive as an example to obtain the formula (12–65) by direct calculation, without the *ad hoc* use of statistical weight factors. Let us suppose that the spins of particles 1 and 2 are initially parallel to the arbitrary directions $\hat{\mathbf{n}}$ and $\hat{\mathbf{n}}'$. If we assume that the interaction between the particles is independent of their spins, these directions are unchanged in the scattering process. We may therefore express the scattering amplitude for the process (a) as

$$f(\mathbf{k}, \mathbf{k}')\chi_n(1)\chi_{n'}(2). \tag{12–68}$$

The scattering amplitude in the antisymmetric state appropriate to the scattering of identical particles is therefore

$$f(\mathbf{k}, \mathbf{k}')\chi_n(1)\chi_{n'}(2) - f(\mathbf{k}, -\mathbf{k}')\chi_n(2)\chi_{n'}(1). \tag{12–69}$$

Denoting the functions $f(\mathbf{k}, \mathbf{k}')$ and $f(\mathbf{k}, -\mathbf{k}')$ temporarily by f_+ and f_-, we can rearrange this expression in the form:

$$\frac{1}{2}\{(f_+ + f_-)[\chi_n(1)\chi_{n'}(2) - \chi_n(2)\chi_{n'}(1)]$$

$$+ (f_+ - f_-)[\chi_n(1)\chi_{n'}(2) + \chi_n(2)\chi_{n'}(1)]\} \tag{12–70}$$

$$= \frac{1}{2}(f_+ + f_-)\,^1\chi + \frac{1}{2}(f_+ - f_-)\,^3\chi,$$

where

$$^1\chi = \chi_n(1)\chi_{n'}(2) - \chi_n(2)\chi_{n'}(1)$$

is antisymmetric, and therefore a singlet function, while

$$^3\chi = \chi_n(1)\chi_{n'}(2) + \chi_n(2)\chi_{n'}(1)$$

is a symmetric triplet function. It follows that the scattering cross section in this state is

$$\sigma(\mathbf{k}, \mathbf{k}', \hat{\mathbf{n}}, \hat{\mathbf{n}}') = \frac{1}{4}|f_+ + f_-|^2(^1\chi, \,^1\chi) + \frac{1}{4}|f_+ - f_-|^2(^3\chi, \,^3\chi). \tag{12–71}$$

The scalar products of spin functions can now be evaluated in terms of $\hat{\mathbf{n}}$ and $\hat{\mathbf{n}}'$, for we have

$$(^{1,3}\chi, \,^{1,3}\chi) = 2 \mp 2\,\mathrm{Re}\,(\chi_n(1)\chi_{n'}(2), \chi_n(2)\chi_{n'}(1))$$

$$= 2[1 \mp |(\chi_n, \chi_{n'})|^2]. \tag{12–72}$$

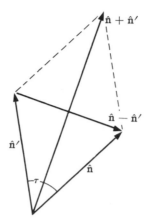

FIG. 12-3. The unit vectors $\hat{\mathbf{n}}$ and $\hat{\mathbf{n}}'$.

The scalar product $(\chi_n, \chi_{n'})$ can be computed conveniently by introducing the coordinates (Fig. 12-3)

$$\hat{\mathbf{i}} = \frac{\hat{\mathbf{n}} + \hat{\mathbf{n}}'}{2 \cos (\tau/2)}, \qquad \hat{\mathbf{j}} = \frac{\hat{\mathbf{n}} - \hat{\mathbf{n}}'}{2 \sin (\tau/2)}, \qquad \hat{\mathbf{k}} = \frac{\hat{\mathbf{n}} \times \hat{\mathbf{n}}'}{\sin \tau}. \qquad (12\text{-}73)$$

Using Eq. (10-132), we find

$$\boldsymbol{\sigma} \cdot \hat{\mathbf{n}} = \begin{pmatrix} 0 & e^{-i\tau/2} \\ e^{i\tau/2} & 0 \end{pmatrix} = (\boldsymbol{\sigma} \cdot \hat{\mathbf{n}}')^*, \qquad (12\text{-}74)$$

and

$$\chi_n = \frac{1}{\sqrt{2}} \begin{pmatrix} 1 \\ e^{i\tau/2} \end{pmatrix}, \qquad \chi_{n'} = \frac{1}{\sqrt{2}} \begin{pmatrix} e^{i\tau/2} \\ 1 \end{pmatrix}, \qquad (12\text{-}75)$$

$$(\chi_{n'}, \chi_n) = \tfrac{1}{2}(e^{i\tau/2} + e^{-i\tau/2}) = \cos (\tau/2). \qquad (12\text{-}76)$$

Thus, Eq. (12-72) becomes

$$(^{1,3}\chi, \,^{1,3}\chi) = 2 \left(1 \mp \cos^2 \frac{\tau}{2} \right) = \begin{cases} 1 - \cos \tau & \text{(singlet)} \\ 3 + \cos \tau & \text{(triplet)}. \end{cases} \qquad (12\text{-}77)$$

Substituting in Eq. (12-71), we have

$$\sigma(\mathbf{k}, \mathbf{k}', \hat{\mathbf{n}}, \hat{\mathbf{n}}') = \tfrac{1}{4}(1 - \cos \tau)|f_+ + f_-|^2 + \tfrac{1}{4}(3 + \cos \tau)|f_+ - f_-|^2. \qquad (12\text{-}78)$$

The cross section for a statistical mixture, in which all directions for $\hat{\mathbf{n}}$ and $\hat{\mathbf{n}}'$ are equally probable, is the average of this expression with respect to the angle τ between $\hat{\mathbf{n}}$ and $\hat{\mathbf{n}}'$. This average is seen immediately to be just the cross section (12–65), namely,

$$\sigma_{1/2}(\mathbf{k}, \mathbf{k}') = \tfrac{1}{4}|f(\mathbf{k}, \mathbf{k}') + f(\mathbf{k}, -\mathbf{k}')|^2 + \tfrac{3}{4}|f(\mathbf{k}, \mathbf{k}') - f(\mathbf{k}, -\mathbf{k}')|^2. \tag{12–79}$$

The statistical weights for the symmetric and antisymmetric spin states for two identical particles of arbitrary spins are easily found from the following consideration: Of the total of $(2s + 1)^2$ states which can be formed from products of spin states $\chi^m(1)\chi^{m'}(2)$, exactly $2s + 1$ have the form $\chi^m(1)\chi^m(2)$, in which $m = m'$; these are symmetric states. Of the remaining $(2s + 1)^2 - (2s + 1) = 2s(2s + 1)$ states with unequal values of m, half are symmetric and half are antisymmetric. They have the forms

$$\chi^m(1)\chi^{m'}(2) + \chi^m(2)\chi^{m'}(1),$$
$$\chi^m(1)\chi^{m'}(2) - \chi^m(2)\chi^{m'}(1), \tag{12–80}$$

respectively. In all, therefore, there are $(s + 1)(2s + 1)$ symmetric and $s(2s + 1)$ antisymmetric states, whence the statistical weights are $(s + 1)/(2s + 1)$ and $s/(2s + 1)$, respectively. If s is an integer, the particles obey Bose statistics, and the cross section is

$$\sigma_B(\mathbf{k}, \mathbf{k}') = \frac{s + 1}{2s + 1}|f(\mathbf{k}, \mathbf{k}') + f(\mathbf{k}, -\mathbf{k}')|^2$$
$$+ \frac{s}{2s + 1}|f(\mathbf{k}, \mathbf{k}') - f(\mathbf{k}, -\mathbf{k}')|^2. \tag{12–81}$$

Similarly, for particles of half-integral spin, obeying Fermi statistics, the cross section is

$$\sigma_F(\mathbf{k}, \mathbf{k}') = \frac{s}{2s + 1}|f(\mathbf{k}, \mathbf{k}') + f(\mathbf{k}, -\mathbf{k}')|^2$$
$$+ \frac{s + 1}{2s + 1}|f(\mathbf{k}, \mathbf{k}') - f(\mathbf{k}, -\mathbf{k}')|^2. \tag{12–82}$$

For systems in which $f(\theta, \phi)$ is independent of ϕ, substitution of $\theta = \pi/2$ yields

$$\sigma_B\left(\frac{\pi}{2}\right) = \frac{2s + 2}{2s + 1}\sigma_c\left(\frac{\pi}{2}\right), \qquad \sigma_F\left(\frac{\pi}{2}\right) = \frac{2s}{2s + 1}\sigma_c\left(\frac{\pi}{2}\right). \tag{12–83}$$

For bosons, the 90° cross section is therefore larger, and for fermions, smaller than the corresponding classical cross section. Accurate measurement of this cross section can, in principle, yield the value of the spin s.

REFERENCES

CONDON, E. U., and G. H. SHORTLEY, *The Theory of Atomic Spectra*. Cambridge: Cambridge University Press, 1951, Chapter VI. Includes detailed consideration of the exclusion principle and its effect upon the structure of atoms.

DIRAC, P. A. M., *The Principles of Quantum Mechanics*. 3rd ed. Oxford: The Clarendon Press, 1947, Chapter IX. The physical basis of the principle of indistinguishability is explained, and the idea of treating permutations as dynamical variables is developed.

MOTT, N. F., and H. S. W. MASSEY, *The Theory of Atomic Collisions*. Oxford: The Clarendon Press, 1949, Chapter V. A careful treatment of the theory of scattering of identical particles.

SCHIFF, L. I., *Quantum Mechanics*. 2nd ed. New York: McGraw-Hill Book Co., Inc., 1955, Chapter IX. Instructive examples in the theory of collisions are worked out.

Problems

12-1. Prove that the exchange integral $K_{\alpha\beta}$ [Eq. (12-39)] is positive. Hint: Use the Fourier transform [cf. Eq. (8-171)]

$$\frac{1}{|\mathbf{r}_1 - \mathbf{r}_2|} = \frac{1}{2\pi^2} \int \exp\left[i\mathbf{k} \cdot (\mathbf{r}_1 - \mathbf{r}_2)\right] \frac{d\mathbf{k}}{k^2}.$$

12-2. Calculate the integral J_{1s^2} [Eq. (12-41)]. Hint: An integral of the form

$$\frac{1}{2} \int \frac{\rho_\alpha(\mathbf{r}_1)\rho_\beta(\mathbf{r}_2)}{|\mathbf{r}_1 - \mathbf{r}_2|} \, d\mathbf{r}_1 \, d\mathbf{r}_2$$

can be interpreted as the electrostatic energy of a distribution of charge of density $\rho_\alpha(\mathbf{r}_1)$ in the field of a charge of density $\rho_\beta(\mathbf{r}_2)$. Hence, the principles of electrostatics can be applied.[1] Alternatively, the integral can be reduced to an elementary one by means of the Fourier transform given above. By this method, $J_{\alpha\beta}$ can be shown to be proportional to the integral

$$\int_0^\infty \frac{dk}{(\kappa^2 + k^2)^4} \qquad \left(\kappa = \frac{2Z}{a_0}\right).$$

12-3. Evaluate the quantities $J_{\alpha\beta}$ and $K_{\alpha\beta}$, given in Eq. (12-43).

12-4. Find the approximate separation of the singlet and triplet $1s3s$ levels in helium. Compare with the experimental value and comment upon the discrepancy.

12-5. Use the variational principle with the trial wave function

$$\psi = \frac{1}{\pi a^3} \exp\left(-\frac{r_1 + r_2}{a}\right),$$

where a is a variational parameter, to find the energy of the helium ground state.

Answer: $\qquad {}^\backprime E^\prime = -\frac{729}{128} \frac{me^4}{2\hbar^2} = -77.5$ ev.

Explain physically why the value of a obtained by the variational calculation is larger than a_0/Z.

12-6. Try to invent a further modification of the trial wave function of Prob. 12-5, so that the calculated ground-state energy will be within $\frac{1}{2}\%$ of the experimental value.

12-7. Does the helium system have metastable states other than the $1s2s$ $^3S_1$ state? Explain.

12-8. Comment upon the statement: "Every observable for a system of identical particles must be symmetric with respect to exchange of the particles."

---

[1] A. Unsöld, *Ann. Physik* **82,** 355 (1926–27).

12–9. Show that the eight eigenfunctions of the total spin S for three electrons can be arranged in three sets: one set of four totally symmetric functions constituting a quartet ($S = \frac{3}{2}$), and two distinct sets of two functions each belonging to $S = \frac{1}{2}$ (doublets). Show also that the doublet functions cannot be "symmetrized" or "antisymmetrized." Such sets of functions are said to belong to an "intermediate symmetry type."

Hint: When Eq. (12–47) is applied to the spin functions for three electrons, it always yields a quartet function. There can be no totally antisymmetric function, because the spin coordinate takes only the two values $\pm\frac{1}{2}$.

12–10. Using a symmetry argument based upon the transformation properties of the p-wave functions, show that the term scheme for three equivalent p-electrons contains the term $^4S_{3/2}$.

12–11. Let F be an operator of the form

$$F(1, 2, \ldots, N) = f(1) + f(2) + \ldots + f(N) = \sum_{i=1}^{N} f(i),$$

in which $f(i)$ operates only upon the coordinates of the ith particle in a system of N identical particles. Show that

$$(\psi_A, F\psi_A) = \sum_{i=1}^{N} (\psi_i, f_i\psi_i),$$

where ψ_A is the antisymmetric wave function (12–45). (It is assumed that the one-particle functions are mutually orthogonal and normalized.)

12–12. Show how Eq. (11–176), Section 11–12, must be generalized to take proper account of the Pauli principle.

12–13. Supply the details of the calculation leading to Eq. (12–76).

12–14. Show that the total cross sections for s-scattering (Section 8–7) of identical particles of spin zero and spin one-half are, respectively,

$$\sigma_0 = \frac{16\pi}{k^2} \sin^2 \delta_0 \qquad \text{and} \qquad \sigma_{1/2} = \frac{4\pi}{k^2} \sin^2 \delta_0.$$

APPENDIX

A-1 Complex integration and the theory of residues. A function $f_1(z)$ of the complex variable $z = x + iy = re^{i\theta}$ is *regular* in the neighborhood of a point a if it has a Taylor expansion

$$f_1(z) = a_0 + a_1(z - a) + a_2(z - a)^2 + \cdots \qquad (|z - a| < R)$$

valid in some circle of radius R with center at a. Consider a second function of this kind, namely,

$$f_2(z) = b_0 + b_1(z - a) + b_2(z - a)^2 + \cdots \qquad (|z - a| < \rho).$$

By substituting $1/(z - a)$ in place of $(z - a)$, a new function is obtained,

$$f_3(z) = b_0 + b_1 \frac{1}{z - a} + b_2 \frac{1}{(z - a)^2} + \cdots,$$

which is defined for $1/|z - a| < \rho$, or what is the same thing,

$$|z - a| > 1/\rho = R'.$$

Now the sum of $f_1(z)$ and $f_3(z)$ is represented by the series

$$a_0 + a_1(z - a) + a_2(z - a)^2 + \cdots + b_0 + \frac{b_1}{z - a} + \frac{b_2}{(z - a)^2} + \cdots$$

within the common region of convergence of the two series above, which is seen to exist provided $R' < R$. One is led in this heuristic way to consider functions defined by an expansion of the form

$$f(z) = \cdots \frac{a_{-2}}{(z - a)^2} + \frac{a_{-1}}{z - a} + a_0 + a_1(z - a) + a_2(z - a)^2 + \cdots,$$

which is called the *Laurent series* for the function $f(z)$. It is valid within an annulus surrounding the point a (Fig. A–1) defined by

$$R' < |z - a| < R.$$

In the theory of complex variables, it is shown that every function which is regular within and on the boundaries of an annulus of this type is represented throughout the annulus by its Laurent expansion.[1]   In

---

[1] E. T. Whittaker and G. N. Watson, *A Course of Modern Analysis.* 4th ed. Cambridge: Cambridge University Press, 1935, Section 5.6, p. 100. K. Knopp, *Theory of Functions.* New York: Dover Publications, 1945, Part I, Chapter 10.

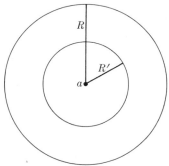

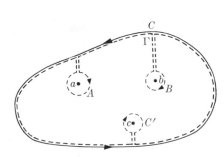

FIG. A–1. Region of convergence   FIG. A–2. Integration contour for
for Laurent series.   the derivation of the residue theorem.

particular, if a is the only singularity of $f(z)$ within R, then R' may be taken
as small as desired. In many important cases, the Laurent expansion
has only a finite number, say m, of terms in negative powers of $(z - a)$,
and $f(z)$ is then said to have *a pole of order m at a.* In the alternative case,
when the series extends to infinitely high negative powers of $(z - a)$, the
point a is an *essential singularity* for $f(z)$. In either case, the cofficient a_{-1}
in the Laurent expansion

$$f(z) = \sum_{n=-\infty}^{\infty} a_n (z - a)^n \qquad (A1\text{–}1)$$

is called the *residue* of $f(z)$ at the point a. Its importance arises from the
following theorem.

THEOREM. *Let $f(z)$ be a function which is analytic within and on a simple
closed contour C except for isolated singularities at the points a, b, c ...
within C. Then*

$$\int_c f(z)\, dz = 2\pi i \cdot (\textit{Sum of residues of } f(z) \textit{ at the singular points.}) \qquad (A1\text{–}2)$$

Proof: By hypothesis, the singularities of $f(z)$ are isolated, and one may
construct a path Γ (shown by the dotted line in Fig. A–2) composed of C,
circular arcs surrounding each singularity, and straight joining parts,
such that $f(z)$ is analytic everywhere within Γ. By Cauchy's theorem,[1]

$$\int_\Gamma f(z)\, dz = 0.$$

Since the straight-line parts of Γ are traversed twice and in opposite

---

[1] E. T. Whittaker and G. N. Watson, *op. cit.*, Section 5.2, p. 85. K. Knopp,
op. cit., Part I, Chapter 4.

directions, their contributions to the integral cancel one another, and one has

$$\int_C f(z)\, dz = -\left[\int_A + \int_B + \int_{C'} + \cdots\right] f(z)\, dz.$$

The integrals on the right-hand side are easily evaluated, for by hypothesis, the radii of the small circles may be chosen so small that $f(z)$ is represented by a (uniformly convergent) Laurent expansion on each of these circles. For example,

$$\int_A f(z)\, dz = \int_A \sum_{n=-\infty}^{\infty} a_n(z-a)^n\, dz = \sum_{n=-\infty}^{\infty} a_n \int_A (z-a)^n\, dz.$$

By writing $z - a = Re^{i\theta}$, $dz = iRe^{i\theta}\, d\theta$, one now has

$$\int_A (z-a)^n\, dz = -R^{n+1} \int_0^{2\pi} e^{i(n+1)\theta}\, d\theta = -\begin{cases} 2\pi i, & n = -1 \\ 0, & n \neq -1. \end{cases}$$

(Note the sense in which the circle A is described.) In summary,

$$\int_A f(z)\, dz = -2\pi i a_{-1},$$

and, after addition of similar contributions from $B, C', \ldots$,

$$\int_c f(z)\, dz = 2\pi i \cdot \text{(Sum of residues of } f(z) \text{ at the singularities within } C.)$$

The utility of the residue theorem in the evaluation of complex integrals will now be illustrated by examples drawn from the text:

(a) $\int_{-\infty}^{\infty} (\sin z/z)\, dz$. The path of integration is the real axis in the z-plane. However, since the function $\sin z/z$ is not singular at any finite point, the contour may be changed to that shown in Fig. A–3 without affecting the value of the integral (Cauchy's theorem). The relation $\sin z = (1/2i)(e^{iz} - e^{-iz})$ may now be used to transform the integral to

$$\frac{1}{2i} \int_C \frac{e^{iz} - e^{-iz}}{z}\, dz = \frac{1}{2i} \int_C \frac{e^{iz}}{z}\, dz - \frac{1}{2i} \int_C \frac{e^{-iz}}{z}\, dz,$$

FIGURE A–3

in which the transformation leading to the right-hand side is now allowed because the path of integration no longer contains the point $z = 0$. *If z has a positive imaginary part,* the function e^{iz} becomes exponentially small as $|z| \rightarrow \infty$, and the contour may be closed in the upper half-plane by a large semicircle (Fig. A–4), and in the limit $|z| \rightarrow \infty$ on this circle, the value of the integral is not affected[11] by this addition to C. The residue of the function

$$\frac{e^{iz}}{z} = \frac{1}{z}\left(1 + iz + \frac{(iz)^2}{2!} + \cdots\right) \qquad \text{at} \qquad z = 0$$

is 1, and since this is the only singularity within C, the residue theorem gives

$$\int_C \frac{e^{iz}}{z}\, dz = 2\pi i.$$

The integral $\int_C (e^{-iz}/z)\, dz$ may be evaluated in a similar way by closing the contour by a large semicircle in the *lower* half-plane (Fig. A–5). Then e^{-iz}/z is regular *everywhere* within this contour and hence $\int_{C'} (e^{-iz}/z)\, dz = 0$. Combining these results, one finds

$$\int_{-\infty}^{\infty} \frac{\sin z}{z}\, dz = \pi.$$

(b) Similar arguments can be used to show that the integral of Eq. (2–45), which can be written

$$a(k) = \frac{1}{\sqrt{\epsilon}}\, \frac{1}{2\pi i}\left[\int_C \frac{\exp\left[i\left(\frac{\epsilon}{2} - k\right)z\right]}{z}\, dz - \int_C \frac{\exp\left[-i\left(\frac{\epsilon}{2} + k\right)z\right]}{z}\, dz\right],$$

is the function defined in Eq. (2–41). Supposing that $|k| < \epsilon/2$, the exponent in the first integral is positive imaginary, and in the second integral, negative imaginary. The path is therefore to be closed in the upper half-plane in the first case, and in the lower half-plane in the second

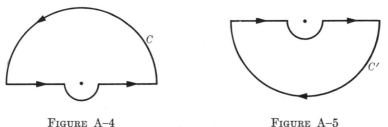

FIGURE A–4   FIGURE A–5

---

[1] Rigorous justification of this procedure amounts to an application of Jordan's lemma. Cf. E. T. Whittaker and G. N. Watson, *op. cit.*, Section 6.222, p. 115.

case. The only pole is, as before, at $z = 0$, and the residue theorem provides the result

$$a(k) = \sqrt{\frac{1}{\epsilon}} \left\{ \operatorname*{Res}_{z=0} \frac{\exp\left[i\left(\frac{\epsilon}{2} - k\right)z \right]}{z} \right\} = \sqrt{\frac{1}{\epsilon}} \qquad \left(|k| < \frac{\epsilon}{2} \right).$$

If $k < -\epsilon/2$, the contours for both integrals are to be closed in the upper half-plane, and since the residues for the two integrals are the same, they cancel each other. If $k > \epsilon/2$, the contours are closed in the lower half-plane, and neither integrand has a singularity within the contour, so that the result is again zero. Hence in both cases,

$$a(k) = 0 \qquad \left(|k| > \frac{\epsilon}{2} \right).$$

(c) The evaluation of the integral in Eq. (2–47) does not involve the residue theorem. It is accomplished as follows: Consider the integral

$$\int_C \exp\left[-\frac{1}{2\sigma^2} (z + ik\sigma^2)^2 \right] dz,$$

in which C is the closed rectangular contour shown in Fig. A–6. By Cauchy's theorem, this integral is zero since the integrand is regular at every point within and on C. Breaking up the path of integration into its straight-line parts, we therefore have

$$\int_{-R}^{R} \exp\left[-\frac{1}{2\sigma^2} (z + ik\sigma^2)^2 \right] dz =$$

$$-\left[\int_{R}^{R-ik\sigma^2} + \int_{R-ik\sigma^2}^{-R-ik\sigma^2} + \int_{-R-ik\sigma^2}^{-R} \right] \exp\left[-\frac{1}{2\sigma^2} (z + ik\sigma^2)^2 \right] dz.$$

If one now allows R to become very large, the integrals along the vertical parts of the path are easily seen to approach zero, and it follows that

$$\int_{-\infty}^{\infty} \exp\left[-\frac{1}{2\sigma^2} (z + ik\sigma^2)^2 \right] dz = \int_{-\infty-ik\sigma^2}^{\infty-ik\sigma^2} \exp\left[-\frac{1}{2\sigma^2} (z + ik\sigma^2)^2 \right] dz.$$

FIGURE A–6

By the substitution $u = (1/\sqrt{2}\sigma)(z + ik\sigma^2)$, this expression is reduced to the real integral[1]

$$\sqrt{2}\,\sigma \int_{-\infty}^{\infty} e^{-u^2}\,du = \sqrt{2\pi}\,\sigma.$$

A–2 Parseval's formula. The relation

$$\int |\psi(x)|^2\,dx = \int |a(p)|^2\,dp$$

can be generalized by applying it to a linear combination of two functions $\psi(x)$ and $\phi(x)$ whose momentum representations are $a(p)$ and $b(p)$. Thus, for all complex values of λ, one has

$$\int |\psi(x) + \lambda\phi(x)|^2\,dx$$
$$= \int [|\psi(x)|^2 + \lambda\psi^*(x)\phi(x) + \lambda^*\phi^*(x)\psi(x) + |\lambda|^2|\phi(x)|^2]\,dx$$
$$= \int |a(p) + \lambda b(p)|^2\,dp$$
$$= \int [|a(p)|^2 + \lambda a^*(p)b(p) + \lambda^*b^*(p)a(p) + |\lambda|^2|b(p)|^2]\,dp,$$

and therefore

$$\lambda \int \psi^*(x)\phi(x)\,dx + \lambda^* \int \phi^*(x)\psi(x)\,dx$$
$$= \lambda \int a^*(p)b(p)\,dp + \lambda^* \int b^*(p)a(p)\,dp,$$

which can only be true for every complex value of λ if

$$\int \psi^*(x)\phi(x)\,dx = \int a^*(p)b(p)\,dp \qquad\qquad\qquad (A2\text{–}1)$$

for any two functions ψ and ϕ. A similar generalization of Bessel's inequality is obviously possible.

Equation (A2–1) may be made the basis of a very thorough treatment of the theory of Fourier transforms. (The interested student may profitably consult P. M. Morse and H. Feshbach, *Methods of Theoretical Physics.* New York: McGraw-Hill Book Co., Inc., 1953, Volume I, Section 4.8, for a readable account of this subject.)

---

[1] R. Courant, *Differential and Integral Calculus.* New York: Nordemann Publishing Company, Inc., 1936, Volume II, p. 561. *Mathematical Tables from Handbook of Chemistry and Physics.* 8th ed. Cleveland: Chemical Rubber Publishing Co., 1947, p. 241, integral No. 352.

A–3 Schwarz's inequality. The simplest form of Schwarz's inequality is obtained by considering two sets of N complex numbers each, for example, $a_1, a_2, \ldots, a_N$ and $b_1, b_2, \ldots, b_N$. Then the sum of absolute squares

$$\sum_{i,j=1}^{n} |a_i b_j - a_j b_i|^2$$

is obviously not negative. Writing

$$|a_i b_j - a_j b_i|^2 = (a_i^* b_j^* - a_j^* b_i^*)(a_i b_j - a_j b_i),$$

and multiplying out the product, one has

$$\sum_{i,j=1}^{N} (a_i^* a_i b_j^* b_j - a_i^* b_i b_j^* a_j - a_j^* b_j b_i^* a_i + a_j^* a_j b_i^* b_i) \geq 0.$$

By suitable renaming of the indices and collection of terms, this is found to be the same as

$$2\left[\left(\sum_i |a_i|^2\right)\left(\sum_i |b_i|^2\right) - \left(\sum_i a_i^* b_i\right)\left(\sum_i b_i^* a_i\right)\right] \geq 0,$$

or

$$\left|\sum_i a_i^* b_i\right|^2 \leq \sum_i |a_i|^2 \sum_i |b_i|^2, \tag{A3-1}$$

which is Schwarz's inequality.

An alternative, frequently encountered proof of Schwarz's inequality follows from the remark that the function

$$\sum_i |a_i + \lambda b_i|^2 = \sum_i |a_i|^2 + 2\lambda\left[\sum_i a_i^* b_i + \sum_i a_i b_i^*\right] + \lambda^2 \sum_i |b_i|^2$$

is not negative for any real value of λ. Consequently, the discriminant of the right-hand side, which is a quadratic expression in λ, must be negative:

$$\left[\text{Re} \sum_i a_i^* b_i\right]^2 \leq \sum_i |a_i|^2 \sum_i |b_i|^2.$$

If $\sum_i a_i^* b_i$ is real, this is Schwarz's inequality. If it is not, one may write

$$\sum_i a_i^* b_i = \left|\sum_i a_i^* b_i\right| e^{i\gamma},$$

and the same reasoning, when applied to the function

$$\sum_i |a_i + \lambda b_i e^{-i\gamma}|^2,$$

results in $\left[\mathrm{Re} \sum_{i} a_i^* b_i e^{-i\gamma} \right]^2 = \left| \sum_i a_i^* b_i \right|^2 \leq \sum_i |a_i|^2 \sum_i |b_i|^2.$

It is obvious that Schwarz's inequality holds for any set of quantities for which a scalar product analogous to $\sum_i a_i^* b_i$ may be defined. For example, if it is applied to integrals of the type $\int \psi^* \phi \, dx$, it takes the form

$$\left| \int \psi^* \phi \, dx \right|^2 \leq \int |\psi|^2 \, dx \int |\phi|^2 \, dx, \qquad (A3\text{--}2)$$

which is used in the text in the proof of the uncertainty relation. In terms of the general scalar product (Section 9–2), it is

$$|(x, y)|^2 \leq (x, x)(y, y). \qquad (A3\text{--}3)$$

From the form of either proof above, it is easy to see that the equality sign holds if and only if x and y are linearly dependent.

A–4 Functional transformations; Dirac delta function. In the equation

$$f(x) = \int G(x, x') g(x') \, dx', \qquad (A4\text{--}1)$$

the dependence of the function $f(x)$ upon the function $g(x)$ is expressed by saying that $f(x)$ is a *linear functional* of $g(x)$. The quantity $G(x, x')$, called a *kernel*, depends upon both variables (x, x') and may be considered to define a *functional transformation* from g to f. It is clear that this transformation is linear: If, by Eq. (A4–1), g_1 and g_2 are transformed into f_1 and f_2, respectively, then by the same transformation, $g_1 + g_2$ is transformed into $f_1 + f_2$. The Fourier transform [Eq. (2–37)] is an example of such a transformation in which the function $\psi(x)$ is a linear functional of the amplitude function $a(k)$ with kernel $(1/\sqrt{2\pi})e^{ikx}$. If one interprets the integral in Eq. (A4–1) as the limit of a sum, then this equation may be compared with the linear transformation (Section 9–5)

$$y_i = \sum_{j=1}^{N} a_{ij} x_j \qquad (A4\text{--}2)$$

of the components of the vector x into those of the vector y in a vector space of N dimensions. In the discussion of the Fourier transform (Section 2–4), the transition from a sum like (A4–2) to an integral was made heuristically. The role of the kernel is taken here by the matrix (a_{ij}).

Linear functional relations of the form (A4–1) play a central role in quantum mechanics, and the formal similarity of Eqs. (A4–1) and (A4–2) is such that it is of considerable advantage to discuss both in the geometrical terms introduced in Chapter 9. Thus the values of the function

$g(x)$ may be regarded as the components (numbered by the *index* x) of a vector g, which now has a continuous range of values instead of the discrete set numbered by the index j in Eq. (A4–2). Similarly, the function G plays the role of a matrix with elements $G(x, x')$, in analogy with the matrix $A = (a_{ij})$, whose elements are numbered by the discrete indices i, j. In one important respect, however, this analogy fails: There is no transformation of the form (A4–1) which corresponds to the identity matrix $a_{ij} = \delta_{ij}$ in (A4–2), i.e., there is no kernel $\delta(x, x')$ for which one may write

$$f(x) = \int \delta(x, x')f(x')\, dx'. \tag{A4–3}$$

This difficulty can be overcome in a rigorous way by modifying the concept of integral (*Stieltjes' integral*). It is, however, more convenient to relax the requirement of mathematical rigor and make use of a "function" which has, by definition, the property (A4–3).

In order to visualize the meaning of Eq. (A4–3), $\delta(x, x')$ may be considered to be a (mathematically nonexistent) limit, such as

$$\delta(x, x') = \delta(x - x') = \lim_{\lambda \to 0} D(x - x', \lambda), \qquad D(x, \lambda) = \frac{1}{\pi}\frac{\lambda}{\lambda^2 + x^2}.$$

$$\tag{A4–4}$$

The function $D(x, \lambda)$, which is shown graphically in Fig. A–7, has a maximum at $x = 0$, and is confined more and more to the neighborhood of this point as λ becomes smaller. Also, $\int_{-\infty}^{\infty} D(x, \lambda)\, dx = 1$, that is, the area under each of the curves in the figure is unity. The meaning of (A4–3) is now clarified by noting that if $f(x')$ is a smooth function in the neighborhood of $x' = x$, then, since only this neighborhood is of importance in the integral, the factor $f(x')$ may be taken equal to $f(x)$, and one has

$$\int_{-\infty}^{\infty} f(x')\, D(x - x', \lambda)\, dx \approx f(x) \int_{-\infty}^{\infty} D(x', \lambda)\, dx' = f(x).$$

The approximation improves with decreasing λ. The reader may show that if $f(x)$ is bounded in $(-\infty, \infty)$ and continuous in the neighborhood of the origin, then

$$\lim_{\lambda \to 0} \int_{-\infty}^{\infty} f(x)\, D(x, \lambda)\, dx = f(0). \tag{A4–5}$$

The precise form (A4–4) for $D(x, \lambda)$ is not important so long as the properties implied in (A4–5) are present. It is possible to devise many similar representations of the delta function, in fact, infinitely many, for example:

$$D(x, \lambda) = \frac{1}{\lambda\sqrt{\pi}}\, e^{-x^2/\lambda^2},$$

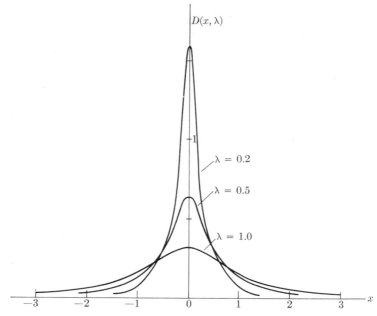

FIG. A-7. The function $D(x, \lambda) = (1/\pi)[\lambda/(\lambda^2 + x^2)]$.

or, in general,

$$D(x, \lambda) = \frac{1}{\lambda(b - a)} \frac{1}{dg/d\theta},$$

where $g(\theta)$ is any monotonic smooth function in (a, b) satisfying

$$g(a) = -\infty, \qquad g(b) = \infty.$$

The mathematically incorrect step involved in the definition of $\delta(x)$ is the interchange of the integration and limit processes in Eq. (A4–5); for this reason, the δ-function is always to be interpreted as a factor in the integrand of an expression like (A4–3).

Another representation of the δ-function is obtained immediately from the Fourier integral theorem if it is written in the form

$$f(x) = \int_{-\infty}^{\infty} \left[\frac{1}{2\pi} \int_{-\infty}^{\infty} e^{ik(x-x')} \, dk \right] f(x') \, dx',$$

which, by comparison with (A4–3), yields

$$\delta(x - x') = \frac{1}{2\pi} \int_{-\infty}^{\infty} e^{ik(x-x')} \, dk. \tag{A4–6}$$

This form has appeared in the discussion of the momentum wave function corresponding to the plane wave e^{ikx} (Section 3–5).

Another representation of $\delta(x' - x)$ is obtained if we consider the unit step function defined by

$$U(x' - x) = \begin{cases} 1, & x' > x, \\ 0, & \text{otherwise.} \end{cases}$$

This function does not have a derivative at $x' = x$, but if it is regarded as the limit of a continuous function which increases sharply from 0 to 1 in a small interval containing the point x', then the derivative may be treated as the limit of a function like $D(x, \lambda)$. Working formally, integrating by parts, and assuming that the interval (a, b) contains the point x, one has

$$\int_a^b f(x') \frac{d}{dx'} U(x' - x)\, dx'$$

$$= f(x') U(x' - x) \Big|_a^b - \int_a^b U(x' - x) \frac{d}{dx'} f(x')\, dx'$$

$$= f(b) - \int_x^b f'(x)\, dx = f(x),$$

whence, by Eq. (A4–3),

$$\frac{d}{dx'} U(x' - x) = \delta(x' - x). \tag{A4–7}$$

Again, consider the expansion of a function

$$f(x) = \sum_n f_n \phi_n(x)$$

in terms of a complete orthonormal set of functions ϕ_n (cf. Chapter 6). The coefficients f_n are

$$f_n = \int \phi_n^*(x') f(x')\, dx',$$

and therefore, by formal interchange of integral and sum,

$$f(x) = \sum_n \int \phi_n^*(x') f(x')\, dx' \phi(x)$$

$$= \int \left[\sum_n \phi_n^*(x') \phi_n(x) \right] f(x')\, dx',$$

whence

$$\delta(x - x') = \sum_n \phi_n^*(x') \phi_n(x). \tag{A4–8}$$

This concept may be readily generalized to three dimensions, for the function

$$\delta(\mathbf{r} - \mathbf{r}') = \delta(x - x')\,\delta(y - y')\,\delta(z - z')$$

has the property

$$\int f(\mathbf{r}')\,\delta(\mathbf{r} - \mathbf{r}')\,d\mathbf{r}'$$

$$= \iiint f(x', y', z')\,\delta(x - x')\,\delta(y - y')\,\delta(z - z')\,dx'\,dy'\,dz'$$

$$= f(x, y, z) = f(\mathbf{r}).$$

The following formal properties of the δ-function,[1] which can be verified by means of the definition

$$f(x) = \int \delta(x - x')f(x')\,dx', \tag{A4-9}$$

are used occasionally in the text:

$$\delta(-x) = \delta(x),$$

$$x\,\delta(x) = 0,$$

$$\delta(ax) = \frac{1}{a}\,\delta(x),\ (a > 0),$$

$$\delta[f(x)] = \frac{1}{df/dx}\,\delta(x - x_0)\qquad f(x_0) = 0,\quad \text{(A4-10)}$$

$$\int \delta(x - x'')\,\delta(x'' - x')\,dx'' = \delta(x - x'),$$

$$f(x)\,\delta(x - x') = f(x')\,\delta(x - x'),$$

$$\frac{d}{dx}\,\delta(x) = \delta'(x) = -\frac{1}{x}\,\delta(x).$$

A–5 The eikonal equation in geometrical optics. A system of rays in an optical medium (Section 4–2) can be specified by means of the *characteristic vector*

$$\boldsymbol{\pi} = \mu\hat{\mathbf{e}},$$

which has, at each point of the medium, a magnitude equal to the index of refraction at that point, and a direction (denoted by the unit vector $\hat{\mathbf{e}}$) parallel to the ray. In terms of $\boldsymbol{\pi}$, the time of propagation of light from a

---

[1] P. A. M. Dirac, *The Principles of Quantum Mechanics.* 3rd ed. Oxford: The Clarendon Press, 1947, p. 60.

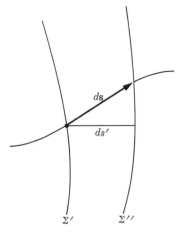

wave surface Σ_0 to another surface Σ (Fig. 4–3) is

$$t == \frac{1}{c}\int \mu\,ds = \frac{1}{c}\int \boldsymbol{\pi}\cdot d\mathbf{s}. \qquad (A5\text{--}1)$$

The second of these integrals is *independent of the path joining the two surfaces*, as may be seen by inspection of Fig. A–8, in which $d\mathbf{s}$ is an element of an arbitrary path of integration, and ds' an element of an optical ray. Thus,

$$\boldsymbol{\pi}\cdot d\mathbf{s} = \mu|d\mathbf{s}|\cos\theta = \mu\,ds' = c\,dt.$$

Clearly, the surface Σ is defined by the equation

$$S = ct = \text{constant} = \text{optical path length from } \Sigma_0 \text{ to } \Sigma,$$

and the equation (A5–1) implies

$$\boldsymbol{\pi} = \nabla S, \qquad (A5\text{--}2)$$

where S is the eikonal defined in the text. Squaring Eq. (A5–2), we have, since $\boldsymbol{\pi}^2 = \mu^2$,

$$(\nabla S)^2 = \mu^2,$$

which is Eq. (4–10). It is the differential equation which characterizes the scalar function S in terms of the given index of refraction of the medium.

A–6 Derivation of Eq. (4–11) from Maxwell's equations. In a source-free medium, the equations governing the propagation of light waves of frequency ω are [1]

$$\nabla \times \mathcal{E} - \frac{i\omega}{c} \mathcal{B} = 0, \qquad \nabla \cdot \mathcal{B} = 0, \tag{A6–1}$$

$$\nabla \times \mathcal{H} + \frac{i\omega}{c} \mathcal{D} = 0, \qquad \nabla \cdot \mathcal{D} = 0, \tag{A6–2}$$

in which the time is contained in each of the field vectors $\mathcal{E}$, $\mathcal{B}$, $\mathcal{D}$, $\mathcal{H}$ in the factor $e^{-i\omega t}$. The constitutive relations are

$$\mathcal{B} = \kappa_m \mathcal{H}, \qquad \mathcal{D} = \kappa_e \mathcal{E}, \tag{A6–3}$$

where the "constants" κ_m and κ_e are now to be regarded as slowly varying functions of spatial coordinates.

The wave equation for the vector $\mathcal{E}$ can be formed by taking the curl of (A6–1) and eliminating the magnetic fields by means of Eqs. (A6–2) and (A6–3). The result is

$$\nabla^2 \mathcal{E} + k^2 \mathcal{E} = -\nabla \left[\mathcal{E} \cdot \frac{\nabla \kappa_e}{\kappa_e} \right] + (\nabla \times \mathcal{E}) \times \frac{\nabla \kappa_m}{\kappa_m}. \tag{A6–4}$$

Similarly, by eliminating the electric fields, we find

$$\nabla^2 \mathcal{H} + k^2 \mathcal{H} = -\nabla \left[\mathcal{H} \cdot \frac{\nabla \kappa_m}{\kappa_m} \right] + (\nabla \times \mathcal{H}) \times \frac{\nabla \kappa_e}{\kappa_e}, \tag{A6–5}$$

in which $k^2 = \omega^2 \kappa_e \kappa_m / c^2$ is the *local* wave number, which depends upon position through κ_e and κ_m. (In any practical case, κ_m is very nearly unity.)

If, as explained in the text, $|\lambda \nabla \kappa_e| \ll \kappa_e$, then the right-hand members of Eqs. (A6–4) and (A6–5) are negligible, whence, if the frequency is sufficiently high, both $\mathcal{E}$ and $\mathcal{H}$ satisfy $\nabla^2 \phi + k^2 \phi = 0$, which is Eq. (4–11).

A–7 Exponential form for unitary matrices and derivation of Eq. (11–21). The exponential matrix [Eq. (9–103)]

$$e^A = 1 + A + \frac{A^2}{2!} + \cdots \tag{A7–1}$$

exists for every matrix A. By multiplication of the defining series, it is easy to show that e^A and e^{-A} are reciprocals:

$$e^A e^{-A} = e^{-A} e^A = 1. \tag{A7–2}$$

---

[1] W. K. H. Panofsky and M. Phillips, *Classical Electricity and Magnetism.* Reading, Mass.: Addison-Wesley Publishing Co., Inc., 1955, Chapter 9.

The matrix e^A is therefore nonsingular. Furthermore, it is evident that

$$\left(e^A\right)^\dagger = e^{(A^\dagger)}. \tag{A7-3}$$

If z is a complex variable, the derivative of the matrix e^{zA}, where A is independent of z, is

$$\frac{d}{dz} e^{zA} = \lim_{h \to 0} \frac{1}{h} \{e^{(z+h)A} - e^{zA}\} = A e^{zA}. \tag{A7-4}$$

The matrix

$$U = e^{iS}, \tag{A7-5}$$

where S is Hermitian ($S = S^\dagger$), is unitary, for

$$U^\dagger = e^{-iS^\dagger} = e^{-iS} = U^{-1}. \tag{A7-6}$$

Conversely, if U is given, a Hermitian matrix S exists for which (A7–5) holds. The proof can be constructed using the representation in which U is diagonal.

Let H be a given matrix, and define $H(z)$ by the equation

$$H(z) = e^{izS} H e^{-izS}, \tag{A7-7}$$

where S is independent of z. By Eq. (A7–4) and the usual rule for differentiation of a product, we have

$$\frac{dH(z)}{dz} = (iS)e^{izS} H e^{-izS} - e^{izS} H e^{-izS}(iS) = i[S, H(z)]. \tag{A7-8}$$

By iteration of this equation, we find the higher derivatives of $H(z)$:

$$\frac{d^2 H(z)}{dz^2} = i\left[S, \frac{dH(z)}{dz}\right] = i^2[S, [S, H(z)]],$$

$$\vdots \tag{A7-9}$$

$$\frac{d^n H(z)}{dz^n} = i^n[S, [S, \dots \, (n), [S, H(z)] \dots \, (n)],].$$

If we now assume that $H(z)$ can be expanded in Taylor's series, we have, since $H(0) = H$,

$$H(z) = H + i[S, H]z + \frac{i^2}{2}[S, [S, H]]z^2 + \cdots \tag{A7-10}$$

The matrix $H' = H(1)$ is the transform

$$H' = e^{iS} H e^{-iS} = UHU^{-1} \tag{A7-11}$$

of H under the unitary transformation U. Setting $z = 1$ in the series (A7–10), which is assumed to be convergent, we obtain Eq. (11–21).

A–8 Physical constants and conversion factors.

Name of Quantity	Symbol	Value
General physical constants:		
Avogadro's number	N_0	6.025×10^{23} molecules/mole
Boltzmann's constant	k	1.3804×10^{-16} erg/deg
Fine-structure constant	α	$e^2/\hbar c = 7.2973 \times 10^{-3} = 1/137.037$
Planck's constant	h	6.6253×10^{-27} erg·sec
	$\hbar$	$h/2\pi = 1.0544 \times 10^{-27}$ erg·sec
Stefan's constant	σ	$2\pi^5 k^4/15c^2 h^3 = 5.668 \times 10^{-5}$ erg/cm^2·deg^4·sec
Velocity of light	c	2.99792×10^{10} cm/sec
Wien's constant	b	$\lambda_{max}T = 0.28979$ cm·deg
Electron:		
Mass	m_e	9.108×10^{-28} gm
Charge	e	-4.803×10^{-10} esu
Classical electron radius	r_e	$e^2/mc^2 = 2.818 \times 10^{-13}$ cm
Compton wave length	λ_c	$\hbar/mc = 3.861 \times 10^{-11}$ cm
Bohr magneton	μ_0	$-e\hbar/2mc = 0.9273 \times 10^{-20}$ erg/gauss
Electron magnetic moment		$-1.0011\,\mu_0$
Nucleons:		
Atomic mass unit	amu	1.6598×10^{-24} gm
Proton mass	M_P	1.00759 amu
Neutron mass	M_N	1.00898 amu
Nuclear magneton	μ_N	5.050×10^{-24} erg/gauss $= (1/1836)\mu_0$
Proton magnetic moment		$2.7928\,\mu_N$
Neutron magnetic moment		$-1.9103\,\mu_N$
Hydrogen:		
Bohr radius	a_0	$\hbar^2/me^2 = 5.29175 \times 10^{-9}$ cm
Rydberg constant:		
for infinite nuclear mass	R_∞	109737.31 cm^{-1}
for finite nuclear mass	R_H	109677.58 cm^{-1}
First ionization potential of hydrogen	E_1	-13.595 ev
Conversion factors:		
Electron volts to ergs		1 ev $= 1.60208 \times 10^{-12}$ erg
Wave length to energy (photons)		$\lambda(A) = 12398/E(ev)$

AUTHOR INDEX

SUBJECT INDEX